MICROBIAL SURFACTANTS
Volume 1: Production and Applications

Books Published in Industrial Biotechnology Series

1. Silkworm Biofactory: Silk to Biology
   Enoch Y Park, Katsumi Maenaka (eds.), 2019

2. Probiotics, the Natural Microbiota in Living Organisms: Fundamentals and Applications
   Hesham El Enshasy, Shang Tian Yang (eds.), 2021

3. Microbial Surfactants—Volume 1: Production and Applications
   Riyazali Zafarali Sayyed, Hesham Ali El Enshasy, Hameeda Bee (eds.), In Press

Series: Industrial Biotechnology

MICROBIAL SURFACTANTS
Volume 1: Production and Applications

Editors

R.Z. Sayyed

Head, Department of Microbiology
PSGVP Mandal's Arts Science & Commerce College
Shahada, India

Hesham Ali El-Enshasy

Institute of Bioproducts Development
University of Technology Malaysia
Johor Bahru, Malaysia

Bee Hameeda

Department of Microbiology
Osmania University
Hyderabad, India

CRC Press
Taylor & Francis Group
Boca Raton London New York

CRC Press is an imprint of the
Taylor & Francis Group, an **informa** business
A SCIENCE PUBLISHERS BOOK

First edition published 2021
by CRC Press
6000 Broken Sound Parkway NW, Suite 300, Boca Raton, FL 33487-2742

and by CRC Press
2 Park Square, Milton Park, Abingdon, Oxon, OX14 4RN

ISBN: 978-0-367-52118-9 (hbk)
ISBN: 978-0-367-52119-6 (pbk)
ISBN: 978-1-003-05663-8 (ebk)

Typeset in Times New Roman
by Radiant Productions

Preface to the Series

Industrial biotechnology has a deep impact on our lives, and is the focus of attention of academia, industry and governmental agencies and become one of the main pillars of knowledge based economy. The enormous growth of biotechnology industries has been driven by our increased knowledge and developments in physics, chemistry, biology, and engineering. Therefore, the growth of this industry in any part of the world can be directly related to the overall development in that region.

The interdisciplinary *Industrial Biotechnology* book series will comprise a number of edited volumes that review the recent trends in research and emerging technologies in the field. Each volume will covers specific class of bioproduct or particular biofactory in modern industrial biotechnology and will be written by internationally recognized experts of high reputation.

The main objective of this work is to provide up to date knowledge of the recent developments in this field based on the published works or technology developed in recent years. This book series is designed to serve as comprehensive reference and to be one of the main sources of information about cutting-edge technologies in the field of industrial biotechnology. Therefore, this series can serve as one of the major professional references for students, researchers, lecturers, and policy makers. I am grateful to all readers and we hope they will benefit from reading this new book series.

Series Editor
Prof. Dr. rer. Nat. Hesham A. El-Enshasy
Johor Bahru, Malaysia

Preface

Biosurfactants are defined as the surface-active biomolecules produced by microorganisms. In recent years, biosurfactants have attracted commercial interest due to their unique properties such as production from renewable sources, high surface activity, high specificity, low toxicity, tolerance to pH, temperature and ionic strength, biodegradability, excellent emulsifying and demulsifying ability, antimicrobial activity, ability to work under extreme conditions, and relative ease of preparation. Biosurfactants have found applications in several industries including organic chemicals, petroleum, petrochemicals, mining, metallurgy (mainly bioleaching), agrochemicals, fertilizers, foods, beverages, cosmetics, pharmaceuticals and many others.

Biosurfactants market would be registering over 5.5% compound annual growth rate exceeding $2.7 billion by 2024 based on Global Market Insights, Inc. The key point of increased demand of biosurfactants is, the drawbacks linked with synthetic surfactants and their ecological hazards. Perse their eco-friendly nature and varied applications like stabilization of aerated systems, agglomeration control of fat globules and preservation of food products of food processing industry make their need in various industrial sectors.

A large variety of microorganisms produce potent surface-active agents referred as biosurfactants. The enormous diversity of biosurfactants make them an interesting group of materials for application in many areas such as agriculture, public health, food, health care, oil recovery, waste utilization and environmental pollution control such as degradation of hydrocarbons present in the soil. Both epiphytic and endophytic plant associate microbes produce biosurfactants and have vital role in motility, signaling and biofilm formation. In addition, biosurfactants can be used to aid germination, act as biocontrol agents, and bioremediation of contaminated soil, replacing chemical surfactants and enhancing plant microbe's interaction.

The enormous diversity of biosurfactants make them an interesting group of materials for different applications in agriculture, food and bioremediation of contaminated soil, replacing chemical surfactants and enhancing plant microbe's interaction. Biosurfactants have proven to be a boon agriculture. Food and in cleaning up of environment contaminated with hydrocarbons, pesticides, heavy metals, etc.

The major advantage of use of biosurfactants for various applications is their environmental feasibility and compatibility vis-à-vis chemical surfactants. These naturally occurring surfactants are advantageous over their corresponding chemically prepared surfactants as they can be produced from natural sources, they are less toxic, they possess high specify and better foaming and emulsifying properties and survive the extreme environments and moreover they are cheaper as they can be produced from the industrial waste and from by-products.

R.Z. Sayyed
Hesham Ali El-Enshasy
Bee Hameeda

Contents

1

Microbial Surfactants
An Overview

Umesh Pravin Dhuldhaj[1,]* and *Chandan R Bora*[2]

1. Introduction

Some of the microbes such as bacteria, yeast and fungi present in an oil field have the capacity to produce biosurfactants with emulsifications of hydrocarbons to increase the availability of petroleum products for effective degradations and uptake (Morikawa et al. 2000, Cameotra and Makkar 2004) and have high biodegradability and low toxicity (Praveesh et al. 2011). Production of biosurfactant is one of the strategies of oil degrading bacteria to survive in unfavorable conditions by adsorption, emulsification and solubilization of oil residues (Nerurkar et al. 2009). Biosurfactants or synthetic surfactants increase the pseudo solubility of oil and petroleum residues in water and makes it a rich source of carbon and energy for living organisms (Pekdemir et al. 2005, Mulligan 2009). These compounds have both hydrophilic and hydrophobic moieties, hence are amphiphilic in nature and reduce interfacial tension between two immiscible solvents (e.g., oil and water) by accumulating at the interface (Nayak et al. 2009, Makkar et al. 2011, Hamzah et al. 2013, Singh and Cameotra 2004). Generally, the hydrophilic moieties are mono-, oligo-, or polysaccharide, peptides and proteins and hydrophobic are fatty alcohols and unsaturated, saturated and hydroxylated fatty acids (Pacwa-Plociniczak et al. 2011). Biosurfactants not only solubilize two immiscible solvent such as an aqueous and non-aqueous solvent, but are highly surface active, lubricating, highly stabilized, detergents with foaming capacity (Cameotra and Makkar 2004, Fenibo et al. 2019a, Pradhan and Bhattacharyya 2018, Sobrinho et al. 2013). At the interface of two immiscible solvents, surfactants or biosurfactants accumulates are hydrophobic head facing to non-aqueous phase (oil phase), and hydrophilic that

[1] School of Life Sciences, Swami Ramanand Teerth Marathwada University, Nanded.
[2] Director, NOVO Cellular Medicine Institute, Fidelity Healthcare Limited, Port of Spain, Trinidad.
* Corresponding author: umeshpd12@gmail.com

are head facing towards hydrophilic aqueous phase which makes the biosurfactant suitable as a foam stabilizer, a good detergent and may be used as food additives (Rodrigues 2015). Biosurfactants present in suspension interact with surface particles and produces the hydrophobic characteristics of minerals' surface and the strong hydrophobic interactions lead to aggregation of particles because of collision amongst hydrophobic molecules. Also these biosurfactant are responsible for stabilization and destabilizations of suspensions; suspension of fine particles which get aggregated and settle down rapidly, they causes destabilizations of suspension, while the stabilization is established when the concentration of surfactant is higher. Broadly these biosurfactant can be categorized into two on the basis of their molecular weight; low molecular weight biosurfactant lowers interfacial and surface tension which increases pseudo solubility while high molecular weight biosurfactant can be used as an emulsion stabilizing agent (Fenibo et al. 2019b).

When two immiscible solvents are exposed to each other, a boundary interface is formed where accumulation of the biosurfactant takes place, because of hydrophobic moieties and aggregation of molecules at very low concentrations (i.e., Critical Micelle Concentration), micelle is formed (Sonawane et al. 2015). Critical Micelle Concentrations (CMC) are the high amount of surfactant monomers present in water with respect to pH, ionic strength and temperature of the system (Seydlová and Svobodová 2008). Micelle in the inner portion contains non-polar constituents and forms a hydrophobic microenvironment (Li et al. 2005). Lesser concentrations of surfactant are needed to reduce surface tension and the efficient one has low CMC (Seydlová and Svobodová 2008, Sivapathasekaran and Sen 2017).

Synthetic surfactant obtained from the petrochemical industry contributes 70–75% of the active tensions used in developed countries (Campos et al. 2013, Akbari et al. 2018). As this synthetic surfactant has a toxic effect on the environment and human beings, petroleum products have to be replaced with renewable, biodegradable and sustainable sources (Akbari et al. 2018, Cortés-Sánchez 2020). Biosurfactants have great importance with respect to synthetic surfactant as it has a high rate of biodegradability and lesser toxicity, hence they are mostly preferred for food processing, pharmaceuticals and bioremediations of environment—often in oil recovery (Praveesh et al. 2011). Biosurfactants are good alternatives to synthetic antioxidant and anti-biofilms chemicals also as they have good antioxidative (Yalcin and Cavusoglu 2010, Takahashi et al. 2012) and antimicrobial properties (Abdollahi et al. 2020). Biosurfactants are surface active molecules produced by microbes having greater significance over their chemical counterpart as they are biodegradable, less toxic and also efficient in extreme conditions such as pH and temperature (Seydlová and Svobodová 2008, Mulligan 2005).

2. Biosurfactant Classification Based on Structure and Functional Groups

Biosurfactant naturally produced by microbes or through the enzymatic processes have diversity in their structural and functional groups. On the basis of their chemical nature, they are classified as the glycolipids, lipopeptides, polysaccharides, proteins and lipoproteins, or mixtures thereof (Table 1) (Ron and Rosenberg 2001, Maier 2003, Mulligan 2005, Muthusamy et al. 2008).

Table 1. Classifications of biosurfactants structure and functional groups and their applications.

Biosurfactants	Sub-types	Variants	Applications
Glycolipid	Rhamnolipids		Uptake of hydrophobic substrates, Antimicrobial properties
	Trehalolipid		Surfactant activity in extreme conditions
	Sophorolipids		Building blocks for biomaterials
Lipo-peptides and Lipoproteins	Surfactin	Esperin	Enhanced microbial oil recovery
		Lichenysin	Enhanced microbial oil recovery
		Pumilacidin	Antiviral antibiotics
	Iturin	Iturin A	Potent antifungal lipopeptide
		Iturin C	Enzyme inhibitor
		Bacillomycin D	Antifungal agent
		Bacillomycin F	Antifungal agent, Antibiotic
		Bacillomycin L	Biocontrol
		Mycosubtilin	Antifungal and hemolytic
	Fengycins	Fengycins A	Antibacterial and antifungal
		Fengycins B	Antifungal lipopeptide
Fatty Acids	Fatty acid ester and Sugar fatty acid esters		Biodiesels
	Sugar Alcohol		Food industry, thickeners
Phospholipids			Key components of cell membranes
Neutral Lipids			Storage lipids
Polymeric Biosurfactants	Emulsan		Structural tailoring and their emulsification
	Alasan		Solubilizations
	Biodispesan		Enhanced microbial oil recovery
	Liposan		Solubilizations and Emulsifications
	Lipomanan		Food industry
Particulate Biosurfactants			Microemulsions
Classifications of Biosurfactant based on head groups			
Anionic			Detergents
Cationic			Cosmetics
Non-ionic			Bioremediations
Zwitterionic			Detergents

2.1 *Glycolipid*

These are the most studied low molecular weight biosurfactant containing carbohydrates along with long chain aliphatic acids. Glycolipids can dissolve contaminating hydrocarbon (e.g., tetradecane, aliphatic and aromatic) during the fermentation and make it available as substrate to the microbes. This group of family can further be categorized into rhamnolipids, sophorolipids, trehalolipids (Fig. 1), etc. and also includes glycopeptides and glycolipopeptides (Das et al. 2009, Das

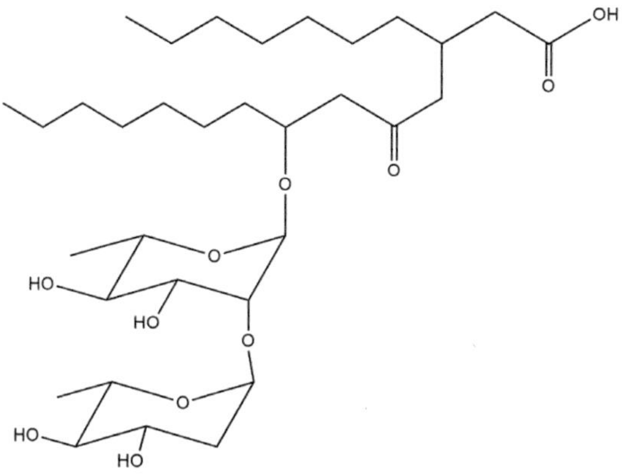

(a) Rhamnolipid

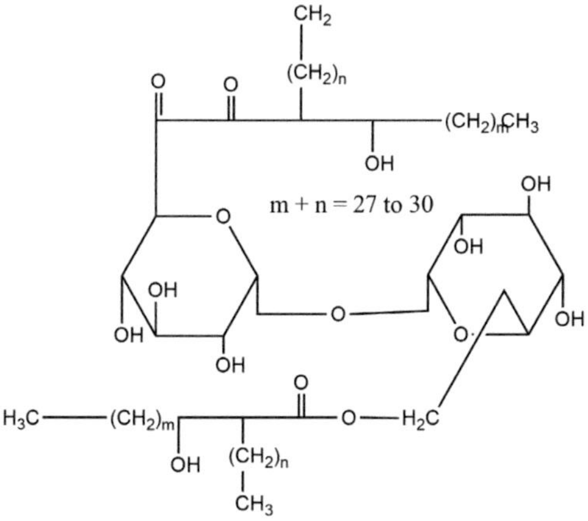

$m + n = 27$ to 30

(b) Trehalolipids

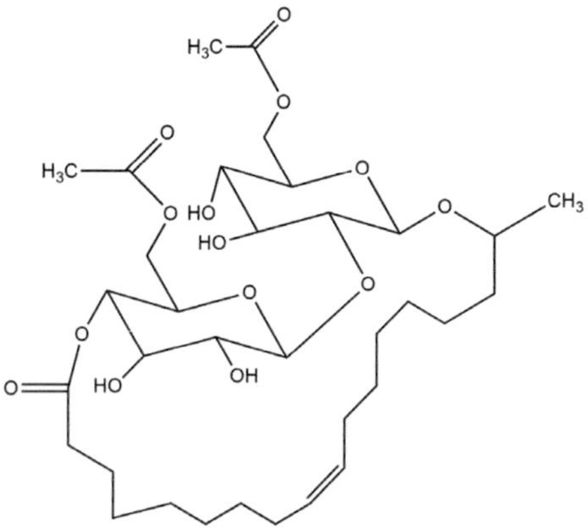

(c) Sophorolipid

Figure 1. Chemical structure of Glycolipids; (a) Rhamnolipid, (b) Trehalolipid and (c) Sophorolipid.

et al. 2008a and b, Ron and Rosenberg 2010). One of the studies recorded good emulsification activity by *Lactobacillus pentosus* because of its secretion of glycopeptides into the system (Vecino et al. 2014, Sivapathasekaran and Sen 2012). Presence of glycolipid sort of biosurfactant in the medium can be detected with the help of orcinol assay.

2.1.1 Rhamnolipids

Rhamnolipids are a well known group of glycolipids in which one or two rhamnose molecules are connected to one or two molecules of β-hydroxydecanoic acid. This sort of glycolipid was firstly observed in *Pseudomonas aeruginosa* (Cameotra and Makkar 2004). *Pseudomonas* species are dominant in the production of rhamnolipids containing L-rhamnose and 3-hydroxy fatty acids.

2.1.2 Trehalolipids

Most of the microbial species including *Mycobacterium*, *Nocardia* and *Corynebacterium* are involved in the production of trehalolipids groups of biosurfactant. This sort of biosurfactant contains mycolic acid which is linked to trehalose disachharide at C-6 and C-69 positions.

2.1.3 Sophorolipids

Sophorolipids, are the groups of biosurfactant dominantly produced by the species of yeast including *Torulopsis bombicola*, *T. petrophilum* and *T. apicola*. Sophorolipids structurally contain dimeric sophorose that are linked to long chain hydroxyl fatty acids.

2.2 Lipopeptides and Lipoproteins

Lipopeptide biosurfactants are high molecular weight biosurfactant, which are amphiphilic in nature and produced not only by fungal species, e.g., *Aspergillus* but also bacterial species, e.g., *Streptomyces*, *Pseudomonas* and *Bacillus*. Similarly a marine *Azotobacter chroococum* produces lipopeptides that shows significant biodegradations (Thavasi et al. 2009) and heavy metal remediation properties (Das et al. 2009). These marine species use aromatic hydrocarbons as the substrate for the productions of such biosurfactant (Das et al. 2008b). Structurally lipopeptides or lipoproteins contain a lipid chain linked to short linear or cyclic oligopeptides. These sorts of biosurfactant can also termed as (amphipathic) lipopolysaccharides or lipoproteins or their mixtures peculiar in net charge of the molecule specifically negative, e.g., surfactin, daptomycin or positive, e.g., polymyxin (Seydlová and Svobodová 2008). Lipopeptides occur alongside with oil dispersion activity, they also have efficient emulsification, biodegradation and remediation properties which makes them suitable for the industrial applications (Sivapathasekaran and Sen 2017, Thavasi et al. 2009, Shaligram and Singhal 2010, Bonilla et al. 2005). These lipopeptides are categorized into three families of cyclic compounds such as surfactin, iturin and fengycin (Raaijmakers et al. 2010) and are also reported in the bacterial species *Bacillus subtilis* MBCU5 (Pandya et al. 2017).

2.2.1 *Surfactin*

Surfactin sort of biosurfactant are reported to be produce by *Bacillus* species for colony spreading in the absence external fluid flow (Kinsinger et al. 2003, Kinsinger et al. 2005, Pornsunthorntawee et al. 2008). It was also observed that *Psuedomonas aeruginosa* BS2 species showed the production of surfactin in the synthetic media and in the industrial effluent of distillery (Dubey and Juwarkar 2001). Surfactin (Fig. 2) can be distinguished as A, B and C on the presence of amino acid sequence such as L-leucine, L-valine and L-isoleucine respectively during formation of the lactone ring

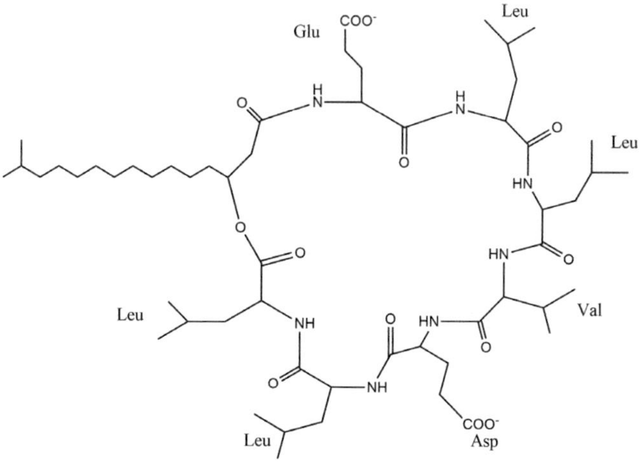

a. General structure of surfactin

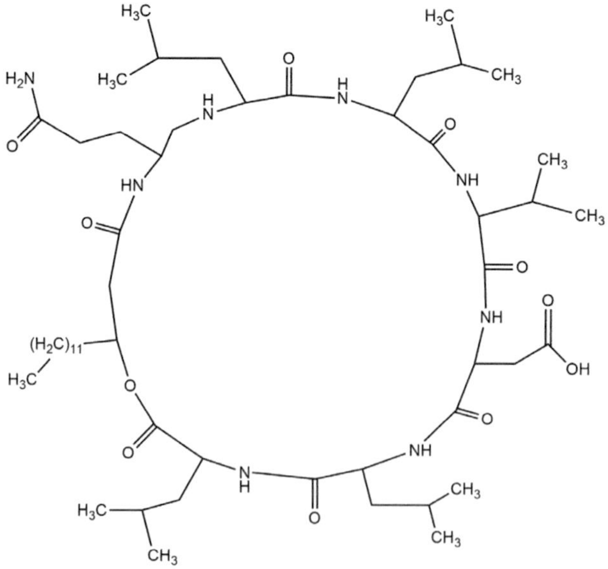

b. Lichenysin

Figure 2. Chemical structure of; (a) Surfactin, (b) Lichenysin.

by using C14–C15 b-hydroxy fatty acid (Singh and Cameotra 2004). Surfactin are well studied biosurfactant and have wide applications such as: (a) prevent clot formation by fibrin, (b) ion channels induction in lipid bilayer membranes, (c) cAMP (cyclic adenosine monophosphate) inhibition, (d) inhibition of platelet along with phospholipase A2 (PLA2) of spleen cytosol and (e) antimicrobial and antitumor properties (Singh and Cameotra 2004). The surfactin biosurfactants have several heptapeptide variants such as esperin, lichenysin, pumilacidin, etc.

2.2.1.1 Esperin

The surfactant comes under surfactin as one the variants of a well established structure brought out by mass spectrophotometry. The revised structure of esperin or esperinic acid is RCHOHCH$_2$CO.Glu.Leu.Leu.Val.Asp.Leu.Leu (Val).OH, where R is C$_{10}$H$_{21}$, C$_{11}$H$_{23}$ and C$_{12}$H$_{25}$ while the other end of the chain, i.e., C-terminal residue has a substitutions of valine around 30%. Esperin as the function of aspartic acid residue has a similar structure with that of OH derived (OH of fatty acids) lactone with carboxylic acid.

2.2.1.2 Lichenysin

This sort of biosurfactant has wide applications in Microbial Enhanced Oil Recovery along with textile, pharmaceutical, cosmetics, etc. *Bacillus licheniformis* growing on hydrocarbon medium supplemented with glucose as the carbon sources produces lichenysins and are anionic cyclic lipoheptapeptide in nature (Nerurkar 2010). The other variants of lichenysins also reported from *Bacillus licheniformis* are lichenysins A, B, C, D, G and surfactant BL86. Lichenysin A has broad spectra of antimicrobial properties. Lichenysin can produced in aerobic as well anaerobic conditions like that of other variants of surfactin while in case of lichenysin B productions there were no need of supplementation with crude oil and can be produced on mineral salts without supplementations of hydrocarbons. Productions of lichenysin G observed on the medium is supplemented with 1-glucose or 1-isoleusine.

2.2.1.3 Pumilacidin

The biosurfactant pumilacidin is produced by the marine bacterial species *Bacillus pumilus* and have antimicrobial properties against *Staphylococcus aureus* with molecular weight less than 3 kDa. Pumilacidin is active at wide range of pH and its productions can be controlled by the growth phase of producing microbes, medium and temperature. Chemically pumilacidin is a mixture of cyclic heptapeptides linked to fatty acids of variable length (Saggese et al. 2018), such cyclic lipopeptides secreted by *Bacillus* sp. strain 176 have significant activity against *Vibrio alginolyticus* strain 178 (Xiu et al. 2017).

2.2.2 Iturin

The production of biosurfactant in the bacterial species *Bacillus subtilis* MBCU5 reported by Pandya et al. (2017). The iturin (Fig. 3) produced by bacterial species are the mixtures of iturin variants having variation at C-length (C14 to C17) of β amino fatty acids and its isomers (Bonmatin et al. 2003, Tilvi and Naik 2007) and having significant antifungal activity (Ongena and Jacques 2008). Lipopeptide biosurfactant

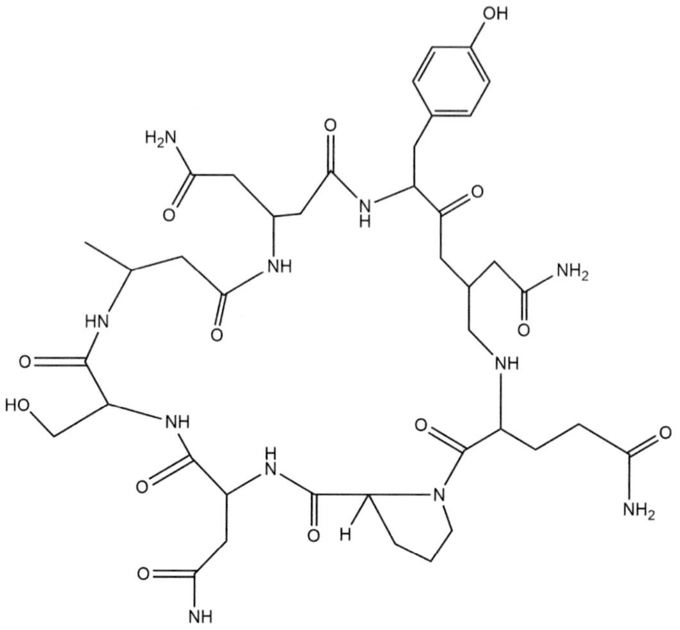

(a) Iturin A

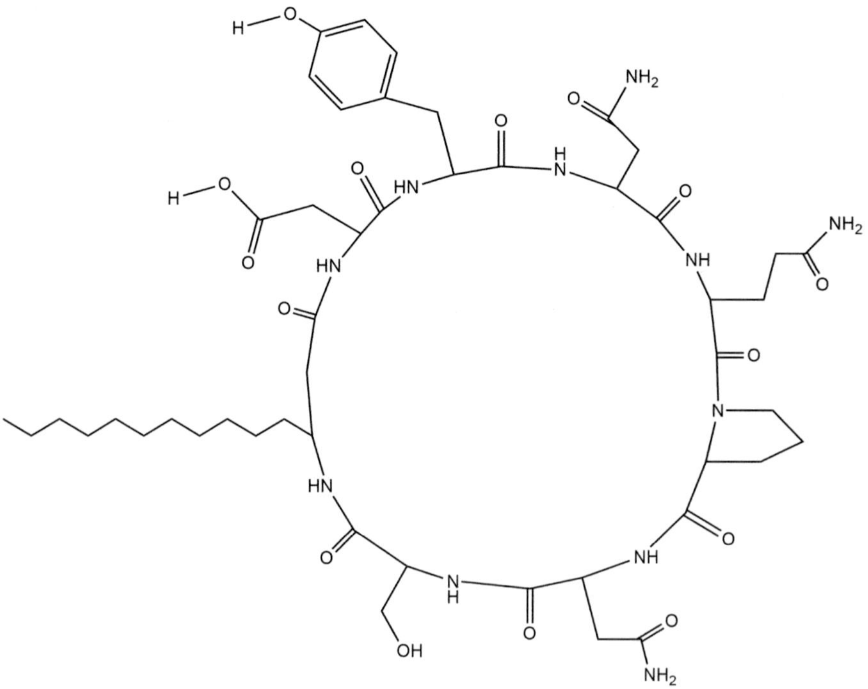

(b) Iturin C

Figure 3 Contd. ...

...Figure 3 Contd.

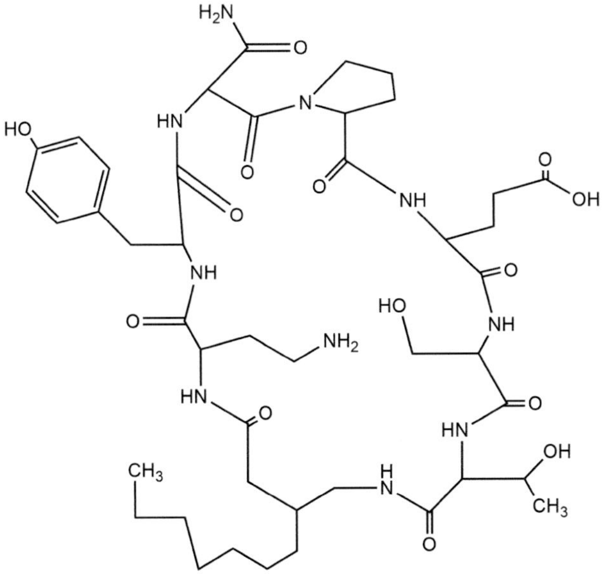

(c) General structure of Bacillomycin

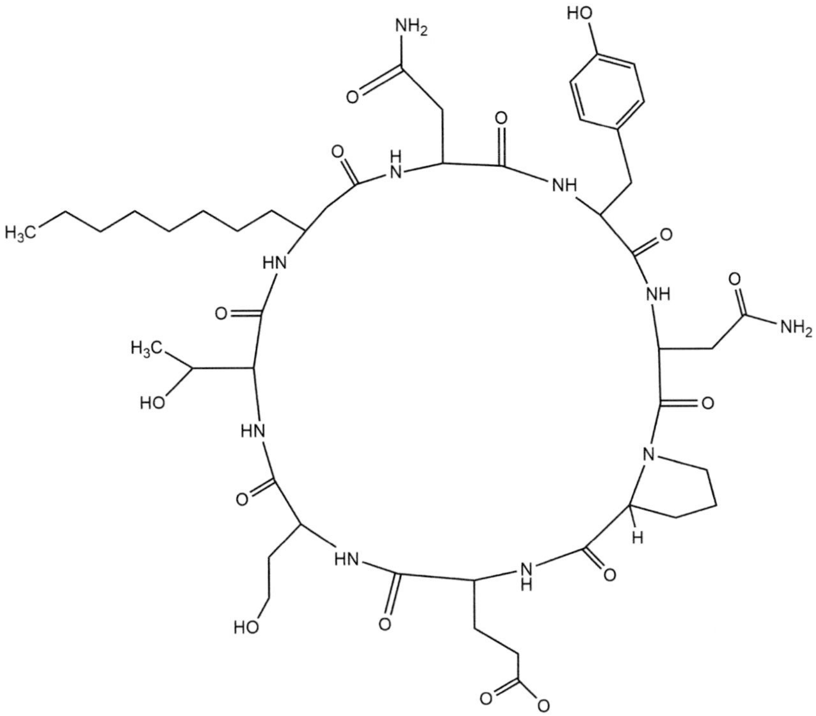

(d) Bacillomycin D

Figure 3 Contd. ...

...Figure 3 Contd.

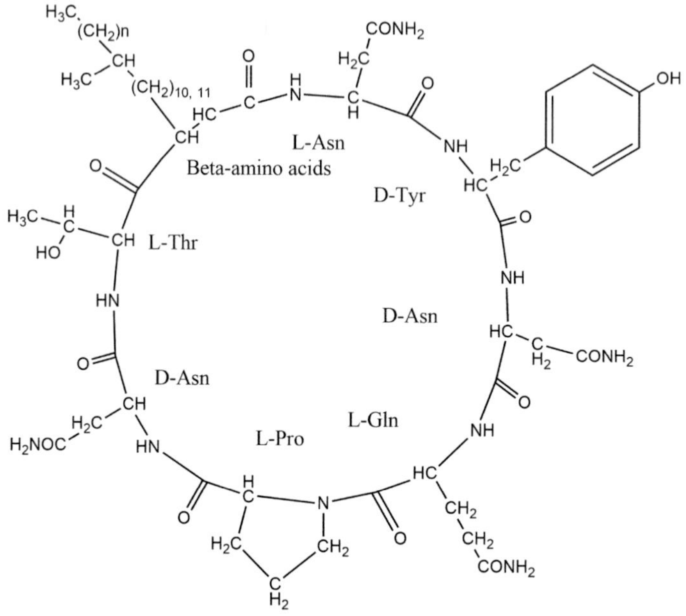

(e) Bacillomycin F

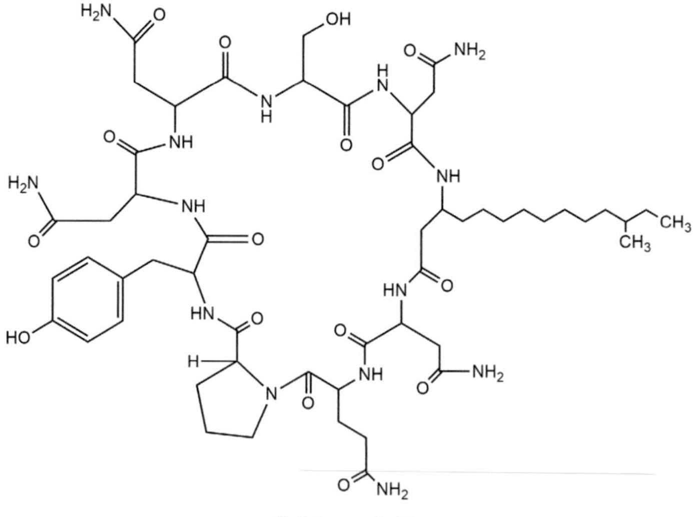

(f) Mycosubtilin

Figure 3. Chemical structure of Iturin group of surfactants; (a) Iturin A, (b) Iturin C, (c) General structure of Bacillomycin, (d) Bacillomycin D, (e) Bacillomycin F and (f) Mycosubtilin.

iturin can be further categorized in to six variants viz. iturin A and C, bacillomycin D, F and L, and mycosubtilin.

2.2.2.1 Iturin A

A well known antimicrobial biosurfactant produced by *Bacillus subtilis* are itutin A that are also counted as the potential antifungal agent for mycosis (Singh and Cameotra 2004). Iturin A is reported to form small vesicles and aggregation of intra-membranous particles which disrupts the plasma membrane and increases the release of electrolytes and products of high molecular mass which results in degradations of phospholipids. Iturin A has a strategic activity against attacking pathogens by increasing the electrical conductance of biomolecular lipid membranes and the formation of pores by lipopeptides. *Bacillus subtilis* K1 has also shown that the production of antifungal lipopeptides contains four variants of iturin A with varying length of β-amino fatty acid (Pathak and Keharia 2014).

2.2.2.2 Iturin C

Iturun C is one of enzyme inhibitor produced by the *Bacillus* strain having peptide moiety with aspartic acid instead of asparagine in iturin A (Jamshidi-Aidjia and Morlock 2018). In one study, it was recorded that *Bacillus subtilis* T-1 have genes for the iturin C along with other biosurfactants like surfactin (Plaza et al. 2006). Lipopeptide biosynthetic genes for the iturin C were also detected in *Bacillus subtilis* B1 and confirmed through the RT-PCR and protonated potassium derivatives of iturin C confirmed through MALDI-TOF-MS (Sajitha et al. 2016). *Bacillus subtilis* K1 produces complex mixture of antifungal lipopeptides and through the mass spectrophotomeric analyses it was observed that it contains three variants of iturin C with varying chain length from C13 to C17 of β-amino fatty acid (Pathak and Keharia 2014).

2.2.2.3 Bacillomycin D

Bacillomycin D sort of iturin biosurfactants are reported to have antimicrobial activity (Moyne et al. 2001) and their lipopeptide biosynthetic genes for the Bacillomycin D also detected in *Bacillus subtilis* B1 and confirmed through the RT-PCR and protonated potassium derivatives of Bacillomycin D confirmed through MALDI-TOF-MS (Sajitha et al. 2016). Sajitha et al. (2016) reported C13 isoform in *Bacillus subtilis* B1, along with reported isoforms of bacillomycin D (C14–C17). The production of bacillomycin D is shown by the bacterial species *Bacillus subtilis* strain Bs49 (Bacillomycin D) is known for the induction of apoptotic activity against cancerous cells and also for having antifungal activity against *Aspergillus ochraceus* in food samples (Lin et al. 2018).

2.2.2.4 Bacillomycin F

Bacillomycin F biosurfactant produced by the bacterial species *Bacillus subtilis* strain I164, have similar structural characteristics like that of iturin A with some differences such as containing L-threonine instead of an L-serine residue which is linked to the β-amino acid and isoC16 and unteisoC1 lipid moiety instead of nC14 and isoC15 of β-amino acid present in bacillomycin F. Bacterial species *Bacillus subtilis* isolated from honey has shown broad spectrum antifungal activity because of bacillomycin F with several variants having varying lengths of aliphatic fatty acids from C14 to C16 (Lee et al. 2008).

2.2.2.5 Bacillomycin L

Bacillomycin L sort of iturin biosurfactants are reported to have antimicrobial activity (Moyne et al. 2001) and biocontrol activity against fungal plant pathogens including *Rhizoctonia solani* Kühn. The bacillomycin L are produced by the bacterial species *Bacillus amyloliquefaciens* K103 growing on lemon (Zhang et al. 2013). Bacillomycin L have a similar structure with that of bacillomycin D with small difference in three amino acids at 1, 3 and 4 among seven with Asp, Ser, and Gln, respectively. For biofilm formation and root colonizations along with surfactin, bacillomycin L is the necessary component to sustain bacteria (*Bacillus subtilis*) in unfavorable conditions (Zeriouh et al. 2014).

2.2.2.6 Mycosubtilin

The biosurfactant mycosubtilin produced by *Bacillus subtilis* helps to form biofilm and colonize roots to challenge against harsh environmental conditions and potential competitors (Ahimou et al. 2000, Ongena and Jacques 2008, Xu et al. 2013). Mycosubtilin are less ecotoxic and help in controlling lettuce downy mildew disease (Deravel et al. 2014) and is considered as the significant antifungal agent (Leclère et al. 2005, Fickers et al. 2009, Fickers 2012, Béchet et al. 2013). The bioactive surfactants like mycosubtilin and surfactin form efficient mixed micelles which are more effective in comparison with simple micelle on biological membranes (Jauregi et al. 2013).

2.2.3 *Fengycin or Plipastatins*

Fengycin (Fig. 4) or Plipastatins is a well known lipopeptide biosurfactant having 16–19 carbon chain with β-hydroxy fatty acid produced by *Bacillus* spp. having antifungal (Chen et al. 2009), antibacterial (Fracchia et al. 2012), antiviral (Huang et al. 2006) and anticancerous (Yin et al. 2013) properties. Fengycins are synthesized through the mechanism of non-ribosomal peptide synthetases (NRPS) (Yaseen et al. 2016). This biosurfactant produced by the *Bacillus subtilis* MBCU5 and are chemically cyclic lipodecapeptides (Pandya et al. 2017). It is found in several isoforms with varying chains of fatty acids and amino acids (Wei et al. 2010, Ramarathnam et al. 2007). Plipastatins produced by *Bacillus cereus* BMG302-fF67 and *Bacillus subtilis* (Volpon et al. 2000) are antifungal antibiotics and play an important role in the inhibition of porcine pancreatic Phospholipase A2. The lipopeptides group of family includes biosurfactant, i.e., fengycin which can be further categorize into variants viz. fengycins A, fengycins B.

2.2.3.1 Fengycins A

Isoforms of fengycin having D-Ala at 6th residue of amino acid composition is fengycin A (Wei et al. 2010). Lipopeptide biosynthetic genes for Fengycins A is also detected in *Bacillus subtilis* B1 and confirmed through the RT-PCR and the protonated potassium derivatives are confirmed through MALDI-TOF-MS (Sajitha et al. 2016). The C16-fengycin A produced by the *Bacillus amyloliquefaciens* fmb60

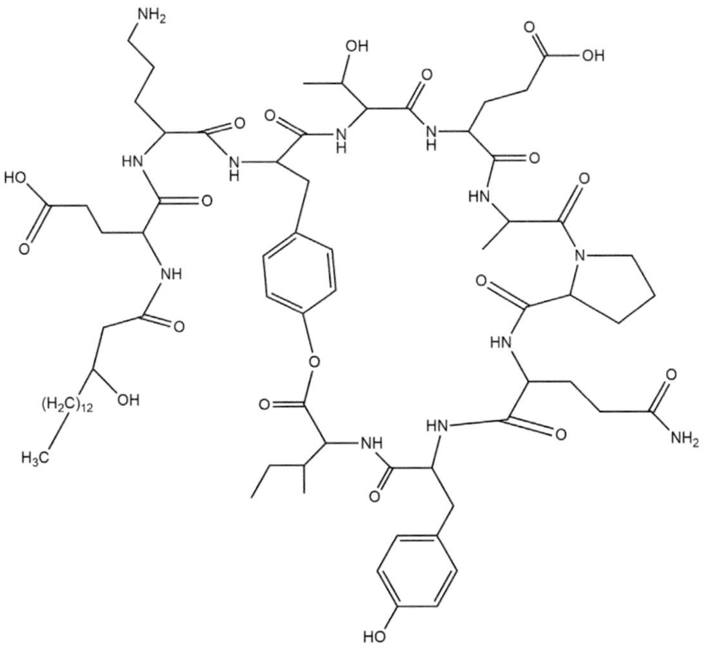

Figure 4. General chemical structure of Fengycins.

has strong antifungal activity against *Candida albicans*. C16-fengycin A destroys the fungal cell wall structure and reduces its hydrophobicity and induces increase in the reactive oxygen species leading to malfunctioning of the mitochondria (Liu et al. 2019).

2.2.3.2 Fengycin B

Fengycin B is another variant of fengycin sort of biosurfactants. Lipopeptide biosynthetic genes for the Fengycin B are detected in *Bacillus subtilis* B1 and confirmed through the RT-PCR and protonated potassium derivatives of Bacillomycin D confirmed through MALDI-TOF-MS (Sajitha et al. 2016). Isoforms of fengycin having D-Val at 6th residue of amino acid composition is fengycin B (Wei et al. 2010).

2.3 Fatty Acids

Fatty acid sort of biosurfactant are important with respect to household application, cosmetics and commercial purposes. This type of biosurfactant can be distinguished by their different derivatives viz. fatty acid ester, fatty acid ester of sugar and sugar alcohol (Gautam and Tyagi 2006).

2.3.1 Fatty Acid Ester and Sugar Fatty Acid Esters

The sugar fatty acid esters biosurfactant such as lactose fatty acid ester can be synthesized in different organic solvents by using lipase (extracted from *Candida antartica* B) from fatty acids and lactose (Enayati et al. 2018). The well known and most developed biosurfactant in this category are sorbitan esters and sucrose esters

(Ruiz 2008). Carbohydrate fatty acids esters have diverse applications in the field of cosmetics, food and pharmaceutical industries. The fatty acid esters of trisaccharides are more hydrophilic due to being in sugar head group; hence these have significant bioactive characteristics (Gonzalez-Alfonso et al. 2018). The sugar fatty acids with fatty acid chain length C14–C16, at minor concentrations 0.001% (w/w) acts as anti-biofilm agent without affecting bacterial growth in case of *Streptococcus mutans* and *Listeria monocytogenes* while in case of *Staphylococcus aureus* prevents adherence to abiotic substratum. At the higher doses (0.1% (w/w)) these sugar fatty acids show similar anti-biofilm activity against *Pseudomonas aeruginosa* (Furukawa et al. 2010).

2.3.2 *Sugar Alcohol*

The biosurfactant substances like sugar alcohols or polyols play a vital role in the physiological and metabolic system of plants as the major sink of photosynthates and can be readily transported (Dumschott et al. 2017). The soluble components of the plants along with leaf and phloem tissues (Dinant 2008, Reidel et al. 2009) contain major portions of sugar alcohols. Sugar alcohols are more or less equally responsible for the primary productions like that of carbohydrates and absence of functional groups (unlike carbohydrates) makes them a suitable candidate for the storage, physiological functions and transport (Merchant and Richter 2011). Sugar alcohols also act as the stress metabolites, the concentrations of these sugar alcohols increases with respect to unfavorable environmental conditions (Williamson et al. 2002).

2.4 *Phospholipids*

Phospholipids (Fig. 5) are the important constitutions of lung surfactant along with protein complex which maintains reduced surface tension at the immiscible interface between liquid and pulmonary air (Bernhard et al. 2016). Phospholipids are also the important substance in several secretions along with bile and lipoproteins and are present in cell membranes. Physiological secretions which contain phospholipids are Eustachian tube (Panaanen 2002), pleura and peritoneum, gastric and intestinal secretions (Bernhard et al. 2001, Ehehalt et al. 2004) and lung surfactants. These secretions, contains unique surfactant at the interface of liquid and pulmonary air (Benson et al. 1983). Phospholipids as Dipalmitoylphosphatidylcholine (DPPC) are the important part of pulmonary surfactant which maintains breathing work and avoids alveolar collapse. This Phospholipid DPPC is lecithin and having two palmitic acids (C16) connected to Phosphotidylcholine (PC) head group.

Lung surfactant phospholipids are the complex mixture of esterified fatty acyl moieties belonging to glycerophosphate at SN-1 and SN-2 position. Phospholipids are synthesized and stored at intracellular lamellar bodies by the alveolar type II epithelial cells (ATII) and targeted for the apical secretions into the extracellular lung lining fluid of alveolus as the lung surfactant (Goss et al. 2013). Phospholipids lung surfactant is also responsible for the maintenance of alveolar homeostasis through their molecular adaptations with lung physiology (Bernhard et al. 2007, Pynn et al. 2010, Gille et al. 2007). Isolated type II pneumocytes (PNII) through the basolateral membrane secretes phospholipids, majorly in the form of phosphotidylcholine which brings systemic phospholipid homeostasis which somehow overcomes their

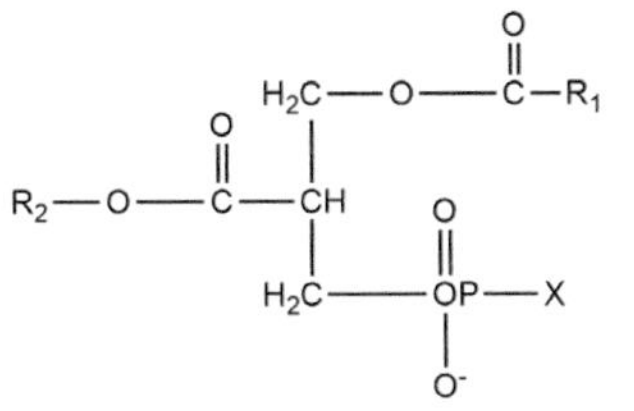

X = -OCH2CH2N+(CH3)3 Phosphatidylcholine (Lecithin), X = -OCH2CH2NH3+ Phosphatidylethanolamine (Cephalin)

X =

Phosphatidylserine

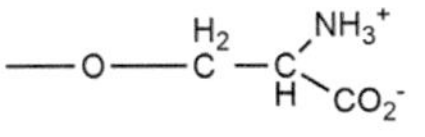

X =

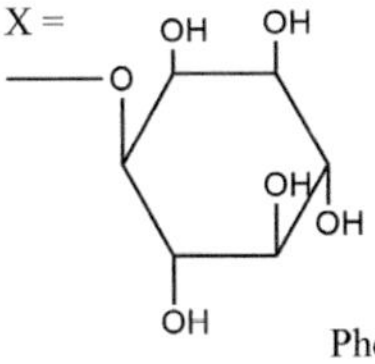

Phosphatidylinositol

X = -OCH2CHOHCH2OH Phosphatidylglycerol

Figure 5. General chemical structure of phospholipids.

secretions of lung surfactant (Bernhard et al. 2016). Phospholipid surfactants are principally glycerophospholipids and phosphotidylcholine contains oleic (with palmitoyl-oleoyl-PC) and linoleic (with palmitoyllinoleoyl-PC) (Bernhard et al. 2001, Bernhard et al. 2016, Gauss et al. 2013).

In the lung physiology and developments, amphilic phospholipids are involved, in which phosphate in combinations with choline, ethanolamine or glycerol present in varying head groups like phosphotidylcholine, phosphotidylethanolamine or phosphotidylglycerol, respectively. The individual molecular species of respective phospholipid class are responsible for the peculiar fluidity, temperature transition of phases and for the precursors of eicosanoids (Dennis and Norris 2015, Wiktorowska-Owczarek 2015).

2.5 *Neutral Lipids*

Neutral lipid are the important component of the pulmonary surfactant and in association with surfactant proteins A (SP-A) aids in the formation DPPC-rich reservoir below the air-water interface. Neutral lipids in presence with SP-A induces the aggregations of DPPC and helps in the transport of DPPC from lipid extract surfactant to the surface. Presence and concentrations of neutral lipids differs with the sort of pulmonary surfactant such as 4% (w/w) in bovine and 7% in canine surfactant. Cholesterol is the major components of neutral lipids and counts around 90% in bovine neutral lipids and acts as the fluidizer that helps for absorption of DPPC which are re-spreading. Cholesterol has negative effect on the surfactant with respect to surface tension reduction in absence of Surfactant Protein A, as it lowers the stabilizations of compressed surface films.

2.6 Polymeric Biosurfactants

Polymeric biosurfactant are extracellular polymers produced by some of the microbes such as *Mesorhizobium, Mucilaginibacter, Paracoccus, Pedobacter, Rahnella,* and *Sphingobium*. These sorts of biosurfactant are involved in the emulsification of hydrophobic substrates (hydrocarbon compounds), enhancement in the viscosity at low pH and microbial interactions (Perfumo et al. 2010). Less information is known regarding surface properties of polymeric biosurfactant (Ozdemir et al. 2003, Johnsen and Karlson 2004, Diggle et al. 2008, Han et al. 2009, Han et al. 2012). Polymeric biosurfactant comprises several variants and best studied polymeric biosurfactants are emulsan, alasan, biodispersan, liposan, lipomanan and some other polysachharide protein and lipopolysachharide complexes (Fenibo et al. 2019a).

2.6.1 Emulsan

Emulsan sort of polymeric biosurfactant, as the name itself suggests, are involved in the emulsifications of hydrocarbons in water at low concentrations—0.001% to 0.01% (Marchant and Banat 2012). These are the high molecular weight biosurfactant (approximately 1 MDa) produced by *Acinetobacter calcoaceticus* and are structurally polysanionic lipopolysachharide in nature, which contains complex of alanine, proteins and polysachharides (Bhattacharya et al. 2017).

2.6.2 Alasan

Alasan biosurfactant such as that of emulsan are involved in the emulsifications of hydrocarbons (paraffins, crude oils, long chain of aliphatic and polyaromatic hydrocarbons) and also aids in solubilizations of polyaromatic hydrocarbons (Bhattacharya et al. 2017) produced by *Acinetobacter radioresistens* KA53 and chemically are a high molecular weight complex structure of polysaccharide and protein. The pure polysaccharide obtained from the complex are termed as the apo-alasan and have lesser activity in a comparison complex structure (Toren et al. 2001). Proteins of the complex are necessary for the activity and structure of the complex while polysaccharides (apo-alasan) do not have emulfying properties, hence do not express the characteristics like that of alasan itself such as change in hydrodynamic shape due to temperature induction. Proteins of the alasan along with the complex together are synergistically active emulsifier (Toren et al. 2001).

2.6.3 Biodispersan

The surfactant biodispersan are heteropolysaccharide in nature with a molecular weight of around 51 kD and are extracellularly produced by *A. calcoaceticus* A2. Similar with that of emulsan, biodispersan have surface active properties and also act as the dispersing and stabilizing agent for several industrial applications. Biodispersan disperses big lime stoneform into µm granules of water suspension. The ions such as Mg^{2+} and PO^{3-} ions have negative impact on the dispersion activity of biodispersan. Biospersan like other similar surfactant, biosynthesize inside the cells and are secreted in external medium as the cell-associated capsules.

2.6.4 Liposan

Liposan sort of polymeric biosurfactants are extracellular, water soluble and can be used as a bioemulsifier. These types of biosurfactant were reported for the first time in the yeast species *Candida lipolytica* and are also produced by *Acinetobacter radioresistens*. Chemically liposan are a complex of protein and carbohydrates in a ratio of approximately 17 and 83% respectively (Chakrabarti 2012, Lahiry and Sinha 2017, Bhattacharya et al. 2017).

2.6.5 Lipomannan

Lipomannan (Fig. 6) are also one of the best studied polymeric biosurfactant which are chemically polysaccharide protein complexes (Santos et al. 2016) produced by *Candida tropicalis* (Galabova et al. 2014). These have a similar structure to that of the microbial whole cell lipid, and constitutes around 50–60 residues which have glycosidic linkages with diglyceride sort of fatty acids. Lipomman most often is present on cell surface of mycobacteria and have a role in the virulence and survival in host cells (Sawettanai et al. 2019).

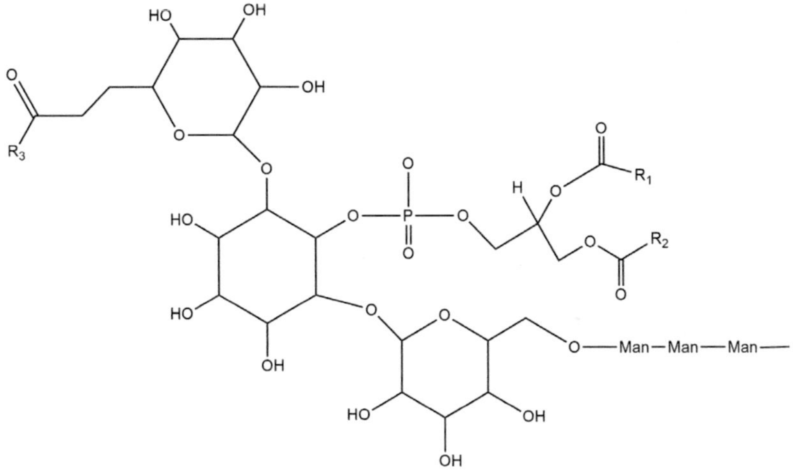

Figure 6. Chemical structure of Lipomannan sort of biosurfactant.

2.7 Particulate Biosurfactant

Particulate biosurfactant plays an important role in the microbial uptake and chemically is the complex of protein, phospholipids and lipo-polysaccharides which forms extracellular membrane vesicles for the compartmentalization of micro-emulsion forming hydrocarbons. This sort of biosurfactant is produced by species of *Acinetobacter* (Chakrabarti 2012). Bhattacharya et al. (2017) reported vesicles of particulate biosurfactants from *Acinetobacter* sp. strain HO1-N with buoyant density of 1.158 g/cm^3 and diameter 20–50 nm.

3. Synthetic Surfactants

Synthetic surfactants are affordable comparative to biosurfactant, because the latter have restricted use and are very expensive due to high productions costs. The production costs can be minimized by using; (a) low cost and waste material as substrate, (b) optimization of production and overproduction of biosurfactant by improvement and development of the strain and (c) optimization of efficient fermentations processes for it (Christova et al. 2013). These synthetic surfactant are similar in functions and structure with some of their biological analogs, but in case of Bioelectrochemical systems, their applications to the living and non-living components are completely different (Pasternak et al. 2020). This bioelectrochemical system has multiple applications with the surfactant—synthetic or natural. Synthetic surfactant in microbial fuel cell (MFC) help to minimize biodeterioration of cathodes at high pH (Pasternak et al. 2016). Presence of a surfactant can improve efficient biofilm formation by interfacing at the surface of electrodes (Zhang et al. 2017) and also increases the availability of hydrophobic substrates to biological systems (Hwang et al. 2019).

In comparison to biosurfactant, synthetic surfactants are not good in degrading organic substrates and that leads to toxic subsequences (Christofi and Ivshina 2002). Synthetic surfactants are used for industrial purposes are derived from petroleum based substrates through the organo-chemical synthesis (Dreja et al. 2012). Applications of synthetic surfactant generate two sorts of pollutants from synthetic surfactants, i.e., its remnants and by-products (Fenibo et al. 2019b). Synthetic surfactants are derived from non-renewable sources and are unable to mediate biodegradation because of their incompatibility and create toxicological effects to living organisms including humans and also lowers functional diversity (Gutierrez 2019, Banat et al. 2010, Fracchia et al. 2012).

3.1 Classification of Surfactant Based on Composition of Head Group

Surfactants classified on the basis of the chemical nature and polarity of their head group are, Anionic, cationic, nonionic and zwitterionic (Dave and Joshi 2017).

3.1.1 Anionic

Linear and non-linear alkylbenzene sulphonate sort of surfactants are considered as the Anionic surfactant (Kowalska et al. 2004, Osadebe et al. 2018) which carry negative charge and are commonly available naturally and synthetically (Bratovcic et al. 2018, Sil et al. 2017). Anionic surfactants having adequate balance of hydrophilic and hydrophobic balance values, good emulsification properties and are efficient in surface tensions reduction, hence very useful in cleansing systems such as personal care products and soaps. Also it has wide applications in oil industry, remediation, pharmaceutics and therapeutics, agriculture and cosmetics (Akbari et al. 2018).

3.1.2 Cationic

The cationic surfactants are effective against opposite charged surfaces, hence very useful as hair conditioners and softeners. Also they have wide applications as anti-corrosive agents, bactericidal, fabric softeners and flotation collectors (Rhein 2002).

The cationic surfactant produced by microbes or its derivations such as alamethicin and mellitin can enhance the immunity of its producers and aids in the conventional antibiotic therapy (Seydlová and Svobodová 2008, Giuliani et al. 2007, Zhang and Falla 2004).

3.1.3 Nonionic

The non-ionic sort of surfactants bear no charge in their head groups but are hydrophilic in nature; hence they are low irritating and effective at low temperatures (Alwadani and Fatehi 2018).

3.1.4 Zwitterionic

Zwitterionic surfactants bear both charges in their head groups and hence are amphoteric in nature and less effective in the cleansing and emulsifications systems. But these surfactants have good dermatological properties and can be effective constituents in cosmetics (Bratovcic et al. 2018). This type of surfactant is skin compatible and used in shampoo preparations (Lukic et al. 2016).

3.2 Industrial Pollutants from the Synthetic Surfactants

The synthetic surfactants have diverse application in the household and in industrial products. Synthetic surfactants are easily available; they are somewhat cheaper in comparison with the biosurfactant. Excess use of these surfactants have a negative effect on eco-toxicity as it releases remnants and toxic byproducts of the surfactants into the aquatic environments. The household or anthropogenic contaminants of surfactants are carried through the waste water into the water bodies like river, lakes and other sediments. Gradual accumulation (micrograms to grams) of surfactants reaches above the concentrations of guideline limits and creates eco-toxic conditions. The most often observed surfactants are linear alkylbenzene sulfonate (LAS), quaternary ammonium compounds (QACs), alkylphenol ethoxylate (APEOs) and alcohol ethoxylate (AEOs) (Jardak et al. 2016). Surfactants released in the form of remnants and pollutants interacted with the biological entities such as peptides, enzymes and DNA by changing the surface of molecule and folding and affected the proper functioning. Cationic surfactants prefer to target the bacterial inner cytoplasmic membrane and disrupt their cellular organization with the chains of alkyl polymers (e.g., Quaternary ammonium compounds). In case of nonionic surfactants, they have antimicrobial activity by binding to the proteins and phospholipids of microbes and increase the permeability of cellular membranes leading to the efflux of necessary nutrients and molecules, resulting in death of the organisms (Ivanković and Hrenović 2010).

4. Application of Biosurfactant

Biosurfactant are a diverse group of chemical compounds having potential industrial and routine applications such as in pharmaceutical/medicine, food, cosmetic, pesticides, oil and biodegradation. The microbes producing biosurfactant are useful for enhanced oil recovery. In comparison to synthetic surfactants, biosurfactant are very particular and needs in trace amount and are efficient in a broad range of oil and reservoir conditions. Biosurfactant have a special role in the food industry as a thickener and emulsifier. The major role of biosurfactant is in the field of environment protection like bioremediations and heavy metal mitigations (Frazetti et al. 2009, Asci et al. 2010) and removal of hydrocarbons from the contaminated sites (Kakugawa et al. 2002). Biosurfactant have anti-adhesive, antiviral, antifungal, etc. like properties hence, they have wide applications in pharmaceutics and medical science (Mukherjee et al. 2006).

5. Conclusion

Biosurfactant are naturally synthesized by microbes such as bacteria, fungi and yeast growing on hydrophobic sources. Biosurfactant have several commercial applications in food industry, cosmetics, agriculture and bioremediations because of their several useful characteristics such as biodegradability, less toxicity, eco-friendly; they can be produced from cheaper renewable substrate and can be active at extreme unfavorable conditions like high temperature, pH and salinity. Biosurfactant in comparison to synthetic surfactant are easy to degrade hence they have application in the waste management, bioremediations and they also manage bioavailability of toxicants in environment. Biosurfactant produced by species of *Bacillus* having enormous application in field of biotechnology and pharmaceutics.

Bioprocess technique for the production biosurfactant observed in the productions of rhamnolipid show significant growth of free bacterial cells in the immobilized system, but as the surfactant produced by the microbes are very expensive, their uses and applications becomes restricted. To overcome such issues, more research should focus on the production of biosurfactant with cheaper substrate or waste products, development and implementations of low cost method for downstream purification and extraction, exploring the recombinant DNA approach to increase the yield and quality, and identification of novel species for the production of biosurfactant from the extreme environments. Biosurfactant synthes is associated with the microbes degrading hydrocarbons, such species should be exploited for more production of quality biosurfactant.

References

Abdollahi, S., Z. Tofighi, T. Babaee, M. Shamsi and G. Rahimzadeh. 2020. Evaluation of antioxidant and anti-biofilm activities of biogenic surfactants derived from *Bacillus amyloliquefaciens* and *Pseudomonas aeruginosa*. Iranian Journal of Pharmaceutical Research 19(2): 115–126.

Ahimou, F., P. Jacques and M. Deleu. 2000. Surfactin and iturin A effects on *Bacillus subtilis* surface hydrophobicity. Enzyme Microb. Technol. 27: 749–754.

Akbari, S., N.H. Abdurahman, R.M. Yunus, F. Fayaz and O.R. Alara. 2018. Biosurfactants—A new frontier for social and environmental safety: a mini review. Biotechnology Research and Innovation 2(1): 81–90.

Alwadani, N. and P. Fatehi. 2018. Synthetic and lignin based surfactants: Challenges and opportunities. Carbon Resour. Convers. 1: 126–138.

Asci, Y., M. Nurbas and Y.S. Acikel. 2010. Investigation of sorption/desorption equilibria of heavy metals ions on/from quartz using rhamnolipid biosurfactant. J. Environ. Manage. 91: 724–731.

Banat, I.M., A. Franzetti, I. Gandolfi, G. Bestetti, M.G. Martinotti, L. Fracchia, T.J. Smyth and R. Marchant. 2010. Microbial biosurfactants production, applications and future potential. Applied Microbiology and Biotechnology 87: 427–444.

Béchet, M., J. Castéra-Guy, J.S. Guez, N.E. Chihib, F. Coucheney, F. Coutte, P. Fickers, V. Leclère, B. Wathelet and P. Jacques. 2013. Production of a novel mixture of mycosubtilins by mutants of *Bacillus subtilis*. Bioresour. Technol. 145: 264–270.

Benson, B.J., J.A. Kitterman, J.A. Clements, E.J. Mescher and W.H. Tooley. 1983. Changes in phospholipid composition of lung surfactant during development in the fetal lamb. Biochim. Biophys. Acta 753: 83–88.

Bernhard, W., A.D. Postle, G.A. Rau and J. Freihorst. 2001. Pulmonary and gastric surfactants. A comparison of the effect of surface requirements on function and phospholipid composition. Comp. Biochem. Physiol. A Mol. Integr. Physiol. 129: 173–182.

Bernhard, W., A. Schmiedl, G. Koster, S. Orgeig, C. Acevedo, C.F. Poets and A.D. Postle. 2007. Developmental changes in rat surfactant lipidomics in the context of species variability. Pediatr. Pulmonol. 42: 794–804.

Bernhard, W. 2016. Lung Surfactant: Function and Composition in the Context of Development and Respiratory Physiology. Annals of Anatomy 208: 146–150.

Bernhard, W., M. Raith, V. Koch, C. Maas, H. Abele, C.F. Poets and A.R. Franz. 2016. Developmental changes in polyunsaturated fetal plasma phospholipids and feto-maternal plasma phospholipid ratios and their association with bronchopulmonary dysplasia. Eur. J. Nutr. 55(7): 2265–74.

Bhattacharya, B., T.K. Ghosh and N. Das. 2017. Application of bio-surfactants in cosmetics and pharmaceutical industry. Sch. Acad. J. Pharm. 6(7): 320–329.

Bonilla, M., C. Olivaro, M. Corona, A. Vazquez and M. Soubes. 2005. Production and characterization of a new bioemulsifier from *Pseudomonas putida* ML2. J. Appl. Microbiol. 98: 456–463.

Bonmatin, J.M., O. Laprevote and F. Peypoux. 2003. Diversity among microbial cyclic lipopeptides: Iturins and surfactins. Activity-structure relationships to design new bioactive agents. Comb. Chem. High Throughput Screening 6: 541–556.

Bratovcic, A., S. Nazdrajic, A. Odobasic and I. Sestan. 2018. The influence of type of surfactant on physicochemical properties of liquid soap. Int. J. Mat. Chem. 8: 31–37.

Cameotra, S.S. and R.S. Makkar. 2004. Recent applications of biosurfactants as biological and immunological molecules. Current Opinion in Microbiology 7: 262–266.

Campos, J.M., T.L. Montenegro Stamford, L.A. Sarubbo, J.M. De Luna, R.D. Rufino and I.M. Banat. 2013. Microbial biosurfactants as additives for food industries. Biotechnology Progress 29(5): 1097–1108.

Chakrabarti, S. 2012. Bacterial Biosurfactant: Characterization, Antimicrobial and Metal Remediation Properties. National Institute of Technology, Surat, India.

Chen, X.H., A. Koumoutsi, R. Scholz, K. Schneider, J. Vater, R. Su"ssmuth, J. Piel and R. Boriss. 2009. Genome analysis of *Bacillus amyloliquefaciens* FZB42 reveals its potential for biocontrol of plant pathogens. J. Biotechnol. 140: 27–37.

Christofi, N. and I.B. Ivshina. 2002. Microbial surfactants and their use in field studies of soil remediation. Journal of Applied Microbiology 93: 915–929.

Christova, N., P. Petrov and L. Kabaivanova. 2013. Biosurfactant production by *Pseudomonas aeruginosa* BN10 cells entrapped in cryogels. Z. Naturforsch. 68c: 47–52.

Cortés-Sánchez, A.J. 2020. Surfactants of microbial origin and its application in foods. Scientific Research and Essays 15(1): 11–17.

Das, P., S. Mukherjee and P. Das. 2008a. Improved bioavailability and biodegradation of a model polyaromatic hydrocarbon by a biosurfactant producing bacterium of marine origin. Chemosphere 72: 1229–1243.

Das, P., S. Mukherjee and R. Sen. 2008b. Genetic regulations of the biosynthesis of microbial surfactants: An overview. Biotechnol. Genet. Eng. Rev. 25: 165–186.

Das, P., S. Mukherjee and R. Sen. 2009. Biosurfactant of marine origin exhibiting heavy metal remediation. Biores. Technol. 100: 4887–4890.

Dave, N. and T.A. Joshi. 2017. Concise review on surfactants and its significance. Int. J. Appl. Chem. 13: 663–672.

Dennis, E.A. and P.C. Norris. 2015. Eicosanoid storm in infection and inflammation. Nat. Rev. Immunol. 15: 511–523.

Deravel, J., S. Lemière, F. Coutte, F. Krier, N. Van Hese, M. Béchet, N. Sourdeau, M. Höfte, A. Leprêtre and P. Jacques. 2014. Mycosubtilin and surfactin are efficient, low ecotoxicity molecules for the biocontrol of lettuce downy mildew. Appl. Microbiol. Biotechnol. 98(14): 6255–64.

Diggle, S.P., S.A. West, A. Gardner and A.S. Griffin. 2008. Communication in bacteria. pp. 11–31. *In*: Hughes, D. and P.D. Ettorre (eds.). Sociobiology of Communication: An Interdisciplinary Perspective. Oxford University Press. U.K.

Dinant, S. 2008. Phloem, transport between organs and long-distance signalling. Comptes Rendus Biol. 331: 334–346.

Dreja, M., I. Vockenroth and N. Plath. 2012. Biosurfactants-exotic specialties or ready for application? Tenside Surfactants Detergents 49.1: 10–17.

Dubey, K. and A. Juwarkar. 2001. Distillery and curd whey wastes as viable alternative sources for biosurfactant production. World J. Microbiol. Biotechnol. 17: 61–69.

Dumschott, K., A. Richter, W. Loescher and A. Merchant. 2017. Post photosynthetic carbon partitioning to sugar alcohols and consequences for plant growth. Phytochemistry 144: 243–252.

Ehehalt, R., C. Jochims, W.D. Lehmann, G. Erben, S. Staffer, C. Reininger and W. Stremmel. 2004. Evidence of luminal phosphatidylcholine secretion in rat ileum. Biochim. Biophys. Acta 1682: 63–71.

Enayati, M., Y. Gong, J.M. Goddard and A. Abbaspourrad. 2018. Synthesis and characterization of lactose fatty acid ester using free and immobilized lipases in organic solvents. Food Chemistry, doi: https://doi.org/10.1016/j.foodchem.2018.06.051.

Fenibo, E.O., S.I. Douglas and H.O. Stanley. 2019a. A review on microbial surfactants: production, classifications, properties and characterization. Journal of Advances in Microbiology Article no.JAMB.51198 18(3): 1–22.

Fenibo, E.O., G.N. Ijoma, R. Selvarajan and Chikere. 2019b. Microbial surfactants: the next generation multifunctional biomolecules for applications in the petroleum industry and its associated environmental remediation. Microorganisms 7: 581.

Fickers, P., J.S. Guez, C. Damblon, V. Leclère, M. Béchet, P. Jacques and B. Joris. 2009. High level biosynthesis of the anteiso-C (17) isoform of the antibiotic mycosubtilin in *Bacillus subtilis* and characterization of its candidacidal activity. Appl. Environ. Microbiol. 75: 4636–4640.

Fickers, P. 2012. Antibiotic compounds from Bacillus: why are they so amazing? Am. J. Biochem. Biotechnol. 8: 40–46.

Fracchia, L., M. Cavallo, M.G. Martinotti and I.M. Banat. 2012. Biosurfactants and bioemulsifiers biomedical and related applications—present status and future potentials. Biomedical Science, Engineering and Technology, Chapter 14, InTech (2012): 325–370.

Frazetti, A., P. Caredda, C. Ruggeri, P.L. Colla, E. Tamburini, M. Papacchinis and G. Bestetti. 2009. Potential applications of surface active compounds by Gordonia sp. strain BS29 in soil remediation technologies. Chemosphere 75: 801–807.

Furukawa, S., Y. Akiyoshi, G.A. O'Toole, H. Ogihara and Y. Morinaga. 2010. Sugar fatty acid esters inhibit biofilm formation by food-borne pathogenic bacteria. International Journal of Food Microbiology 138: 176–180.

Galabova, D., A. Sotirova, E. Karpenkoy and O. Karpenk. 2014. Role of microbial surface-active compounds in environmental protection. pp. 41–83. *In*: Fanum, M. (ed.). The Role of Colloidal Systems in Environmental Protection. Elsevier B.V. All rights reserved.

Gauss, A., R. Ehehalt, W.D. Lehmann, G. Erben, K.H. Weiß, Y. Schaefer, P. Kloeters-Plachky, A. Stiehl, W. Stremmel, P. Sauer and D.N. Gotthardt. 2013. Biliary phosphatidylcholine and lysophosphatidylcholine profiles in sclerosing cholangitis. World J. Gastroenterol. 19: 5454–5463.

Gautam, K.K. and V.K. Tyagi. 2006. Microbial surfactants: A review. Journal of Oleo Science 55(4): 155–166.

Gille, C., B. Spring, W. Bernhard, C. Gebhard, D. Basile, K. Lauber, C.F. Poets and T.W. Orlikowsky. 2007. Differential effect of surfactant and its saturated phosphatidylcholines on human blood macrophages. J. Lipid Res. 48: 307–317.

Giuliani, A., G. Pirri and S. Fabiole Nicoletto. 2007. Antimicrobial peptides: an overview of a promising class of therapeutics. Central European Journal of Biology 2: 1–33.

Gonzalez-Alfonso, J.L., L. Casas-Godoy, J. Arrizon, D. Arrieta-Baez, A.O. Ballesteros, G. Sandoval and F.J. Plou. 2018. Lipase-catalyzed synthesis of fatty acid esters of trisaccharides. *In*: Georgina Sandoval (ed.). Lipases and Phospholipases: Methods and Protocols, Methods in Molecular Biology. Springer Science+Business Media, LLC, part of Springer. Nature 1835: 287–296.

Goss, V., A.N. Hunt and A.D. Postle. 2013. Regulation of lung surfactant phospholipid synthesis and metabolism. Biochimica et Biophysica Acta 1831: 448–458.

Gutierrez, T. 2019. Marine microbial surfactants: searching for needles in a Haystack. EC Microbiology 15.4: 239–241.

Hamzah, A., N. Sabturani and S. Radiman. 2013. Screening and optimization of biosurfactant production by the hydrocarbon-degrading bacteria. Sains Malaysiana 42(5): 615–623.

Han, C., S. Spring, A. Lapidus, T.G. del Rio, H. Tice, A. Copeland, J.F. Cheng, S. Lucas, F. Chen, M. Nolan, D. Bruce, L. Goodwin, S. Pitluck, N. Ivanova, K. Mavromatis, N. Mikhailova, A. Pati, A. Chen and J.C. Detter. 2009. Complete genome sequence of *Pedobacter heparinus* type strain (HIM 762-3T). Standards in Genomic Sciences 1: 54–62.

Han, S.I., H.J. Lee, H.R. Lee, K.K. Kim and K.S. Whang. 2012. *Mucilaginibacter polysacchareus* sp. nov., an exopolysaccharide-producing bacterial species isolated from the rhizoplane of the herb *Angelica sinensis*. International Journal of Systematic and Evolutionary Microbiology 62: 632–637.

Huang, Q.X., Z.X. Lu, H.Z. Zhao, X.m. Bie, F.X. Lu and S.J. Yang. 2006. Antiviral activity of antimicrobial lipopeptide from *Bacillus subtilis* fmbj against Pseudorabies virus. Porcine Parvovirus, Newcastle disease virus and infectious bursal disease virus *in vitro*. Int. J. PeptRes. Therap. 12: 373–7.

Hwang, J.H., K.Y. Kim, E.P. Resurreccion and W.H. Lee. 2019. Surfactant addition to enhance bioavailability of bilge water in single chamber microbial fuel cells (MFCs). J. Hazard. Mater. 368: 732–738.

Ivanković, T. and J. Hrenović. 2010. Surfactants in the environment. Arh Hig Rada Toksikol 61: 95–110.

Jamshidi-Aidjia, M. and G.E. Morlock. 2018. Fast equivalency estimation of unknown enzyme inhibitors *in situ* the effect-directed fingerprint, shown for *Bacillus* lipopeptide extracts. Anal. Chem. 90(24): 14260–14268.

Jardak, K., P. Drogui and R. Daghrir. 2016. Surfactants in aquatic and terrestrial environment: occurrence, behavior, and treatment processes. Environ. Sci. Pollut. Res. 23: 3195–3216.

Jauregi, P., F. Coutte, L. Catiau, D. Lecouturier and P. Jacques. 2013. Micelle size characterization of lipopeptides produced by *Bacillus subtilis* and their recovery by the two step ultrafiltration process. Sep. Purif. Technol. 104: 175–182.

Johnsen, A.R. and U. Karlson. 2004. Evaluation of bacterial strategies to promote the bioavailability of polycyclic aromatic hydrocarbons. Applied Microbiology and Biotechnology 63: 452–459.

Kakugawa, K., M. Tamai, K. Imamura, K. Miyamoto and S. Miyoshi. 2002. Isolation of yeast Kurtzmanomyces sp. I-11, novel producer of mannosylerythriotol lipid. Biosci. Biotechnol. Biochem 66: 188–191.

Kinsinger, R., M.C. Shirk and R. Fall. 2003. Rapid surface motility in *Bacillus subtilis* is dependent on extracellular surfactin and potassium ion. J. Bacteriol. 185: 5627–5631.

Kinsinger, R.F., D.B. Kearns, M. Hale and R. Fall. 2005. Genetic requirements for potassium ion-dependent colony spreading in *Bacillus subtilis*. J. Bacteriol. 187: 8462–8469.

Kowalska, I., M. Kabsch-Korbutowiez, K. Majewska-Nowak and T. Winnicki. 2004. Separation of anionic surfactants on ultrification membranes. Desalination. 126: 33–40.

Lahiry, S. and R. Sinha. 2017. Biosurfactant: Pharmaceutical perspective. J. Anal. Pharm. Res. 4(3): 1–3.

Leclère, V., M. Béchet, A. Adam, J.S. Guez, B. Wathelet, M. Ongena, P. Thonart, F. Gancel, M. Chollet-Imbert and P. Jacques. 2005. Mycosubtilin overproduction by *Bacillus subtilis* BBG 100 enhances the organism's antagonistic and biocontrol activities. Appl. Environ. Microbiol. 8: 4577–4584.

Lee, H., J.J. Churey and R.W. Worobo. 2008. Purification and structural characterization of bacillomycin F produced by a bacterial honey isolate active against Byssochlamys fulva H25. Journal of Applied Microbiology 105: 663–673.

Li, D., L. Sun and M. Lian. 2005. Application of surfactants in soil remediation. Journal of Chemical and Pharmaceutical Research 7(3): 364–366.

Lin, F., J. Yang, U. Muhammad, J. Sun, Z. Huang, W. Li, F. Lv and Z. Lu. 2018. Bacillomycin D-C16 triggers apoptosis of gastric cancer cells through the PI3K/Akt and FoxO3a signaling pathways. Anti-Cancer Drugs 0(0): 1–10.

Liu, Y., J. Lu, J. Sun, X. Zhu, L. Zhou, Z. Lu and Y. Lu. 2019. C16-Fengycin A affect the growth of *Candida albicans* by destroying its cell wall and accumulating reactive oxygen species. Applied Microbiology and Biotechnology https://doi.org/10.1007/s00253-019-10117-5.

Lukic, M., I. Pantelic and S. Savic. 2016. An overview of novel surfactants for formulation of cosmetics with certain emphasis on acidic active substances. Tenside Surfact. Det. 53: 7–19.

Maier, R.M. 2003. Biosurfactants: evolution and diversity in bacteria. Adv. Appl. Microbiol. 52: 101–121.

Makkar, R.S., S.S. Cameotra and I.M. Banat. 2011. Advances in utilization of renewable substrates for biosurfactant production. AMB (Applied Microbiology Biotechnology) Express 1: 5.

Marchant, R. and I.M. Banat. 2012. Microbial biosurfactants: challenges and opportunities for future exploitation. Trends Biotechnol. 30(11): 558–565.

Merchant, A. and A.A. Richter. 2011. Polyols as biomarkers and bioindicators for 21st century plant breeding. Funct. Plant Biol. 38: 934–940.

Morikawa, M., Y. Hirata and T. Imanaka. 2000. A study on the structure–function relationship of the lipopeptide biosurfactants. Biochimica et Biophysica Acta 1488: 211–218.

Moyne, A.L., R. Shelby, T.E. Cleveland and S. Tuzunl. 2001. Bacillomycin D: an iturin with antifungal activity against *Aspergillus flavus*. J. Appl. Microbiol. 90: 622–629.

Mukherjee, S., P. Das and R. Sen. 2006. Towards commercial production of microbial surfactants. Trends Biotechnol. 24: 509–515.

Mulligan, C.N. 2005. Environmental applications for biosurfactants. Environ. Pollut. 133: 183–198.

Mulligan, C.N. 2009. Recent advances in the environmental applications of biosurfactants. Current Opinion in Colloid and Interface Science 14: 372–378.

Muthusamy, K., S. Gopalakrishnan, T.K. Ravi and P. Sivachidambaram. 2008. Biosurfactants: properties, commercial production and application. Curr. Sci. 94: 736–747.

Nayak, A.S., M.H. Vijaykumar and T.B. Karegoudar. 2009. Characterization of biosurfactant produced by *Pseudoxanthomonas* sp. PNK-04 and its application in bioremediation. International Biodeterioration and Biodegradation 63: 73–79.

Nerurkar, A.S., K.S. Hingurao and H.G. Suthar. 2009. Bioemulsifiers from marine microorganisms. Journal of Scientific and Industrial Research 68: 273–277.

Nerurkar, A.S. 2010. Structural and molecular characteristics of lichenysin and its relationship with surface activity. pp. 304–315. *In*: Sen, R. (ed.). Biosurfactants, Landes Bioscience and Springer Science+Business Media.

Ongena, M. and P. Jacques. 2008. Bacillus lipopeptides: versatile weapons for plant disease biocontrol. Trends Microbiol. 16: 115–125.

Osadebe, A.U., C.A. Onyiliogwu, B.M. Suleiman and G.C. Okpokwasili. 2018. Microbial degradation of anionic surfactants from laundry detergents commonly discharged into a riverine ecosystem. Journal of Applied Life Sciences International Article no. JALSI.40131, 16(4): 1–11.

Ozdemir, G., T. Ozturk, N. Ceyhan, R. Isler and T. Cosar. 2003. Heavy metal biosorption by biomass of *Ochrobactrum anthropi* producing exopolysaccharide in activated sludge. Bioresource Technology 90: 71–74.

Paananen, R., D.P.A.D. Postle, G. Clark, V. Glumoff and M. Hallman. 2002. Eustachian tube surfactant is different from alveolar surfactant: determination of phospholipid composition of porcine eustachian tube lavage fluid. J. Lipid Res. 43(1): 99–106.

Pacwa-Plociniczak, M., G.A. Plaza, Z. Piotrowska-Seget and S.S. Cameotra. 2011. Environmental applications of biosurfactants: Recent advances. International Journal Molecular Science 12: 633–654.

Pandya, U., S. Prakash, K. Shende, U.P. Dhuldhaj and M. Saraf. 2017. Multifarious allelochemicals exhibiting antifungal activity from *Bacillus subtilis* MBCU5. 3Biotech 7: 175.

Pasternak, G., J. Greenman and I. Ieropoulos. 2016. Regeneration of the power performance of cathodes affected by biofouling. Appl. Energy 173: 431–437.

Pasternak, G., T.D. Askitosari and M.A. Rosenbaum. 2020. Biosurfactants and synthetic surfactants in bioelectrochemical systems: A mini-review. Front. Microbiol. 11: 358.

Pathak, K.V. and H. Keharia. 2014. Identification of surfactins and iturins produced by potent fungal antagonist, *Bacillus subtilis* K1 isolated from aerial roots of banyan (Ficus benghalensis) tree using mass spectrometry. 3. Biotechnology 4: 283–295.

Pekdemir, T., M. Copur and K. Urum. 2005. Emulsification of crude oil–water systems using biosurfactants. Process Safety Environmental Protection 83(B1): 38–46.

Perfumo, A., T.J.P. Smyth, R. Marchant and I.M. Banat. 2010. Production and roles of biosurfactants and bioemulsifiers in accessing hydrophobic substrates. pp. 1501–1512. *In*: Timmis, K.N. (ed.). Handbook of Hydrocarbon and Lipid Microbiology. Springer-Verlag, Berlin, Germany.

Plaza, G.A., I. Zjawiony and I.M. Banat. 2006. Use of different methods for detection of thermophilic biosurfactant-producing bacteria from hydrocarbon-contaminated and bioremediated soils. J. Petrol. Sci. Eng. 50: 71–77.

Pornsunthorntawee, N., N. Arttaweeporn, S. Paisanjit, P. Somboonthanate, M. Abe, R. Rujiravanit and S. Chavadej. 2008. Isolation and comparison of biosurfactants produced by *Bacillus subtilis* PT2 and *Pseudomonas aeruginosa* SP4 for microbial surfactant-enhanced oil recovery. Biochemical Engineering Journal 42: 172–179.

Pradhan, A. and A. Bhattacharyya. 2018. An alternative approach for determining critical micelle concentration: Dispersion of ink in foam. Journal Surfactants Detergent 21(5): 745–50.

Praveesh, B.V., A.R. Soniyamby, C. Mariappan, M. Palaniswamy and S. Lalitha. 2011. Microbial surfactants: an overview. International Journal of Current Research and Review 3(10): 38–46.

Pynn, C.J., M.V. Picardi, T. Nicholson, D. Wistuba, C.F. Poets, E. Schleicher, J. Perez-Gil and W. Bernhard. 2010. Myristate is selectively incorporated into surfactant and decreases dipalmitoylphosphatidylcholine without functional impairment. Am. J. Physiol. Regul. Integr. Comp. Physiol. 299: R1306–1316.

Raaijmakers, J.M., I. de Bruijin, O. Nybroe and M. Ongena. 2010. Natural functions of lipopeptides from *Bacillus* and *Pseudomonas*: more than surfactants and antibiotics. FEMS Microbiol. Rev. 34: 1037–1062.

Ramarathnam, R., S. Bo, Y. Chen, W.G.D. Fernando, G. Xuewen and T. de Kievit. 2007. Molecular and biochemical detection of fengycin- and bacillomycin D producing *Bacillus* spp., antagonistic to fungal pathogens of canola and wheat. Can. J. Microbiol. 53: 901–11.

Reidel, E.J., E.A. Rennie, V. Amiard, L. Cheng and R. Turgeon. 2009. Phloem loading strategies in three plant species that transport sugar alcohols. Plant Physiol. 149: 1601–1608.

Rhein, L. 2002. Surfactant action on skin and hair: Cleansing and skin reactivity mechanisms. pp. 305–369. *In*: Johansson, I. and P. Somasundaran (eds.). Handbook for Cleaning/Decontamination of Surfaces. 1st ed. Elsevier: Amsterdam, The Netherlands; Boston, MA, USA.

Rodrigues, L.R. 2015. Microbial surfactants: Fundamentals and applicability in the formulation of nano-sized drug delivery vectors. Journal of Colloid and Interface Science 449: 304–316.

Ron, E.Z. and E. Rosenberg. 2001. Natural role of biosurfactants. Environ. Microbiol. 3: 229–236.

Ron, E.Z. and E. Rosenberg. 2010. Role of biosurfactants. pp. 2515–2520. *In*: Timmis, K.N. (ed.). Handbook of Hydrocarbon and Lipid Microbiology.

Ruiz, C.C. 2008. Sugar-based Surfactants: Fundamentals and Applications (Vol. 143): CRC Press.

Saggese, A., R. Culurciello, A. Casillo, M.M. Corsaro, E. Ricca and L. Baccigalupi. 2018. A marine isolate of *Bacillus pumilus* secretes a pumilacidin active against *Staphylococcus aureus*. Mar. Drugs 16: 180.

Sajitha, K.L., S.A. Dev and E.J. Maria Florence. 2016. Identification and characterization of lipopeptides from *Bacillus subtilis* B1 against Sapstain fungus of rubberwood through MALDI-TOF-MS and RT-PCR. Current Microbiology 73: 46–53.

Santos, D.K.F., R.D. Rufino, J.M. Luna, V.A. Santos and L.A. Sarubbo. 2016. Biosurfactants: multifunctional biomolecules of the 21st century. International Journal of Molecular Sciences 17(3): 401.

Sawettanai, N., H. Leelayuwapan, N. Karoonuthaisiri, S. Ruchirawat and S. Boonyarattanakalin. 2019. Synthetic lipomannan glycan microarray reveals the importance of α(1,2) mannose branching in DC-SIGN binding. J. Org. Chem. 84(12): 7606–7617.

Seydlová, G. and J. Svobodová. 2008. Review of surfactin chemical properties and the potential biomedical applications. Cent. Eur. J. Med. 3(2): 123–133.

Shaligram, N.S. and R.S. Singhal. 2010. Surfactin—A review on biosynthesis, fermentation, purification and applications. Food Technol. Biotechnol. 48: 119–134.

Sil, J., P. Dandapat and S. Das. 2017. Health care applications of different biosurfactants: Review. Int. J. Sci. Res. 6: 41–50.

Singh, P. and S.S. Cameotra. 2004. Potential applications of microbial surfactants in biomedical sciences. Trends in Biotechnology 22.3: 142–146.

Sivapathasekaran, C. and R. Sen. 2012. Performance evaluation of batch and unsteady state fedbatch reactor operations for the production of a marine microbial surfactant. J. Chem. Technol. Biotechnol. 88: 719–726.

Sivapathasekaran, C. and R. Sen. 2017. Origin, properties, production and purification of microbial surfactants as molecules with immense commercial potential. Tenside Surf. Det. 54(2): 92–107.

Sobrinho, H.B., J.M. Luna, R.D. Rufino, A.L.F. Porto and L.A. Sarubbo. 2013. Biosurfactants: classification, properties and environmental applications. Recent Developments in Biotechnology 11: 1–29.

Sonawane, S., S. Pal, S. Tayade and N. Bisht. 2015. Application of surfactant in various fields a short review. International Journal of Scientific & Engineering Research 6(12): 30–32.

Takahashi, M., T. Morita, T. Fukuoka, T. Imura and D. Kitamoto. 2012. Glycolipid biosurfactants, mannosylerythritol lipids, show antioxidant and protective effects against H_2O_2-induced oxidative stress in cultured human skin fibroblasts. J. Oleo. Sci. 61: 457–64.

Thavasi, R., V.R.M.S. Nambaru, S. Jayalakshmi, T. Balasubramanian and I.M. Banat. 2009. Biosurfactant production form Azotobacter chroococcum isolated from the marine environment. Mar. Biotechnol. 11: 551–556.

Tilvi, S. and C.G. Naik. 2007. Tandem mass spectrometry of Kahalaides: identification of new cyclic depsipeptides, kahlide R and S from Elysia grandifolia. J. Mass Spectro. 1: 70–80.

Toren, A., S. Navon-Venezia, E.Z. Ron and E. Rosenberg. 2001. Emulsifying activities of purified alasan proteins from acinetobacter radioresistens KA53. Applied and Environmental Microbiology 67(3): 1102–1106.

Vecino, X., R. Devesa-Rey, J.M. Dominguez, J.M. Cruz and A.B. Moldes. 2014. Effect of soil loading and pH during batch solvent extraction of fluorene from soil using a lipopetide biosurfactant aqueous solution. IMETI, 7th International Multi-Conference on Engineering and Technological Innovation, Proceedings. pp. 11–13.

Volpon, L., F. Besson and J.M. Lancelin. 2000. NMR structure of antibiotics plipastatins A and B from *Bacillus subtilis* inhibitors of phospholipase A(2). FEBS Lett. 485(1): 76–80.

Wei, Y.H., L.C. Wang, W.C. Chen and S.Y. Chen. 2010. Production and characterization of fengycin by indigenous *Bacillus subtilis* F29-3 originating from a potato farm. Int. J. Mol. Sci. 11: 4526–38.

Wiktorowska-Owczarek, A., M. Berezińska and J.Z. Nowak. 2015. PUFAs: Structures, metabolism and functions. Adv. Clin. Exp. Med. 24: 931–941.

Williamson, J.D., D.B. Jennings, W.W. Guo, D.M. Pharr and M. Ehrenshaft. 2002. Sugar alcohols, salt stress, and fungal resistance: polyols - Multifunctional plant protection? J. Am. Soc. Hortic. Sci. 127: 467–473.

Xiu, P., R. Liu, D. Zhang and C. Sun. 2017. Pumilacidin-like lipopeptides derived from marine bacterium *Bacillus* sp. strain 176 suppress the motility of *Vibrio alginolyticus.* Applied and Environmental Microbiology 83(12): e00450–17.

Xu, Z., J. Shao, B. Li, X. Yan, Q. Shen and R. Zhang. 2013. Contribution of bacillomycin D in *Bacillus amyloliquefaciens* SQR9 to antifungal activity and biofilm formation. Appl. Environ. Microbiol. 79: 808–815.

Yalcin, E. and K. Cavusoglu. 2010. Structural analysis and antioxidant activity of a biosurfactant obtained from *Bacillus subtilis* RW-I. Turk. J. Biochem. 35: 243–7.

Yaseen, Y., F. Gancel, D. Drider, M. Bechet and P. Jacques. 2016. Influence of promoters on the production of fengycin in *Bacillus* spp. Research in Microbiology xx: 1–10.

Yin, H., C. Guo, Y. Wang, D. Liu, Y. Lv, F. Lv and Z. Lu. 2013. Fengycin inhibits the growth of the human lung cancer cell line 95D through reactive oxygen species production and mitochondria-dependent apoptosis. Anticancer Drugs 24: 587–98.

Zeriouh, H., A. de Vicente, A. Pérez-García and D. Romero. 2014. Surfactin triggers biofilm formation of *Bacillus subtilis* in melon phylloplane and contributes to the biocontrol activity. Environ. Microbiol. 16: 2196–2211.

Zhang, B., C. Dong, Q. Shang, Y. Cong, W. Kong and P. Li. 2013. Purification and partial characterization of Bacillomycin L produced by *Bacillus amyloliquefaciens* K103 from Lemon. Appl. Biochem. Biotechnol. 171: 2262–2272.

Zhang, L. and T.J. Falla. 2004. Cationic antimicrobial peptides—An update. Expert. Opin. Invest. Drugs 13: 97–106.

Zhang, Y., J. Jiang, Q. Zhao, Y.Z. Gao, K. Wang, J. Ding, H. Yu and Y. Yao. 2017. Accelerating anodic biofilms formation and electron transfer in microbial fuel cells: role of anionic biosurfactants and mechanism. Bioelectrochemistry 117: 48–56.

2

Rhamnolipid Biosurfactants
Structure, Biosynthesis, Production, and Applications

Dimple S Pardhi,[1] Rachana Bhatt,[2]
Rakeshkumar R Panchal,[1] Vikram H Raval[1] and
Kiransinh N Rajput[1,*]

1. Introduction

Biosurfactants or microbial surfactants are surface-active, extracellular, secondary metabolites produced during the stationary phase of microbial growth. Most strains of bacteria and a few yeasts and molds are recorded as the biosurfactant producing microorganisms. These amphiphilic biomolecules usually contain carbohydrates, amino acids, and phosphate groups as a hydrophilic domain, whereas long-chain fatty acids as a hydrophobic domain. Biosurfactants' nature generally depends on their microbial origin and carbon : nitrogen (C : N ratio). Biosurfactants are broadly classified in two ways, according to their molecular weights and chemical structures. Based on molecular weights, they are divided into low molecular weight and high molecular weight biosurfactants. Depending on the structural diversity, low molecular weighted biosurfactants include glycolipids, lipopeptides or lipoproteins, and phospholipids whereas high molecular weighted biosurfactants includes polymeric surfactants and particulate surfactants.

Rhamnolipid (RL), an anionic-surface-active member of the glycolipid family is one of the extensively studied biosurfactants. It possesses a hydrophilic domain containing one or two rhamnose units and a hydrophobic fatty acid domain. Jarvis

[1] Department of Microbiology and Biotechnology, University School of Sciences, Gujarat University, Ahmedabad-380009, Gujarat, India.
[2] Department of Biomedical Engineering, Rutgers University-New Brunswick, 590 Taylor Road, Piscataway, NJ, USA-08854.
* Corresponding author: rajputkn@yahoo.com

and Johnson (1949) firstly discovered the rhamnolipids as the extracellular product of *Pseudomonas aeruginosa* in presence of water immiscible compounds as a carbon source. Amongst the microorganisms studied, the maximum number of *Pseudomonas* strains are reported as potential rhamnolipid producers. Rhamnolipids' commercial applications are limited because of their higher production cost and comparatively lower yield. Hence, it is crucial to improve the efficiency of bioprocesses for increased yield and decreased production cost. The medium composition and cultivation conditions positively affect fermentation processes. To accomplish cost-effective bioprocess, inexpensive carbon sources like crude oil, waste frying oil, soybean oil, petroleum products are to be included in the culture medium. Another approach to reduce the production cost involves strain development or designing a recombinant strain with higher rhamnolipids yield. It will significantly decrease the purification steps in downstream processing.

The biosynthetic pathway carried out by *Pseudomonas aeruginosa* in submerged fermentation, primarily produces two forms of rhamnolipids; mono- and di-rhamnolipid. Two sequential reactions catalyze these biomolecules; each of them uses a specific enzyme, i.e., rhamnosyltransferase I and rhamnosyltransferase II. Several analytical methods viz., emulsification index, drop collapsed test, oil displacement activity, hemolytic activity, surface tension measurement, oil agar plate, blue agar plate, bacterial adhesion to the hydrocarbon, orcinol assay, and emulsification assay are used to screen the potential biosurfactant producing microorganisms. Mass spectrometry is the most effective and widely used technique for analyzing the functional groups of rhamnolipids. Fourier transform infrared spectroscopy, high-performance liquid chromatography, rhamnose test, and nuclear magnetic resonance are the other techniques used for structural characterization.

As a biosurfactant, rhamnolipid shows exclusive properties like biodegradability, low toxicity, high surface/interfacial activity, structural diversity, biocompatibility, and better stability towards the range of pH, temperature, and salt concentrations compared to the synthetic one. Such properties make them a worthwhile alternative to the synthetic surfactants in various applications like cleaning of oil storage tanks, microbial enhanced oil recovery, removal of dyes/metals/hydrocarbons from the environment, making antimicrobial-antiadhesive products, and food additives, etc. In this chapter, rhamnolipid structure, biosynthesis, production, improvement, and characterization approaches are discussed. Moreover, it also discusses the global market and applications of rhamnolipids in agriculture, bioremediation, petroleum, pharmaceuticals, cosmetics, foods, etc.

2. Structure and Properties of Rhamnolipids

Amid the various classes of biosurfactants explored, **'rhamnolipids'** the member of glycolipid biosurfactants stand apart. Primarily, rhamnolipids are crystalline acids composed of β-hydroxy fatty acids attached to the carboxyl ends of rhamnose sugar molecules. They consist of one or two molecules of L-rhamnose and a long-chain β-hydroxy fatty acid containing 6–18 carbon atoms, as a hydrophilic and hydrophobic group, respectively. However, *Pseudomonas aeruginosa* produces rhamnolipids predominantly containing molecular chain of 10-C. The first OH group of the β-hydroxy fatty acid is involved in the glycosidic linkage with the reducing end of the L-rhamnose disaccharide. In contrast, the second -OH group is involved with ester formation.

Based on the presence of rhamnose molecules, rhamnolipids are categorized as mono-rhamnolipids (RL1) and di-rhamnolipids (RL2) where one and two rhamnose molecules are attached with two molecules of β-hydroxydecanoic acid, respectively. The respective molecular weight of mono- and di-rhamnolipids are 504 and 650 Da (Mulligan and Gibbs 2004). Apart from the mono- and di-rhamnolipids, recently a further two structurally diverse isoforms of rhamnolipids have been classified as tri-rhamnolipids (RL3) and tetra-rhamnolipids (RL4) containing one and two rhamnose molecules attached with only one molecule of β-hydroxydecanoic acid, respectively (Kamal-Ahmed 2015). The chemical structures of different isoforms of rhamnolipids are presented in Fig. 1. Amongst these four, isoforms mono- and di-rhamnolipids are principal biosurfactants produced by *P. aeruginosa*.

The microbial bioprocess produces a mixture of rhamnolipid congeners with divers' chain length, degree of unsaturated fatty acid chains, and several rhamnose molecules. It has been determined that approximately 60 rhamnolipid congeners and homologues occur in the production broth. In contrast, the predominant class

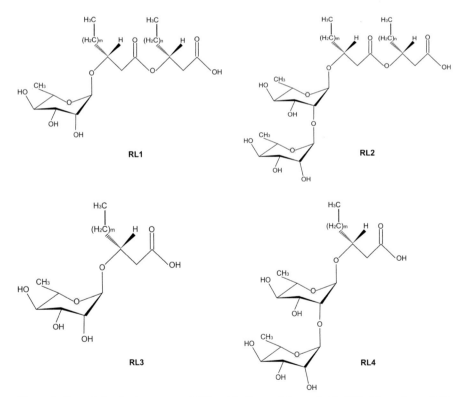

Figure 1. General chemical structures of different isoforms of rhamnolipids. RL1: Mono-rhamnolipid-di-lipidic, RL2: Di-rhamnolipid-di-lipidic, RL3: Tri-rhamnolipid-mono-lipidic, RL4: Tetra-rhamnolipid-mono-lipidic; m, n: 6–18.

of rhamnolipid and congeners concentration depends on the strain used and culture conditions. A small change in the congener's composition can significantly affect the physicochemical properties of the rhamnolipids. The hydrophilic-lipophilic-balance (HLB) values of rhamnolipids are generally obtain in the range of 4 to 6, which reflects good and stable water-oil emulsifier activity.

Production of rhamnolipids enables *P. aeruginosa* to grow on hydrophobic carbon sources. About < 100 mg/L concentration of hydrophobic substrate is sufficient to support their growth (Müller et al. 2012). Depending on the various substrates used, produced rhamnolipids have different homologues. For example, a rhamnolipid mixture produced using vegetable oil from *Pseudomonas aeruginosa* LBI contains mono-rhamnolipid (Rha-C10-C10) and di-rhamnolipid (Rha-Rha-C10-C10). However, *P. aeruginosa* MN1 synthesizing the mixture of rhamnolipids using glycerol as a carbon source showed the presence of different homologues (Rha-C10, Rha-C8-C10, Rha-C10-C8, Rha-C10-C12: 1, Rha C10-C10, Rha-C10-C10, Rha-C10-C10, Rha-C10-C10, RhaRha-C8-C10, Rha-Rha-C10-C8, Rha-Rha-C8-C12: 1, Rha-Rha-C10-C10, Rha-Rha-C10-C12: 1, Rha-Rha-C12: 1-C10, Rha-Rha-C10-C12, RhaRha-C12-C10, Rha-Rha-C8-C8) containing approximately 35% of mono-rhamnolipids (Araújo et al. 2018). Moreover, greater the homologue quantity; better the critical micelle concentration (CMC) and surface tension reduction as compared to the isolated di-rhamnolipid and mixture of only two homologues.

Rhamnolipids have some valuable properties over the chemically synthesized counterparts, which makes them a better choice over synthetic surfactants. These include the ability of surface tension reduction, biodegradability, low toxicity, non-mutagenicity, emulsification, de-emulsification, dispersion, wetting, dissolution, and decontamination. Rhamnolipids remarkably decrease the surface tension of water from 72 to 28 mN/m, and interfacial tension of water/oil system from 43 to < 1 mN/m (Liang-Ming et al. 2008). They are easily biodegradable compounds and hence, principally suitable for environmental applications such as bioremediation and dispersion of hydrocarbons. Rhamnolipids have lower toxicity and non-mutagenicity than the synthetically derived surfactants. They can exhibit excellent emulsification as well as de-emulsification ability with variety of vegetable oils and hydrophobic hydrocarbons. The highest emulsification index of 86.4% has been reported with rhamnolipids against toluene (Kamal-Ahmed 2015). Thus, rhamnolipids have been extensively used in petroleum, cosmetics, food, medical, detergents, and environmental cleaning owing to their high specificity, biodegradability, bioavailability, and biocompatibility.

3. Rhamnolipid Producing Microorganisms

Hydrocarbon-degrading microbial species mainly produce rhamnolipids, and their amphiphilic nature essentially helps microorganisms to uptake and utilize the hydrophobic hydrocarbons. Some microorganisms showed rhamnolipid production only if hydrocarbons are available as a carbon source. Furthermore, several species are reported with only mono-rhamnolipid production. The biosurfactants' structural diversity and yield widely depend on the origin site of producing microorganisms along with the growth components provided. Oil contaminated sites such as petrochemical industries, crude oil contaminated localities, used edible oils, tannery effluents, and oil reservoirs are the prime source for isolation of potential rhamnolipid producers.

3.1 Rhamnolipid Producing Pseudomonas sp.

Rhamnolipid-producing strains mainly belong to the genus *Pseudomonas*, and the most reported species is *P. aeruginosa*. It contributes approximately 50–60% of the total rhamnolipid production. Some rhamnolipid producing strains of *Pseudomonas* sp. are listed in Table 1. Researchers have also studied the recombinant or mutant strains of *Pseudomonas* sp. for enhanced rhamnolipid production. Some mutant strains like *Pseudomonas putida, P. aeruginosa* PTCC1637, and *P. aeruginosa* MIG-N146 (Raza et al. 2007, Tahzibi et al. 2004, Guo et al. 2009) were investigated for increased rhamnolipid yield.

Furthermore, marine samples are also reported as potential rhamnolipid producers with operational performance capability in extreme environments. They produce highly stable products which are generally useful in pharmaceutical preparations. They can utilize a diverse range of substrates, which are usually harmful to other microorganisms. A hydrocarbon-degrading thermophilic strain *Pseudomonas aeruginosa* AP02-1 was reported with better rhamnolipid production (Perfumo et al. 2006). Varjani and Upasani (2016) studied a thermophilic and halo-tolerant rhamnolipid producer *P. aeruginosa* NCIM 5514. An oil-contaminated mangrove wetlands' isolate *P. aeruginosa* KVD-HR42 also showed rhamnolipid production (Deepika et al. 2016). It exhibited vast stability in a wide range of pH (2.0–12.0), temperature (20–120°C), and NaCl concentrations (0–2.0 M).

Pseudomonas aeruginosa is still the most proficient rhamnolipid producer, but its pathogenic nature may cause safety and health problems during large-scale production and applications. Some rhamnolipid-producers can cause nosocomial

Table 1. Rhamnolipids producing *Pseudomonas* sp.

Microorganisms	Reference
Pseudomonas guguanensis	Ramya Devi et al. 2019
Pseudomonas azotoformans AJ15	Das and Kumar 2018
Pseudomonas pachastrellae LOS20	Kaskatepe et al. 2017
Pseudomonas aeruginosa OG1	Ozdal et al. 2017
Pseudomonas indica MTCC 3714	Bhardwaj et al. 2015
Pseudomonas putida BD2	Janek et al. 2013
Pseudomonas nitroreducens	Onwosi and Odibo 2013
Pseudomonas fluorescens	Vasileva-Tonkova et al. 2011
Pseudomonas alcaligenes	Oliveira et al. 2009
Pseudomonas luteola	Onbasli and Aslim 2009
Pseudomonas cepacian	Onbasli and Aslim 2009
Pseudomonas stutzeri	Celik et al. 2008
Pseudomonas chlororaphis	Gunther et al. 2006

infections, specifically in individuals with low immunity; perhaps rhamnolipids may be significantly contributing to the pathogenicity. Thus, extensive investigations should be carried out to discover safe and economical methods for rhamnolipid production. Some metabolic engineering efforts have been made to improve rhamnolipid production and reducing its pathogenicity. The critical genes rhl*A*, *rhlB*, and *rhlC* for rhamnolipid biosynthesis, are exclusively found in *Pseudomonas* sp. and *Burkholderia* sp., but they have shown successful expression after metabolic insertion in few of the non-pathogenic host bacteria to produce safe rhamnolipids. The rhlB and rhlC genes can be altered at the expression level to modify the composition of mono- and di-rhamnolipids.

Additionally, cell-free rhamnolipid synthesis should be designed using enzymes and precursors from non-pathogenic microorganisms; this may eliminate the pathogenic effects and simplify the purification by reducing the complexity of the quorum-sensing mechanism. Another approach to reduce or eliminate the pathogenicity of *P. aeruginosa* may involve the *in vivo* or *in vitro* enzymatic degradation of the toxins like pyocyanin during rhamnolipid production. The production costs of rhamnolipids can be significantly reduced by direct utilization of fermentation broth without rhamnolipid purification or the rhamnolipid-producing strains may directly utilize the alternative substrates.

3.2 Other Genera Producing Rhamnolipids

The rhamnolipid biosynthesis is not limited to *Pseudomonas*, other genera including *Bacillus, Pantoea, Pseudoxanthomonas, Burkholderia, Ralstonia,* and *Alcaligenes* are also reported for rhamnolipid productions. Some rhamnolipid producing strains other than *Pseudomonas* sp. are listed in Table 2.

The main reason behind interest in such new rhamnolipid producing genera and strains is their non-pathogenic nature unlike the opportunistic pathogen *Pseudomonas* sp. Hošková et al. (2013) studied the comparative rhamnolipid production by non-pathogenic bacteria *Acinetobacter calcoaceticus* and *Enterobacter asburiae*. The rhamnolipids formed by *A. calcoaceticus* and *E. asburiae* both showed excellent

Table 2. Rhamnolipid producing other genera.

Microorganisms	Reference
Stenotrophomonas maltophilia IITR87	Tripathi et al. 2020
Achromobacter sp. (PS1)	Joy et al. 2019
Serratia rubidaea SNAU02	Nalini and Parthasarathi 2018
Ochrobactrum anthropi HM-1	Ibrahim 2017
Citrobacter freundii HM-2	Ibrahim 2017
Burkholderiathailandensis E264	Rienzo et al. 2016
Virgibacillussalaries	Elazzazy et al. 2015
Bacillus Lz-2	Li et al. 2015
Nocardia otitidiscaviarum MTCC 6471	Vyas and Dave 2011
Burkholderia plantarii DSM 9509[T]	Hörmann et al. 2010
Renibacterium salmoninarum 27BN	Christova et al. 2004

emulsification activity against aromatic and aliphatic hydrocarbons as well as several plant oils.

The rhlA, rhlB, and rhlC genes responsible for rhamnolipid biosynthesis in *Pseudomonas* sp. are present in natural non-infectious *B. thailandensis* and the genetically similar pathogen *B. pseudomallei* (Dubeau et al. 2009). Both bacterial strains contain all the three genes in a single gene cluster and produce rhamnolipids with 3-hydroxy fatty acid moieties which is normally not observed in the *Pseudomonas* sp. Likewise, Andrä et al. (2006) discovered a rhamnolipid containing trimer of hydroxyfatty acids from *B. plantarii*. Apart from these naturally occurring rhamnolipid producers, researchers have also genetically modified some non-producing strains. They artificially inserted the *rhlA*, *rhlB*, and *rhlC* genes into non-pathogenic, suitable host microorganism for the production of selective rhamnolipids.

4. Metabolic Pathway and Regulation of Rhamnolipid Production

4.1 Metabolic Pathway of Pseudomonas aeruginosa

The pathway followed by microorganisms to synthesize rhamnolipids are responsible for their diverse chemical composition and make them applicable in various fields. Similar to the other biomolecules, the culture conditions of producing microorganisms affect the rhamnolipids biosynthesis too. The biosynthetic pathway for rhamnolipid production in model bacteria *Pseudomonas aeruginosa* is shown in Fig. 2.

P. aeruginosa utilizes the *de novo* fatty acid pathway as the source of substantial fatty acids for rhamnolipid synthesis. The cell wall lipopolysaccharide (LPS) of *P. aeruginosa* is composed of rhamnose. Glycerol replaces acetate and provides carbon to the rhamnose molecules by the condensation of two carbon units without any reorganization of their C-C bonds. The rhamnolipids derive all the significant carbons from these glycerol carbon, however carbon required for the β-oxidation intermediate "β-hydroxydecanoic acid" is supplied by the acetate. The fatty acid synthesis for rhamnolipid differs from the typical biosynthesis at the level of ketoacyl reduction. Rhamnolipids are sequentially catalyzed as mono- and di-rhamnolipids by rhamnosyltransferase I and rhamnosyltransferase II, the glycosyltransferase units. The bicistronic operon genes *rhlA* and *rhlB* products showed independent activities of both proteins (Wittgens et al. 2017). The rhamnosyltransferase II encoding gene *rhlC*, localized separately from gene *rhlA* and *rhlB* at an alternative chromosomal site. Both *rhlA* and *rhlC* genes roughly bound to the inner-membrane, while *rhlB* represented as a membrane-bound gene. The 3-(3-hydroxyalkanoyloxy) alkanoic acid (HAA), an essential component of mono-rhamnolipid (L-rhamnosyl-3-hydroxydecanoyl-3-hydroxydecanoic acid) is synthesized using activated hydroxy fatty acid by *rhlA* gene.

Moreover, *rhlB* glycosyltransferase catalyze deoxy thymidine diphosphate-L-rhamnose (dTDP-L-rhamnose) and HAA condensation to form mono-rhamnolipids. In contrast, *rhlC* uses mono-rhamnolipid as a substrate combined with dTDP-L-rhamnose to synthesize di-rhamnolipid (L-rhamnosyl-L-rhamnosyl-3-hydroxydecanoyl-3-hydroxydecanoic acid) and also exhibited sequence homology

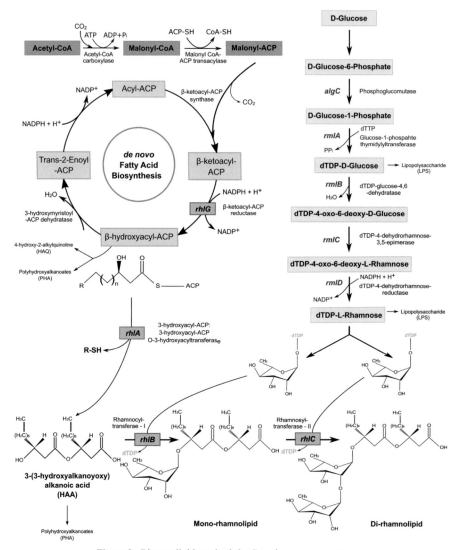

Figure 2. Rhamnolipid synthesis by *Pseudomonas aeruginosa.*

with rhamnosyltransferases associated with LPS synthesis (Tiso et al. 2017). The *rhlA, rhlB,* and *rhlC* are the critical genes involved in rhamnolipid synthesis and are also reported in *Burkholderia thailandensis*, *B. paseudomallei*, and *Escherichia coli.*

P. aeruginosa can release HAA in the cell environment as a biosurfactant because it also holds surface-active properties. This is an essential step of rhamnolipid production, even though its function is unidentified. The gene *rhlG* is especially involved in the fatty acid precursor draining for rhamnolipid production, and consistently affects the polyhydroxyalkanoates (PHA) synthesis. It also provides a fatty acid precursor, acyl carrier protein (ACP) to form the 4-hydroxy-2-alkylquinolines (HAQs) containing QS-related *Pseudomonas* quinolone signal

(PQS). Rhamnolipid primary developments and PHA synthesis involve HAA as a common molecule, but PHA synthesis is not compulsory for rhamnolipid production.

Figure 2 showed *algC* as a principal gene for the biosynthetic pathway of dTDP-D-glucose, D-rhamnose, and dTDP-L-rhamnose. It converts the D-glucose-6-phosphate to D-glucose-1-phosphate, a precursor of dTDP-D-glucose and dTDP-L-rhamnose used for LPS and exopolysaccharide alginate production. An operon *rml*ABCD holds genes *rmlA*, *rmlB*, *rmlC*, and *rmlD* which catalyzed the dTDP-L-rhamnose pathway in *P. aeruginosa*. Kubicki et al. (2019) stated that marine biosurfactants' synthesis originates from similar pathways followed by non-marine bacteria.

4.2 Role of Quorum Sensing in Regulation

The rhamnolipid biosynthesis in *P. aeruginosa* is regulated by environmental factors and the quorum sensing (QS) system. The QS system regulates approximately 10% genes of the *P. aeruginosa*. It corresponds to a bacterial signalling mechanism which affect the production throughout the cellular growth phase of autoinducers, the mediating molecules. The autoinducers interact with the transcriptional regulators when the concentration limit is reached and it allows the specific expression of the group genes. An interspecies highly studied autoinducer N-acyl homoserine lactones (AHL) was released by more than 70 Gram-negative species found to be used as a signalling molecule (Araújo et al. 2018).

P. aeruginosa can grow inside the host cells and refrain from damaging them until their population concentration reaches to the significant level necessary for biofilm formation. Then, microorganisms become hostile to the host's immune system and cause diseases. The QS system of *P. aeruginosa* regulates the production of some essential compounds of biofilm formation and acts on the release of extracellular DNA. *P. aeruginosa* contains three recognized systems for the QS detection, LasI/ LasR, RhI/RhRR and a *Pseudomonas* quinolone signal (PQS) system (Tielker et al. 2005). In QS, *las* and *rhl* are the central regulating systems. The catalytic enzymes Las and RhlI produce the homoserine lactones 3OC12-HSL and C4-HSL signalling molecules in the synthesis process that binds and modulates the corresponding transcriptional regulators LasR and RhRR, respectively. The system requires the RsaL protein and LasR as well, to negatively regulate the expression of both genes and affect the rhamnolipid biosynthesis indirectly (Rasamiravaka et al. 2015).

The 2-heptyl-3-hydroxy-4-quinolone based third signalling system, selects the signal *Pseudomonas* quinolone (PQS), as a part of *P. aeruginosa* quorum-sensing regulatory network. The pqsABCD gene products promote PQS biosynthesis and bind the LysR-type regulator PqsR (or MvfR). In a typical complexed regulatory network, las direct the expression of *pqsR* while the *rhl* QS system represses. The profile of PQS production is similar to the rhamnolipid profiles, as it reaches its maximal level in the stationary phase. Mutant genes of *pqsR* and *pqsE* reduced the levels of rhamnolipid synthesis, even when supplied with exogenous C4-HSL and indicated a direct relation of *pqsR* and PQS in the rhamnolipid biosynthesis.

An understanding of the QS system and the rhamnolipid biosynthetic pathway helps to solve the problems that occur during their production. Hence, some studies have established an anti-QS property of natural herbal medicinal substances recently (Bouyahya et al. 2017). The inhibition of QS-molecules requires specific screening of individual molecules with diverse chemical natures.

5. Screening of Potential Biosurfactant Producers

It is necessary to determine the biosurfactant producing ability of microorganisms to select the potential one. Various methods can be employed for screening and assess the rate of biosurfactant production. Each technique works on a different principle according to the specificity between the substrates/chemicals used and cell components. Some of these techniques are modified for effective performance and to obtain efficient biosurfactant producers. Some commonly used detection methods for biosurfactant producing microorganism are compared in Table 3.

Table 3. Screening methods for biosurfactants/rhamnolipids.

Screening Methods	Advantages	Limitations	References
Drop collapse test	Qualitative, rapid and easy; small volume of samples; microplates can be used	Less sensitive; required significant concentration of biosurfactant	Pardhi et al. 2020
Oil displacement test	Qualitative, rapid and easy; small volume of samples used; significant for less active biosurfactants	Positive response can disappear within seconds hence continuous attention required to	Pardhi et al. 2020
Emulsification index test	Qualitative, simple and readily used method	Indicates only presence of biosurfactants	Pardhi et al. 2020
Surface tension measurement	Qualitative, quantitative, easy and accurate	Special equipment required; multiple samples can't be measured simultaneously; high volume of samples required	Barakat et al. 2017
Blue agar plate method	Qualitative and easy; different culture conditions and factors can also be applied directly on the agar plates	Specific for only anionic biosurfactants; CTAB is toxic and inhibits the growth of some microbes	Vyas and Dave 2011
Oil agar plate technique	Qualitative and easy	Time consuming	Satpute et al. 2010
Hemolytic activity	Qualitative and simple	Not specific; some biosurfactants don't show hemolytic activity	Pardhi et al. 2020
Bacterial adhesion to the hydrocarbon assay	Qualitative and simple	Indirect	Satpute et al. 2010
Emulsification assay	Qualitative	Less consistent	Saravanan and Vijayakumar 2012
Orcinol assay	Qualitative and quantitative	Not convenient for biosurfactants other than glycolipid	Saravanan and Vijayakumar 2012

The drop collapsed and oil displacement test are the oldest methods for the screening of biosurfactant producing microorganisms. Both are qualitative, rapid, and easy to handle methods; besides this, they require a small amount of sample. A limitation of these tests is their less sensitivity towards the low biosurfactant concentrations. Some modified versions of drop collapsed test are parafilm M, tilting glass slide, microplate, and penetration assay (Satpute et al. 2010). They work on the same principle, only parafilm or microplates are used rather than the glass slide. Biosurfactant can mix two immiscible liquids by dispersing one liquid into another. It represents micellular solubilization with large solubilized particles, called emulsions. Based on this property, biosurfactant producers can be determined by a simple and most widely used method emulsification index (E_{24}%). Another approach to screen a potential producer is to measure the blood hemolysis formed by biosurfactant production. Although it is a simple and easy method, it is not recommended as it limits the specificity of some biosurfactant producing microorganisms. Pardhi et al. (2020) have compared and summarized the methods used to screen a potential biosurfactant producer.

Besides, several other procedures like surface tension measurement, oil agar plate, and blue agar plate, bacterial adhesion to the hydrocarbon (BATH) assay, emulsification assay, and orcinol assay are recognized as a probable screening method for biosurfactants. Surface tension measurement depends on the reduction of surface and interfacial tensions occur due to the production of biosurfactants. It is well known for its accuracy, but the main drawback is the use of a special equipment which increases the ultimate cost and is not able to measure multiple samples simultaneously (Barakat et al. 2017). The oil and blue agar plate depend on the fact that biosurfactants can utilize the oil and glucose as a carbon and energy source, respectively. A clear zone formed around the cultures by utilization of oil/glucose is directly proportional to the concentration of biosurfactant produced. Blue agar plate is specially used to detect anionic surfactants like rhamnolipids but found quite risky as its main component cetyl trimethyl ammonium bromide (CTAB) is toxic for some microorganisms' growth (Satpute et al. 2010, Vyas and Dave 2011).

Bacterial adhesion to the hydrocarbon (BATH) assay is used to evaluate the microbial cells' surface hydrophobicity. Usually, the microorganisms that directly uptake hydrocarbons are demonstrated with high surface hydrophobicity. Similarly, replica plate assay and solubilization of crystalline anthracene also measures the hydrophobicity of the cells (Satpute et al. 2010). The colorimetric assays, emulsification and orcinol estimates the sugar and related elements for determining the microbial ability to produce biosurfactant. Saravanan and Vijayakumar (2012) proposed orcinol assay as a direct assessment of glycolipids, especially the rhamnolipids. It is used for both qualitative as well as quantitative analysis of biosurfactant production.

6. Strategies to Improve Rhamnolipid Production

Rhamnolipid production depends on the nutritional components utilized along with the cultural conditions provided to the producing strain. These nutritional and physiochemical factors affect the biosurfactants' production at genetic levels. Although

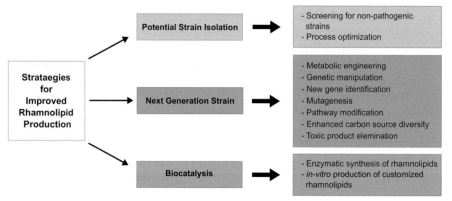

Figure 3. Strategies for development of rhamnolipid producing strain.

rhamnolipids are applicable in different fields, their production is not promoted at the industrial level because of its higher production cost. The use of low-cost substrates is one of the important factors that favourably affects the rhamnolipid production cost and increases the yield with reasonable cost. The main problem in using such renewable substrates is the rhamnolipid producing strain. Hence, several strategies are introduced to obtain a high yield rhamnolipid producing strains (Fig. 3).

The strategies define an ideal rhamnolipid producing strain should: (i) utilize cheap and renewable substrates to avoid high-substrate cost, (ii) minimal or no use of chemical defoamers, (iii) produce highly pure products instead of mixtures to reduce the purification steps, and (iv) non-pathogenic nature.

6.1 Strain Development

The rhamnolipid production can be optimized through development of the wild-type strains. It involves the isolation of a potential rhamnolipid producing strain from the environment where conditions are favourable for biosurfactant-producers. A wide range of substrates that are utilized by different microorganisms are studied and screened for the favourable conditions assisting the high yield. The strain improvement may be carried out with random mutagenesis by physical or chemical mutagens leading to chance phenomenon can probably increase the rhamnolipid yield.

Raza et al. (2006) obtained a mutant strain *Pseudomonas aeruginosa* EBN-8 by gamma radiation aided mutagenesis which gave 8.5 g/L of rhamnolipids with vegetable oil refinery wastes. A UV irradiated mutant *P. fluorescens* 29L showed capability of degrading pyrene (Husain 2008). Similarly, Dobler et al. (2016) also carried out UV mutagenesis on rhamnolipid producing microorganisms and obtained two-fold increased yield of rhamnolipids than the parent strain. The chemically induced mutagenesis using N-methyl-N'-nitro-N-nitrosoguanidine (MNNG) created a mutant *P. aeruginosa* MIG-N146 showed improved rhamnolipid production (Guo et al. 2009).

P. aeruginosa MR01, a gamma irradiated mutant exhibited almost 1.5-fold greater rhamnolipid production than the parent strain (Lotfabad et al. 2016). Another UV assisted mutant strain *P. aeruginosa* 15GR reported the yield of 5.21 g/L rhamnolipid, which was approximately two times higher than the wild-type production (2.5 g/L) (El-Housseiny et al. 2019). The chance mutagenesis is extensively used for increased production of microbial compounds, but sometimes it is observed that the mutants cannot hold their productivity over a period of time.

6.2 Recombinant Strains

The recombinant strains are supposed to be designed with balanced metabolic fluxes towards the products, remove/alter the unwanted autologous genes, and heterologous genes capable of utilizing alternative substrates are inserted to enhance the metabolism. *Pseudomonas aeruginosa* is reported as the highest rhamnolipid producing strains amongst the other bacterial producers studied. Since they are pathogenic, it is necessary to develop a technique to overcome pathogenicity. Hence, metabolic engineering is to be introduced to obtain the potential and safe rhamnolipid-producing strains. This technique involves the identification of critical genes for rhamnolipid synthesis and insert them into the suitable host cells.

The precursors HAA and dTDP-L-rhamnose molecules are essential for rhamnolipid synthesis in *P. aeruginosa* (discussed in section 4). HAA is formed by β-hydroxyacyl-ACP involving *rhlA* gene. Some microorganisms showed the metabolic pathways for both β-hydroxyacyl-ACP and dTDP-L-rhamnose, thus the important genes of rhamnolipid biosynthesis are probably transformed in the non-pathogenic host to form a recombinant bacterial strain. As shown in Fig. 2, the recombinants with *rhlA*, and *rhlB* gene are able to produce only mono-rhamnolipids. However, insertion of *rhlA*, *rhlB*, and *rhlC* genes showed heterologous expression and allows the host to produce both mono-rhamnolipids as well as di-rhamnolipids. The approach to construct rhamnolipid producing recombinant strain can possibly avoid the complicated regulatory mechanism related with the pathogenic strain *P. aeruginosa*. *Pseudomonas putida* and *Escherichia coli* are mainly preferred as a bacterial host for recombinant rhamnolipid production. A summary of recently studied recombinant strains for rhamnolipid production are shown in Table 4.

Wang et al. (2007) successfully introduced the *rhlAB* gene in *E. coli* TnERAB. It showed 0.08 g/L of rhamnolipids with 4 g/L of glucose while, it gradually increased the yield upto 0.18 g/L when glucose (4 g/L) supplemented lysogeny broth is used. The production of di-rhamnolipid can be predominantly increased by the recombinant strain if high number of *rhlC* genes are expressed in the host cells. Han et al. (2014) used pET30a(+) (Novagen) plasmid to construct the rhamnolipid producing recombinant *E. coli* BL21(DE). Tavares et al. (2013) investigated the overexpression of *rhlAB* under the control of tac promoter in *Burkholderia kururiensis*, resulting in increased rhamnolipid production from 0.78 to 5.76 g/L. Similarly, *P. aeruginosa* SG produced 20.98 g/L of rhamnolipid with increased number of *rhlAB* copies under an indigenous promotor of oprl (Zhao et al. 2015). A mono-rhamnolipid producing non-pathogenic recombinant *P. chlororaphis* also showed di-rhamnolipid production

Table 4. Recombinant bacterial strains and rhamnolipid production.

Recombinant Strain	Carbon Source (g/L)	Yield (g/L)	Inserted Genes	Reference
P. aeruginosa PrhlAB	Crude glycerol (60)	2.87	*rhlAB*	Zhao et al. 2019
P. stutzeri Rhl	Crude glycerol (60)	0.87	*rhlABRI*	Zhao et al. 2019
P. aeruginosa DAB	Crude oil (10)	17.30	*rhlAB*	He et al. 2017
P. putida KT2440	Glucose (35)	14.90	*rhlAB*	Beuker et al. 2016
P. putida KT2440	Glucose (20)	0.57	*rhlAB*	Setoodeh et al. 2014
P. putida KT2440	Glucose (N)	1.70	*rhlAB*	Henkel et al. 2014
P. putida KT42C1	Glucose (N)	1.50	*rhlAB*	Wittgens et al. 2011
P. putida T2440	Glucose (N)	1.50	*rhlAB*	Wittgens et al. 2011
P. putida KCTC1067	Soybean oil (20)	7.30	*rhlAB*	Cha et al. 2008
E. coli W3110	Glucose (5)	0.12	*rhlAB*; *rmlBDAC*	Cabrera-Valladares et al. 2006
E. coli HB101	Oleic acid (4)	0.05	*rhlAB*; *rmlBDAC*	Cabrera-Valladares et al. 2006

* **N**, Not mentioned

by overexpressing the *rhlC* gene under a constitutive promoter (Solaiman et al. 2015). Approximately 50 rhamnolipid homologues are produced by recombinant *B. kururiensis*; showing characteristics of both *Pseudomonas* sp. as well as *Burkholderia* sp. (Tavares et al. 2013). This indicates that genetic engineering can modify the composition and congener types of rhamnolipids, responsible for alteration of physicochemical properties of the wild-type rhamnolipid. Moreover, introducing site directed mutagenesis on *rhlB* can also alter the rhamnolipid composition (Han et al. 2014).

Sometimes rhamnolipid producing recombinant strains also form by-products like PHA that covers approximately 50% of the cell dry weight and lowers the rhamnolipid production (Costa et al. 2009). The biochemical network reconstruction by quantitative investigation at process level provides information that may be helpful to remove or reduce undesirable by-products. This flux balance analysis (FBA) did not require any data regarding the quantitative metabolites (Müller et al. 2012). In a comparative study using 13C NMR, it was found that the ratio of mono- and di-rhamnolipid produced directly depends on the carbon source (Choi et al. 2011). Furthermore, the mutants with no rhamnolipid producing ability promoted more PHA accumulation than the parent strains while the PHA-deficient strains showed less HAA, resulting in rhamnolipid formation.

The classical glycolysis and degradation of lactose like sugars is not found in *P. aeruginosa*. Thus, it is suggested to develop rhamnolipid producing recombinant strains that can easily utilize arabinose, galactose, lactose, and xylose by introducing the autologous genes (Müller et al. 2012). The investigations distinctly showed the possibilities of rhamnolipid production through non-pathogenic strains. It is difficult for heterologous hosts to overproduce the rhamnolipids because both the precursors required for synthesis are derived from the central metabolic pathways. Thus, a strategy that increases the precursor availability and balancing flux must be applied along with recombinant approaches to get better rhamnolipid yield.

6.3 Low-cost Carbon Source

Microorganisms generally convert complex organic molecules into simpler forms. These compounds are utilized as a substrate and provide carbon and energy required for microbial growth and production of secondary metabolites. Rhamnolipids are one of the interesting secondary metabolites; hence the selection of suitable substrates for their commercially and economically viable production is necessary. Since the substrates and downstream processing of rhamnolipid are costly; researchers have replaced them with inexpensive and readily available compounds. Such low-cost materials involved agro-industrial wastes as well as some by-products of food and petrochemical industries that eventually helps to reduce the production cost. Besides, utilization of these waste resources microorganisms cleans the environment and produce valuable products. Some low-cost substrate utilizing *Pseudomonas* sp. and their rhamnolipid yields are listed in Table 5.

Researchers have used vegetable oils like olive, palm, soybean, grapeseed, sunflower, rapeseed, coconut, and canola from mill effluents as an alternative of substrates for rhamnolipid productions. Industrial by-products such as molasses, buttermilk, and whey are studied as a low-cost substrate. Some investigators used corn, potato, wheat, and tapioca wastes from starch industries and reported them useful for high yield rhamnolipid.

Among such agro-industrial residues, soybean oil (37 g/L) showed promising rhamnolipid production from *P. aeruginosa* E03-40 compared to the glycerol (20 g/L) (Araújo et al. 2018). Similarly, 1% (v/v) olive oil was successfully utilized by *Pseudomonas aeruginosa* W10 and produced 2 g/L of rhamnolipid (Chebbi et al. 2017). Camilios-Neto et al. (2008) investigated the mixture of sugarcane bagasse, sunflower seed meal, and glycerol for rhamnolipid production by *Pseudomonas aeruginosa* UFPEDA614 and reported maximum yield, 46 g/L. Furthermore, *P. aeruginosa* strain ATCC, H1 and SY1 showed rhamnolipid production with kefir media as 11.7, 10.8, and 3.2 g/L and with fish meal media as 12.3, 9.3, and 10.3 g/L, respectively (Kaskatepe et al. 2015). Rhamnolipid producing natural bacterial species (other than *Pseudomonas*) are listed in Table 6.

Ibrahim (2017) reported 4.9 g/L and 4.1 g/L of rhamnolipid by *Ochrobactrum anthropi* HM-1 and *Citrobacter freundii* HM-2 using 2% (v/v) of waste frying oil. This low-cost substrate utilization developed as a successful approach for enhanced rhamnolipid production and reduced approximately 20–30% of the production costs at the industrial level. As discussed, earlier rhamnolipid production is predominantly obtained by bacterial species hence mostly they undergo the submerged fermentation (SmF). Several researchers have studied the effect of solid-state fermentation on bacterial rhamnolipid producers, but its production rates are comparatively lower than the SmF.

Table 5. Carbon sources used with different *Pseudomonas* and rhamnolipids yield.

Microorganisms	Carbon Source g/L (w/v; v/v)	Rhamnolipid Type	Yield (g/L)	Reference
Pseudomonas guguanensis	Vegetable oil (N)	RL1	0.04–0.05	Ramya Devi et al. 2019
Pseudomonas aeruginosa OG1	Waste frying oil (52)	Crud	13.31	Ozdal et al. 2017
Pseudomonas aeruginosa KVD-HR42	Karanja oil (23.85)	Crud	5.90	Deepika et al. 2016
Pseudomonas aeruginosa NCIM 5514	Glucose (10)	RL1 and RL2	3.178	Varjani and Upasani 2016
Pseudomonas aeruginosa ATCC 9027	Glucose (5)	RL1	2.1	Grosso-Becerra et al. 2016
Pseudomonas aeruginosa ATCC	Kefir (N)	Crud	11.7	Kaskatepe et al. 2015
Pseudomonas aeruginosa ATCC	Fish meal (N)	Crud	12.3	Kaskatepe et al. 2015
Pseudomonas indica MTCC 3714	Mixture of rice bran, de-oiled rice bran and glucose (40)	RL2	9.6	Bhardwaj et al. 2015
Pseudomonas aeruginosa	Corn steep liquer (100) + Molasses (100)	RL1 and RL2	3.2	Gudiña et al. 2015
Pseudomonas putida BD2	Glucose (20)	RL1 and RL2	0.15	Janek et al. 2013
Pseudomonas aeruginosa MTCC 7815	*Mesuaferrea* seed oil (N)	Crude	3.54	Singh et al. 2013
Pseudomonas aeruginosa RS29	Glycerol (N)	Crude	6.0	Saikia et al. 2012
Pseudomonas nitroreducens	Glucose (40)	Crude	5.46	Onwosi and Odibo 2013
Pseudomonas aeruginosa MSIC02	Glycerol (18)	RL1 and RL2	1.26	de Sousa et al. 2011
Pseudomonas aeruginosa OCD1	N-octadecane (20)	Crude	0.98	Datta et al. 2011
Pseudomonas alcaligenes	Palm oil (N)	Crude	2.3	Oliveira et al. 2009
Pseudomonas luteola B17	Molasses (50)	Crude	0.53	Onbaslil and Aslim 2009
Pseudomonas aeruginosa UFPEDA614	Glycerol (30)	Crude	46.0	Camilios-Neto et al. 2008
Pseudomonas stutzeri G11	Crude oil (10)	Crude	0.5	Celik et al. 2008
Pseudomonas fluorescens HW-6	Hexadecane (15)	Crude	2.0	Vasileva-Tonkova et al. 2006
Pseudomonas chlororaphis NRRL B-30761	Glucose (20)	Crude	1.0	Gunther et al. 2006
Pseudomonas sp. DSM 2874	Rapeseed oil (N)	RL1 and RL2	45.0	Trummler et al. 2003

* **N**, Not mentioned; **RL**, Rhamnolipid; **RL1**, Mono-rhamnolipid; **RL2**, Di-rhamnolipid

Table 6. Different carbon sources used with other genera and rhamnolipids yield.

Microorganisms	Carbon source g/L (w/v; v/v)	Rhamnolipid Type	Yield (g/L)	Reference
Virgibacillus salaries	Sunflower frying oil (N)	Crud	2.80	Elazzazy et al. 2015
Burkholderia kururiensis KP23ᵀ	Glycerol (30)	RL1	0.78	Tavares et al. 2013
Acinetobacter calcoaceticus NRRL B-59190	Glycerol (10)	RL1 and RL2	2.00	Hošková et al. 2013
*Burkholderia glumae*AU6208	Canola oil (N)	RL1 and RL2	1.00	Costa et al. 2011
Thermus aquaticus CCM 3488	Sunflower oil (2)	RL1 and RL2	2.79	Řezanka et al. 2011
Meiothermus ruber CCM 2842	Sunflower oil (2)	RL1 and RL2	1.50	Řezanka et al. 2011
Enterobacter hormaechei NRRL B-59185	Glycerol (10)	RL1 and RL2	2.40	Rooney et al. 2009
Pantoea stewartii	Glycerol (10)	RL1 and RL2	2.20	Rooney et al. 2009
Pseudoxanthomonas sp. PNK-04	Mannitol (20)	RL1 and RL2	0.28	Nayak et al. 2009
Burkholderia thailandensis E264	Glycerol (40)	RL2	2.79	Funston et al. 2016
Burkholderia thailandensis E264	Canola oil (40)	RL1 and RL2	1.47	Dubeau et al. 2009
Renibacterium salmoninarum 27BN	*n*-hexadecane (20)	RL1 and RL2	0.92	Christova et al. 2004

* **N**, Not mentioned; **RL**, Rhamnolipid; **RL1**, Mono-rhamnolipid; **RL2**, Di-rhamnolipid

6.4 Rhamnolipid Production Conditions

The choice of a bioprocess directly influences the productivity of target biomolecule. It helps the microorganisms to convert the complex organic molecules into easily utilizable form and consume them for self-benefit. Various biochemical reactions are followed by microorganisms to produce products such as amino acids, vitamins, ethanol, carbon dioxide, etc. Fermentation is an important step that requires a lot of efforts to optimize production. Hence, some possible factors (carbon source, nitrogen source, fermentation duration, temperature, inducer, foam management, use of mixed strains, fermentation styles, cell recycles) that affect the fermentation cost can be improved to enhance the production. At different stages of a fermentation process, other compounds like antibiotics, enzymes, growth factors and biosurfactants are also produced. These compounds known as secondary metabolites are synthesized along with the main product of interest. The conditions to be applied for fermentation process must be optimized and these controlled conditions must be scaled up for higher rhamnolipid production to compete the synthetic surfactants. Based on the culture conditions, fermentation can be divided into submerged state and solid-state fermentation, while based on production mode, it can be divided as continuous and batch.

Submerged (SmF) or liquid state fermentation utilizes liquid media for microbial processes, hence suitable for secondary metabolite producing microbes which show optimal growth under moist conditions. In this method, simple and easy purification of biomolecules is possible as microorganisms secret them in the fermentation broth. The disadvantage of SmF includes sometimes leaching of valuable products from the liquid portion at each recovery step.

Generally, researchers design mineral salt medium with one or two major substrates and use the shake flask method for SmF, which provide maximum agitation intended for optimal growth of rhamnolipid producing microorganisms. Liu et al. (2011) studied rhamnolipid production by *Pseudomonas* sp. BS1 using the shake flask method and collected 0.9 g/L yield. Rienzo et al. (2016) had increased the scale and used a 10 L bioreactor for rhamnolipid by *Pseudomonas aeruginosa* ATCC 9027 and *Burkholderia thailandensis* E264 using SmF.

Solid or surface-state fermentation (SSF) carried out with solid substrates like a banana peel, cassava dregs, cassava bagasse, coffee husk, molasses, rice husk, tapioca peel, and wheat bran. These are generally low-cost, renewable, and readily available materials. SSF involves low moisture content environments, hence suitable for biosurfactant producing fungi, unlike the bacteria. As it utilizes low-cost or renewable substrates, it is a reasonable, economical, and upfront method for biosurfactant production. Moreover, concentrated products with less effluent production can be collected at large scale production. The only disadvantage of SSF includes difficulty in recovery and purification. Successful approaches for optimal rhamnolipid production from *Pseudomonas aeruginosa* UFPEDA 614 and *Serratia rubidaea* SNAU02 using solid-state cultivations was reported (Neto et al. 2008, Nalini and Parthasarthi 2014).

6.5 Advances in Downstream Processing

The initial stages of rhamnolipid downstream processing involve economic recovery and qualitative purification. After the production, these two are the vital steps used to determine the bioprocess probability at a large scale. Rhamnolipid recovery accounts for approximately 60–80% of the total production cost, like any other biotechnological products. The surface and micelle forming physicochemical properties of rhamnolipids make their recovery easy. Sometimes based on the producing microorganisms, rhamnolipids are extracted using combination of two or more methods.

The downstream processing involves purification techniques based on the rhamnolipid purity needed for particular applications like pharmaceutical grade or detergent grade. The versatility of rhamnolipids has an important place in various applications but its complex mixture makes the purification quite challenging. The producing strain is also responsible for the cost of purification involved. Yet, some applications do not require highly pure rhamnolipids and thereby the products with less purity and moderate costs may be applied.

Some common analytical techniques viz., precipitation, filtration, solvent extraction, centrifugation, and foam fractionation are reported for biosurfactant recoveries. One of the simplest methods for extraction of rhamnolipid from

Pseudomonas sp. is acid precipitation and ammonium precipitation. The rhamnolipid precipitation can be performed using suitable solvents like acetone, ethyl acetate, methanol, chloroform, butanol, pentane, acetic acid, hexane, ether, and dichloromethane. Sometimes the solvents used for purification and recovery of the biosurfactants are found toxic for the producing microorganisms; hence approaches are carried out to replace them with low toxic, cheap, and readily available solvents that eventually reduces the recovery costs.

Foam fractionation involves the precipitation of biosurfactants in foam using their surface activity mainly in the continuous mode of the bioprocess. It gives products with high purity levels, hence reliable for purification purpose. A rhamnolipid extraction from *Pseudomonas aeruginosa* ATCC 9027 and *Burkholderia thailandensis* E264 was successfully reported using this method (Rienzo et al. 2016). Generally, ammonium sulphate precipitation is used to recover the biosurfactants with higher molecular weight. The organic solvent extraction is also a widely accepted method for recovery. Besides these, some other techniques like crystallization, ion-exchange, adsorption-desorption, membrane ultrafiltration, and tangential flow filtration are also used for purification of rhamnolipids (Stanburry et al. 2016). The biosurfactant's properties study is essential to choose the downstream processes for efficient recovery of active and cost-effective rhamnolipids.

6.6 *Biocatalysis*

Besides the microbial rhamnolipid production, some modified rhamnolipids are also produced through biocatalysis with large quantities of L-rhamnose. Zho and Rock (2008) showed experimental evidence that HAA can be formed *in vitro* using the purified *rhlA*. Twelve naturally occurred rhamnolipids (Rha2-C14-C14) were synthesised *in vitro* from *B. plantarii* (Howe et al. 2006). The *rhlA*, *rhlB*, and *rhlI* genes are successfully purified from the recombinant *E. coli* strain, showing the possibilities of *in vitro* rhamnolipid production (Kiss et al. 2017). Although the catalytic efficiency is a concern at certain extent, this is accepted as a promising procedure to achieve the rhamnolipids.

The rhamnolipids obtained from *in vitro* processes involves easy purification methods in downstream stage. Apart from this, it also eliminates the pathogenic consequence by acquiring the dTDP-L-rhamnose and HAA from a non-pathogenic strain. The efforts spend on *in vitro* rhamnolipid synthesis established as a worthwhile investigation as both mono- and di-rhamnolipids are obtained using this technique. It is sometimes found challenging for *E. coli* as a host to express and purify the *rhlC* gene. However, non-purified *rhlC* genes may be applied for *in vitro* rhamnolipid synthesis. Gutsmann et al. (2000) showed an effective simultaneous rhamnolipid synthesis accomplished with hydrophobically aided switching phase synthesis. The energy balances related with the substrate conversion yields depends on the assumption that respective strain will be a recombinant rhamnolipid producing strain. This suggests that the metabolic pathways such as classical glycolysis and pentose-sugar degradation are modified for the peculiar substrate, hence no growth or by-products occur during the production phase. It will help to overproduce the specific rhamnolipid.

7. Characterization of Rhamnolipids

Several modern methods are used for characterization and identification of rhamnolipids and other biosurfactants. These involve chromatographic and spectrophotometric methods for determining the chemical composition of unknown products like crude biosurfactants. The widely used analytical methods viz., rhamnose test, thin layer chromatography, high-performance liquid chromatography, Fourier transform infrared spectrometry, and mass spectrometry are discussed here. These methods can be applied individually or collectively, depending on the type of the compound to be characterized. The structural and functional characterization of the biosurfactant will help to discover their applications in various fields.

7.1 *Rhamnose Test*

The carbohydrate groups present in the biosurfactant molecule can be assayed by rhamnose test. A volume of 0.5 ml of cell-free supernatant mixed with 0.5 ml of 5% phenol solution and 2.5 ml of sulfuric acid, and incubated for 15 min. After incubation, the absorbance of the solution was measured at 490 nm. This test is mainly used to screen or characterize the glycolipid type biosurfactants (Aboulsould et al. 2007).

7.2 *Thin-Layer Chromatography (TLC)*

The characterization and identification of the crude biosurfactants traditionally rely on the determination of relative affinity of present compounds towards both mobile and stationary phase by the thin-layer chromatography (TLC). It is a simple and oldest method used for structural characterization of biosurfactants present in the mixture by separation. Based on biosurfactants, different solvent systems are used for detection. Typically, the non-volatile organic and inorganic solvents like diethyl ether, acetic acid, n-hexane, ethyl acetate, and pyridine are preferred. The solvent system, containing chloroform : methanol : water, 65 : 15 : 2 (v/v/v) used for rhamnolipid analysis produced by *Psuedomonas putida* BD2 (Janek et al. 2013). Similarly, Kaskatepe et al. (2015) used chloroform : methanol : acetic acid (65 : 15 : 2; v/v/v) to evaluate the rhamnolipid isolated from *P. aeruginosa* ATCC 9027. Comparatively, TLC is a quick, easy, and inexpensive method amongst the methods used for biosurfactant detection.

7.3 *High-Performance Liquid Chromatography (HPLC)*

High-performance liquid chromatography (HPLC) is an advanced chromatographic method involving a mobile phase, a stationary phase, and a detector. The mobile phase moves the solution introduced through the injector port and continuously flows the components on a solid stationary phase. The migration of compounds depends on their noncovalent interactions with the HPLC column. The response allocated by elution of the sample displayed by the detector and subsequently a peak generated on the chromatogram.

HPLC usually help to separate the lipopeptide type biosurfactants, but some glycolipids are successively separated and identified using HPLC device coupled with an evaporative light scattering detector (ELSD) or mass spectrometry (MS). This technique separates various components based on their polarity. The fractions can be collected for individual peaks to analyze their structural moiety. HPLC-MS is a significant method for investigating the molecular mass of each fraction. A composition of rhamnolipid produced by *Pseudomonas aeruginosa,* RLAT10 and RL47T2 strains identified using LC-MS (Haba et al. 2003).

7.4 Fourier Transform Infrared Spectroscopy (FT-IR)

Fourier transform infrared spectroscopy (FT-IR) reveals the chemical bonds and functional groups present in the biosurfactants. This technique relies on the fact that most of the molecules absorb light in the infrared (IR) region of the electromagnetic spectrum. This absorption corresponds precisely to the bonds present in the molecule. The most common type of FT-IR analysis is the pressed pellet method using a KBr disk. A rhamnolipid detection produced by *Pseudomonas fluorescens* Migula 1895-DSMZ was reported by KBr liquid cell method (Abouseoud et al. 2007). Nowadays, an advanced approach introduced in the FT-IR, i.e., attenuated total reflectance (ATR) crystal accessory, that helps to get rapid and more effective results. It became popular because of its accurate results and low time requirements.

7.5 Nuclear Magnetic Resonance (NMR)

The accurate structural information of interested compounds can be derived from NMR using solid-state and high-resolution practices. Its parameters of significance for structural characterization of glycolipids are (a) the chemical shifts of the absorption frequency, (b) coupling constants, and (c) the integral height of the NMR signals. Furthermore, NMR procedures can be applied as the correlation spectroscopy and heteronuclear multiple quantum coherence, or one- and two-dimensional 1H and 13C NMR spectroscopy. Lotfabad et al. (2016) have reported two structurally distinguished rhamnolipids by 1H and 13C NMR chemical shifts. A series of FTIR, NMR (1H and 13C), and LC/MS was carried out to analyze the rhamnolipid produced from *P. indica* MTCC 3714 (Bhardwaj et al. 2015). The structurally characteristics revealed the existing compound as a di-rhamnolipid.

7.6 Mass Spectrophotometry (MS)

Mass spectrometry (MS) is generally coupled with supplementary techniques for improved performance such as gas chromatography-MS (GC-MS), electrospray ion-MS (ESI-MS), secondary ion-MS (SIMS), liquid chromatography-ESI-MS (LC-ESI-MS), ultra-high-performance liquid-high-resolution-MS (UHPLC-HRMS), and matrix-assisted laser desorption/ionization-time of flight-MS (MALDI TOF-MS). The molecular mass per charge (m/z) value of valuable compounds determined using the ESI-MS and SIMS. The structural identification of rhamnolipid produced by *Pseudomonas aeruginosa* UKMP14T is analyzed by ESI-MS/MS (Sabturani et al. 2016). A structural illumination of rhamnolipids produced by *B. plantarii* DSM9509[T] showed Rha2-C14-C14 congeners by MS. It can decrease the water surface tension to 29.4 mN/m and have the critical micelle concentration (CMC) between 15 and

20 mg/L (Hörmann et al. 2010). Dubeau et al. (2009) investigated the functional groups of rhamnolipids produced by *B. thailandensis* and *B. pseudomallei* using LC/MS. Both have exhibited the identical long-chain containing congeners including Rha-Rha-C14-C14 as dominant. *B. thailandensis* rhamnolipid reduces the water surface tension to 42 mN/m and shows CMC of 225 mg/L. Similarly, Costa et al. (2011) reported a mono- and di-rhamnolipid producing strain *B. glumae* AU 6208. The structural analysis using LC/MS revealed the presence of Rha2-C14-C14 congeners. Its CMC was estimated to be between 25 to 27 mg/L and the interfacial tension of $C16/H_2O$ was decreased from 40 to 1.8 mN/m.

8. Patents and Global Market of Rhamnolipids

Rhamnolipids are established as a significant and applicable product for many different preparations. Therefore, some investigators have scaled up the production technology from laboratory scale to large-scale. Though, limited number of companies produce the rhamnolipid biosurfactants on an industrial scale, it is the most marketable among all other biosurfactants. The list of rhamnolipid producing companies around the world are shown in Table 7. The bacterial rhamnolipids are sold under the market name BioFuture and EC601 are supplied by BioFuture Ltd. (Dublin) and Ecochem Ltd. (Canada), respectively. The diluted and CMC based product has a relative cost of 0.02 and 0.23 £ per litres of the solution (Randhawa and Rahman 2014).

Rhamnolipids are well-characterized amongst all the biosurfactants discovered and hence, their demand is progressively growing as the most desirable of Biosurfactants. Comparatively, rhamnolipids have the maximum number of patents (Table 8) and research publications amongst the biosurfactants investigated. But the higher production costs are holding rhamnolipids a step back from becoming the champions of their field.

Table 7. Rhamnolipid producing companies in the world.

Company	Location	Application Field
Unilever and Evonik	UK	Household cleaning products
AGAE Technologies	USA	Cosmetics, pharmaceuticals
NatSurFact Laboratories	USA	Personal care and cleaning
TeeGene Biotech	UK	Pharmaceuticals, antimicrobial and anti-cancer components, cosmetics
AGAE Technologies LLC	USA	Pharmaceutical, cosmetics, enhanced oil recovery, personal care, bioremediation (*in situ* and *ex situ*)
Jeneil Biosurfactant Co. LLC	USA	Cleaning products, enhanced oil recovery
Paradigm Biomedical Inc.	USA	Pharmaceuticals
Rhamnolipid Companies, Inc.	USA	Agriculture, pharmaceuticals, cosmetics, enhanced oil recovery, bioremediation, food products

Table 8. Patents for rhamnolipid productions and applications.

Patent No.	Patent Title	References
US 20190029250A1	Preventing and destroying citrus greening and citrus canker using rhamnolipid	DeSanto 2019
US 20140148588A1	Process for the isolation of rhamnolipids	Schilling et al. 2014
US 20140080771B2	Method for treating rhinitis and sinusitis by rhamnolipids	Leighton 2014a
US 8765694B2	Method for treating obesity	Leighton 2014b
US 20130296461B2	Aqueous coatings and paints incorporating one or more antimicrobial biosurfactants and methods for using same	Sadasivan 2015
EP 2410039A1	Rhamnolipids with improved cleaning	Unilever PLC London 2012
WO 20120255918 A1	Use of rhamnolipids in the water treatment industry	DeSanto and Keer 2012
CA 20130130319A1	Cells and methods for producing rhamnolipids	Schaffer et al. 2013
US 8592381B2	Method for treating rhinitis and sinusitis by rhamnolipids	Leighton 2013
US 8183198B2	Rhamnolipid-based formulations	DeSanto 2012
WO 20120322751A9	Use of rhamnolipids as a drug of choice in the case of nuclear disasters in the treatment of the combination radiation injuries and illnesses in humans and animals	Piljac 2012
US 20110123623A1	Rhamnolipid mechanism	DeSanto 2011a
US 20110270207A1	Rhamnolipid-based formulations	DeSanto 2011b
US 7985722 B2	Rhamnolipid-based formulations	DeSanto 2011c
US 20110306569A1	Rhamnolipid biosurfactant from *Pseudomonas aeruginosa* strain NY3 and methods of use	Yin et al. 2011
US 20110257115A1	Method for treating rhinitis and sinusitis by rhamnolipids	Leighton 2011
US 7968499B2	Rhamnolipid compositions and related methods of use	Gandhi and Skebba 2011
US 20100249058A1	Feed additive and feed	Ito et al. 2010
US 20090220603A1	Use of rhamnolipids in wound healing, treating burn shock, atherosclerosis, organ transplants, depression, schizophrenia and cosmetics	Piljac et al. 2009b
EP 1889623A3	Use of rhamnolipids in wound healing, treating burn shock, atherosclerosis, organ transplants, depression, schizophrenia and cosmetics	Piljac and Piljac 2009a
US 20090126948A1	Use of rhamnolipid based formulations for fire suppression and chemical and biological hazards	DeSanto 2009
US 20080213194A1	Rhamnolipid-based formulations	DeSanto 2008
US 20080261891A1	Compositions and methods for using syringopeptin 25A and rhamnolipids	Weimer 2008
TW 200812634A	Dermatological anti-wrinkle agent	Eiko and Toshi 2008

Table 8 Contd. ...

...Table 8 Contd.

Patent No.	Patent Title	References
US 20070207930A1	Rhamnolipid compositions and related methods of use	Gandhi et al. 2007a
US 20070191292A1	Antimycotic rhamnolipid compositions and related methods of use	Gandhi et al. 2007b
WO 2007095259A3	Antimycotic rhamnolipid compositions and related methods of use	Bensaci et al. 2007
US 7202063B1	Processes for the production of rhamnolipids	Gunther et al. 2007
US 7262171B1	Use of rhamnolipids in wound healing, treating burn shock, atherosclerosis, organ transplants, depression, schizophrenia and cosmetics	Piljac and Piljac 2007
US 20060233935A1	Rhamnolipids in bakery products	Van Haesendonck et al. 2006
US 7129218 B2	Use of rhamnolipids in wound healing, treatment and prevention of gum disease and periodontal regeneration	Stipcevic et al. 2006
US 20040224905A1	Use of rhamnolipids in wound healing, treatment and prevention of gum disease and periodontal regeneration	Stipcevic et al. 2004
WO 2004040984A1	Rhamnolipids in bakery products	Van Haesendonck and Vanzeveren 2004

Therefore, researchers have to focus on the ideal rhamnolipid producing strain, alternative low-cost substrate, and nominal bioreactor process strategies. The present market value of rhamnolipid (95%, R-95) is around $200/10 mg (AGAE technologies, USA) and $227/10 mg (Sigma-aldrich) (Randhawa and Rahman 2014). Rhamnolipids are established as the most applicable biomolecule in different sectors, if they become economically viable than nothing can stop them to rule the market of surface-active compounds.

9. Applications of Rhamnolipid

Amongst various biosurfactants, rhamnolipids are being extensively studied with considerable interest in recent years. Rhamnolipids have commercial and potential applications in different industrial sectors as shown in Fig. 4.

9.1 Environment

The environmental problems are increasing day by day because of excessive use of chemical compounds. To combat these problems, they are to be replaced with eco-friendly microbial compounds. Rhamnolipids are helpful to remove the chemical contaminants from soils and improve its fertility. They are also used to control the oil spillage and detoxification of oil-contaminated industrial effluents and soils. The hydrocarbon degradation and removal is enhanced in presence of rhamnolipids, as it solubilizes the oil/water emulsions and raise the hydrocarbon solubility in soil.

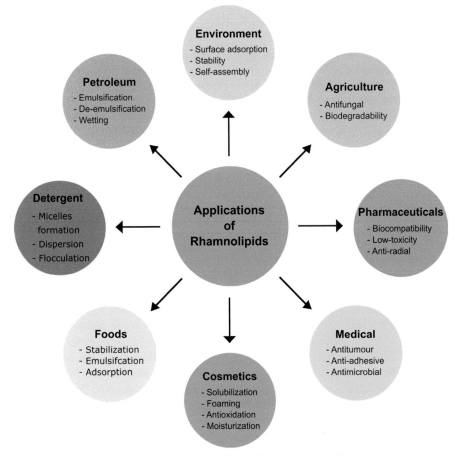

Figure 4. Applications of rhamnolipids in various fields.

Rhamnolipids are reported to remove Ni and Cd from soils with 80–100% effectiveness under controlled conditions and 20–80% in field (Mulligan and Wang 2004).

Deepika et al. (2016) revealed that addition of rhamnolipids along with the pure culture of *P. aeruginosa* showed increased biodegradation of n-paraffins, octadecane, hexadecane, phenanthrene, tetradecane, hexadecane, creosote, pristine or hydrocarbon mixtures present in soil. Dahrazma and Mulligan (2007) reported reduction in cadmium toxicity, removal of Cu and Ni, and increased naphthalene biodegradation in the presence of rhamnolipids. Rhamnolipid obtained from *P. aeruginosa* J4 reduced the surface tension up to 30 dynes/cm from 72 dynes/cm with CMC of 50 mg/L and remidiated 40 to 100% of diesel from diesel/water system (Whang et al. 2008). *P. stutzeri* strain G11 produced 0.003 g/L of rhamnolipid with 69% crude oil degradation rate (Celik et al. 2008).

9.2 Agriculture

The antimicrobial activity and biodegradability of rhamnolipids make them suitable in agricultural fields for increasing soil quality, inhibiting plant pathogens, supporting beneficial plant-microbe interactions, and preparing the pesticides. It generally removes the water-soluble pollutants from soil to augment its health. Some heavy metals like Cd, Ba, Ca, Li, Cu, Mg, Ni, Mn, and Zn are successfully removed from the soil using this anionic rhamnolipids-foaming technique (Mulligan and Gibbs 2004). Rhamnolipids are also found useful in eliminating the polyhydroxyalkanoates (PAH) (Poggi-Varaldo and Rinderknecht-Seijas 2003) and pentachlorophenol (Mulligan and Eftekhari 2003) from the soil.

A rhamnolipid with 500 mg/L concentration can remove approximately 60% of the hexadecane under laboratory conditions. Comparative studies showed that rhamnolipids could increase the four-ring PHA solubilization almost five times more effectively than sodium dodecyl sulfate (SDS). Similarly, SDS was found less efficient in removing the hydrocarbons from sandy loam and silt loam soil while rhamnolipids removed approximately 10% more hydrocarbons from it (Mulligan and Gibbs 2004). Rhamnolipids can be used as an effective biopesticides in agriculture. They induce resistance in plants by non-specific immunity. Rhamnolipids can stimulate the defence genes in tobacco and act as a potent protector in monocotyledonous plants against the biotrophic fungi. They are suitable participant for treating the wastewater and preventing overwatering during irrigation.

Rhamnolipids also facilitate the absorption of essential nutrients and fertilizers available in the root area and breaks the impermeable barriers allowing water to spread throughout the soil (Sinumvayo and Ishimwe 2015). *Alternaria alternata*, causing post-harvest black rot in tomatoes can be controlled using rhamnolipids integrated with *Rhodotorula glutinis* (Yan et al. 2014). Renfro et al. (2014) have investigated the *Phytophtora* blight caused in *Capsicum annuum* which can be regulated by continuous application of rhamnolipid (150 mg/L). The zoospore forming plant pathogens that have acquired the resistance against commercial chemical pesticides were also inhibited by rhamnolipids (Sinumvayo and Ishimwe 2015).

9.3 Petroleum

Rhamnolipids can enhance the biodegradation and exclusion of oil through mobilization, emulsification, de-emulsification, and solubilization. The variety of their applications in petroleum industries involve microbial enhanced oil recovery (MEOR), cleaning of oil tankers, waste oil-water treatments, and transportation of crude oil (Banat et al. 2000). Additionally, they also play the role of biocatalyst and anti-corrosion agent for up-gradation of petroleum.

Lai et al. (2009) reported removal of total petroleum hydrocarbons (TPH) from contaminated soil (3000 mg TPH/kg dry soil) using 0.2 mass% of rhamnolipids. *Ochrobactrum anthropic* HM-1 and *Citrobacter freundii* HM-2 rhamnolipids recovered 70% and 67% of residual oil from oil-saturated sand packed-columns (Ibrahim 2017). Płaza et al. (2008) reported rhamnolipids from *Ralstonia picketti* SRS and *Alcaligenes piechaudii* SRS that can degrade crude oil up to 80% in 20 days. A rhamnolipid produced by *P. aeruginosa* reduced the water surface tension from 72 to 35.26 mN/m at CMC of 127 mg/L, hence is suggested to be use in microbial enhanced oil recovery (MEOR) (Câmara et al. 2019).

9.4 Detergents

Investigators are developing interest in biosurfactants for replacing the chemical laundry detergents with an eco-friendly substitute. Rhamnolipids are able to form micelles that attract the hydrophilic moieties of oily stains and remove them from desired material. Benincasa et al. (2004) reported a detergent-like rhamnolipid from *P. aeruginosa* showing excellent foaming properties. A rhamnolipid produced by *P. aeruginosa* actively removed the hydrophilic stains from cotton cloth at neutral conditions (7 pH, 60–80°C and 40 min of washing) and as they are easily degraded by soil microflora, they can be used in detergent formulation (Bafghi and Fazaelipoor 2012).

9.5 Medical

The use of biosurfactants for medical preparations is not completely accepted yet, but the significant biological activities of rhamnolipids including antimicrobial, anti-adhesive, antitumor, anti-mycoplasma, and hemolytic activities makes them potential components in the medical field and may be promoted for medicine preparations in the future. Abalos et al. (2001) used different concentrations of rhamnolipid to inhibit the growth of *Aspergillus niger* and *Gliocadium virens* (6 μg/mL); *Chaetonium globosum*, *Panicillium crysogenum* and *Aureobasidium pullulans* (32 μg/mL) and *Botrytis cinerea* and *Rhizoctonia solani* at (18 μg/mL). Rhamnolipid with antitumor activity is reported from *P. aeruginosa* BN10 (Christova et al. 2014). The biofilm architecture in *P. aeruginosa* PAO1 is affected by rhamnolipid production (Davey et al. 2003).

9.6 Cosmetics

Cosmetic applications involves oil-based substances that stabilized by emulsifiers or surfactants to obtain emulsions. Rhamnolipids owed skin compatibility and low skin-irritations, hence replaced the synthetic surfactants as emulsifiers, cleaning agents, foaming agents, wetting agents, solubilizers, antimicrobial agents, as well as enzymatic mediators in various forms such as creams, gels, lotions, films, liquids, powders, pastes, sprays, and sticks (Araújo et al. 2018). Patents have been approved for cosmetics containing rhamnolipids for anti-wrinkle and anti-aging products (Table 8), released as a commercial cosmetic for skin care.

9.7 Pharmaceuticals

Rhamnolipids have shown the applicability in pharmaceuticals due to their biocompatible nature, high emulsification activity and relatively low toxicity. They also exhibited antimicrobial activities against several pathogenic microorganisms like *Micrococcus luteus*, *Bacillus cereus*, *Staphylococcus aureus*, and *Listeria monocytogenes*. *Pseudomonas* sp. MK5, a seashore soil isolate showed great antibacterial and antifungal activities; possibly applicable in pharmaceuticals (Murugan et al. 2018). Several rhamnolipid producing companies (Table 7) and patents (Table 8) listed have pharmaceutical use of the biosurfactant produced.

However, though rhamnolipids show many positive responses, their use is limited in this field assuming the fact that being a biomolecule it may found antigenic in nature.

9.8 Food Processing

Rhamnolipids produced by *Pseudomonas aeruginosa* are classified as promising biosurfactants for food processing since the United States Environmental Protection Agency approved them for making food products. A combination of niacin and rhamnolipids extended the shelf life of salad by inhibiting mold growth. Natamycin, nisin, and rhamnolipid mixture increased the shelf life of cottage cheese and inhibited the growth of gram-positive bacteria (Van Haesendonck and Vanzeveren 2004). Rhamnolipids help in rhamnose synthesis that can be used as a food flavourer (Sinumvayo and Ishimve 2015). Van Haesendonck and Vanzeveren (2004) used 0.01% (w/w) of rhamnolipid to improve the loaf volume, texture and shelf-life of the bread. Rhamnolipids can disrupt the biofilm formation of food-born pathogens and control their growth on food products. They also regulate the retard staling, consistency, fat stabilization, and flavour oil solubilization in ice cream and bakery preparations (Sinumvayo and Ishimve 2015).

10. Conclusion

Rhamnolipids are getting a lot of attention as surface active agents from the industries as a novel renewable substrate utilizer. Among the publications and patents filed for glycolipids, rhamnolipids stands at the top with maximum research reports. Being a biomolecule, rhamnolipids are biodegradable and comparatively low-toxic than the synthetic surfactants. Moreover, they exhibit anti-adhesive and antioxidant activity which make them valuable products for cosmetics. Rhamnolipid synthesis is essential for the *Pseudomonas* sp. to grow on a hydrophobic hydrocarbon source. The large-scale rhamnolipid production faces some problems due to low-productivity strains, high-cost substrate, and downstream processing. Hence forward, new agro-industrial wastes and vegetable oils are developed as a potential carbon source for future rhamnolipid producing strains. The future strain can be naturally screened and optimized or metabolically engineered by inserting the gene for key enzyme rhamnosyltransferases. A recombinant strain will be able to produce a desired product, i.e., either mono- or di-rhamnolipids. Furthermore, biocatalysis involves the modification and *in vitro* production of rhamnolipids using HAA precursor and L-rhamnose that decreases the downstream processing cost. Current reports showed rhamnolipids have the highest economical potentials, amongst the biosurfactant investigated. The rhamnolipids have enough potential to replace the synthetic surfactants used in bioremediation, agriculture, petroleum, detergents, pharmaceuticals, medicals, and food processing. Hence, economically sustainable rhamnolipid production would make them a principal surface-active compound in the coming years.

References

Abalos, A., A. Pinazo, M.R. Infante, M. Casals, F. García and A. Manresa. 2001. Physicochemical and antimicrobial properties of new rhamnolipids produced by *Pseudomonas aeruginosa* AT10 from soybean oil refinery wastes. Langmuir 17(5): 1367–1371. DOI: https://doi.org/10.1021/la0011735.

Abouseoud, M., R. Maachi and A. Amrane. 2007. Biosurfactant production from olive oil by *Pseudomonas fluorescence*. Curr. Res. Educat. Topic. Tren. App. Microbiol. 340–347.

Andrä, J., J. Rademann, J. Howe, M.H. Koch, H. Heine, U. Zähringer and Brandenburg. 2006. Endotoxin-like properties of a rhamnolipid exotoxin from *Burkholderia* (*Pseudomonas*) *plantarii*: immune cell stimulation and biophysical characterization. Biol. Chem. 387(3): 301–310. DOI: 10.1515/BC.2006.040.

Araújo, A., J. Rocha, M.O. Filho, S. Matias, S.O. Júnior, C. Padilha and E. Santos. 2018. Rhamnolipids biosurfactants from *Pseudomonas aeruginosa*—a review. Biosci. Biotechnol. Rese. ASIA 15(4): 767–781. DOI: http://dx.doi.org/10.13005/bbra/2685.

Bafghi, M.K. and M.H. Fazaelipoor. 2012. Application of rhamnolipid in the formulation of a detergent. J. Surf. Deterg. 15: 679–684. DOI: https://doi.org/10.1007/s11743-012-1386-4.

Banat, I.M., R.S. Makkar and S.S. Cameotra. 2000. Potential commercial applications of microbial surfactants. Appl. Microbiol. Biotechnol. 53: 495–508. DOI: 10.1007/s002530051648.

Barakat, K., S. Hassan and O. Darwesh. 2017. Biosurfactant production by haloalkaliphilic *Bacillus* strains isolated from red sea, Egypt. Egyptian J. Aquat. Res. 1–6. DOI: https://doi.org/10.1016/j.ejar.2017.09.001.

Benincasa, M., A. Abalos, I. Oliveira and A. Manresa. 2004. Chemical structure, surface properties and biological activities of the biosurfactant produced by *Pseudomonas aeruginosa* LBI from soapstock. Antonie Van Leeuwenhoek 85: 1–8. DOI: https://doi.org/10.1023/B:ANTO.0000020148.45523.41.

Bensaci, M., P. Skebba, L. Victoria and N.R. Gandhi. 2007. Antimycotic rhamnolipid compositions and related methods of use. WO 2007095259A3. World Intellectual Property Organization (PCT).

Beuker, J., A. Steier, A. Wittgens, F. Rosenau, M. Henkel and R. Hausmann. 2016. Integrated foam fractionation for heterologous rhamnolipid production with recombinant *Pseudomonas putida* in a bioreactor. AMB Expr. 6: 11. DOI 10.1186/s13568-016-0183-2.

Bhardwaj, G., S.S. Cameotra and H.K. Chopra. 2015. Utilization of oil industry residues for the production of rhamnolipids by *Pseudomonas indica*. J. Surf. Deterg. 18: 887–893. DOI: 10.1007/s11743-015-1711-9.

Bouyahya, A., N. Dakka, A. Et-touys, J. Abrini and Y. Bakri. 2017. Medicinal plant products targeting quorum sensing for combating bacterial infections. Asian Pac. J. Trop. Med. 10: 729–743. DOI: 10.1016/j.apjtm.2017.07.021.

Cabrera-Valladares, N., A.P. Richardson, C. Olvera, L.G. Treviño, E. Déziel, F. Lépine and G. Soberón-Chavez. 2006. Monorhamnolipids and 3-(3-hydroxyalkanoyloxy) alkanoic acids (HAAs) production using *Escherichia coli* as a heterologous host. Appl. Microbiol. Biotechnol. 73: 187–194. DOI: https://doi.org/10.1007/s00253-006-0468-5.

Câmara, J.M.D.A., M.A.S.B. Sousa, E.L. Barros Neto and M.C.A. Oliveria. 2019. Application of rhamnolipid biosurfactant produced by *Pseudomonas aeruginosa* in microbial-enhanced oil recovery (MEOR). J. Petrol Explor. Prod. Technol. 9: 2333–2341. DOI: https://Doi.Org/10.1007/S13202-019-0633-X.

Camilios-Neto, D., A. Meira, J.M. de Araújo, D.A. Mitchell and C. Krieger. 2008. Optimization of the production of rhamnolipids by *Pseudomonas aeruginosa* UFPEDA 614 in solid-state culture. Appl. Microbiol. Biotechnol. 81: 441–448. DOI 10.1007/s00253-008-1663-3.

Celik, G.Y., B. Aslim and Y. Beyatli. 2008. Enhanced crude oil biodegradation and rhamnolipid production by *Pseudomonas stutzeri* strain G11 in the presence of Tween-80 and Triton X-100. J. Environ. Biol. 29: 867–70.

Cha, M., N. Lee, M. Kim and S. Lee. 2008. Heterologous production of *Pseudomonas aeruginosa* EMS1 biosurfactant in *Pseudomonas putida*. Biores. Technol. 99: 2192–2199. DOI: https://doi.org/10.1016/j.biortech.2007.05.035.

Chebbi, A., M. Elshikh, F. Haque, S. Ahmed, S. Dobbin, R. Marchant, S. Sayadi, M. Chamkha and I.M. Banat. 2017. Rhamnolipids from *Pseudomonas aeruginosa* strain W10; as antibiofilm/antibiofouling products for metal protection. J. Basic Microbiol. DOI: http://dx.doi.org/10.1002/jobm.201600658.

Choi, M.H., J. Xu, M. Gutierrez, T. Yoo, Y-H. Cho and S.C. Yoon. 2011. Metabolic relationship between polyhydroxyalkanoic acid and rhamnolipid synthesis in *Pseudomonas aeruginosa*: comparative 13C NMR analysis of the products in wildtype and mutants. J. Biotechnol. 151: 30–42. DOI: https://doi. org/10.1016/j.jbiotec.2010.10.072.

Christova, A., P. Petrovb and L. Kabaivanova. 2014. Biosurfactant production by *Pseudomonas aeruginosa* BN10 cells entrapped in cryogels. Verlag der Zeits. fürNaturfors. 68(1-2): 47–52. DOI: 10.5560/ ZNC.2013.68c0047.

Christova, N., B. Tuleva, Z. Lalchev, A. Jordanova and B. Jordanov. 2004. Rhamnolipid biosurfactants produced by *Renibacterium salmoninarum* 27BN during growth on n-hexadecane. Zeits. fürNatur. C 59(1-2): 70–74. DOI: https://doi.org/10.1515/znc-2004-1-215.

Costa, S.G., E. De´ziel and F. Le´pine. 2011. Characterization of rhamnolipid production by *Burkholderia glumae*. Lett. Appl. Microbiol. 53(6): 620–627. DOI: https://doi.org/10.1111/j.1472-765X.2011.03154.x.

Costa, S.G.V.A.O., F. Lépine, S. Milot, E. Déziel, M. Nitschke and J. Contiero. 2009. Cassava wastewater as substrate for the simultaneous production of rhamnolipid and polyxydroxyalkanoates by *Pseudomonas aeruginosa*. J. Ind. Microbiol. Biotechnol. 36: 1063–1072. DOI: 10.1007/s10295-009-0590-3.

Dahrazma, B. and C.N. Mulligan. 2007. Investigation of the removal of heavy metals from sediments using rhamnolipid in a continuous flow configuration. Chemosphere 69: 705–711. DOI: 10.1016/j. chemosphere.2007.05.037.

Das, A.J. and R. Kumar. 2018. Utilization of agro-industrial waste for biosurfactant production under submerged fermentation and its application in oil recovery from sand matrix. Biores. Technol. 260: 233–240. DOI: https://doi.org/10.1016/j.biortech.2018.03.093.

Datta, S., S. Sahoo and D. Biswas. 2011. Optimization of culture conditions for biosurfactant production from *Pseudomonas aeruginosa* OCD. J. Adv. Sci. Res. 2(3): 32–36.

Davey, M.E., N.C. Caiazza and G.A. O'Toole. 2003. Rhamnolipid surfactant production affects biofilm architecture in *Pseudomonas aeruginosa* PAO1. J. Bacteriol. 185(3): 1027–1036. DOI: 10.1128/ JB.185.3.1027-1036.2003.

de Saousa, J.R., J.A. de Costa Correia, J.G.L. de Almeida, S. Rodriguies, O.D.L. Passoa, V.M.M. Melo and L.R.B. Goncalvis. 2011. Evaluation of a co-product of biodiesel production as carbon source in the production of biosurfactant by *P. aeruginosa* MSIC02. Proc. Biochem. 46(9): 1831–1839. DOI: https://doi.org/10.1016/j.procbio.2011.06.016.

Deepika, K.V., S. Kalam, P.R. Sridhar, A.R. Podile and P.V. Bramhachari. 2016. Optimization of rhamnolipid biosurfactant production by mangrove sediment bacterium *Pseudomonas aeruginosa* KVD-HR42 using response surface methodology. Biocatal. Agri. Biotechnol. 5: 38–47. DOI: https:// doi.org/10.1016/j.bcab.2015.11.006.

DeSanto, K. 2008. Use of rhamnolipid-based formulations for fire suppression and chemical and biological hazards. US 20080213194A1, United States Patent Application Publication.

DeSanto, K. 2009. Use of rhamnolipid-based formulations for fire suppression and chemical and biological hazards. US 20090126948A1, United States Patent.

DeSanto, K. 2011a. Rhamnolipid mechanism. US 20110123623A1, United States Patent Application Publication.

DeSanto, K. 2011b. Rhamnolipid-based formulations. US 20110270207A1, United States Patent Application Publication.

DeSanto, K. 2011c. Rhamnolipid-based formulations. US 7985722 B2, United States Patent.

DeSanto, K. 2012. Rhamnolipid-based formulations. US 8183198B2, United States Patent.

DeSanto, K. and D.R. Keer. 2012. Use of rhamnolipids in the water treatment industry. WO 20120255918A1, World Intellectual Property Organization (PCT).

DeSanto, K. 2019. Preventing destroying citrus greening and citrus canker using rhamnolipid. US 20190029250A1, United States Patent Application Publication.

Dobler, L., L.F. Vilela, R.V. Almeida and B.C. Neves. 2016. Rhamnolipids in perspective: gene regulatory pathways, metabolic engineering, production and technological forecasting. New Biotechnol. 33: 123–135. DOI: https://doi.org/10.1016/j.nbt.2015.09.005.

Dubeau, D., E. Déziel, D.E. Woods and F. Lépine, 2009. *Burkholderia thailandensis* harbors two identical *rhl* gene clusters responsible for the biosynthesis of rhamnolipids. BMC Micro. 9: 263–270. DOI:10.1186/1471-2180-9-263.

Eiko, K. and T. Toshi. 2008. Dermatological anti-wrinkle agent. TW 200812634A, World Patent 2008/001921, Tokyo.

Elazzazy, A.M., T.S. Abdelmoneim and O.A. Almaghrabi. 2015. Isolation and characterization of biosurfactant production under extreme environmental conditions by alkali-halo-thermophilic bacteria from Saudi Arabia. Saudi J. Biol. Sci. 22(4): 466–475. DOI: 10.1016/j.sjbs.2014.11.018.

El-Housseiny, G.S., K.M. Aboshanab, M.M. Aboulwafa and N.A. Hassouna. 2019. Rhamnolipid production by a gamma ray-induced *Pseudomonas aeruginosa* mutant under solid state fermentation. AMB Expr. 9(1): 7. DOI: https://doi.org/10.1186/s13568-018-0732-y.

Funston, S.J., K. Tsaousi, M. Rudden, T.J. Smyth, P.S. Stevenson, R. Marchant and I.M. Banat. 2016. Characterising rhamnolipid production in *Burkholderia thailandensis* E264, a non-pathogenic producer. Appl. Microbiol. Biotechnol. 100: 7945–7956. DOI: https://doi.org/10.1007/s00253-016-7564-y.

Gandhi, N.R. and V.L.P. Skebba. 2007a. Process for the production of rhamnolipids. US 7202063A1, United States Patent.

Gandhi, N.R. and V.L.P. Skebba. 2007b. Antimycotic rhamnolipid compositions and related methods of use. US 20070191292A1, United States Patent Application Publication.

Gandhi, N.R. and V.L.P. Skebba. 2011. Rhamnolipid compositions and related methods of use. US 7968499B2, United States Patent.

Grosso, V., A. Gonzalez-Valdez, M.-J. Granados-Martínez, E. Morales Ruiz, L. Servín-González, J.-L. Méndez, G. Delgado, R. Morales-Espinosa, G. Ponce-Soto, M. Cocotl-Yañez and G. Soberón-Chávez. 2016. *Pseudomonas aeruginosa* ATCC 9027 is a non-virulent strain suitable for mono-rhamnolipids production. Appl. Microbiol. Biotechnol. 100: 1–10. DOI: 10.1007/s00253-016-7789-9.

Gudiña, E.J., A.I. Rodrigues, E. Alves, M.R. Domingues, J.A. Teixeira and L.R. Rodrigues. 2015. Bioconversion of agroindustrial by-products in rhamnolipids toward applications in enhanced oil recovery and bioremediation. Bioresour. Technol. 177: 87–93. DOI: https://doi.org/10.1016/j.biortech.2014.11.069.

Gunther, N.W., A. Nuñez, L. Fortis and D.K. Solaiman. 2006. Proteomic based investigation of rhamnolipid production by *Pseudomonas chlororaphis* strain NRRL B-30761. J. Ind. Microbiol. Biotechnol. 33: 914–920. DOI: https://doi.org/10.1007/s10295-006-0169-1.

Gunther, N.W., D.K.Y. Soleiman and W.A. Fett. 2007. Process for the production of rhamnolipids. US 7202063B1, United State Patents.

Guo, Y.P., Y.Y. Hu, R.R. Gu and H. Lin. 2009. Characterization and micellization of rhamnolipidic fractions and crude extracts produced by *Pseudomonas aeruginosa* mutant MIG-N146. J. Coll. Inter. Sci. 331: 356–363. DOI: https://doi.org/10.1016/j.jcis.2008.11.039.

Gutsmann, T., A.B. Schromm, M.H.J. Koch, S. Kusumoto, K. Fukase, M. Oikawa, U. Seydel and K. Brandenburg. 2000. Lipopolysaccharide-binding protein-mediated interaction of lipid A from different origin with phospholipid membranes. Phys. Chem. Chem. Phys. 2: 4521–4528. DOI: https://doi.org/10.1039/B004188M.

Haba, E., A. Abalos, O. Jauregui, M.J. Espuny and A. Manresa. 2003. Use of liquid chromatography-mass spectroscopy for studying the composition and properties of rhamnolipids produced by different strains of *Pseudomonas aeruginosa*. J. Surf. Deter. 6: 155–161. DOI: https://doi.org/10.1007/s11743-003-0260-7.

Han, L., P. Liu, Y. Peng, J. Lin, Q. Wang and Y. Ma. 2014. Engineering the biosynthesis of novel rhamnolipids in *Escherichia coli* for enhanced oil recovery. J. Appl. Microbiol. 117: 139–50. DOI: https://doi.org/10.1111/jam.12515.

He, C., W. Dong, J. Li, Y. Li, C. Huang and Y. Ma. 2017. Characterization of rhamnolipid biosurfactants produced by recombinant *Pseudomonas aeruginosa* strain DAB with removal of crude oil. Biotechnol. Lett. 39: 1381–1388. DOI: 10.1007/s10529-017-2370-x.

Henkel, M., A. Schmidberger, M. Vogelbacher, C. Kühnert, J. Beuker, T. Bernard, T. Schwartz, C. Syldatk and R. Hausmann. 2014. Kinetic modelling of rhamnolipid production by *Pseudomonas aeruginosa* PAO1 including cell density-dependent regulation. Appl. Microbiol. Biotechnol. 98(16): 7013–7025. DOI: https://doi.org/10.1007/s00253-014-5750-3.

Hörmann, B., M. Muller, C. Syldatk and R. Hausmann. 2010. Rhamnolipid production by *Burkholderia plantarii* DSM 9509(T). European J. Lipid Sci. Technol. 112: 674–680.

Hošková, M., O. Schreiberová, R. Ježdík, J. Chudoba, J. Masák, K. Sigler and T. Řezanka. 2013. Characterization of rhamnolipids produced by non-pathogenic *Acinetobacter* and *Enterobacter* bacteria. Biores. Technol. 130: 510–516. DOI: https://doi.org/10.1016/j.biortech.2012.12.085.

Howe, J., J. Bauer, J. Andra, A.B. Schromm, M. Ernst, M. Rossle, U. Zahringer, J. Rademann and K. Brandenburg. 2006. Biophysical characterization of synthetic rhamnolipids. FEBS J. 273: 5101–5112. DOI: https://doi.org/10.1111/j.1742-4658.2006.05507.x.

Husain, S. 2008. Effect of surfactants on pyrene degradation by *Pseudomonas fuorescens* 29L. World J. Microbiol. Biotechnol. 24: 2411–2419. DOI: https://doi.org/10.1007/s11274-008-9756-9.

Ibrahim, H. 2017. Characterization of biosurfactants produced by novel strains of *Ochrobactrum anthropi* HM-1 and *Citrobacter freundii* HM-2 from used engine oil-contaminated soil. Egyptian J. Petr. 1–9. DOI: https://doi.org/10.1016/j.ejpe.2016.12.005.

Ito, S., M. Suzuki, K. Suzuki and Y. Kobayashi. 2010. Feed additive and feed. US 20100249058A1, United States Patent Application Publication.

Janek, T., M. Lukaszewicza and A. Krasowska. 2013. Identification and characterization of biosurfactants produced by the arctic bacterium *Pseudomonas putida* BD2. Coll. Surf. B: Biointer. 110: 379–386. DOI: 10.1016/j.colsurfb.2013.05.008.

Jarvis, F.G. and M.J. Johnson. 1949. A glyco-lipide produced by *Pseudomonas aeruginosa*. J. American Chem. Soci. 71: 4124–4126. DOI: https://doi.org/10.1021/ja01180a073.

Joy, S., P. Rahman, S. Khare and S. Sharma. 2019. Production and characterization of glycolipid biosurfactant from *Achromobacter* sp. (PS1) isolate using one-factor-at-a-time (OFAT) approach with feasible utilization of ammonia-soaked lignocellulosic pretreated residues. Bioproc. Biosys. Eng. 42. DOI: 10.1007/s00449-019-02128-3.

Kamal-Alahmad. 2015. The definition, preparation and application of rhamnolipids as biosurfactants. Inter. J. Nutr. Food Sci. 4(6): 613–623. DOI: 10.11648/j.ijnfs.20150406.13.

Kaskatepe, B., S. Yildiz, M. Gumustas and S.A. Ozkan. 2015. Biosurfactant production by *Pseudomonas aeruginosa* in kefir and fish meal. Brazilian J. Microbiol. 46(3): 855–859. DOI: 10.1590/S1517-838246320140727.

Kaskatepe, B., Y. Sulhiye, M. Gumustas and S. Ozkan. 2017. Rhamnolipid production by *Pseudomonas putida* IBS036 and *Pseudomonas pachastrellae* LOS20 with using pulps. Curr. Pharma. Anal. 13: 138–144. DOI: 10.2174/1573412912666161018144635.

Kiss, K., W.T. Ng and Q. Li. 2017. Production of rhamnolipids-producing enzymes of *Pseudomonas* in *E. coli* and structural characterization. Front. Chem. Sci. Eng. 11: 133–138. DOI: https://doi.org/10.1007/s11705-017-1637-z.

Kubicki, S., A. Bollinger, N. Katzke, K.-E. Jaeger, A. Loeschcke and S. Thies. 2019. Marine biosurfactants: biosynthesis, structural diversity and biotechnological applications. Mar. Dru. 17(408): 1–30. DOI: 10.3390/md17070408.

Lai, C., Y. Huang, Y. Wei and J. Chang. 2009. Biosurfactant-enhanced removal of total petroleum hydrocarbons from contaminated soil. J. Haz. Mat. 167: 609–614. DOI: https://doi.org/10.1016/j.jhazmat.2009.01.017.

Leighton, A. 2011. Methods for treating rhinitis sinusitis by rhamnolipids. US 20110257115A1, United States Patent Application Publication.

Leighton, A. 2013. Methods for treating rhinitis sinusitis by rhamnolipids. US 8592381B2, United States Patent.

Leighton, A. 2014a. Method for treating rhinitis and sinusitis by rhamnolipids. US 20140080771B2, United States Patent.

Leighton, A. 2014b. Methods for treating obesity. US 8765694B2, United States Patent.

Li, S., Y. Pi, M. Bao, C. Zhang, Y. Li, P. Sun and J. Lu. 2015. Effect of rhamnolipid biosurfactant on solubilization of polycyclic aromatic hydrocarbons. Mar. Poll. Bull. 101(1): 219–225. DOI: https://doi.org/10.1016/j.marpolbul.2015.09.059.

Liang-Ming, W., P.W.G. Liu, M. Chih-Chung and C. Sheng-Shung. 2008. Application of biosurfactants, rhamnolipid, and surfactin, for enhanced biodegradation of diesel-contaminated water and soil. J. Haz. Mat. 151: 155–163. DOI: 10.1016/j.jhazmat.2007.05.063.

Liu, T., J. Hou, Y. Zuo, S. Bi and J. Jing. 2011. Isolation and characterization of a biosurfactant-producing bacterium from daqing oil-contaminated sites. African J. Microbiol. Res. 5(21): 3509–3514. DOI: 10.5897/AJMR11.684.

Lotfabad, T.B., N. Ebadipour and R. Roostaazad. 2016. Evaluation of a recycling bioreactor for biosurfactant production by *Pseudomonas aeruginosa* MR01 using soybean oil waste. J. Chem. Technol. Biotechnol. 91: 1368–1377. DOI: https://doi.org/10.1002/jctb.4733.

Müller, M.M., J.H. Kugler, M. Henkel, M. Gerlitzki, B. Hormann, M. Pohnlein, C. Syldatk and R. Hausmann. 2012. Rhamnolipids—next generation surfactants? J. Biotechnol. 162: 366–380. DOI: https://doi.org/10.1016/j.jbiotec.2012.05.022.

Mulligan, C. and B. Gibbs. 2004. Types, production and applications of biosurfactants. Proc. Indian. Natn. Sci. Acad. B70(1): 31–55.

Mulligan, C. and S. Wang. 2004. Rhamnolipid foam enhanced remediation of cadmium and nickel contaminated soil. J. Water Air Soil Poll. 157: 315–330. DOI: https://doi.org/10.1023/B:WATE.0000038904.91977.f0.

Mulligan, C.N. and F. Eftekhari. 2003. Remediation with surfactant foam of PCP-contaminated soil. Eng. Geol. 70(3-4): 269–279. DOI: https://doi.org/10.1016/S0013-7952(03)00095-4.

Murugan, T., M. Murugan and W.J. Albino. 2018. Rhamnolipid biosurfactants produced by *Pseudomonas* sp. MK5 and its efficacy on pharmaceutical application. Res. J. Pharm. Technol. 10(8): 2645–2649. DOI: 10.5958/0974-360X.2017.00470.X.

Nalini, S. and R. Parthasarathi. 2014. Production and characterization of rhamnolipids produced by *Serratia rubidaea* SNAU02 under solid-state fermentation and its application as biocontrol agent. Biores. Technol. 173: 231–238. DOI: 10.1016/j.biortech.2014.09.051.

Nalini, S. and R. Parthasarathi. 2018. Optimization of rhamnolipid biosurfactant production from *Serratia rubidaea* SNAU02 under solid-state fermentation and its biocontrol efficacy against *Fusarium* wilt of eggplant. Ann. Agrar. Sci. DOI: https://doi.org/10.1016/j.aasci.2017.11.002.

Nayak, A., M. Vijaykumar and T. Karegoudar. 2009. Characterization of biosurfactant produced by *Pseudoxanthomonas* sp. PNK-04 and its application in bioremediation. International Biodeter. Biodegr. 63(1): 73–79. DOI: https://doi.org/10.1016/j.ibiod.2008.07.003.

Neto, D.C., J.A. Meira, J.M. Araújo, D.A. Mitchell and N. Krieger. 2008. Optimization of the production of rhamnolipids by *Pseudomonas aeruginosa* UFPEDA 614 in solid-state culture. Appl. Microbiol. Biotechnol. 81: 441–448. DOI: https://doi.org/10.1007/s00253-008-1663-3.

Oliveira, F., L. Vazquez, N. De Campos and F. De Franca. 2009. Production of rhamnolipids by a *Pseudomonas alcaligenes* strain. Proc. Biochem. 44: 383–389. DOI: http://dx.doi.org/10.1016/j.procbio.2008.11.014.

Onbasli, D. and B. Aslim. 2009. Biosurfactant production in sugar beet molasses by some *Pseudomonas* spp. J. Environ. Biol. 30(1): 161–163. DOI: http://www.jeb.co.in info@jeb.co.in.

Onwosi, C.O. and F.J.C. Odibo. 2013. Use of response surface design in the optimization of starter cultures for enhanced rhamnolipid production by *Pseudomonas nitroreducens*. African J. Biotechnol. 12(19): 2611–2617. DOI: 10.5897/AJB12.2635 ISSN 1684-5315.

Ozdal, M., S. Gurkok and O.G. Ozdal. 2017. Optimization of rhamnolipid production by *Pseudomonas aeruginosa* OG1 using waste frying oil and chicken feather peptone. 3 Biotech (Springer), 7(117): 1–8. DOI: 10.1007/s13205-017-0774-x.

Pardhi, D., R. Panchal and K. Rajput. 2020. Screening of biosurfactant producing bacteria and optimization of production conditions for *Pseudomonas guguanensis* D30. Biosci. Biotechnol. Res. Commun. 13(1): 170–179. DOI: http://dx.doi.org/10.21786/bbrc/13.1special issue/28.

Perfumo, A., I.M. Banat, F. Canganella and R. Marchant. 2006. Rhamnolipid production by a novel thermophilic hydrocarbon-degrading *Pseudomonas aeruginosa* APO2-1. Appl. Microbiol. Biotechnol. 72: 132–138. DOI: 10.1007/s00253-005-0234-0.

Piljac, G. 2012. The use of rhamnolipids as a drug of choice in the case of nuclear disasters in the treatment of the combination radiation injuries and illnesses in humans and animals. WO 20120322751A9, World Intellectual Property Organization.

Piljac, T. and G. Piljac. 2009a. Use of rhamnolipids in wound healing, treating burn shock, atherosclerosis, organ transplants, depression, schizophrenia and cosmetics. EP 1889623A3, European Patent Application.

Piljac, T. and O. Piljac. 2009b. Use of rhamnolipids in wound healing, treating burn shock, atherosclerosis, organ transplants, depression, schizophrenia and cosmetics. US 20090220603A1, United States Patent Application Publication.

Płaza, G., K. Lukasik, J. Wypych, G. Nalecz-Jawecki, C. Berry and R. Brigmon. 2008. Biodegradation of crude oil and distillation products by biosurfactant-producing bacteria. J. Environ. Stu. 17(1): 87–94.

Poggi-Varaldo, H.M. and N. Rinderknecht-Seijas. 2003. A differential availability enhancement factor for the evaluation of pollutant availability in soil treatments. Act. Biotechnol. 23: 271–280. DOI: 10.1002/abio.200390034.

Ramya Devi, K.C., R.L. Sundaram, S. Vajiravelu, V. Vasudevan and G.K.M. Elizabeth. 2019. Structure elucidation and proposed *de novo* synthesis of an unusual mono-rhamnolipid by *Pseudomonas guguanensis* from Chennai Port area. Sci. Rep. 9: 5992. DOI: https://doi.org/10.1038/s41598-019-42045-9.

Randhawa, K.K.S. and P.K.S.M. Rahman. 2014. Rhamnolipid biosurfactants—past, present, and future scenario of global market. Front. Microbiol. 5: 454. DOI: https://doi.org/10.3389/fmicb.2014.00454.

Rasamiravaka, T., Q. Labtani, P. Duez and M. El Jaziri. 2015. The formation of biofilms by *Pseudomonas aeruginosa*: a review of the natural and synthetic compounds interfering with control mechanisms. Biomed Res. Int. DOI:10.1155/2015/759348.

Raza, Z., M.S. Khan and Z.M. Khalid. 2007. Evaluation of distant carbon sources in biosurfactant production by a gamma ray-induced *Pseudomonas putida* mutant. Proc. Biochem. 42(4): 686–692. DOI: 10.1016/j.procbio.2006.10.001.

Raza, Z.A., M.S. Khan, Z.M. Khalid and A. Rehman. 2006. Production of biosurfactant using different hydrocarbons by *Pseudomonas aeruginosa* EBN-8 Mutant. Zeits. Fur. Naturfors. C 61(1-2): 87–94. DOI: 10.1515/znc-2006-1-216.

Renfro, T.D., W. Xie, G. Yang and G. Chen. 2014. Rhamnolipid surface thermodynamic properties and transport in agricultural soil. Coll. Surf. B Biointer. 115: 317–322. DOI: https://doi.org/10.1016/j.colsurfb.2013.12.021.

Řezanka, T., T. Kyselová and K. Sigler. 2011. Rhamnolipid-producing thermophilic bacteria of species *Thermus* and *Meiothermus*. Extremophil. 15(6): 697–709. DOI: 10.1007/s00792-011-0400-5.

Rienzo, M.A., I.D. Kamalanathan and P.J. Martin. 2016. Comparative study of the production of rhamnolipid biosurfactants by *B. thailandesis* E264 and *P. aeruginosa* ATCC 9027 using foam fractionation. Proc. Biochem. 51: 820–827. DOI: https://doi.org/10.1016/j.procbio.2016.04.007.

Rooney, A.P., N.P. Price, K.J. Ray and T.M. Kuo. 2009. Isolation and characterization of rhamnolipid-producing bacterial strain from a biodiesel facility. FEMS Microbiol. Lett. 295: 82–87. DOI: 10.1111/j.1574-6968.2009.01581.x.

Sabturani, N., J. Latif, S. Radiman and A. Hamzah. 2016. Spectroscopic analysis of rhamnolipid produced by produced by *Pseudomonas aeruginosa* UKMP14T. Malaysian J. Analyt. Sci. 20(1): 31–43.

Sadasivan. 2015. Aqueous coatings and paints incorporating one or more antimicrobial biosurfactants and methods for using same. US 20130296461B2, United States Patent.

Saikia, R.R., S. Deka, M. Deka and H. Sarma. 2012. Optimization of environmental factors for improved production of rhamnolipid biosurfactant by *Pseudomonas aeruginosa* RS29 on glycerol. J. Basic Microbiol. 52: 446–457. DOI: https://doi.org/10.1007/s13213-011-0315-5.

Saravanan, V. and S. Vijayakumar. 2012. Isolation and screening of biosurfactant producing microorganisms from oil contaminated soil. J. Acad. Ind. Res. 1(5): 264–268.

Satpute, S.K., A.G. Banpurkar, P.K. Dhakephalkar, I.M. Banat and B.A. Chopade. 2010. Methods for investigating biosurfactants and bioemulsifiers. Crit. Revie. Biotechnol. 1–18. DOI: 10.3109/07388550903427280.

Schaffer, S., M. Mirja, T. Anja and S. Nadine. 2013. Cells and methods for producing rhamnolipids. CA 20130130319A1. InternationalesVeroffentulichungsdatum.

Schilling, M., M. Ruetering, V. Dahl and F. Cabirol. 2014. Process for the isolation of rhamnolipids. US 20140148588A1. United State Patent Application Publication.

Setoodeh, P., A. Jahanmiri, R. Eslamloueyan, A. Niazi, S.S. Ayatollahi, F. Aram, M. Mahmoodi and A. Hortamani. 2014. Statistical screening of medium components for recombinant production of

Pseudomonas aeruginosa ATCC 9027 rhamnolipids by non-pathogenic cell factory *Pseudomonas putida* KT2440. Mol. Biotechnol. 56: 175–191. DOI 10.1007/s12033-013-9693-1.

Singh, S.P., P. Bharali and B.K. Konwar. 2013. Optimization of nutrient requirements and culture conditions for the production of rhamnolipid from *Pseudomonas aeruginosa* (MTCC 7815) using Mesua ferrea seed oil. Indian J. Micrbiol. 53(4): 467–476, http://dx.doi.org/10.1007/s12088-013-0403-2.

Sinumvayo, J.P. and N. Ishimve. 2015. Agriculture and food applications of rhamnolipids and its production by *Pseudomonas aeruginosa*. J. Chem. Eng. Proc. Tech. 6(2): 223. DOI: 10.4172/2157-7048.1000223.

Solaiman, D.K.Y., R.D. Ashby, N.W. Gunther and J.A. Zerkowski. 2015. Dirhamnose-lipid production by recombinant non-pathogenic bacterium *Pseudomonas chlororaphis*. Appl. Microbiol. Biotechnol. 99: 4333–4342. DOI: https://doi.org/10.1007/s00253-015-6433-4.

Stanburry, P., A. Whitaker and S. Hall. 2016. Principles of Fermentation Technology. Chap. 10: 619–686, 3rd Edt., Elsevier, USA.

Stipcevic, T., T. Piljac, J. Piljac, T. Dujmic and G. Piljac. 2004. Use of rhamnolipid in wound healing, treatment and preservation of gum disease and periodontal regeneration. United States Patent US 20040224905A1, United States Patent Application Publications.

Stipcevic, T., T. Piljac, J. Piljac, T. Dujmic and G. Piljac. 2006. Use of rhamnolipids in wound healing, treatment and prevention of gum disease and periodontal regeneration. US 7129218B2, United States Patent.

Tahzibi, A., F. Kamal and M.M. Assadi. 2004. Improved production of rhamnolipids by a *Pseudomonas aeruginosa* mutant. Iran Biomed. J. 8(1): 25–31. DOI: 10.4172/2157-7048.1000223.

Tavares, L.F.D., P.M. Silva, M. Junqueira, D.C.O. Mariano, F.C.S. Nogueira, G.B. Domont, D.M.G. Freire and B.C. Neves. 2013. Characterization of rhamnolipids produced by wild-type and engineered *Burkholderia kururiensis*. Appl. Microbiol. Biotechnol. 97(5): 1909–1921. DOI: 10.1007/s00253-012-4454-9.

Tielker, D., S. Hacker, R. Loris, M. Strathmann, J. Wingender, S. Wilhelm, F. Rosenau and K.E. Jaeger. 2005. *Pseudomonas aeruginosa* lectin LecB is located in the outer membrane and is involved in biofilm formation. Microbiol. 151: 1313–1323. DOI:10.1099/mic.0.27701-0.71.

Tiso, T., S. Thies, M. Müller, L. Tsvetanova, L. Carraresi, S. Bröring, K.-E. Jaeger and L.M. Blank. 2017. Rhamnolipids: production, performance, and application. In consequences of microbial interactions with hydrocarbons, oils, and lipids: production of fuels and chemicals. Handbook of Hydrocarbon and Lipid Microbiology. Springer, Cham. DOI: https://doi.org/10.1007/978-3-319-31421-1_388-1.

Tripathi, V., V.K. Gaur, N. Dhiman, K. Gautam and N. Manickam. 2020. Characterization and properties of the biosurfactant produced by PAH-degrading bacteria isolated from contaminated oily sludge environment. Environ. Sci. Pollut. Res. Int. 27(22): 27268–27278. DOI: 10.1007/s11356-019-05591-3.

Trummler, K., F. Effenberger and C. Syldatk. 2003. An integrated microbial/enzymatic process for production of rhamnolipids and L-(+)-rhamnose from rapeseed oil with *Pseudomonas* sp. DSM 2874. European J. Lipid Sci. Technol. 105: 563–571. DOI: 10.1007/s00253-012-4454-9.

Unilever PLC London. 2012. Rhamnolipids with improved cleaning. EP 2410039A1, European Patent Application.

Van Haesendonck, I.P.H. and E.C.A. Vanzeveren. 2004. Rhamnolipids in bakery products. W. O. 2004/040984, International Application Patent (PCT).

Van Haesendonck, I.P.H., E. Claude and A. Vanzeveren. 2006. Rhamnolipids in bakery products. US 20060233935A1, Patent Application Publication, United States.

Varjani, S.J. and V.N. Upasani. 2016. Carbon spectrum utilization by an indigenous strain of *Pseudomonas aeruginosa* NCIM 5514: production, characterization and surface-active properties of biosurfactant. Bioresour. Technol. 221: 510–516. DOI: https://doi.org/10.1016/j.biortech.2016.09.080.

Vasileva-Tonkova, E., A. Sotirova and D. Galabova. 2011. The effect of rhamnolipid biosurfactant produced by *Pseudomonas fluorescens* on model bacterial strains and isolates from industrial wastewater. Curr. Microbiol. 62: 427–433. DOI: 10.1007/s00284-010-9725-z.

Vasileva-Tonkova, E.S., D. Galabova, E. Stoimenova and Z. Lalchev. 2006. Production and properties of biosurfactants from a newly isolated *Pseudomonas fluorescens* HW-6 growing on hexadecane. Zeits. Fur. Naturfors. C 61(7-8): 553–559. DOI: 10.1515/znc-2006-7-814.

Vyas, T. and B. Dave. 2011. Production of biosurfactant by *Nocardia otitidiscaviarum* and its role in biodegradation of crude oil. International J. Environ. Sci. Technol. 8(2): 425–432.

Wang, Q., X. Fang, B. Bai, Z. Liang, P.J. Shuler, W.A. Goddard and Y. Tang. 2007. Engineering bacteria for production of rhamnolipid as an agent for enhanced oil recovery. Biotechnol. Bioeng. 98(4): 842–853. DOI: https://doi.org/10.1002/bit.21462.

Weimer, B.C. 2008. Compositions and methods for using syringopeptin 25A and rhamnolipids. US 20080261891A1, Patent Application Publication, United States.

Whang, L., P. Liu, C. Ma and S. Cheng. 2008. Application of biosurfactants, rhamnolipid and surfactin, for enhanced biodegradation of diesel-contaminated water and soil. J. Haz. Mat. 151: 155–163. DOI: http://dx.doi.org/10.1016/j.jhazmat.2007.05.063.

Wittgens, A., T. Tiso, T.T. Arndt, P. Wenk, J. Hemmerich, C. Müller, R. Wichmann, B. Küpper, M. Zwick, S. Whilhelm, R. Hausmann, C. Syldatk, F. Rosenau and L.M. Blank. 2011. Growth independent rhamnolipid production from glucose using the non-pathogenic *Pseudomonas putida* KT2440. Microb. Cell. Fact. 10: 80. https://doi.org/10.1186/1475-2859-10-80.

Wittgens, A., F. Kovacic, M.M. Müller, M. Gerlitzki, B. Santiago-Schübel, D. Hofmann, T. Tiso, L.M. Blank, M. Henkel and R. Hausmann. 2017. Novel insights into biosynthesis and uptake of rhamnolipids and their precursors. Appl. Microbiol. Biotechnol. 101: 2865–2878. DOI: https://doi.org/10.1007/s00253-016-8041-3.

Yan, F., S. Xu, Y. Chen and X. Zheng. 2014. Effect of rhamnolipids on *Rhodotorula glutinis* biocontrol of *Alternaria alternata* infection in cherry tomato fruit. Postharvest Biol. Tec. 97: 32–35. DOI: https://doi.org/10.1016/j.postharvbio.2014.05.017.

Yin, X., O.R. Corvallis, M. Nie and Q. Shen. 2011. Rhamnolipid biosurfactant from *Pseudomonas aeruginosa* strain NY3 and methods of use. US 2011/0306569 A1, United States Patent Application Publication.

Zhao, F., Q.F. Cui, S.Q. Han, H.P. Dong, J. Zhang, F. Ma and Y. Zhang. 2015. Enhanced rhamnolipid production of *Pseudomonas aeruginosa* SG by increasing copy number of *rhlAB* genes with modified promoter. RSC Adv. 5: 70546–7052. DOI: 10.1039/c5ra13415c.

Zhao, F., H. Jiang, H. Sun, C. Liu, S. Hana and Y. Zhang. 2019. Production of rhamnolipids with different proportions of mono-rhamnolipids using crude glycerol and a comparison of their application potential for oil recovery from oily sludge. RSC Adv. 9: 2885–2891. DOI: 10.1039/c8ra09351b.

Zhu, K. and C.O. Rock. 2008. RhlA converts β-hydroxyacyl-acyl protein intermediates in fatty acid synthesis to the β-hydroxydecanoyl-β-hydroxydecanoate component of rhamnolipids in *Pseudomonas aeruginosa*. J. Bacteriol. 190(9): 3147–3154. DOI: 10.1128/JB.00080-08.

3

Lipopeptides Biosurfactants Production, and Applications in Bioremediation and Health

Hebatallah H Abo Nahas,[1,*] *Fatma A Abo Nouh,*[2]
Sh Husien,[3] *Sara A Gezaf,*[4] *Safaa A Mansour,*[2]
Ahmed M Abdel-Azeem[2] *and Essa M Saied*[5,6]

1. Introduction

Biosurfactant are divided into five broad groups: glycolipids, lipopeptides and lipoproteins, phospholipids, hydroxylated and crossed-linked and fatty acids, polymeric surfactants and particulate surfactants based on their chemical composition and types of microbes producing them (Maier 2003). Compared to chemical surfactants, these compounds have several advantages such as lower toxicity, higher biodegradability and effectiveness at extreme temperatures or pH values (Mukherjee et al. 2006). Although these compounds present interesting features as compared with their chemical counterparts, many of the envisaged applications depend considerably on whether they can be produced economically. Most biosurfactants are considered secondary metabolites; nevertheless, some may play essential roles for the survival of the producing-microorganisms either through facilitating nutrient transport, microbe-host interactions or as biocide agents (Van et al. 2006). Biosurfactant roles include increasing the surface area and bioavailability of hydrophobic water-insoluble

[1] Zoology Department (Physiology), Faculty of Science, University of Suez Canal, Ismailia, Egypt.
[2] Botany and Microbiology Department, Faculty of Science, Suez Canal University, Ismailia 41522, Egypt.
[3] Egyptian Petroleum Research Institute (EPRI), Nasr City, Cairo 11727, Egypt.
[4] Botany Department, Faculty of Science, AL-Arish University, North Sinai, Egypt.
[5] Chemistry Department, Faculty of Science, Suez Canal University, Ismailia, Egypt.
[6] Institute for Chemistry, Humboldt Universität zu Berlin, Brook-Taylor-Str. 2, 12489 Berlin, Germany.
Email: saiedess@hu-berlin.de
* Corresponding author: hebatallah_hassan@science.suez.edu.eg

substrates, heavy metal binding, bacterial pathogenesis, quorum sensing and biofilm formation (Singh and Cameotra 2004). Biosurfactants are amphipatic molecules with both hydrophilic and hydrophobic moieties that partition preferentially at the interface between fluid phases that have different degrees of polarity and hydrogen bonding, such as oil and water, or air and water interfaces (Banat et al. 2000). Lipopeptides biosurfactants are characterized by diverse functional properties (emulsification/de-emulsification, dispersing, foaming, viscosity reducers, solubilizing and mobilizing agents, pore forming capacity) permitting their use in many domains. Besides, lipopeptides biosurfactants have more advantages over synthetic emulsifiers; low toxicity, higher biodegradability, and higher efficiency toward extreme temperature, pH, and salinity offering great opportunities as replacements for chemical surfactants (Sanket and Yagnik 2013). Lipopeptides can be classified to *Bacillus*-related lipopeptides, *Pseudomonas*-related lipopeptides, other bacteria-related lipopeptides, actinomycete-related lipopeptides, and fungal-related lipopeptides (Mnif and Ghribi 2015).

2. Historical Perspectives and Structural Diversity of Lipopeptides

Lipopeptides are among the most popular biosurfactants, and are predominantly synthesized by *Bacillus* spp. (Hathout et al. 2000). The group of lipopeptides isolated during 1950s and 1960s was from *Bacillus* sp. These contain more than 20 different peptides linked to various fatty acid chains. At present, more than 100 different compounds can so be described (Walia and Cameotra 2015). Structurally, they are constituted by a fatty acid in combination with a peptide moiety and correspond to an isoform group that differs by the composition of the peptide moiety, the length of the fatty acid chain, and the link between the two parts. Lipopeptides are defined as cyclic, low molecular weight compounds with antimicrobial potential and are largely produced by *Bacillus* and *Pseudomonas* sp. (Cai et al. 2013). In general, the molecular weight of lipopeptides ranges from 1000–2000 Da. They are synthesized by specific gene clusters, namely nonribosomal peptides synthetase (NRPs) via a multi-enzyme biosynthesis pathway (Stein 2005). Lipopeptides lowers the surface and interfacial tension more efficiently than glycolipids. Lipopeptides which are synthesized by *Bacillus* sp. have a good heterogeneity in accordance with the type and sequence of amino acid moiety and nature, length and branching of fatty acid chain and their moiety (Hathout et al. 2000). Lipopeptides are mainly classified into: Iturin biosurfactants, Surfactin biosurfactants and Fengycin biosurfactants (Malviya et al. 2020). Table 1 shows a comparison between the three families in their structure, biosynthesis, their different isomeric forms, and mechanism in a more conclusive way (Ali et al. 2014, Meena and Kanwar 2015, Patel et al. 2015, Fira et al. 2018). Surfactin was first discovered in 1968 by Arima et al. (1968) from the supernatant of culture of *Bacillus subtilis* an exocellular compound was isolated having an excellent biosurfactant activity. This compound was named surfactin and its structure was explicated as that of a lipopeptide (Sachdev and Cameotra 2013). They are cyclic lipopeptides and are heptapeptides interlinked with a β-hydroxy fatty acid. These are a mixture of isoforms A, B, C and D and are classified in accordance with the

Table 1. A comparison between the three lipopeptide families in their structure, biosynthesis, their different isomeric forms, and mechanism with more conclusively way (Ali et al. 2014, Cawoy et al. 2013, Deleu et al. 2008, Fira et al. 2018, Meena and Kanwar 2015, Patel et al. 2015).

	Surfactin	Iturin	Fengycin
Chemical structure	Seven amino acids linked to β-hydroxy fatty acid with 13–15 carbon atoms	Seven amino acids bind to β-hydroxy fatty acid with 14-18 carbon atoms	Ten amino acidsconnect to β-hydroxy fatty acid with 14-17 carbon atoms
Chemical formula			
Mechanism	It interacts with the lipid bilayers and further interface with biological membrane until complete disruption and solubilization of it	Their action is based on formation of ion-conducting pores	It interacts with the cell membrane causing changes in its structure and permeability
Pathogens effect	It has hemolytic, antiviral, and antibacterial abilities; however, it has limited antifungal activity	It has strongly hemolytic, antifungal, antibacterial and antiviral activity	It has less hemolytic, strong antifungal activity, especially against filamentous fungi
Isomeric forms	Surfactin, lychenisin, halobacilin, and umilacidin	Iturin A, C, D and E, bacillomycin, and mycosubtilin	Fengycin, maltacin, and plipastatin

difference in their fatty acid sequence (Peypoux et al. 1999). They possess a very potent haemolytic, antiviral, antimicrobial activity. Surfactin has a wide range of therapeutic applications than other biosurfactants, Iturin illustrated in 1950, a second similar compound to Mycosubtilin that was the first antifungal compound from *Bacillus subtilis*. Its name is associated to Ituri, a region from Congo, where the compound was isolated from a soil sample. Iturin was first found to be a strong anti-fungal agent with constrained anti-bacterial activity against *micrococcus* and *sarcina* strains (Peypoux et al. 1981). Iturin was the first compound discovered of the iturin group and its best-known member, was isolated from a *B. subtilis* strain (Peypoux et al. 1978). Iturinare heptapeptides are linked to β-amino fatty acid chain with a length of C14 to C17. There are many types of iturin: Iturin A, C, bacillomycin D, F, L, Lc and mycosubtilin. Iturin A is the only lipopeptide which is produced by all *B. subtilis* strains (Kim et al. 2010). Iturin possess a very high antifungal activity against many yeast and fungal species whereas they have very limited antibacterial activity but no antiviral activity (Walia and Cameotra 2015). Fengycin represents the third family of lipopeptides after the Surfactin and Iturin and is also called Plipastatin (Deleu et al. 2008). Discovery of Fengycin and plipastatin was concomitant. In 1986, German and Japanese teams simultaneously discovered a third family of lipopetides: fengycin produced by *Bacillus subtilis* and plipastatin produced from *Bacillus cereus* (Volpon et al. 2000).

Fengycins are classified into Fengycins A and B. Fengycin A contains Ala at position 6 which is replaced by Val in case of Fengycin B (Meena and Kanwar 2015). They really interact to lipid layers and to a certain extent have the ability to change cell membrane structure and permeability in a dose dependent way (Hathout et al. 2000).

3. Biosynthesis of Lipopeptides Biosurfactants

Lipopeptides were synthesized from *Bacillus substitutes* by non-ribosomal peptide synthetizes (NRPSs) or hybrid polyketide synthases and non-ribosomal peptide synthetizes (PKSs/NRPSs). Several hundred of bioactive compounds were synthesized by those modular proteins, which were considered as a mega enzymes that catalyzed the different reactions to form polyketide or peptide transformation (Awan et al. 2017, Reimer et al. 2018). Each functional unit was called a module that was subdivided into several domains, which works as a catalytic agent for the biochemical reaction. A typical NRPS module contain 1000 residues of amino acids that is usually responsible for one reaction cycle to recognize the selective substrate and activate as adenylate "A-domain", peptide-bound formation "condensation, C-domain", and covalent bond tethering as an enzyme-bound thioester "Peptidyl-Carrier-Protein PCP-domain" (Baldim et al. 2017, Farag et al. 2019). Domains basic set through the module can be expanded by substrate modifying domains for instance, substrate epimerization (E-domain), methylation, hydroxylation, and the formation of heterocyclic ring. Furthermore, A thioesterase domain (Tedomain) is usually found in the last module to confirm the thioester bond cleavage between the last PCP domain and the nascent peptide and in several cases, it is responsible for the peptide cyclization (Ongena and Jacques 2008).

In the state of surfactin synthetases, three large open reading frames (ORFs) coding; srfA-A, srfA-B and srfA-C were responsible for surfactin synthetases that are designated to be in seven modules linear array (MEI Yu-wei 2020). srfA-A and srfA-B products contain three modules, and srfA-C contain the last one, after that, the fatty acid chain was added to the amino acid activated in the first module. Additionally, the first thioesterase that was fused with the C-terminal end of the last activation PCP domain was responsible for the synthesized product release from the enzymatic template. Whereas, the fourth gene srfA-D that is known as second thioesterase/acyltransferase (Te/At-domain) stimulates the biosynthesis process initiation (Qian et al. 2017, Steller et al. 2004).

Plipastatin or Fengycin were synthesized by the same way as surfactin "NRPSs" but are encoded by five open reading frames operons (ORFs) that is called ppsA–E (or fenA–E) (Yaseen et al. 2018, Yaseen et al. 2017). Two modules were recorded in the first three enzymes, three modules in the fourth, and one module in the last enzyme. Nevertheless, Iturine derivatives were synthesized with NRPS hybrid complex (Crowe-McAuliffe et al. 2018, Liu et al. 2018).

Iturine operon composed of four open reading frames (ORFs) that were known as fenF, mycA, mycB and mycC or ituD, ituA, ituB and ituC for mycosubtilin or iturin, respectively. Where, the last three genes code for the NRPSs were responsible for the incorporation between the first residue for mycA (or ituA). The following four residues were for mycB (or ituB) and the two last residues were for mycC (or ituC) (Aron et al. 2005, Zhang et al. 2020). Structural difference between Iturin A and Mycosubtilin (in which the last amino acids are inverted) can be illustrated by the intragenic domain change that is noticed in mycC and ituC. Furthermore, a malonyl-CoA transacylase (MCT-domain) was a code for FenF (ituD) and the mycA also contains genes related to polyketide synthases (Moreira et al. 2020). These genes are very important in the biosynthesis of the fatty acids chain's last steps (last elongation and b-amination) before its transfer to the first amino acid of the peptidic moiety [acyl-CoA ligase (AL-domain)], acyl carrier protein (ACP-domain), β-keto acyl synthetase (KS-domain), amino transferase (AMT domain) (Aron et al. 2005).

Moreover, the lipopeptides biosynthesis process can be explained in another way, where the process starts with phosphopantetheinyl group transfer to the PCP (peptidyl carrier protein), which was catalyzed by the phosphopantetheinyl transferase (Dejong et al. 2016). After that, amino acids were activated by the adenylation (A-domain) where it included a different module that was composed of several domains, including PCP and starting with A domain. Afterwards, Activated amino acid transferred into the 4′-phosphopantetheine group of PCP led to a thioester bond formation. Whereas, the C-domain or condensation catalyzed the peptide bond formation bond between amino acids during the peptide biosynthesis process (Fischbach and Walsh 2006). Additionally, modules may contain the E-domain that is responsible for epimerization of particular amino acids. As a result, many non-ribosomal peptides may contain D and L amino acids stereoisomers, this feature improved their resistance towards the proteolytic enzymes action (Fira et al. 2018).

4. Lipopeptides Producing Strains

4.1 Bacillus-related Lipopeptides

Broadly there are three types of lipopeptides, namely, surfactin, iturin, and fengycin that are produced by various *Bacillus* species.

4.1.1 Bacillus-related Surfactin

Surfactin, a cyclic lipopeptide is one of the best known biosurfactants up until this point, which was first detailed in *Bacillus subtilis* (Arima et al. 1968). Late investigations demonstrate that surfactin shows intense antiviral, antimycoplasma, antitumoral, and anticoagulant exercises just as inhibitors of enzymes (Mukherjee and Das 2005). Such properties of surfactins qualify them for expected applications in medication or biotechnology. Under characteristic conditions, the Surfactin is produce with a combination of its isomers. A portion of this blend relies upon outside components like the development medium and physico-synthetic factors and the kind of culture conditions. This Surfactin lipopeptide particle presents high protection from warmth, cold and stearic influences (Tsan et al. 2007).

4.1.2 Bacillus-related Iturin

Iturin lipopeptide is a cyclic peptide of seven amino acids (heptapeptides) connected to a fatty acid (β-amino) chain that can shift from C-14 to C-17 carbon molecules. Such molecules are incredibly intriguing due to their biological and physicochemical properties, which can be utilize in food, pharmaceutical industries and oil (Aranda et al. 2005). The iturin group of compounds are cyclic lipoheptapeptides, which contain a-amino unsaturated fat in its side chain. Lipopeptides having a place with the iturin family are intense antifungal specialists which can thus be utilized as biopesticides for plant defense (Romero et al. 2007).

4.1.3 Bacillus-related Fengycin

It is a bioactive lipopeptide extracted from a few strains of *Bacillus subtilis* and has antifungal properties against the filamentous fungi (Akpa et al. 2001). Fengycins are known to show strong antifungal activity but are less haemolytic compare to iturins and surfactins (Hathout et al. 2000). The lipopetide was first determined to be an antifungal specialist and later as a phospholipase A2 inhibitor. Extraction of fengycin was also obtain by *Bacillus thuringiensis* (Nishikiori et al. 1986).

4.2 Pseudomonas-related Lipopeptides

4.2.1 Pseudomonas-related Viscosinamide

Pseudomonas fluorescens DR54 are produced Viscosinamide that show antagonistic properties against pathogenic fungi that grow on plants like *Pythium ultimum* and *Rhizoctonia solani*. Bio-surfactant properties of viscosinamide contrast from the known bio-surfactant viscosin that contains glutamine as opposed to glutamate at the second amino corrosive position. Detachment and assurance of structure of this new compound with anti-microbial and antifungal properties has been uncovered. Tests in plants also demonstrated that purified viscosinamide decreased the aerial mycelium advancement of both *P. ultimum* and *R. solani* (Lee et al. 2010, De Faria et al. 2011).

4.2.2 *Pseudomonas-related Tolaasin*

There are many types of Tolaasin (Tolaasin I, tolaasin II, and tolaasins A, B, C, D, and E are the accessible analogs) and barely any different metabolites are extracted by *P. tolaasii*. Every one of these analogs show differences in their amino acid structure. All observed lipodepsipeptides of bacterial origin are kept up by the β-hydroxy octanoyl phi chain at the N-end, aside from tolaasin A, where the acyl moiety is a gamma-carboxy butanoyl phi moiety (Mazzola et al. 2009). *P. tolaasii* is the causal agent of brown blotch disease of *Agaricus bisporus* and causes yellowing of *Pleurotus ostreatus* too (Chao et al. 2007).

4.2.3 *Pseudomonas-related Tensin*

P. fluorescens strain 96.578 are producing antifungal compound known as tensin which is an antifungal cyclic lipopeptide. The NMR information acquired for tensin anticipated is as a cyclic lipopeptide made out of a 3-hydroxydecanoyl buildup in blend with 11 amino acid residues that include 5 leucine (Leu), 1 isoleucine (Ile), 1 aspartic acid (Asp), 1 glutamine (Gln), 1 glutamic acid (Glu), 1 threonine (Thr), and 1 serine (Ser) amino acids (Rainey et al. 1991). *P. fluorescens* is already utilize as a biocontrol operator against parasites, for example, *R. solani*. Different investigations recommended that tensin is effective against radial growth of *R. solani* (Cho et al. 2010).

4.2.4 *Pseudomonas-related Syringotoxin and Syringopeptin*

Syringotoxin is extracted from *P. syringae* pv. *syringae* which is a known as pathogen of various species of citrus trees (Nielsen et al. 1999). Peptide moiety of a bioactive molecule contains amino acids arranged in the sequence Ser-Dab-Gly-Hse-Orn-aThr-Dhb-(3-OH) Asp-(4-Cl) Thr with the terminal carboxy group closing a macrocyclic ring on the OH group of the N-terminal Ser (Nielsen et al. 2002). *P. syringae* pv. Is produced Syringopeptin is another major phytotoxic antibiotic. The amino acid sequence is Ser-Ser-Dab-Dab-Arg-Phe-Dhb-4(Cl) Thr-3(OH) Asp with the betacarboxy group of the C-terminal residue closing a macrocyclic ring on the OH group of the N-terminal Ser (Couillerot et al. 2009).

4.2.5 *Pseudomonas-related Pseudophomins*

The pseudophomins are isolated from *P. fluorescens* strain BRG100 that are bacterium with potential application as a biocontrol agent for agricultural weeds and plant pathogens. The pseudophomins are cyclic lipodepsipeptides classify into two group pseudophomins A and B. Pseudophomin A are utilized as stronger inhibition toward green foxtail (Setariaviridis) and play a role in induced root germination interestingly, Pseudophomin B appear higher antifungal activity against the phytopathogens for example *Sclerotinia sclerotiorum* and *Phoma lingam/Leptosphaeria maculans* (Galonic et al. 2007).

4.2.6 Pseudomonas-related Massetolide A

Massetolide A is extracted from *P. fluorescens* SS101, known as a biocontrol agent. By using *P. infestans, P. fluorescens* SS101 was effective in preventing infection in leaves of tomato (*Lycopersicon esculentum*) and essentially diminished the expansion of existing late blight lesions. Purified mass etolide A displayed essentially control of *P. infestans* both locally and systemically by induced resistance (Sun et al. 2001).

4.3 Actinomycetes-related Lipopeptides

Actinomycetes are commonly known for the production of CLPs antibiotics. These compounds such as friulimicin are isolated from *Actinoplanesfriuliensis* (Aretz et al. 2000), amphomycins isolated from *Streptomyces canus* (Schneidera et al. 2013), and laspartomycins isolated from *Streptomyces viridochromogenes* (Wang et al. 2011). This compound are structurally composed of 11 amino acid residues, of which 10 residues form a ring structure and an exocyclic Asp1 residue related with an acyl group (Wang et al. 2011). The laspartomycins discovered for first time in 1968 by Naganawa et al. from the soil bacterium *Streptomyces viridochomogenes* var. komabensis (ATCC29814) were reported to be active against *Staphylococcus aureus* (Naganawa et al. 1968). They were extracted as a combination of at least three peptide compounds that differ in their attached fatty acid side chains (Borders et al. 2002). The main component of this mixture is Laspartomycin C and its structure was recently fully clarified as a cyclic lipopeptide with a 2, 3-unsaturated C15-fatty acid side chain (Borders et al. 2007). The laspartomycins are comprised of 11 amino acid residues of which ten residues form a ring structure and an exocyclic Asp1 residue related with an acyl group (Muller et al. 2007).

In comparison with the other lipopeptides, e.g., friulimicin and amphomycins, the laspartomycins contain a unique cyclic peptide core and 2, 3-unsaturated acyl side chains instead of 3, 4-unsaturated acyl groups (Borders et al. 2007). Recent studies have shown that they are active against a broad spectrum of Gram-positive pathogens such as MRSA, VRSA, vancomycin-intermediate *S. aureus* (VISA) and VRE (Curran et al. 2007). The amphomycin is already used in the veterinary industry as a topical antibacterial agent (Dini 2005). Amphomycin appear o have the ability to inhibit the formation of dolicholphosphate-mannose (dol-P-Man) in eukaryotic N-linked glycoprotein biosynthesis (Banerjee 1989). The friulimicins were isolated as a mixture of four lipopeptides (Vertesy et al. 2000). They were described as novel lipopeptide antibiotics with peptidoglycan synthesis-inhibiting activity (Aretz et al. 2000). The friulimicins vary from the laspartomycins in their fatty acid side chains and amino acid composition at positions 2, 4, 9 and 10 (Wecke et al. 2009). Friulimicin is an effective compound against Gram-positive pathogens including MRSA and methicillin-resistant *S. epidermidis* (Heinzelmann et al. 2005). Recent studies have reported that friulimicin has only mode of action, which is different from that of daptomycin, a peptide antibiotic in current clinical use (Schneider et al. 2009).

4.4 Fungal-related Lipopeptide

The echinocandins are compounds produced by fungi involve a new group of antifungal lipopeptide agents. Echinocandins isolated from *Aspergillus rugulosus* and A. nidulans showed antifungal and anti-yeast activities (Tran et al. 2007). This group are including compounds such as amphiphilic, cyclic hexapeptides with an N-linked acyl lipid side chain and the molecular weight is around 1,200 Da (Tffoth et al. 2012). The most advantage these lipopeptides have is a cyclic peptide and an N-terminal fatty acid group but the fatty acid group is sometime either branched or unbranched with a chain length of 14–18 carbon atoms. Both the cyclic structure (Kurokawa and Ohfune 1993) and an acyl side chain are basically required for the biological activity of the compound. This confirms the significance of the acyl group for the mechanism of antimicrobial action (Rodriguez et al. 1999). Echinocandins show antimicrobial activity by blocking the synthesis of 1-3-β-D-glucan, a critical component of the fungal cell wall, through noncompetitive inhibition of the enzyme 1-3-β-D-glucan synthase. The echinocandin antifungal spectrum is limited with few exceptions to *Candida* spp. and *Aspergillus* spp. (Baindara and Korpole 2016).

5. Surfactin, Iturin, Fengycin and Plipastatin: A Significant Candidate in Lipopeptides Biosurfactant

5.1 Surfactin

One of the most powerful biosurfactants known to date, surfactin, is a cyclic lipopeptide (Sen 2010). The type of surfactin differs according to the order of the amino acids present in the molecule and the size of the lipid portion. The hydrophobic amino acids of the Surfactin molecule are found in the cyclic structure at 2, 3, 4, 6 and 7 positions from right below, while the residues of Glu and Asp, found at 1 and 5 positions, add two negative charges to the molecule respectively (Korenblum et al. 2012). Typically, Surfactin isoforms coexist as a mixture of many peptide variants of a distinct aliphatic fatty acid chain length in the cell of a bacterium. The pattern of amino acids and β-hydroxy fatty acids in the molecule of Surfactin depends not only on the bacterial strain of the producer but also on the kind of conditions of the bacterial culture (Seydlová et al. 2011). An intramolecular hydrogen bond may form the β-turn, while the β-sheet may rely on an intermolecular hydrogen bond (Zou et al. 2010). Surfactin is made with a mixture of its isomers under natural conditions. The composition of this mixture depends on external factors such as physico-chemical and growth medium factors and conditions of culture. This lipopeptide surfactin molecule is extremely resistant to heat, cold, and stearic effects. Surfactin amino acid chain can differ in sequence, while the molecules of Surfactin can be classified into four isoforms, namely Surfactin A, Surfactin B, Surfactin C, and Surfactin D (Meena et al. 2017).

The residues number 2 and 6 of the Surfactin molecule facing each other in the vicinity of the acidic side chains Glu-1 and Asp-5, which determine the minor polar character of the molecule. Residue 4 is associated with a long lipid chain consisting of a major hydrophobic domain, which to a lesser extent includes residue 3 and 7 side chains, giving it amphiphilic characteristics and powerful surfactant properties (Tsan et al. 2007). The surfactin groups of compounds are seen to be a cyclic Lipoheptapeptides, containing seven D- and L-amino acid residues and one β-hydroxy fatty acid residue with an amino acid sequence totally different from that of the iturin group (Shaligram and Singhal 2010). The diversity of the peptide moiety allows surfactin, lichenysin, esperin, and pumilacidin to be distinguished from the surfactin family (Ongena and Jacques 2008). It has the potential to minimize water surface tension from 72 to 27 mN/m at a concentration as low as 0.005% (Shaligram and Singhal 2010). It is also called surfactin because of this extraordinary surfactant action (Mnif and Ghribi 2015). Surfactins are effective biosurfactants that exhibit antibacterial activity but no defined fungi toxicity (with certain exceptions). Recent studies have shown that surfactin exhibits potent antiviral, antymicoplasma, antitumoral, anticoagulant, and enzyme inhibitor activity (Sen 2010).

5.2 *Iturin*

Iturin lipopeptide is a cyclic peptide of 7 amino acids (heptapeptide) connected to a chain of fatty acids (β-amino) that can differ from carbon molecules of C14 to C17, exhibiting strong in vitro antifungal activity via the formation of fungal membrane ion-conducting pores (Ongena and Jacques 2008). They show structural variation in the residues of amino acids and in the length and branching of the fatty acid chain. Iturins A, C, D and E, Bacillomycins D, F and L, Bacillopeptin and Mycosubtilin, all of which are arranged in an LDDLLDL amino acid configuration sequence, are some examples of these amphiphilic compounds (Aranda et al. 2005). Iturin A consists in particular of up to eight isomers (Iturin A1–A8) of varying lengths (10–14 carbons) and fatty acid chain branching. Due to their biological and physico-chemical properties, which can be exploited in the food, oil and pharmaceutical industries, these lipopeptide molecules are of appreciable interest. Both *Bacillus subtilis* strains produce Iturin-family lipopeptides (Tsuge et al. 2001).

Iturin is a special pore-forming lipopeptide class and is well known for its antifungal activity against a wide range of pathogenic yeasts and fungi, but are limited to antibacterial activities to certain bacteria such as *Micrococcus luteus* (Mnif and Ghribi 2015). Indeed, the ability of iturin compounds to increase the permeability of membrane cells is due to the formation of ion-conducting pores, the properties of which rely on both the membrane lipid composition and the peptide cycle structure. The antifungal activity is therefore linked to the interaction of iturin lipopeptides with the target cytoplasmic cell membrane, whose permeability to K1 is greatly increased (Mnif and Ghribi 2015). Iturins are well-known biocontrol agents for plant pathogens such as *Pectobacterium carotovorum* subsp., *Xanthomonas campestris* Cucurbitae (Zeriouh et al. 2011, Gong et al. 2015), *Carotovorum, Rhizoctonia solani, Fusarium graminearum,* etc. (Patel et al. 2015).

5.3 *Fengycin and Plipastatin*

A third family of lipopeptides, fengycin is produced by *Bacillus subtilis* and plipastatin is produced by *Bacillus cereus.* Fengycin was known as the first lipopeptide with an anti-fungal agent and then an inhibitor of phospholipase A2. There was only a minor structural distinction between these two compounds, and both of them and their biological activities are still in question today (Volpon et al. 2000). Fengycins are categorized into A and B Fengycins. These are referred to as plisplastatins as well. It is found that fengycins have a powerful antifungal activity, but they are less heamolytic than iturins and surfactins. They actually interact with lipid layers and have the capacity to alter the structure and permeability of the cell membrane in a dose-dependent manner (Hathout et al. 2000). These bioactive molecules are lipopeptides in the β-hydroxy fatty acid chain containing a lactone ring that can be saturated or unsaturated. The structure of Fengycin includes a 10 amino acid peptide chain connected to a chain of fatty acids (Akpa et al. 2001). By inducing lipid bilayer perturbations and local electrostatic driven remodelling, it works on the plasma membrane (Horn et al. 2013). It works extensively against filamentous fungi. Inhibiting *Rhizoctonia solani* in the cotton rhizosphere was discovered (Guo et al. 2014). While it has limited therapeutic implications at present, its milder hemolytic activity suggests potential medical applications relative to other lipopeptides. It was marketed as a powerful candidate for dermatomycoses (candidiasis and ringworm) treatment (Eeman et al. 2014).

6. Emulsification and De-emulsification Capacity of Lipopeptides

Emulsification refers to the dispersion of one liquid into another (as microscopic droplets), between the functional characteristics of biosurfactants, leading to the mixing of two immiscible liquids. This property is particularly useful for the development of emulsions of oil and water for the environment, cosmetics, and food (Pathak and Keharia 2014). Surfactants may have both surface tension reduction and emulsification function. Emulsifiers are actually known as molecules that are surface-active and that can form the emulsion of two immiscible liquids (Satpute et al. 2010). Emulsifiers, however, can only bind water-insoluble substrates together to form an emulsion, high-molecular mass biosurfactants are typically better emulsifiers than biosurfactants with low molecular mass, low-molecular weight biosurfactants, however, have been broadly defined as effective bioemulsifiers (Mnif and Ghribi 2015). As mentioned by Khopade et al. (2012) and Sarafin et al. (2014), B4 and *Kocuria marina* BS-15 displayed greater emulsifying activity against hydrocarbons, the emulsion of hexane, heptane, and octane formed using 48-h-old culture supernatant of *B. subtilis* K1 producing lipopeptide biosurfactant held stable for up to 2 days, whereas the four-stroke engine oil emulsion remained stable for more than a year.

The lipopeptide derived from *P. aeruginosa* displayed an interesting emulsification effect against motor lubricant oil, crude oil, peanut oil, gasoline, diesel, xylene, naphthalene and anthracene, higher than that observed with Triton X-100 at the same concentration (1 mg/ml), allowing its potential use in the bioremediation of hydrocarbon-polluted environments (Thavasi et al. 2011). The emulsification activity and stability of the pseudofactin developed by *P. fluorescens* BD5 was higher than that of the synthetic surfactants Tween 20 and Triton X100 and therefore had a high potential for use in industrial fields such as bioremediation and biomedicine (Janek et al. 2010).

The emulsion can be stabilized (emulsifiers) or destabilized (de-emulsifiers) by biosurfactants. De-emulsification is the ability to separate emulsions between the bulk phase and the internal phase by the destruction of the stable surface. De-emulsification can be of interest in the processing of foodstuffs, especially with regard to fat and oil products, as well as in the treatment of waste (Nitschke and Costa 2007). According to Bosch and Axcell (2005), this is significant in the processes of oil production, where natural emulsifying agents hinder the production processes. It is performed by affecting the conditions of thermodynamics at the interface. In fact, as described by Das (2001), de-emulsifiers are largely consumed in the treatment of emulsions generated in the treatment and processing of crude petroleum. Water-oil emulsions actually occur during the production, transportation, and processing of oil. De-emulsification is becoming more troublesome as the water content in these emulsions rises. Water/oil types are most of the emulsions formed in oil fields and can almost be destabilized by chemically synthesized de-emulsifiers that can cause toxicity to the environment and living organisms. Therefore, microbial derived bio-de-emulsifiers provide a great opportunity to prevent harmful threats to the environmental ecosystem when handling such emulsions before discharge. Also, Liu et al. (2009) reported the production of a *Dietzia* sp. for this reason. S-JS-1 bio-de-emulsifier known as a mixture of lipopeptide homologues from waste frying oil and paraffin, which achieved 88.3% of the water/oil emulsion oil separation ratio and 76.4% of the oil/water emulsion water separation ratio within 5 h. A cell-wall combined lipopeptide bio-de-emulsifier developed by *Alcaligenes* sp. In W/O and O/W kerosene emulsions, 96.5% and 49.8% of the emulsion breaking ratio were achieved within 24 h, respectively, and 95 percent of the water separation ratio in oilfield petroleum emulsion was also shown within 2 h (Huang et al. 2009). In another analysis, an outstanding demulsifying capability of crude oil emulsion with a satisfactory operation compared to a chemical demulsifier, polyether with an increase in the consistency of separated water, was shown (Liu et al. 2010). In the presence of relatively high NaCl concentrations (up to 25 percent), a *P. nitroreducens*-derived lipopeptide bioemulsifier that is stable over a pH range of 5–11 and a temperature range of 20–90°C showed emulsifying activity, while SDS showed no activity beyond 10 percent NaCl. It produced stable emulsions with aliphatic (hexadecane, n-heptane, cyclohexane), aromatic (xylene, benzene, toluhexane) and aromatic (xylene, benzene, toluhexane), and petroleum (gasoline, diesel, kerosene, crude oil) compounds and displayed maximum emulsification behavior with weathered crude oil (97%) (Mnif and Ghribi 2015). A bio-de-emulsifier lipopeptide developed by *Paenibacillus alvei* showed a high ability to split heavy crude oil emulsion with the potential to be used as an environmentally friendly and non-toxic material in the petroleum industry. For water-in-oil (W/O) and oil-in-water (O/W) model emulsions, an extracellular lipopeptide bio-de-emulsifier developed by *Bacillus cereus* achieved 95.61 and 95.40 percent de-emulsifying ratios within 12 h, respectively (Mnif and Ghribi 2015).

7. Biomedical Lipopeptides as Anti-Microbial Agents and as Immunomodulator

Lipopeptides appear as strong anti-microbial, anti-fungal, anti-viral, anti tumour and immune modulator properties. Practically the entirety of the *Bacillus* species produce an antimicrobial agent known as lipopeptides and many strains of *B. subtilis* and *B. amyloliquefaciens* have demonstrated to produce lipopeptides. Gong et al.

(2006) and Qian et al. (2009) demonstrated that the crude lipopeptides are stable at heat, pH, and appear to have high ability as biocontrol agents against different pathogenic microorganism (Romero et al. 2007). Surfactin is likewise active against several types of viruses such as (the Semliki Forest virus, Simian immunodeficiency virus, Herpes simplex virus (HSV-1 and HSV 2), Vesicular stomatitis virus, Feline calicivirus, and the Murine encephalomyocarditis virus). The length of the carbon chain in cyclic Surfactin lipopeptide affects its ability for viral inactivation (Huang et al. 2006).

The inactivation of enveloped viruses, particularly herpes viruses and retroviruses by Surfactin, is essentially more effective than that of nonenveloped viruses (Seydlova et al. 2011). This proposes that the antiviral activity of Surfactin is principally because of the physicochemical interaction between the membrane active surfactant property of Surfactin and the virus lipid membrane (Vollenbroich et al. 1997). It has likewise been observed that antimicrobial lipopeptides containing Surfactin inactivate cell-free viruses of the *Porcine parvovirus, Pseudo rabies* virus, Bursal disease virus, and Newcastle disease virus (Singla et al. 2014).

Iturin and Fengycin are the main lipopeptides having strong antifungal activities, while Surfactin has antibacterial activity (Gordillo and Maldonado 2012). The involvement of antifungal lipopeptides, Iturins and Fengycins, was found to show biocontrol activity against *Bacillus* strains as well as against many kinds of plant pathogens (Ongena and Jacques 2008). Recently, a new lipopeptide referred to as "Kinnurin" extracted from *Bacillus cereus* has been found to show a good antifungal activity (Ajesh et al. 2013). Fungi toxicity of Iturins closely depend on their ability to penetrate the cell membrane of the target organism (Deleu et al. 2008).

Surfactins that were produced from *Bacillus circulans* were also found to be active against many drug-resistant bacteria such as *Alcaligenes faecalis, Proteus vulgaris, Escherichia coli, Pseudomonas aeruginosa* and methicillin-resistant *Staphylococcus aureus*. The minimal inhibitory and minimal bactericidal concentrations of the Surfactin were found to be less than those of the ordinary antibiotics tested alongside (Das et al. 2008). Plaza et al. (2013) have exhibit that surfactin, a type of lipopeptide isolated from *Bacillus subtilis*, growing on molasses have anti-bacterial and anti-fungal properties. Nielson et al. (1999), discovered a new cyclic depsipeptide named viscosinamide that are extracted from *Pseudomonas fluorescens* DR54 which exhibit anti-fungal properties against *Phythiumultimum* and *Rhizoctonia solani*. The authors concluded that the new CLPs possess both biosurfactant and antibiotic properties, hence can be used as a potent biological control agent. Nielson et al. (2002) reported a new cyclic compound produced from *Pseudomonas fluorescens* strain 96.578 having anti-fungal activity against plant pathogenic fungus *Rhizoctonia solani* called tensin. Kruijt et al. (2009) also discovered a putisolvin like cyclic lipopeptide biosurfactant produced from *P. putida* 267 having zoosporicidal and anti-fungal activity. Das et al. (2008) reported the anti-microbial activity of a lipopeptide biosufactant isolated from a marine *Bacillus circulans*. The produced biosurfactant showed a potent anti microbial activity against both gram positive and gram negative pathogenic and semi pathogenic microbial strains. Furthermore, this biosurfactant was also found to be non haemolytic.

7.1 Antiobesity Activity of the Lipopeptides

Obesity is considered as a surpassing lifestyle disorder particularly in developing countries. It predominate in new world countries as a result of junk food intake, including lack of physical activity and high fructose corn syrup added foodstuff consumption (Bray 2013). Inhibitory activity of pancreatic lipase has been extensively used for the exploration of potentially effective natural products (Lunagariya et al. 2014). Lipopeptide as an exclusive class of biosurfactants have recently appeared as promising molecules because of their structural novelty, versatility and varied properties that are valuable for advanced therapeutic applications (Gudiña et al. 2013). *Bacillus subtilis* SPB1 lipopeptide may be a major drug of the future to treat the obesity-related metabolic disorders. *B. subtilis* lipopeptides can be administered orally, in order to achieve an effective control on body weight (Zouari et al. 2016). *Bacillus subtilis* SPB1 crude Lipopeptide biosurfactant has both protective and healing action on overweight persons and it has diminished the body weight of hefty rats and thus can treat hyperlipidemia without side effects. *Bacillus subtilis* lipopeptide reduced the body weight of rats by diminishing the pancreatic lipase activity (Zouari et al. 2016). A brilliant study conducted by (Chen et al. 2020) revealed that Lipopeptide show anti-obesity effects. In this study, *Bacillus velezensis* strain FJAT-52631 coproduce iturins, fengycins, and surfactins. Results appeared that the FJAT-52631 crude Lipopeptide, purified iturin, fengycin, and surfactin standards have shown potential inhibition against lipase with dose-dependence manners (half maximal inhibitory concentration (IC50) = 0.011, 0.005, 0.056, and 0.005 mg/mL, respectively). Additionally, surfactin and fengycin had the comparable activities with orlistat, but iturin not. This study revealed that the type of the lipopeptides and inhibition mechanism were reversible and competitive. The docking analysis displayed that fengycin and surfactin could directly interact with the active amino acid residues (oAsp or Ser) of lipase, but not with iturin. The quenching mechanism of lipase was static and only one binding site between lipase and lipopeptide was inferred from the fluorescence analysis. Chen and coworker recommended that *B. velezensis* lipopeptides would have the potential to act as lipase inhibitors.

References

Ajesh, K., S. Sudarslal, C. Arunan and K. Sreejith. 2013. Kannurin, a novel lipopeptide from *Bacillus cereus* strain AK1: isolation, structural evaluation and antifungal activities. Journal of Applied Microbiology 115(6): 1287–1296.

Ali, S., S. Hameed, A. Imran, M. Iqbal and G. Lazarovits. 2014. Genetic, physiological and biochemical characterization of *Bacillus* sp. strain RMB7 exhibiting plant growth promoting and broad spectrum antifungal activities. Microbial Cell Factories 13(1): 144. https://doi.org/10.1186/s12934-014-0144-x.

Aron, Z.D., P.C. Dorrestein, J.R. Blackhall, N.L. Kelleher and C.T. Walsh. 2005. Characterization of a new tailoring domain in polyketide biogenesis: the amine transferase domain of MycA in the mycosubtilin gene cluster. Journal of the American Chemical Society 127(43): 14986–14987.

Aranda, F.J., J.A. Teruel and A. Ortiz. 2005. Further aspects on the hemolytic activity of the antibiotic lipopeptide iturin A. Biochimica et Biophysica Acta—Biomembranes 1713(1): 51–56. https://doi.org/https://doi.org/10.1016/j.bbamem.2005.05.003.

Aretz, W., J. Meiwes, G. Seibert, G. Vobis and J. Wink. 2000. Friulimicins: novel lipopeptide antibiotics with peptidoglycan synthesis inhibiting activity from *Actinoplanesfriuliensis* sp. nov. I. Taxonomic studies of the producing microorganism and fermentation. J. Antibiot. (Tokyo) 53: 807–15.

Arima, K., A. Kakinuma and G. Tamura. 1968. Surfactin, a crystalline peptide lipid surfactant produced by *Bacillus subtilis*: isolation, characterization and its inhibition of fibrin clot formation. Biochem. Biophys. Res. Commun. 31: 488–494.

Aron, Z.D., P.C. Dorrestein, J.R. Blackhall, N.L. Kelleher and C.T. Walsh. 2005. Characterization of a new tailoring domain in polyketide biogenesis: the amine transferase domain of MycA in the mycosubtilin gene cluster. Journal of the American Chemical Society 127(43): 14986–14987.

Asaka, O. and M. Shoda. 1996. Biocontrol of *Rhizoctonia solani* damping-off of tomato with *Bacillus subtilis* RB14. Applied and Environmental Microbiology 62(11): 4081.

Awan, A.R., B.A. Blount, D.J. Bell, W.M. Shaw, J.C.H. Ho, R.M. McKiernan and T. Ellis. 2017. Biosynthesis of the antibiotic nonribosomal peptide penicillin in baker's yeast. Nature Communications 8(1): 15202. https://doi.org/10.1038/ncomms15202.

Bais, H.P., R. Fall and J.M. Vivanco. 2004. Biocontrol of *Bacillus subtilis* against infection of arabidopsis roots by *Pseudomonas syringae* is facilitated by biofilm formation and surfactin production. Plant Physiology 134(1): 307.

Bais, H.P., T.L. Weir, L.G. Perry, S. Gilroy and J.M. Vivanco. 2006. The role of root exudates in rhizosphere interactions with plants and other organisms. Annual Review of Plant Biology 57(1): 233–266.

Banat, I.M., R. Makkar and S. Cameotra. 2000. Potential commercial applications of microbial surfactants. Applied Microbiology and Biotechnology 53: 495–508.

Banerjee, D.K. 1989. Amphomycin inhibits mannosylphosphoryldolichol synthesis by forming a complex with dolichylmonophosphate. J. Biol. Chem. 264: 2024–8 [PubMed: 2464586].

Besson, F., F. Peypoux, G. Michel and L. Delcambe. 1977. Structure de la *bacillomycine* L, antibiotique de Bacillus subtilis Eur. J. Biochem. 77.

Blée, E. 2002. Impact of phyto-oxylipins in plant defense. Trends in Plant Science 7(7): 315–322.

Borders, D.B., W.V. Curran, A.A. Fantini, N.D. Francis, H. Jarolmen and R.A. Leese. 2002. Derivatives of laspartomycin and preparation and use thereof. U.S. Pat. Pub. No.: WO/2002/005838.

Borders, D.B., R.A. Leese, H. Jarolmen, N.D. Francis, A.A. Fantini, T. Falla, J.C. Fiddes and A. Aumelas. 2007. Laspartomycin, an acidic lipopeptide antibiotic with a unique peptide core. J. Nat. Prod. 70: 443–6.

Bosch, R. and E. Axcell. 2005. Produced-water chemical treatments enable environmental compliance. World Oil 226(10): 75–78.

Branda, S.S., J.E. González-Pastor, S. Ben-Yehuda, R. Losick and R. Kolter. 2001. Fruiting body formation by *Bacillus subtilis*. Proc. Natl. Acad. Sci. U.S.A. 98(20): 11621–6. doi: 10.1073/pnas.191384198. PMID: 11572999; PMCID: PMC58779.

Baldim, J.L., B.L. da Silva, D.A. Chagas-Paula, J.H.G. Lago and M.G. Soares. 2017. A strategy for the identification of patterns in the biosynthesis of nonribosomal peptides by Betaproteobacteria species. Scientific Reports 7(1): 10400. https://doi.org/10.1038/s41598-017-11314-w.

Bray, G.A. 2013. Energy and fructose from beverages sweetened with sugar or high-fructose corn syrup pose a health risk for some people12. Advances in Nutrition 4(2): 220–225. https://doi.org/10.3945/an.112.002816.

Cai, X.C., H. Li, Y.R. Xue and C.H. Liu. 2013. Study of endophytic *Bacillus amyloliquefaciens* CC09 and its antifungal CLPs. J. Appl. Biol. Biotechnol. 1: 1–5.

Cawoy, H., M. Mariutto, G. Henry, C. Fisher, N. Vasilyeva, P. Thonart, J. Dommes and M. Ongena. 2013. Plant defense stimulation by natural isolates of *Bacillus* depends on efficient surfactin production. Molecular Plant-Microbe Interactions® 27(2): 87–100. https://doi.org/10.1094/MPMI-09-13-0262-R.

Chen, K., Z. Tian, Y. Luo, Y. Cheng and C.-a. Long. 2018. Antagonistic activity and the mechanism of *Bacillus amyloliquefaciens* DH-4 against citrus green mold. Phytopathology® 108(11): 1253–1262. https://doi.org/10.1094/phyto-01-17-0032-r.

Chen, M., T. Liu, J. Wang, Y. Chen, Q. Chen, Y. Zhu and B. Liu. 2020. Strong inhibitory activities and action modes of lipopeptides on lipase. Journal of Enzyme Inhibition and Medicinal Chemistry 35(1): 897–905. https://doi.org/10.1080/14756366.2020.1734798.

Cho, K.-H., S.-T. Kim and Y.-K. Kim. 2007. Purification of a pore-forming peptide toxin, tolaasin, produced by *Pseudomonas tolaasii* 6264. J. Biochem. Mol. Biol. 40: 113–118.

Cho, K.H., H.S. Wang and Y.K. Kim. 2010. Temperature dependent hemolytic activity of membrane poreforming peptide toxin, tolaasin. J. Pept. Sci. 16: 85–90.

Conrath, U., G.J.M. Beckers, V. Flors, P. García-Agustín, G. Jakab, F. Mauch, M.-A. Newman, C.M.J. Pieterse, B. Poinssot, M.J. Pozo, A. Pugin, U. Schaffrath, J. Ton, D. Wendehenne, L. Zimmerli and B. Mauch-Mani. 2006. Priming: Getting ready for battle. Molecular Plant-Microbe Interactions® 19(10): 1062–1071.

Couillerot, O., C. Prigent-Combaret, J. Caballero-Mellado and Y. Moe˜nne-Loccoz. 2009. *Pseudomonas fluorescens* and closely-related fluorescent pseudomonads as biocontrol agents of soil-borne phytopathogens. Lett. Appl. Microbiol. 48: 505–512.

Crowe-McAuliffe, C., M. Graf, P. Huter, H. Takada, M. Abdelshahid, J. Nováček, V. Murina, G.C. Atkinson, V. Hauryliuk and D.N. Wilson. 2018. Structural basis for antibiotic resistance mediated by the *Bacillus subtilis* ABCF ATPase VmlR. Proceedings of the National Academy of Sciences 115(36): 8978–8983. https://doi.org/10.1073/pnas.1808535115.

Curran, W.V., R.A. Leese, H. Jarolmen, D.B. Borders, D. Dugourd, Y. Chen and D.R. Cameron. 2007. Semisynthetic approaches to laspartomycin analogues. J. Nat. Prod. 70: 447–50.

Das, P., S. Mukherjee and R. Sen. 2008. Antimicrobial potential of a lipopeptide biosurfactant derived from a marine *Bacillus circulans*. J. Appl. Microbiol. 104: 1675–1684.

Das, M. 2001. Characterization of de-emulsification capabilities of a *Micrococcus* species. Bioresource Technology 79(1): 15–22.

De Faria, A.F., D. Ste´fani, B.G. Vaz, I´S Silva, J.S. Garcia, M.N. Eberlin, M.J. Grossman, O.L. Alves and L.R. Durrant. 2011. Purification and structural characterization of fengycin homologues produced by *Bacillus subtilis* LSFM-05 grown on raw glycerol. J. Ind. Microbiol. Biotechnol. 38: 863–871.

Dejong, C.A., G.M. Chen, H. Li, C.W. Johnston, M.R. Edwards, P.N. Rees, M.A. Skinnider, A.L. Webster and N.A. Magarvey. 2016. Polyketide and nonribosomal peptide retro-biosynthesis and global gene cluster matching. Nature Chemical Biology 12(12): 1007–1014.

Deleu, M., M. Paquot and T. Nylander. 2008. Effect of fengycin, a lipopeptide produced by *Bacillus subtilis*, on model biomembranes. Biophysical Journal 94(7): 2667–2679.

Dini, C. 2005. MraY inhibitors as novel antibacterial agents. Curr. Top Med. Chem. 5: 1221–36.

Dixon, R.A., L. Achnine, P. Kota, C.-J. Liu, M.S.S. Reddy and L. Wang. 2002. The phenylpropanoid pathway and plant defence—a genomics perspective. Molecular Plant Pathology 3(5): 371–390. https://doi.org/10.1046/j.1364-3703.2002.00131.x.

Duitman, E.H., L.W. Hamoen, M. Rembold, G. Venema, H. Seitz, W. Saenger, F. Bernhard, R. Reinhardt, M. Schmidt and C. Ullrich. 1999. The mycosubtilin synthetase of *Bacillus subtilis* ATCC6633: a multifunctional hybrid between a peptide synthetase, an amino transferase, and a fatty acid synthase. Proceedings of the National Academy of Sciences 96(23): 13294–13299.

Eeman, M., G. Olofsson, E. Sparr, M.N. Nasir, T. Nylander and M. Deleu. 2014. Interaction of fengycin with stratum corneum mimicking model membranes: a calorimetry study. Colloids Surf B Biointerfaces 121: 27–35. doi:10.1016/ j.colsurfb.2014.05.019.

Fan, H., J. Ru, Y. Zhang, Q. Wang and Y. Li. 2017. Fengycin produced by *Bacillus subtilis* 9407 plays a major role in the biocontrol of apple ring rot disease. Microbiological Research 199: 89–97.

Farace, G., O. Fernandez, L. Jacquens, F. Coutte, F. Krier, P. Jacques, C. Clément, E.A. Barka, C. Jacquard and S. Dorey. 2015. Cyclic lipopeptides from *Bacillus subtilis* activate distinct patterns of defence responses in grapevine. Molecular Plant Pathology 16(2): 177–187.

Farag, S., R.M. Bleich, E.A. Shank, O. Isayev, A.A. Bowers and A. Tropsha. 2019. Inter-modular linkers play a crucial role in governing the biosynthesis of non-ribosomal peptides. Bioinformatics 35(19): 3584–3591. https://doi.org/10.1093/bioinformatics/btz127.

Finking, R. and M.A. Marahiel. 2004. Biosynthesis of nonribosomal peptides. Annual Review of Microbiology 58(1): 453–488.

Fira, D., I. Dimkić, T. Berić, J. Lozo and S. Stanković. 2018. Biological control of plant pathogens by *Bacillus* species. Journal of Biotechnology 285: 44–55.

Fischbach, M.A. and C.T. Walsh. 2006. Assembly-line enzymology for polyketide and nonribosomal peptide antibiotics: logic, machinery, and mechanisms. Chemical Reviews 106(8): 3468–3496.

Galonic´, D.P., E.W. Barr, C.T. Walsh, J.M. Bollinger and C. Krebs. 2007. Two interconverting Fe(IV) intermediates in aliphatic chlorination by the halogenase CytC3. Nat. Chem. Biol. 3: 113–116.

Gong, A.D., H.P. Li, Q.S. Yuan, X.S. Song, W. Yao, W.J. He, J.B. Zhang and Y.C. Liao. 2015. Antagonistic mechanism of iturin A and plipastatin A from Bacillus amyloliquefaciens S76-3 from wheat spikes against Fusarium graminearum. PloS one 10(2): e0116871. https://doi.org/10.1371/journal. pone.0116871.

Gong, M., J.D. Wang, J. Zhang, H. Yang, X.F. Lu, Y. Pei and J.Q. Cheng. 2006. Study of the antifungal ability of *Bacillus subtilis* strain PY-1 *in vitro* and identification of its antifungal substance (iturin A). Acta Biochem. Biophysiol. Sin. 38: 233–240.

Gordillo, A. and M.C. Maldonado. 2012. Purification of peptides from *Bacillus* strains with biological activity. Chromatography and Its Applications 11: 201–225.

Gudiña, E.J., V. Rangarajan, R. Sen and L.R. Rodrigues. 2013. Potential therapeutic applications of biosurfactants. Trends in Pharmacological Sciences 34(12): 667–675.

Guo, Q., W. Dong, S. Li, X. Lu, P. Wang, X. Zhang, Y. Wang and P. Ma. 2014. Fengycin produced by Bacillus subtilis NCD-2 plays a major role in biocontrol of cotton seedling damping-off disease. Microbiol Res. 169(7-8): 533–40. doi: 10.1016/j.micres.2013.12.001. Epub 2013 Dec 12. PMID: 24380713.

Hamdache, A., R. Azarken, A. Lamarti, J. Aleu and I.G. Collado. 2013. Comparative genome analysis of *Bacillus* spp. and its relationship with bioactive nonribosomal peptide production. Phytochemistry Reviews 12(4): 685–716. https://doi.org/10.1007/s11101-013-9278-4.

Hathout, Y., Y.P. Ho, V. Ryzhov, P. Demirev and C. Fenselau. 2000. Kurstakins: a new class of lipopeptides isolated from *Bacillus thuringiensis*. J. NatProd. 63(11): 1492–1496.

Heinzelmann, E., S. Berger, C. Muller, T. Hartner, K. Poralla, W. Wohlleben and D. Schwartz. 2005. An acyl-CoA dehydrogenase is involved in the formation of the Delta cis3 double bond in the acyl residue of the lipopeptide antibiotic friulimicin in Actinoplanesfriuliensis. Microbiology 151: 1963–74.

Hiradate, S., S. Yoshida, H. Sugie, H. Yada and Y. Fujii. 2002. Mulberry anthracnose antagonists (iturins) produced by *Bacillus amyloliquefaciens* RC-2. Phytochemistry 61(6): 693–698.

Hmidet, N., H. Ben Ayed, P. Jacques and M. Nasri. 2017. Enhancement of Surfactin and Fengycin production by *Bacillus mojavensis* A21: Application for diesel biodegradation. BioMed. Research International 2017: 5893123. https://doi.org/10.1155/2017/5893123.

Horn, J.N., A. Cravens and A. Grossfield. 2013. Interactions between fengycin and model bilayers quantified by coarse-grained molecular dynamics. Biophys. J. 105: 1612–1623. doi:10.1016/j. bpj.2013.08.034.

Hossain, M.M., F. Sultana and S. Islam. 2017. Plant growth-promoting fungi (PGPF): Phytostimulation and induced systemic resistance. pp. 135–191. Singh, D. P., H.B. Singh and R. Prabha (eds.). Plant-Microbe Interactions in Agro-Ecological Perspectives: Volume 2: Microbial Interactions and Agro-Ecological Impacts. Singapore, Springer Singapore.

Huang, X., Z. Lu, H. Zhao, X. Bie, F. L"u and S. Yang. 2006. Antiviral activity of antimicrobial lipopeptide from *Bacillus subtilis* fmbj against Pseudorabies Virus, Porcine Parvovirus, Newcastle Disease Virus and Infectious Bursal Disease Virus *in vitro*. International Journal of Peptide Research and Therapeutics 12(4): 373–377.

Huang, X., Z. Wei, X. Gao, X.S. Yang and Y. Cui. 2008. Optimization of inactivation of endospores of *Bacillus cereus* in milk by surfactin and fengycin using a response surface method. International Journal of Peptide Research and Therapeutics 14(2): 89–95.

Huang, X.F., J. Liu, L.J. Lu, Y. Wen, J.C. Xu, D.H. Yang and Q. Zhou. 2009. Evaluation of screening methods for demulsifying bacteria and characterization of lipopeptide bio-demulsifier produced by *Alcaligenes* sp. Bioresource Technology 100(3): 1358–1365.

Janek, T., M. Łukaszewicz, T. Rezanka and A. Krasowska. 2010. Isolation and characterization of two new lipopeptide biosurfactants produced by *Pseudomonas fluorescens* BD5 isolated from water from the Arctic Archipelago of Svalbard. Bioresource Technology 101(15): 6118–6123. doi: 10.1016/j. biortech.2010.02.109.

Jiang, J., Y. Zu, X. Li, Q. Meng and X. Long. 2020. Recent progress towards industrial rhamnolipids fermentation: Process optimization and foam control. Bioresource Technology 298: 122394.

Jin, P., H. Wang, Z. Tan, Z. Xuan, G.Y. Dahar, Q.X. Li, W. Miao and W. Liu. 2020. Antifungal mechanism of bacillomycin D from *Bacillus velezensis* HN-2 against Colletotrichum gloeosporioides Penz. Pesticide Biochemistry and Physiology 163: 102–107. https://doi.org/https://doi.org/10.1016/j. pestbp.2019.11.004.

Kannojia, P., K.K. Choudhary, A.K. Srivastava and A.K. Singh. 2019. Chapter Four - PGPR Bioelicitors: Induced Systemic Resistance (ISR) and Proteomic Perspective on Biocontrol. PGPR Amelioration in Sustainable Agriculture. Singh, A.K., A. Kumar and P.K. Singh (eds.). Woodhead Publishing: 67–84.

Khopade, A., B. Ren, X.Y. Liu, K. Mahadik, L. Zhang and C. Kokare. 2012. Production and characterization of biosurfactant from marine *Streptomyces* species B3. Journal of Colloid and Interface Science 367(1): 311–318.

Kim, P., J. Ryu, Y. Kim and Y. Chi. 2010. Production of biosurfactant lipopeptides Iturin A, Fengycin and Surfactin A from *Bacillus subtilis* CMB32 for control of Colletotrichum gloeosporioides. Journal of Microbiology and Biotechnology 20(1): 138–145. doi: 10.4014/jmb.0905.05007.

Kinsinger, R.F., M.C. Shirk and R. Fall. 2003. Rapid surface motility in *Bacillus subtilis* is dependent on extracellular surfactin and potassium ion. Journal of Bacteriology 185(18): 5627.

Kloepper, J.W., C.-M. Ryu and S. Zhang. 2004. Induced systemic resistance and promotion of plant growth by *Bacillus* spp. Phytopathology® 94(11): 1259–1266.

Kruijt, M., H. Tran and J.M. Raaijmakers. 2009. Functional, genetic and chemical characterization of biosurfactants produced by plant growth-promoting *Pseudomonas putida* 267. J. Appl. Microbiol. 107: 546–556.

Kurokawa, N. and Y. Ohfune. 1993. Synthetic studies on antifungal cyclic-peptides, echinocandins – stereoselective total synthesis of echinocandin-d via a novel peptide coupling. Tetrahedron 49: 6195–6222.

Leclère, V., R. Marti, M. Béchet, P. Fickers and P. Jacques. 2006. The lipopeptides mycosubtilin and surfactin enhance spreading of *Bacillus subtilis* strains by their surface-active properties. Archives of Microbiology 186(6): 475–483. https://doi.org/10.1007/s00203-006-0163-z.

Lee, S.C., S.H. Kim, I.H. Park, S.Y. Chung, M. Subhosh Chandra and Y.L. Choi. 2010. Isolation, purification, and characterization of novel fengycin S from *Bacillus amyloliquefaciens* LSC04 degrading-crude oil. Biotechnol. Bioproc. Eng. 15: 246–253.

Liu, J., X.F. Huang, L.J. Lu, J.C. Xu, Y. Wen, D.H. Yang and Q. Zhou. 2010. Optimization of biodemulsifier production from *Alcaligenes* sp. S-XJ-1 and its application in breaking crude oil emulsion. Journal of Hazardous Materials 183(1-3): 466–473.

Liu, J., X.F. Huang, L.J. Lu, J.C. Xu, Y. Wen, D.H. Yang and Q. Zhou. 2009. Comparison between waste frying oil and paraffin as carbon source in the production of biodemulsifier by *Dietzia* sp. S-JS-1. Bioresour Technol. 100(24): 6481–7. 0.1016/j.biortech.2009.07.006. Epub 2009 Jul 29. PMID: 19643603.

Liu, Y., S. Kyle and P.D. Straight. 2018. Antibiotic stimulation of a *Bacillus subtilis* migratory response. mSphere 3(1): e00586–00517. https://doi.org/10.1128/mSphere.00586-17.

Lugtenberg, B.J.J., L. Dekkers and G.V. Bloemberg. 2001. Molecular determinants of rhizosphere colonization by *Pseudomonas*. Annual Review of Phytopathology 39(1): 461–490.

Lunagariya, N.A., N.K. Patel, S.C. Jagtap and K.K. Bhutani. 2014. Inhibitors of pancreatic lipase: state of the art and clinical perspectives. EXCLI Journal 13: 897.

Maier, R.M. 2003. Biosurfactant: Evolution and diversity in bacteria. Adv. Appl. Microbiol. 52: 101–121.

Malviya, D., P.K. Sahu, U.B. Singh, S. Paul, A. Gupta, A.R. Gupta, S. Singh, M. Kumar and D. Paul. 2020. Lesson from ecotoxicity: Revisiting the microbial lipopeptides for the management of emerging diseases for crop protection. Int. J. Environ. Res. Public Health 17: 1434; doi:10.3390/ijerph17041434.

Mazzola, M., I. De Bruijn, M.F. Cohen and J.M. Raaijmakers. 2009. Protozoan-induced regulation of cyclic lipopeptide biosynthesis is an effective predation defense mechanism for *Pseudomonas fluorescens*. Appl. Environ. Microbiol. 75: 6804–6811.

Meena, K.R. and S.S. Kanwar. 2015. Lipopeptides as the antifungal and antibacterial agents: Applications in food safety and therapeutics. BioMed. Research International, 473050. https://doi.org/10.1155/2015/473050.

Meena, M., P. Swapnil, K. Divyanshu, S. Kumar, Harish, Y.N. Tripathi, A. Zehra, A. Marwal and R.S. Upadhyay. 2020. PGPR-mediated induction of systemic resistance and physiochemical alterations in plants against the pathogens: Current perspectives. Journal of Basic Microbiology 60(10): 828–861. https://doi.org/10.1002/jobm.202000370.

MEI Yu-wei, Y., Z.-y. YU Fan and LONGXu-wei. 2020. Recent progress on fermentation and antibacterial applications of surfactin. China Biotechnology 40(5): 105–116. https://doi.org/10.13523/j.cb.1912023.

Mnif, I. and D. Ghribi. 2015. Potential of bacterial derived biopesticides in pest management. Crop Protection 77: 52–64. https://doi.org/https://doi.org/10.1016/j.cropro.2015.07.017.

Mnif, I. and D. Ghribi. 2015. Lipopeptides biosurfactants: Mean classes and new insights for industrial, biomedical, and environmental applications. Peptide Science 104(3): 129–147.

Molina-Santiago, C., J.R. Pearson, Y. Navarro, M.V. Berlanga-Clavero, A.M. Caraballo-Rodriguez, D. Petras, M.L. García-Martín, G. Lamon, B. Haberstein, F.M. Cazorla, A. de Vicente, A. Loquet, P.C. Dorrestein and D. Romero. 2019. The extracellular matrix protects *Bacillus subtilis* colonies from *Pseudomonas* invasion and modulates plant co-colonization. Nature Communications 10(1): 1919. https://doi.org/10.1038/s41467-019-09944-x.

Moreira, S.M., T.A. de Oliveira Mendes, M.F. Santanta, S.A. Huws, C.J. Creevey and H.C. Mantovani. 2020. Genomic and gene expression evidence of nonribosomal peptide and polyketide production among ruminal bacteria: a potential role in niche colonization? FEMS Microbiology Ecology 96(2). https://doi.org/10.1093/femsec/fiz198.

Mukherjee, A.K. and K. Das. 2005. Correlation between diverse cyclic lipopeptides production and regulation of growth and substrate utilization by *Bacillus subtilis* strains in a particular habitat. FEMS Microbiol. Eco. 54: 479–489.

Mukherjee, S., P. Das and R. Sen. 2006. Towards commercial production of microbial surfactants. Trends in Biotechnology 24: 509–515.

Muller, C., S. Nolden, P. Gebhardt, E. Heinzelmann, C. Lange, O. Puk, K. Welzel, W. Wohlleben and D. Schwartz. 2007. Sequencing and analysis of the biosynthetic gene cluster of the lipopeptide antibiotic Friulimicin in *Actinoplanesfriuliensis*. Antimicrob. Agents Chemother. 51: 1028–37.

Naganawa, H., M. Hamada, K. Maeda, Y. Okami and T. Takeushi. 1968. Laspartomycin, a new anti-staphylococcal peptide. J. Antibiot. (Tokyo) 21: 55–62.

Nielsen, T.H,, C. Christophersen, U. Anthoni and J. Sørensen. 1999. Viscosinamide, a new cyclic depsipeptide with surfactant and antifungal properties produced by *Pseudomonas fluorescens* DR54. J. Appl. Microbiol. 87: 80–90.

Nielsen, T.H., D. Sørensen, C. Tobiasen, J.B. Andersen, C. Christophersen, M. Givskov and J. Sørensen. 2002. Antibiotic and biosurfactant properties of cyclic lipopeptides produced by fluorescent *Pseudomonas* spp. from the sugar beet rhizosphere. Applied and Environmental Microbiology, 68(7): 3416–3423. https://doi.org/10.1128/aem.68.7.3416-3423.2002.

Nishikiori, T., H. Naganawa, Y. Muraoka, T. Aoyagi and H. Umezawa. 1986. Plipastatins: new inhibitors of phospholipase A2, produced by *Bacillus cereus* BMG302-fF67. II. Structure of fatty acid residue and amino acid sequence. J. Antibiot. (Tokyo) 39: 745–754.

Nitschke, M. and S.G.V.A.O. Costa. 2007. Biosurfactants in food industry. Trends in Food Science & Technology 18(5): 252–259.

O'Brien, P.A. 2017. Biological control of plant diseases. Australasian Plant Pathology 46(4): 293–04.

Ongena, M., P. Jacques, Y. Touré, J. Destain, A. Jabrane and P. Thonart. 2005. Involvement of fengycin-type lipopeptides in the multifaceted biocontrol potential of *Bacillus subtilis*. Applied Microbiology and Biotechnology 69(1): 29–38.

Ongena, M. and P. Jacques. 2008. *Bacillus* lipopeptides: versatile weapons for plant disease biocontrol. Trends in Microbiology 16(3): 115–125. https://doi.org/https://doi.org/10.1016/j.tim.2007.12.009.

Ongena, M., E. Jourdan, A. Adam, M. Paquot, A. Brans, B. Joris, J.L. Arpigny and P. Thonart. 2007. Surfactin and fengycin lipopeptides of *Bacillus subtilis* as elicitors of induced systemic resistance in plants. Environmental Microbiology 9(4): 1084–1090. https://doi.org/10.1111/j.1462-2920.2006.01202.x.

Patel, S., S. Ahmed and J.S. Eswari. 2015. Therapeutic cyclic lipopeptides mining from microbes: latest strides and hurdles. World Journal of Microbiology and Biotechnology 31(8): 1177–1193. https://doi.org/10.1007/s11274-015-1880-8.

Pathak, H. 2012. Serratia-The 4T Engine oil degrader. Journal of Petroleum & Environmental Biotechnology 01(S1). doi: 10.4172/scientificreports.117.

Pathak, K.V. and H. Keharia. 2014. Application of extracellular lipopeptide biosurfactant produced by endophytic *Bacillus subtilis* K1 isolated from aerial roots of banyan (*Ficus benghalensis*) in microbially enhanced oil recovery (MEOR). 3 Biotech 4(1): 41–48.

Pérez-García, A., D. Romero and A. de Vicente. 2011. Plant protection and growth stimulation by microorganisms: biotechnological applications of Bacilli in agriculture. Current Opinion in Biotechnology 22(2): 187–193. https://doi.org/https://doi.org/10.1016/j.copbio.2010.12.003.

Peypoux, F., M. Guinand, G. Michel, L. Delcambe, B. Das and E. Lederer. 1978. Structure of iturine A, a peptidolipid antibiotic from *Bacillus subtilis*. Biochemistry 17(19): 3992–3996. doi: 10.1021/bi00612a018.

Peypoux, F., F. Besson, G. Michel and L. Delcambe. 1981. Structure of bacillomycin D, a new antibiotic of the iturin group. European Journal of Biochemistry 118(2): 323–327.

Piyush Baindara and Suresh Korpole. 2016. Lipopeptides: Status and strategies to control fungal infection. pp. 97–121. Basak, A. et al. (eds.). Recent Trends in Antifungal Agents and Antifungal Therapy. Springer India DOI 10.1007/978-81-322-2782-3_4.

Plaza, G.A., A. Turek, E. Król and R. Szczyglowska. 2013. Antifungal and antibacterial properties of surfactin isolated from *Bacillus subtilis* growing on molasses. African Journal of Microbiology Research 7: 3165–3170.

Qian, C.D., B.Q. Li, T. Zhao, Q.G. Guo, X.Y. Lu, S.Z. Li and P. Ma. 2009. Isolation and stability analysis of lipopeptides produced by *Bacillus subtilis* strain BAB21. China J. Agric. Sci. Technol. 11: 69–74.

Qian, S., J. Sun, H. Lu, F. Lu, X. Bie and Z. Lu. 2017. L-glutamine efficiently stimulates biosynthesis of bacillomycin D in *Bacillus subtilis* fmbJ. Process Biochemistry 58: 224–229. https://doi.org/https://doi.org/10.1016/j.procbio.2017.04.026.

Rainey, P., C. Brodey and K. Johnstone. 1991. Biological properties and spectrum of activity of tolaasin, alipodepsipeptide toxin produced by the mushroom pathogen *Pseudomonas tolaasii*. Physiol. Mol. Plant 39: 57–70.

Ramey, B.E., M. Koutsoudis, S.B.V. Bodman and C. Fuqua. 2004. Biofilm formation in plant–microbe associations. Current Opinion in Microbiology 7(6): 602–609. https://doi.org/https://doi.org/10.1016/j.mib.2004.10.014.

Reimer, J.M., A.S. Haque, M.J. Tarry and T.M. Schmeing. 2018. Piecing together nonribosomal peptide synthesis. Current Opinion in Structural Biology 49: 104–113. https://doi.org/https://doi.org/10.1016/j.sbi.2018.01.011.

Rodriguez, M.J., V. Vasudevan, J.A. Jamison, P.S. Borromeo and W.W. Turner. 1999. The synthesis of water soluble prodrugs analogs of echinocandin B. Bioorg. Med. Chem. Lett. 9: 1863–1868.

Romano, A., D. Vitullo, A. Di Pietro, G. Lima and V. Lanzotti. 2011. Antifungal lipopeptides from *Bacillus amyloliquefaciens* strain BO7. Journal of Natural Products 74(2): 145–151. https://doi.org/10.1021/np100408y.

Romero, D., A. de Vicente, R.H. Rakotoaly, S.E. Dufour, J.-W. Veening, E. Arrebola, F.M. Cazorla, O.P. Kuipers, M. Paquot and A. Pérez-García. 2007. The Iturin and Fengycin families of lipopeptides are key factors in antagonism of *Bacillus subtilis* toward Podosphaerafusca. Molecular Plant-Microbe Interactions® 20(4): 430–440. https://doi.org/10.1094/MPMI-20-4-0430.

Sachdev, D.P. and S.S. Cameotra. 2013. Biosurfactants in agriculture. Appl. Microbiol. Biotechnol. 97: 1005–1016.

Sanket, G.K. and B.N. Yagnik. 2013. Current trend and potential for microbial biosurfactants. Asian J. Exp. Biol. Sci. 4(1): 1–8.

Sarafin, Y., M.B.S. Donio, S. Velmurugan, M. Michaelbabu and T. Citarasu. 2014. Kocuria marina BS-15 a biosurfactant producing halophilic bacteria isolated from solar salt works in India. Saudi Journal of Biological Sciences 21(6): 511–519.

Satpute, S.K., A.G. Banpurkar, P.K. Dhakephalkar, I.M. Banat and B.A. Chopade. 2010. Methods for investigating biosurfactants and bioemulsifiers: a review. Critical Reviews in Biotechnology 30(2): 127–144.

Schneider, T., K. Gries, M. Josten, I. Wiedemann, S. Pelzer, H. Labischinski and H.G. Sahl. 2009. The lipopeptide antibiotic Friulimicin B inhibits cell wall biosynthesis through complex formation with bactoprenol phosphate. Antimicrob. Agents Chemother. 53: 1610–8.

Schneidera, T., A. Mullera, H. Miess and H. Gross. 2013. Cyclic lipopeptides as antibacterial agent-potent antibiotic activity mediated by intriguing mode of actions. Int. J. Med. Microbiol. 304(1): 37–43. doi: org/10.1016/j.ijmm.2013.08.009.

Sen, R. 2010. Biosurfactants. New York, N.Y.: Springer Science+Business Media.

Seydlova, G., R. Cabala and J. Svobodova. 2011. Biomedical engineering, trends, research and technologies. *In*: Surfactin—Novel Solutions for Global Issues. 13: 306–330. InTech, Rijeka, Croatia.

Shaligram, N.S. and R.S. Singhal. 2010. Surfactin—a review on biosynthesis, fermentation, purification and applications. Food Technology and Biotechnology 48(2): 119–134.

Singh, P. and S. Cameotra. 2004. Potential applications of microbial surfactants in biomedical sciences. Trends in Biotechnology 22: 142–146.

Singla, R.K., H.D. Dubey and A.K. Dubey. 2014. Therapeutic spectrum of bacterial metabolites. Indo Global Journal of Pharmaceutical Sciences 2(2): 52–64.

Stein, T. 2005. *Bacillus subtilis* antibiotics: structures, syntheses and specific functions. Molecular Microbiology 56(4): 845–857.

Steller, S., D. Vollenbroich, F. Leenders, T. Stein, B. Conrad, J. Hofemeister, P. Jacques, P. Thonart and J. Vater. 1999. Structural and functional organization of the fengycin synthetase multienzyme system from *Bacillus subtilis* b213 and A1/3. Chemistry and Biology 6(1): 31–41.

Steller, S., A. Sokoll, C. Wilde, F. Bernhard, P. Franke and J. Vater. 2004. Initiation of surfactin biosynthesis and the role of the SrfD-thioesterase protein. Biochemistry 43(35): 11331–11343. https://doi.org/10.1021/bi0493416.

Sun, X., D.J. Zeckner, W.L. Current, R. Boyer, C. McMillian, N. Yumibe and S.H. Chen. 2001. N-acyloxymethyl carbamate linked prodrugs of pseudomycins are novel antifungal agents. Bioorg. Med. Chem. Lett. 11: 1875–1879.

Sur, S., T.D. Romo and A. Grossfield. 2018. Selectivity and mechanism of fengycin, an antimicrobial lipopeptide, from molecular dynamics. The Journal of Physical Chemistry B 122(8): 2219–2226. https://doi.org/10.1021/acs.jpcb.7b11889.

Tffoth, V., C.T. Nagy, I. Pffocsi and T. Emri. 2012. The echinocandin B producer fungus *Aspergillus nidulans* var. *roseus* ATCC 58397 does not possess innate resistance against its lipopeptide antimycotic. Appl. Microbiol. Biotechnol. 95: 113–122.

Thavasi, R., V.S. Nambaru, S. Jayalakshmi, T. Balasubramanian and I.M. Banat. 2011. Biosurfactant production by *Pseudomonas aeruginosa* from renewable resources. Indian Journal of Microbiology 51(1): 30–36.

Tran, H., A. Ficke, T. Asiimwe, M. Ho¨fte and J.M. Raaijmakers. 2007. Role of the cyclic lipopeptide massetolide a in biological control of *Phytophthora infestans* and in colonization of tomato plants by *Pseudomonas fluorescens*. New Phytol. 175: 731–742.

Tsan, P., L. Volpon, F. Besson and J.M. Lancelin. 2007. Structure and dynamics of surfactin studied by NMR in micellar media. J. Am. Chem. Soc. 129(7): 1968–1977.

Van Hamme, J.D., A. Singh and O.P. Ward. 2006. Physiological aspects. Part 1 in a series of papers devoted to surfactants in microbiology and biotechnology. Biotechnology Advances 24: 604–620.

Vertesy, L., E. Ehlers, H. Kogler, M. Kurz, J. Meiwes, G. Seibert, M. Vogel and P. Hammann. 2000. Friulimicins: novel lipopeptide antibiotics with peptidoglycan synthesis inhibiting activity from *Actinoplanesfriuliensis* sp. nov. II. Isolation and structural characterization. J. Antibiot. (Tokyo) 53: 816–27.

Volpon, L., F. Besson and J.M. Lancelin. 2000. NMR structure of antibiotics plipastatins A and B from *Bacillus subtilis* inhibitors of phospholipase A(2). FEBS Lett. 485: 76–80.

Vollenbroich, D., M. ¨Ozel, J. Vater, R.M. Kamp and G. Pauli. 1997. Mechanism of inactivation of enveloped viruses by the biosurfactant surfactin from *Bacillus subtilis*. Biologicals 25(3): 289–297.

Walia, N.K. and S.S. Cameotra. 2015. Lipopeptides: Biosynthesis and applications. J. Microb. Biochem. Technol. 7: 103–107. doi:10.4172/1948-5948.1000189.

Walsh, C.T. 2004. Polyketide and nonribosomal peptide antibiotics: modularity and versatility. Science 303(5665): 1805.

Wang, Y., Y. Chen, Q. Shen and X. Yin. 2011. Molecular cloning and identification of the laspartomycin biosynthetic gene cluster from *Streptomyces viridochromogenes*. Gene 483(1): 11–21.

Wecke, T., D. Zuhlke, U. Mader, S. Jordan, B. Voigt, S. Pelzer, H. Labischinski, G. Homuth, M. Hecker and T. Mascher. 2009. Daptomycin versus Friulimicin B: in-depth profiling of *Bacillus subtilis* cell envelope stress responses. Antimicrob. Agents Chemother. 53: 1619–23.

Yaseen, Y., F. Gancel, M. Béchet, D. Drider and P. Jacques. 2017. Study of the correlation between fengycin promoter expression and its production by *Bacillus subtilis* under different culture conditions and the impact on surfactin production. Archives of Microbiology 199(10): 1371–1382. https://doi.org/10.1007/s00203-017-1406-x.

Yaseen, Y., A. Diop, F. Gancel, M. Béchet, P. Jacques and D. Drider. 2018. Polynucleotide phosphorylase is involved in the control of lipopeptide fengycin production in *Bacillus subtilis*. Archives of Microbiology 200(5): 783–791. https://doi.org/10.1007/s00203-018-1483-5.

Yu, G.Y., J.B. Sinclair, G.L. Hartman and B.L. Bertagnolli. 2002. Production of iturin A by *Bacillus amyloliquefaciens* suppressing *Rhizoctonia solani*. Soil Biology and Biochemistry 34(7): 955–963.

Zeriouh, H., D. Romero, L. Garcia-Gutierrez, F.M. Cazorla, A. de Vicente and A. Perez-Garcia. 2011. The iturin-like lipopeptides are essential components in the biological control arsenal of *Bacillus subtilis* against bacterial diseases of cucurbits. Mol. Plant Microbe. Interact. 24(12): 1540–52. doi: 10.1094/MPMI-06-11-0162. PMID: 22066902.

Zhang, Y., Y. Wang, Y. Qin and P. Li. 2020. Complete genome sequence of *Bacillus velezensis* LPL-K103, an antifungal cyclic lipopeptide bacillomycin L producer from the surface of lemon. 3 Biotech 10(1): 8.

Zouari, R., K. Hamden, A.E. Feki, K. Chaabouni, F. Makni-Ayadi, C. Kallel, F. Sallemi, S. Ellouze-Chaabouni and D. Ghribi-Aydi. 2016. Protective and curative effects of *Bacillus subtilis* SPB1 biosurfactant on high-fat-high-fructose diet induced hyperlipidemia, hypertriglyceridemia and deterioration of liver function in rats. Biomed Pharmacother. 84: 323–329. doi: 10.1016/j.biopha.2016.09.023. Epub 2016 Sep 22. PMID: 27665478.

4

Recent Developments in Biomedical and Therapeutic Application of Biosurfactants

Hebatallah H Abo Nahas,[1,]* *Hanaa F Abd EL-kareem,*[2]
Yousra A El-Maradny,[3] *Yousef H Abo Nahas,*[4]
Sara A Gezaf,[5] *Ahmed M Abdel-Azeem*[6] and
Essa M Saied[7,8]

1. Introduction

Biosurfactants are natural surfactant as they are produced from a natural renewable resource. The source could range from micro-organism to plant, or an animal and the manufactured product ought to be acquired by separation processes as extraction, precipitation or distillation (Holmberg 2001). As hydrocarbon particles, the biosurfactants began to use in 1960s with many applications especially in food processing, medical and oil manufacturing due to their low toxicity whatever the strain of the microorganisms used to produce them (Roy 2018). Due to their great biodegradability and low toxicity, they can be simply manufactured from renewal energy resources and used in medicinal, foodstuff, cosmetic, textile, oil and agricultural industries (Khan et al. 2014). Because of the pathogenicity of the producer of some biosurfactants such as *Pseudomonas* and *Bacillus*, Naughton et al.

---

[1] Zoology Department, Faculty of Science, Suez Canal University, Ismailia, Egypt.
[2] Zoology Department, Faculty of Science Ain Shams University, Egypt.
[3] Microbiology Department, High Institute of Public Health, Alexandria University, Egypt.
[4] Microbiology Department, Faculty of Medicine, Helwan University, Cairo11795, Egypt.
[5] Botany Department, Faculty of Science, Arish University, North Sinai, Egypt.
[6] Botany and Microbiology. Department, Faculty of Science, University of Suez Canal, Ismailia 41522, Egypt.
[7] Chemistry Department, Faculty of Science, Suez Canal University, Ismailia, Egypt.
[8] Institute for Chemistry, Humboldt Universität zu Berlin, Brook-Taylor-Str. 2, 12489 Berlin, Germany.
* Corresponding author: hebatallah_hassan@science.suez.edu.eg

(2019) stated that they were interested in yeast and yeast-like fungi non-pathogenic bacteria in producing biosurfactants that are not toxic or pathogenic.

The amphiphoteric property of the biosurfactant give them the opportunity to be a good intermediate in the emulsification process in different applications such as in oil contaminated samples, as the oil increase the production of the biosurfactant from the microorganisms that integrate the oil into small droplets and then use these droplets for their energy production (Nurfarahin et al. 2018).

There are many factors affecting the type and the amount of surfactant produced as: the strain of the bacteria, nature of the carbon source, the nitrogen source and as well as the carbon: nitrogen proportion, temperature, air circulation and salinity. Biosurfactants have anti-adhesive properties so they are suggested to be of use as versatile fixings or added substances (Pardhi et al. 2018). In addition to the medical use of the biosurfactant as antibiotics, antifungal and antiviral, they are also used in food processing, in the agricultural industry and for cosmetics (Chiewpattanakul et al. 2010). Some biosurfactants can be adjuvants for antigens, can inhibit the formation of clot and activate clot lysis (Gudiña et al. 2013). Moreover, Gudiña et al. (2013) stated that biosurfactant can intermingle with cell membranes of the organisms with the adjacent environments and so they can act as therapeutics for cancer diseases.

Though different types of biosurfactants have been discovered, lipopeptides and glycolipids came to the fore because of their inherent properties that make it possible to apply them in the food industry, such as antibacterial and anti-adhesive activity against numerous species (*Pseudomonas aeruginosa, Escherichia coli, Bacillus subtilis,* and *Staphylococcus aureus*), antioxidant activity and low cytotoxicity (Ribeiro et al. 2020).

Biosurfactants seem to be vital in the dietary management in the field of farm animal production. They have an impact on nutrition as they effect the physiological and production parameters; especially the biosurfactants that are derived from plants as alkyl polyglucosides (Naughton et al. 2019).

To produce economically viable biosurfactants, the cost of the fermentation media has to be lower, so agro-industrial waste products may be added as substrates to increase the viability of the large-scale production of biosurfactants and make these natural products more economical (Ribeiro et al. 2020). Biosurfactants are also used in solubilizing vegetable oils, stabilizing fats through cooking, and enhancing the organoleptic properties of bread. They can be used also in preparation of ice cream, and used to substitute the baking powder and synthetic additives (Ribeiro et al. 2020).

Research on the applications of biosurfactants in modifying environmental pollution started over two decades ago, Biosurfactants also have a significant advantage in terms of resource usage. Biosurfactants are manufactured by microorganisms as secondary metabolites, mainly if they grow on water-insoluble substrates as hydrocarbons. Making of biosurfactants can be either extracellular or as cell-bound molecules. When biosurfactants are manufactured extracellularly, they cause emulsification to the substrate (i.e., hydrocarbon), while, if they are manufactured as a part of the cell membrane of the microorganism, this cause a passage of substrates through the membrane. So, the use of biosurfactant-producing microorganisms in the biodegradation of hydrocarbons and biopreparation of polluted sites is an effective microbiological treatment method of environmental pollutants (Olasanmi and Thring 2018). Moreover, biosurfactants can be used in enhancement decay availability of the pollutants and increase the rate of biodegradation. The usage of biosurfactants to clear-out the environmental pollutants has been discovered due to their capacity to elevate the bioavailability of contaminants to microorganisms, and the solubilization and mobilization of hydrocarbons in soils by biosurfactant-enhanced washing resulting in removal of the contaminants. Thus, Mazaheri Assadi and Tabatabaee (2010) mentioned that the data of the chemical and physical nature of the contaminated area is needed if biosurfactants are used in petroleum bioremediation.

Biosurfactants have become an exciting and noticeable trend in industries, principally in cleaning up environmental pollutants in petroleum wastes. Their usefulness and advantages give these microbial products a hopeful future in industrial applications (Olasanmi and Thring 2018).

2. General Biosurfactants Properties and Functions

Biosurfactants, or bio-surface active compounds are compounds derived from microorganisms rather than manufactured by chemical substances. Moreover, these microorganisms yield less toxic biosurfactants (Santos et al. 2016). Biosurfactants can act as a molecular interfacial film that decrease the surface tension in liquid and the interfacial tension between different liquid phases at the interfacial boundary between different liquid phases. As the surfactants has 2 parts in its molecule (hydrophilic and hydrophobic), the hydrophilic part can counterpart with the hydrophilic phase while the hydrophobic part can counterpart with the hydrophobic phase (Pacwa-Płociniczak et al. 2011).

Biosurfactants can decrease the surface tension of water by producing critical micelle concentration (CMC), they also have biodegradability, various structures, effect at high temperature and pH. Moreover, biosurfactants can be manufactured from low-priced raw materials, so they are used for engineering, cultivated, nutrition,

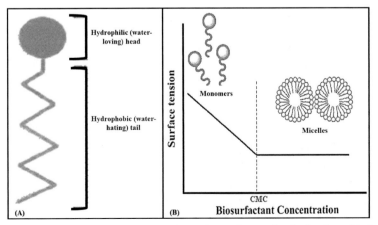

Figure 1. (A) Structure of biosurfactants, (B) Decreasing surface tension by forming critical micelle concentration of biosurfactants (Santos et al. 2016).

cosmetics and medicinal applications (De et al. 2015). Biosurfactants are categorized according to their molecular weight, or according to their chemical composition.

2.1 Classification Based on Molecular Weight (Pacwa-Płociniczak et al. 2011)

i) **Low molecular weight biosurfactants:** These biosurfactants decrease the surface and interfacial tension at the air/water boundaries as: glycolipids, lipopeptides, flavolipids, corynomycolic acids and phospholipids.

ii) **High molecular weight biosurfactants:** They also called the bioemulsans and are more effective in stabilizing oil in water emulsions as: emulsan, alasan, liposan, polysaccharides and protein complexes.

2.2 Classification Based on Chemical Structure (Santos et al. 2016)

i) **Glycolipids:** These types of biosurfactants as carbohydrates constituent of mono-, di-, tri and tetrasaccharides attached to a long-chain aliphatic acid or lipopeptide.

ii) **Lipopeptides and lipoproteins:** These types involve a lipid combined with a polypeptide chain.

iii) **Fatty acids and neutral lipids:** When alkanes were exposed to microbial oxidations, this resulted in the construction of straight chain fatty acids known as a surfactant.

iv) **Phospholipids:** They are main constituents of microbial membranes, when specific hydrocarbon degrading bacteria are developed on alkane substrates their level rises greatly.

v) **Polymeric microbial surfactants:** They are polymeric heterosaccharide having proteins.

vi) **Particulate biosurfactant:** There are some bacteria that produce extracellular membrane vesicles of hydrocarbons that form micro-emulsion, which is important for the uptake of alkane by microbial cells.

There are several kinds of original microorganisms, and new ones producing diseases have appeared constantly every year while there is no discovery of new chemical antibiotics. Because the diseases turned out to be resistant to the antibiotics, so, the biosurfactants become unique substitutions and can be used as an effective therapeutic agent (Rodrigues et al. 2006).

2.3 Microorganisms Generating Biosurfactants

The biosurfactants that generating from the microorganisms have different chemical composition according to the site of isolation of microorganisms and the nutritional factors for their development. Several microorganisms, which are isolated from various sources for industrial use, can grow on substances that can be harmful to other microorganisms, and the field of manufacture of the biosurfactants by these microorganisms is well known (Lígia R. Rodrigues and Teixeira 2010). Table 1 and

Table 1. Microorganisms producing potential biosurfactants with their applications (Pardhi et al. 2018).

Microorganism	Biosurfactant produced	Applications
Pseudomonas sp.	Rhamnolipid	Processing of food Treatment of water waste Bioremediation of oil-contaminated sites, hydrocarbon contaminated sites, marine and soil environments Insecticidal effect
Bacillus subtilis	Iturin	Anti-fungal agent
Bacillus subtilis	Surfactin	Anti-mycoplasma agent Anti-tumor agents Anti-fungal agent Anti-viral agents
Serratia marcescens	Lipopeptide	Bioremediation of marine oil pollution, hydrocarbon contaminated sites Prompting the activity of Granulocyte colony stimulating factor (G-CSF), Granulocyte-macrophage colony stimulating factor (GM-CSF)
Candida sp. SY-16	Mannosylerythritol (Glycolipid)	Recovery of oil Bioremediation of marine environment Anti-tumor agents

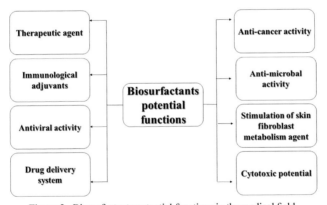

Figure 2. Biosurfactants potential functions in the medical field.

Fig. 2 show different biosurfactants that produced from microorganisms and the potential functions of the biosurfactants in the medical field, respectively.

As shown in Table 1, the applications of the biosurfactants produced from microorganisms includes processing of food, bioremediation of different contaminated sites and anti-tumor, anti-fungal, anti-viral, and anti-mycoplasma agents (Pardhi et al. 2018).

3. Cytotoxic Potential and Anti-tumor Performance of Biosurfactants

Biosurfactants are very promising due to their less toxic and more degradable properties (Akbari et al. 2018). The biosurfactants are considered by various cell functions as: differentiation, immune response, and signal transduction (Santos et al. 2016). Moreover, (Duarte et al. 2014) stated that some types of biosurfactants promote apoptosis to cancer cells. In addition, the extracellular glycolipid of the microbes may help in the treatment of diseased cells as the biosurfactant that has a fatty acid chain may act as a natural antioxidant compound and prevent the formation of free radicals (Vecino et al. 2017), scavenge these free radicals (Vecino et al. 2017).

In the cellular respiration, there is superoxide ions formed as a result of this respiration that inactivate the antioxidant enzymes and result in the formation of free radicals which cause harmful effects. Christova et al. (2014) state that some biosurfactants have a high percentage of total antioxidant capacity, trehalose lipid which is one of the biosurfactants component has cytotoxic effect against the activity of some malignant cells as differentiation-prompting action in some cancer cell line. They deduced that this effect relayed on the structure of the hydrophobic moiety of the biosurfactants. Moreover, some biosurfactants stimulated the formation of interleukin-1β and tumor necrosis factor-α. Wu et al. (2017) suggested that the biosurfactants triggered the increase in the expression of the tumor suppressor p53 gene. As shown in Fig. 3, we summarize the anti-tumor activity of biosurfactants.

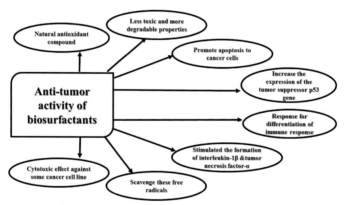

Figure 3. The anti-tumour activity of the biosurfactants.

4. Antimycoplasma Performance of Biosurfactants

Mycoplasmas are causative agents of extreme human and animal diseases, including acute respiratory inflammation (such as pneumonia) and urogenital tract diseases, and tend to be co-factors of AIDS pathogenesis (WHO report 1993, Blanchard et al. 1994). These tiny free-living species are eukaryotic cell parasites and are one of the main contaminants influencing the culture of tissue cells. The most widespread agents that cause this contamination are the mollicute species *Mycoplasma orale* (a human species), *Mycoplasma hyorhinis* (a porcine species), *Mycoplasma arginini* (a bovine species), and *Acholeplasma laidlawii* (a bovine species) (McGarrity et al. 1978). Contaminating mycoplasmas impair a range of cellular processes and cell morphology, replenish the nutrients in the growth medium, and disrupt virus replication (Stanbridge and Doersen 1978, Chowdhury et al. 1994). It is necessary to remove these agents from cell cultures used for basic research, diagnosis, and biotechnological development for both biological and ecological purposes. Treatment with antibiotics is the most effective technique for removing, inactivating, or suppressing mycoplasmas in cell cultures (Schmidt and Erfle 1984, Uphoff et al. 1992, Fleckenstein et al. 1994). Generally, antibiotic therapies don't lead to long-lasting effective decontamination, and due to cytotoxic effects and the growth of resistant mycoplasma strains, unintended side effects on eukaryotic cells have been reported (Schmidt and Erfle 1984, Uphoff et al. 1992). Mycoplasmas lack a cell wall but are surrounded by a three-layer cytoplasmic membrane, therefore antibiotics such as the penicillin, which are widely accepted additives in cell culture media and disrupt murein formation in cell walls, are not impactful against mycoplasmas.

Some of the most valuable sources of new antibiotics are the secondary metabolites of numerous bacteria, fungi, and yeast. The soil bacterium *Bacillus subtilis* creates a variety of substances with complex structures and antibiotic properties (Zuber et al. 1993). One of these, surfactin, is a 1.036 molecular weight cyclic lipopeptide antibiotic. Surfactin is a potent biosurfactant with high surface activity and numerous important biological properties (Arima 1968, Lin et al. 1994, Ishigami et al. 1995). It displays antifungal properties, mild antibacterial properties (Bernheimer and AVIGAD 1970, Tsukagoshi et al. 1970), and hemolysis; prevents fibrin clot formation (Arima 1968, Bernheimer and AVIGAD 1970); triggers the formation of ion channels in lipid bilayer membranes (Sheppard et al. 1991); inhibits enzymes such as cyclic AMP phosphodiesterase (Hosono and Suzuki 1983); demonstrates antiviral and antitumor activities (Kameda et al. 1974, Vollenbroich et al. 1997b); and suppresses starfish oocyte maturation (Toraya et al. 1995).

Vollenbroich et al. (1997b) calculated the cytotoxicity of surfactin with a 50% cytotoxic concentration of 30 to 64 mM for a variety of human and animal cell lines in vitro to evaluate the applicability of this antiviral and antibacterial medication. At the same time, increases in proliferation rates and changes in the morphology of mycoplasma-contaminated mammalian cells have been reported following treatment with this drug. A single treatment over one passage resulted in complete removal of viable *Mycoplasma hyorhinis* cells from numerous adherent cell lines, and *Mycoplasma orale* was removed from non-adherent human T-lymphoid cell lines by double treatment. A DNA fluorescence test, an enzyme-linked immunosorbent assay, and two distinct PCR techniques monitored this effect. The mode of action of surfactin was demonstrated by the disintegration of the mycoplasma membranes as detected by electron microscopy. Disintegration is obviously attributable to the physicochemical interaction of the membrane-active surfactant with the outer part of the bilayer lipid membrane, which induces changes in permeability and eventually contributes to the disintegration of the mycoplasma membrane system by a detergent reaction at higher concentrations. The limited cytotoxicity of surfactin for mammalian cells facilitates specific inactivation of mycoplasmas without considerable harmful effects on cell metabolism and the proliferation rate in cell culture (Vollenbroich et al. 1997a).

Vollenbroich et al. (1997b) treated a confluent ML cell culture that was heavily infected with *M. hyorhinis* with surfactin at several concentrations below 80% cytotoxic concentration (12.5, 25, and 50 mM) to clarify the mode of action of the drug and examined the effects of the drug through transmission electron microscopy.

Mycoplasmas incubated without surfactin in a cell culture were visible on the surfaces of the ML cells as intact particles. The creation of small holes in the mycoplasma membrane and swelling of the particles have been observed after incubation of mycoplasmas with 12.5 mM surfactin at 378C for 1 hr. Mycoplasmas attached to the cell surface were particularly directly affected by the drug. At a concentration of surfactin of 25 mM, particles were induced to burst. Mycoplasmas, which were embedded in cell membrane pockets and clefts, also began to disintegrate. Finally, at a surfactin concentration of 50 mM disruption of the mycoplasma lipid bilayer included complete disintegration of the membrane systems, which caused bursting of all microorganisms. This behaviour was considered to be due to inhibition by surfactin of cyclic AMP phosphodiesterase (Hosono and Suzuki 1983). Electron microscopic experiments have provided proof that surfactin influences the envelopes of contaminating mycoplasmas. Surfactin, evidently, disrupts the plasma membrane, which is its main activity site. It induces leakage, contributes to complete disintegration of the membrane systems at higher concentrations, and eventually causes the mycoplasmas to burst (Vollenbroich et al. 1997a).

Previous experiments with artificial membranes (Sheppard et al. 1991, Maget-Dana and Ptak 1995), protoplasts (Tsukagoshi et al. 1970), and eukaryotic cells (Hosono and Suzuki 1985) have shown that, depending upon the composition of the membrane lipid, surfactin easily attaches to cell membranes with a high degree of selectivity. The fatty acid portion of surfactin is anchored in the lipid bilayer, exhibiting excellent affinities to cholesterol and phospholipids (McGarrity et al. 1978, Hosono and Suzuki 1985). Cholesterol was present in the Mycoplasma species at levels comparable to those found in eukaryotic cell plasma membranes (25 to 30% [wt/wt] of total membrane lipids) (Razin 1978, Rottem 1980). Compared to eukaryotic cells, higher levels of membrane phospholipids (particularly phosphatidylglycerol, phosphatidylcholine, and phosphatidylethanolamine) found in Mycoplasma cells can lead to higher susceptibility of mycoplasma membranes to surfactin (Razin 1978, Rottem 1980). Evidently, surfactin interacts with the mycoplasma membrane in micellar form, causing an osmotic influx of medium and finally total disruption of the cells.

5. Antimicrobial Performance of Biosurfactants

5.1 Antibacterial Activity

Lipopeptides are known as possibly the most useful antimicrobial agents for the ability of their molecules to self-associate and create a pore-bearing tube, or micellular aggregation, inside a lipid membrane upon which their antibiotic activity depends (Deleu et al. 2008, Biniarz et al. 2017). Surfactin, for instance, possesses many biological and physical actions, including antibacterial, antifungal, antitumor, antiviral, antimycoplasma and inflammatory effects, and properties of anti-platelet and hemolytic activity (Seydlová and Svobodová 2008). The pattern of the hydrocarbon chains is influenced by their involvement in penetrating the membrane through hydrophobic interactions, and the membrane thickness is changed (Bonmatin et al. 2003). These actions are nonspecific modes of action, and are effective for acting on various cell membranes, whether in Gram-positive or Gram-negative bacteria

(Lu et al. 2007). This suggests that the success of surfactin as a lipopeptide is on the integrity of the membrane rather than other essential cellular processes and may make up the next generations of antibiotics (Rodrigues and Teixeira 2010).

It is also possible to rely on not only surfactin, but marine *B. circulans* biosurfactant, against several Gram-positive and Gram-negative pathogenic and semi-pathogenic bacteria for antimicrobial output. The productive performers also include *Micrococcus flavus*, *B. pumilus*, *Mycobacterium smegmatis*, *Escherichia coli*, *Serratia marcescens*, and *Proteus vulgaris* (Das et al. 2008). Overlapping patterns, such as those of surfactin lipopeptides and lichenysin, are the chemical structures of these types of biosurfactant fractions. In addition, moderate antimicrobial activity has been observed against methicillin-resistant *Staphylococcus aureus* (MRSA) and other pathogenic multidrug-resistant (MDR) strains (Cameotra and Makkar 2004). This confirms that, due to their non-hemolytic nature, biosurfactants may be utilized as drugs in antimicrobial chemotherapy (Fracchia et al. 2012).

A recent report by Huang and coworkers, has found that the utilization of surfactin and polylysine against *Salmonella enteritidis* in dairy products may be important. It was observed that the minimum inhibitory concentrations of 6.25 surfactin and 31.25 ug/ml polylysine affected *S. enteritidis*, after having utilized a response surface methodology. They noticed, too, that *S. enteritidis* decreased at a temperature of 4.45°C by 6 orders of magnitude, with an action time of 6.9 h, and a concentration of 10.03 µg/ml at a surfactin/polylysine ratio of 1:1 (Huang et al. 2011).

Other antimicrobial lipopeptides have been developed by *Bacillus licheniformis*, *B. pumilus*, and *B. polymyxa*. Polymyxin B from *B. polymyxa*, for instance, provides antibacterial activity towards a broad spectrum of Gram-negative pathogenic bacteria. It is a cationic agent which attaches to the outer membrane of anionic bacteria and contributes to a micelle reaction that disrupts the integrity of the membrane (Naruse et al. 1990, Landman et al. 2008). *Escherichia coli*, *Klebsiella* spp., Enterobacter spp., *Pseudomonas aeruginosa*, *Enterococcus* spp., and Acinetobacter spp. are common nosocomial pathogenic bacteria (Fiore et al. 2017). They are affected by polymyxins, and significant features of polymyxins have been identified against *Acinetobacter baumannii*, *Haemophilus* sp., *Pasteurella* sp., *Salmonella* sp., *Shigella* sp., and *Vibrio cholera* (Landman et al. 2008).

Viscosin, a cyclic lipopeptide, generated by *Pseudomonas*, possesses the property of antimicrobial action and other significant biological properties (Saini et al. 2008). The group of glycolipids includes either rhamnolipids or sophorolipids, which can demonstrate important antimicrobial activities (Mnif and Ghribi 2016). This was utilized very well against *B. subtilis* with a MIC of 8 ug/ml (Benincasa et al. 2004), conducted by a combination of rhamnolipid homologues. There have been observations that *Candida antarctica* strains have developed Mannosylerythritol lipids (MEL-A and MEL-B) that demonstrate antimicrobial activity against Gram-positive bacteria (Kitamoto et al. 1993).

Massetolides A-H, also cyclic depsipeptides, were extracted from the marine habitat organisms of *Pseudomonas* and were observed to possess in vitro antimicrobial activity towards *Mycobacterium tuberculosis* and *Mycobacterium avium*-intracellulare (Gerard et al. 1997).

Precursors and degeneration products of sphingolipids biosurfactants were reported to prevent the interaction of *Streptococcus mitis* with buccal epithelial cells and of *Staphylococcus aureus* with nasal mucosal cells (Bibel et al. 1992).

Mannosylerythritol lipid (MEL), a yeast glycolipid biosurfactant derived from vegetable oils by Candida strains, has been reported to show several antimicrobials, immunological and neurological effects (Shah et al. 2007, Van Bogaert et al. 2007). Kitamoto et al. (1993) have shown that MEL demonstrates antimicrobial activity mainly against gram-positive bacteria.

Antibacterial activity against *Klebsiella pneumonia, Escherichia coli, Vibrio cholera, B. subtilis* and *Staphylococcus aureus* has been demonstrated by biosurfactants developed by *Staphylococcus saprophyticus* SBPS 15 (Mani et al. 2016).

5.2 Antifungal Activity

It has been reported for a long time that biosurfactants can execute antifungal activities, but only small numbers of performance against human pathogenic fungi have been investigated (Chung et al. 2000). Direct lysis of zoospores produced by the intercalation of biosurfactants through zoospore plasma membranes that are independent of a cell wall, is the main mode of action of biosurfactants on fungal pathogens (Vatsa et al. 2010).

In vitro antifungal activity of biosurfactants strongly inhibited various pathogenic yeasts, along with human mycoses, including Candida spp., *Colletotrichum gloeosporioides, Corynespora cassiicola, Cryptococcus neoformans, Fusarium* spp., *Fusarium oxysporum, Rhizoctonia* spp., *Trichophyton rubrum*, and *Trichosporon asahii* (Mnif and Ghribi 2016, Fariq and Saeed 2016).

Another lipopeptide, iturin A, generated by *B. subtilis* was observed to have useful antifungal properties which influences the morphology and membrane system of yeast cells (Thimon et al. 1995, Ahimou et al. 2000). With the development of small vesicles and the aggregation of intramembranous particles, this lipopeptide has been shown to move through the cell wall and damage the plasma membrane. Iturin also travels through the plasma membrane and interferes with the nuclear membrane and likely with membranes of other cytoplasmic organelles. This lipopeptide has been suggested as an important antifungal agent for profound mycosis (Tanaka et al. 1997). Also observed to have antimicrobial activity against *Aspergillus flavus* were other members of the iturin group, including bacillomycin D and bacillomycin Lc, but the different lipid chain lengths evidently influenced lipopeptide activity against other fungi (Moyne et al. 2001). Thereby, the members of the iturin-like biosurfactant group are considered potential alternative antifungal agents.

Viscosinamide, a cyclic depsipeptide, was identified by Nielsen and coworkers as a novel antifungal surface active agent generated by *Pseudomonas fluorescens* and with different characteristics as compared to the biosurfactant viscosin, known to be derived from the same species and to have antibiotic action (Neu et al. 1990, Nielsen et al. 1999).

5.3 Antiviral Activity

An effective antiviral action can be performed by surfactin, which is regarded as the best in medical field applications (Naruse et al. 1990, Biniarz et al. 2017). As reported in experiments *in vitro*, surfactin and fengycin from *B. subtilis* could significantly diminish the activities of cell-free virus stocks of several viruses, such as the porcine parvovirus, the pseudorabies virus, the Newcastle disease virus, and the bursal disease virus. They were also capable of inhibiting infections and the replication of these viruses (Huang et al. 2006).

Due to various physicochemical interactions between the lipid membrane virus and the membrane-active surfactant, surfactin was reported to demonstrate antiviral action. This induces changes in permeability and eventually results in the membrane system disintegration via a micelle effect at higher concentrations (Vollenbroich et al. 1997b). In addition, surfactin was found to affect feline calicivirus, herpes simplex virus, murine encephalomyocarditis virus, Semliki Forest virus, simian immunodeficiency virus, and vesicular stomatitis virus. Gram-positive *B. pumilus* cells were reported to generate pumilacidin A, B, C, D, E, F, and G. These pumilacidins have important antiviral activity against herpes simplex virus 1 (HSV-1), inhibitory activity against H+, K+-ATPase, and great protection against gastric ulcer (Naruse et al. 1990), possibly by inhibiting the microbial activity that leads to these ulcers (Rodrigues and Teixeira 2010).

There are also studies of sophorolipids acting against the human immunodeficiency virus (Shah et al. 2005). Trehalose dimycolate and therapeutic drug monitoring in the trehalose lipids group have been shown to be highly resistant to intranasal influenza virus infection in mice, though T-lymphocyte proliferation with gamma/delta T-cell receptors has been induced, along with the preservation of acquired infection resistance (Franzetti et al. 2010). The complex of rhamnolipid alginate is another lower molecular mass compound that acts against viruses. The rhamnolipid alginate complex affected HSV types 1 and 2, particularly the herpes virus cytopathic effect in the Madin-Darby bovine kidney cell line, by a concentration lower than the critical micelle concentration (Remichkova et al. 2008).

5.4 Anti-Adhesive Activity

It has been found that biosurfactants prevent the adhesion of pathogenic organisms to solid surfaces or infection sites, so that prior adhesion of biosurfactants to solid surfaces of implant materials could be a new and successful means of countering pathogenic microorganism colonization (Singh and Cameotra 2004). Precoating vinyl urethral catheters by running the surfactin solution through them prior to inoculation with media led to the reduction of the amount of biofilm produced by *Salmonella typhimurium, Salmonella enterica, Escherichia coli* and *Proteus mirabilis* (Mireles et al. 2001). These findings have tremendous potential for practical applications, considering the significance of opportunistic infections by Salmonella species, including urinary tract infections of AIDS patients.

The biosurfactant surlactin, developed by several Lactobacillus isolates (Velraeds et al. 1996), was mentioned as an effective anti-adhesive coating for catheter materials. There is significant interest in the role of the Lactobacillus species in the female urogenital tract as a barrier to infection (Boris and Barbés 2000). It is believed that these species contribute to vaginal microbiota regulation by competing with other microorganisms for epithelial cell adherence and by generating biosurfactants. There are observations of uro-pathogens and yeast inhibition of biofilm formation on silicone rubber performed by *Lactobacillus acidophilus* biosurfactants (Velraeds et al. 1998, Reid 2000). *Lactobacillus fermentum* RC-14 has been shown by Heinemann and co-workers to produce surface-active components that can prevent adhesion of uro-pathogenic bacteria, including *Enterococcus faecalis* (Heinemann et al. 2000).

6. Biosurfactants in Health-related Applications

6.1 Dermatological Applications

Biosurfactant constitutes proteins and lipids that are safe to be used as an alternative to chemical preservatives in cosmetic and skin products to overcome the problem of skin irritation from chemical agents (Tamara Stipcevic et al. 2013, Bujak et al. 2015, Sałek and Euston 2019). Drakontis and Amin (2020) recommended the use of biosurfactants as eco-friendly and non-toxic agents in high-value products such as cosmetics and pharmaceutical preparations. In addition, biosurfactants such as glycolipids with anti-irritating effects on the skin have an advantage over chemical surfactants when used in topical preparations and cosmetics (Ramisse et al. 2000).

A study by (Rodríguez-López et al. 2019) compared the irritant effect and antimicrobial activity of two biosurfactants from *Lactobacillus pentosus* and the corn stream. In the same study they found that the biosurfactant from corn stream showed promising results in inhibiting pathogenic bacteria without any irritating effect on the biological membrane. On the other hand, research by (Vecino et al. 2018) found that the biosurfactant produced by *Lactobacillus pentosus* (PEB) exhibited significant antiadhesive and antimicrobial properties against various tested microorganisms that are normal skin microbiota. Additionally, they stated that any minor change in the content of sugar, protein, and/or lipid moieties found in the biosurfactant polymer will lead to great changes in the biosurfactant characteristics and thus in their use in skin care products and cosmetics.

Vecino et al. (2018) investigated the antimicrobial and anti-adhesive properties of cell-bound biosurfactants produced by *Lactobacillus pentosus* (PEB), which are characterized as glycolipid molecules against several micro-organisms found amongst human skin flora. Vecino et al. (2018) investigated the antimicrobial and anti-adhesive properties of cell-bound biosurfactants produced by *Lactobacillus pentosus* (PEB), which are characterized as glycolipid molecules against several micro-organisms found amongst human skin flora. Besides, rhamnolipid biosurfactant showed a potential *in vitro* and *in vivo* inhibitory effect against *Trichophyton rubrum* that causes most dermatophytes diseases (Sen et al. 2019).

Further study proved that, rhamnolipids produced from *Pseudomonas aeruginosa* are characterized by unique structure, large scale production, and had an important role in the cosmetic industries instead of synthetic surfactant (Maier and Soberón-Chávez 2000). Rhamnolipids have been used as an anti-wrinkle agent and they are also added to nail care products, deodorants, toothpaste, and many cosmetics (Irfan-Maqsood and Seddiq-Shams 2014, Varvaresou and Iakovou 2015). Other research found that rhamnolipid absorbed through the bloodstream after their topical application on the skin and this mechanism was used in the treatment of burns by radiation (DeSanto 2011).

6.2 Applications in Wound Healing

Wound healing is an important process involving the repair of damaged skin barriers. There are limited products that stimulate wound healing (Boateng et al. 2015). Lipoprotein biosurfactants are characterized by their antioxidant and moisturizing activity that give them an advantage to be used in wound healing preparations (Sun et al. 2006). The biosurfactant *B. subtilis* SPB1 demonstrated significant epithelial tissue reformation and wound healing effect when applied topically to a wound of the rat model and compared with untreated and CICAFLORA™-treated models (Zouari et al. 2016). Another biosurfactant BS15 isolated from *B. stratosphericus* A15 showed *in vivo* wound closure and *in vitro* antimicrobial and antioxidant activity that protects the wound during healing (Sana et al. 2017). A safe biosurfactant was also extracted from *B. licheniformis* SV1 that enhanced the *in vivo* wound healing process when applied topically in an ointment formula (Gupta et al. 2017).

Sana et al. (2018) found that rhamnolipid biosurfactant produced from *Pseudomonas aeruginosa* C2 was a promising agent for promoting wound closure and as an antibiotic by stimulating collagen production, antimicrobial effect, and re-epithelialized. Furthermore, rhamnolipid coated with silver nanoparticles was recommended to be incorporated in non-woven bioresorbable dressings to enhance and speed wound healing (Khalid et al. 2019).

Sophorolipids is a lipid biosurfactant obtained from yeast *Starmerella bombicola* which upon its combination with wound healing cream could be an added value to inhibit the wound infection during healing but without any healing effect (Lydon et al. 2017). Rhamnolipids and sophorolipids biosurfactants were combined with polymers in wafer dressings show promising treatment for the chronic wound (Akiyode and Boateng 2018). The topical application of an ointment containing di-rhamnolipid has accelerated wound healing and showed beneficial therapeutic applications in the treatment of skin ulcers, chronic wounds, and burns (Tamara Stipcevic et al. 2006).

Despite the advantages of the use of biosurfactants over synthetic surfactants, especially in cosmetic products, the high cost of production of biosurfactants has limited their commercial use. Moutinho et al. (2020) recommended considering the substrate and electricity cost during the production process of biosurfactant as they constituted the highest percentage of production cost to make biosurfactants available for commercial use.

Should also focus on lowering substrate costs as well as the identification of energy-efficient unit operations to lower electricity cost.

6.3 Oral Care

Rhamnolipids from non-pathogenic *Burkholderia thailandensis* was recommended to be added in oral hygiene products as these rhamnolipids showed inhibitory action towards various normal oral microbiota (Elshikh et al. 2017).

Recently, Resende et al. (2019) investigated the effect of a combination of biosurfactant derived from *P. aeruginosa, B. methylotrophicus,* and *C. bombicola,* with fungal chitosan in toothpaste formulations to inhibit *Streptococcus mutans* biofilm which is the most predominant of the oral microflora forming biofilms. In addition, they had tested the toxicity, pH, foaming effect, and antibacterial activity of these formulations in comparison with available toothpaste formulations in the market. Finally, they recommended the use of these non-toxic and foaming formulations in oral hygiene products to inhibit dental biofilm especially those formed by *S. mutans.*

Similarly (Farias et al. 2019), tested the same combination of biosurfactant from the same sources in (Resende et al. 2019) study with chitosan from fungus Mucorales and peppermint essential oil in mouthwash preparations. These formulations were safe and showing potential antimicrobial activity against oral cariogenic microorganisms in comparison to commercial mouthwashes.

Rhamnolipids produced by probiotic bacteria showed a significant endodontic antimicrobial activity that plays an important role in the control of the oral infection by *Streptococcal pulpal* (Amer 2019).

Biosurfactant produced by *Lactobacillus casei* showed antibacterial activity against the oral *Staphylococcus aureus* strain and thus prevent biofilm formation and decreasing oral infections. Additionally, this study also demonstrated the antioxidant and antiproliferative activity of the extracted biosurfactant (Merghni et al. 2017). Stipcevic et al. (2004) discovered in their patent the potential effect of rhamnolipid in the treatment of gum diseases and their benefits in periodontal regeneration.

6.4 *Drug Delivery Systems, Including Vaccine*

The poor oral bioavailability of hydrophobic drugs is a major problem in pharmaceutical industries. Biosurfactants were recommended to be used instead of chemical surfactants in the microemulsion formulation to ensure drug safety and minimize its cytotoxicity and irritating effects on the gut (Rodrigues 2015). The selection of suitable biosurfactants to be used in the microemulsion is a challenging process because of the lack of adequate information about available microbial surfactants (Rodrigues 2015). MEL, rhamnolipids, and sophorolipids were known for their use in microemulsions as a result of their emulsifying, foaming, and detergent activity (Kitamoto et al. 2009, Nguyen et al. 2010). Xie et al. (2005) found that rhamnolipids can significantly be used in the formation of microemulsions with the co-surfactant medium-chain alcohols.

In addition to the use of biosurfactants in microemulsion, they are promising agents that are incorporated in the production of stable nanoparticles targeted for drug delivery (Sengupta et al. 2005). The use of a non-toxic and eco-friendly process in the formation of nanoparticles is one of the important aspects of nanotechnology (Chen et al. 2007). The lipopeptide biosurfactant surfactin from *B. subtilis* was used as a safe and biodegradable stabilizer in the synthesis of silver nanoparticles (Reddy et al. 2009). Another study used rhamnolipid biosurfactant, formed by *Pseudomonas aeruginosa* as an environmentally compatible and simple process in the synthesis of nickel oxide nanoparticles using the microemulsion technique (Palanisamy and Raichur 2009). Furthermore, rhamnolipids and sophorolipids have been used in the stabilization and formulation of metal-bound silver nanoparticles that showed potential antibacterial and antifungal activity (Kasture et al. 2008). In another study, sophorolipids exhibited antibacterial activity against both Gram-positive and -negative bacteria when used in the coating of silver and gold nanoparticles (Singh et al. 2009). Where gold nanoparticles coated with sophorolipids showed significant cyto and geno-compatibility in comparison to the silver ones (Singh et al. 2010). Additionally, glycolipid biosurfactant from *Brevibacterium casei* MSA19 was considered as a "green" stabilizer during the production of uniform silver nanoparticles (Kiran et al. 2010).

Novel liposomes for drug delivery were synthesized using rhamnolipid, where these patented liposomes are non-toxic, eco-friendly, stable and with extended shelf life (Chen et al. 2007). Mannosylerythritol lipids A (MEL-A) biosurfactant exhibits self-assembly property that enables it to increase the gene transfection efficiency through cationic liposomes (Igarashi et al. 2006). In addition, MEL-A-containing liposomes showed promising DNA encapsulation and membrane fusion which play an important role in gene transfection and gene therapy (Ueno et al. 2007a,b).

Biosurfactant extracted from *B. cereus* strain was tested for its cytotoxicity and immunogenicity effect. This study concluded that the extracted lipopeptide biosurfactant was non-toxic and activated the humoral immune response when added as an adjuvant in an inactive AIV-H9N2 vaccine and was compared with oil-based vaccine (Basit et al. 2018). Another research suggested the use of WH1 fungin, a surfactin lipopeptide produced by *B. amyloliquefaciens* as an oral immunoadjuvant. WH1 fungin exhibited strong oral immunogenicity through their protection of the antigenic protein from gastric acidity and proteolytic degradation (Gao et al. 2013).

Smith et al. (2020) demonstrated the benefits of biosurfactants and their application in COVID-19 pandemics as they can be used in cleansing agents, drug delivery, and treatment of acute respiratory distress syndrome (ARDS), immunomodulatory, and vaccine production.

6.5 Others

Surfactin biosurfactant produced by *B. subtilis* was incorporated in the functional pore-forming process by producing a channel in lipid bilayer membranes (Sheppard et al. 1991).

Further clinical application of glycolipid, particularly di-rhamnolipid, was in the treatment of neuropsychiatric disorders and control of neurodegenerative diseases by stimulating the proliferation of neural stem cells (Tamara Stipcevic et al. 2013).

Rhamnolipids were active against many food pathogenic bacteria such as *Listeria monocytogenes* (Magalhães and Nitschke 2013). *B. cereus, Staphylococcus aureus, Escherichia coli,* and *Salmonella enteritidis* which potentiate the use of rhamnolipids in food processing to control foodborne pathogens (de Freitas Ferreira et al. 2019).

References

Ahimou, F., P. Jacques and M. Deleu. 2000. Surfactin and iturin A effects on *Bacillus subtilis* surface hydrophobicity. Enzyme and Microbial Technology 27(10): 749–754.

Akbari, S., N.H. Abdurahman, R.M. Yunus, F. Fayaz and O.R. Alara. 2018. Biosurfactants—a new frontier for social and environmental safety: A mini review. Biotechnology Research and Innovation 2(1): 81–90. https://doi.org/10.1016/j.biori.2018.09.001.

Akiyode, O. and J. Boateng. 2018. Composite biopolymer-based wafer dressings loaded with microbial biosurfactants for potential application in chronic wounds. Polymers 10(8): 918. https://doi.org/10.3390/polym10080918.

Amer Hashim, Z. 2019. Investigating the Potential of Biosurfactants in the Control of Tooth Infections (Doctoral dissertation, Cardiff University).

Arima, K. 1968. Surfactin, acrystalline peptidelipid surfactant produced by *Bacillus subtilis*: isolation, characterization and its inhibition of fibrin clot formation. Biochem. Biophys. Res. Commun. 31: 488–494.

Basit, M., M.H. Rasool, M.F. Hassan, M. Khurshid, N. Aslam, M.F. Tahir, M. Waseem and B. Aslam. 2018. Evaluation of lipopeptide biosurfactants produced from native strains of *Bacillus cereus* as adjuvant in inactivated low pathogenicity avian influenza H9N2 vaccine. International Journal of Agriculture and Biology 20(6): 1419–1423. https://doi.org/10.17957/IJAB/15.0676.

Benincasa, M., A. Abalos, I. Oliveira and A. Manresa. 2004. Chemical structure, surface properties and biological activities of the biosurfactant produced by *Pseudomonas aeruginosa* LBI from soapstock. Antonie Van Leeuwenhoek 85(1): 1–8.

Bernheimer, A.W. and L.S. Avigad. 1970. Nature and properties of a cytolytic agent produced by *Bacillus subtilis*. Microbiology 61(3): 361–369.

Bibel, D.J., R. Aly and H.R. Shinefield. 1992. Inhibition of microbial adherence by sphinganine. Canadian Journal of Microbiology 38(9): 983–985.

Biniarz, P., M. Łukaszewicz and T. Janek. 2017. Screening concepts, characterization and structural analysis of microbial-derived bioactive lipopeptides: a review. Critical Reviews in Biotechnology 37(3): 393–410.

Blanchard, A. and L. Montagnier. 1994. AIDS-associated mycoplasmas. Annual Review of Microbiology 48(1): 687–712.

Boateng, J., R. Burgos-Amador, O. Okeke and H. Pawar. 2015. Composite alginate and gelatin based bio-polymeric wafers containing silver sulfadiazine for wound healing. International Journal of Biological Macromolecules 79: 63–71. https://doi.org/10.1016/j.ijbiomac.2015.04.048.

Bonmatin, J.M., O. Laprévote and F. Peypoux. 2003. Diversity among microbial cyclic lipopeptides: iturins and surfactins. Activity-structure relationships to design new bioactive agents. Combinatorial Chemistry & High Throughput Screening 6(6): 541–556.

Boris, S. and C. Barbés. 2000. Role played by lactobacilli in controlling the population of vaginal pathogens. Microbes and Infection 2(5): 543–546.

Bujak, T., T. Wasilewski and Z. Nizioł-Łukaszewska. 2015. Role of macromolecules in the safety of use of body wash cosmetics. Colloids and Surfaces B: Biointerfaces 135: 497–503. https://doi.org/10.1016/j.colsurfb.2015.07.051.

Cameotra, S.S. and R.S. Makkar. 2004. Recent applications of biosurfactants as biological and immunological molecules. Current Opinion in Microbiology 7(3): 262–266.

Chen, C.Y., S.C. Baker and R.C. Darton. 2007. The application of a high throughput analysis method for the screening of potential biosurfactants from natural sources. Journal of Microbiological Methods 70(3): 503–510. https://doi.org/10.1016/j.mimet.2007.06.006.

Chiewpattanakul, P., S. Phonnok, A. Durand, E. Marie and B.W. Thanomsub. 2010. Bioproduction and anticancer activity of biosurfactant produced by the dematiaceous fungus *Exophiala dermatitidis* SK80. J. Microbiol. Biotechnol. 20(12): 1664–1671.

Chowdhury, M.I.H., T. Munakata, Y. Koyanagi, S. Arai and N. Yamamoto. 1994. Mycoplasma stimulates HIV-1 expression from acutely-and dormantly-infected promonocyte/monoblastoid cell lines. Archives of Virology 139(3-4): 431–438.

Christova, N., S. Lang, V. Wray, K. Kaloyanov, S. Konstantinov and I. Stoineva. 2014. Production, structural elucidation, and *in vitro* antitumor activity of trehalose lipid biosurfactant from *Nocardia farcinica* strain. Journal of Microbiology and Biotechnology 25. https://doi.org/10.4014/jmb.1406.06025.

Chung, Y.R., C.H. Kim, I. Hwang and J. Chun. 2000. Paenibacillus koreensis sp. nov., a new species that produces an iturin-like antifungal compound. International Journal of Systematic and Evolutionary Microbiology 50(4): 1495–1500.

Das, P., S. Mukherjee and R. Sen. 2008. Antimicrobial potential of a lipopeptide biosurfactant derived from a marine *Bacillus circulans*. Journal of Applied Microbiology 104(6): 1675–1684.

de Freitas Ferreira, J., E.A. Vieira and M. Nitschke. 2019. The antibacterial activity of rhamnolipid biosurfactant is pH dependent. Food Research International 116(August 2018): 737–744. https://doi.org/10.1016/j.foodres.2018.09.005.

De, S., S. Malik, A. Ghosh, R. Saha and B. Saha. 2015. A review on natural surfactants. RSC Advances 5(81): 65757–65767. https://doi.org/10.1039/C5RA11101C.

Deleu, M., M. Paquot and T. Nylander. 2008. Effect of fengycin, a lipopeptide produced by *Bacillus subtilis*, on model biomembranes. Biophysical Journal 94(7): 2667–2679.

DeSanto, K. 2011. Rhamnolipid Mechanism.

Drakontis, C.E. and S. Amin. 2020. Biosurfactants: Formulations, properties, and applications. Current Opinion in Colloid and Interface Science 48: 77–90. https://doi.org/10.1016/j.cocis.2020.03.013.

Duarte, C., E.J. Gudiña, C.F. Lima and L.R. Rodrigues. 2014. Effects of biosurfactants on the viability and proliferation of human breast cancer cells. AMB Express 4: 40. https://doi.org/10.1186/s13568-014-0040-0.

Elshikh, M., S. Funston, A. Chebbi, S. Ahmed, R. Marchant and I.M. Banat. 2017. Rhamnolipids from non-pathogenic *Burkholderia thailandensis* E264: Physicochemical characterization, antimicrobial and antibiofilm efficacy against oral hygiene related pathogens. New Biotechnology 36: 26–36. https://doi.org/10.1016/j.nbt.2016.12.009.

Farias, J.M., T.C.M. Stamford, A.H.M. Resende, J.S. Aguiar, R.D. Rufino, J.M. Luna and L.A. Sarubbo. 2019. Mouthwash containing a biosurfactant and chitosan: An eco-sustainable option for the control of cariogenic microorganisms. International Journal of Biological Macromolecules 129: 853–860. https://doi.org/10.1016/j.ijbiomac.2019.02.090.

Fariq, A. and A. Saeed. 2016. Production and biomedical applications of probiotic biosurfactants. Current Microbiology 72(4): 489–495.

Fiore, M., A.E. Maraolo, I. Gentile, G. Borgia, S. Leone, P. Sansone and M.C. Pace. 2017. Nosocomial spontaneous bacterial peritonitis antibiotic treatment in the era of multi-drug resistance pathogens: A systematic review. World Journal of Gastroenterology 23(25): 4654.

Fleckenstein, E., C.C. Uphoff and H.G. Drexler. 1994. Effective treatment of mycoplasma contamination in cell lines with enrofloxacin (Baytril). Leukemia 8(8): 1424–1434.

Fracchia, L., M. Cavallo, M.G. Martinotti and I.M. Banat. 2012. Biosurfactants and bioemulsifiers biomedical and related applications–present status and future potentials. Biomedical Science, Engineering and Technology 14: 326–335.

Franzetti, A., I. Gandolfi, G. Bestetti, T.J. Smyth and I.M. Banat. 2010. Production and applications of trehalose lipid biosurfactants. European Journal of Lipid Science and Technology 112(6): 617–627.

Gao, Z., X. Zhao, S. Lee, J. Li, H. Liao, X. Zhou, J. Wu and G. Qi. 2013. WH1fungin a surfactin cyclic lipopeptide is a novel oral immunoadjuvant. Vaccine 31(26): 2796–2803. https://doi.org/10.1016/j.vaccine.2013.04.028.

Gerard, J., R. Lloyd, T. Barsby, P. Haden, M.T. Kelly and R.J. Andersen. 1997. Massetolides A–H, antimycobacterial cyclic depsipeptides produced by two pseudomonads isolated from marine habitats. Journal of Natural Products 60(3): 223–229.

Gudiña, E.J., V. Rangarajan, R. Sen and L.R. Rodrigues. 2013. Potential therapeutic applications of biosurfactants. Trends in Pharmacological Sciences 34(12): 667–675. https://doi.org/10.1016/j.tips.2013.10.002.

Gupta, S., N. Raghuwanshi, R. Varshney, I.M. Banat, A.K. Srivastava, P.A. Pruthi and V. Pruthi. 2017. Accelerated *in vivo* wound healing evaluation of microbial glycolipid containing ointment as a transdermal substitute. Biomedicine and Pharmacotherapy 94: 1186–1196. https://doi.org/10.1016/j.biopha.2017.08.010.

Heinemann, C., J.E. van Hylckama Vlieg, D.B. Janssen, H.J. Busscher, H.C. van der Mei and G. Reid. 2000. Purification and characterization of a surface-binding protein from *Lactobacillus fermentum* RC-14 that inhibits adhesion of *Enterococcus faecalis* 1131. FEMS Microbiology Letters 190(1): 177–180.

Holmberg, K. 2001. Natural surfactants. Current Opinion in Colloid & Interface Science 6(2): 148–159. https://doi.org/10.1016/S1359-0294(01)00074-7.

Hosono, K. and H. Suzuki. 1983. Acylpeptides, the inhibitors of cyclic adenosine 3', 5'-monophosphate phosphodiesterase. III. Inhibition of cyclic AMP phosphodiesterase. The Journal of Antibiotics 36(6): 679–683.

Hosono, K. and H. Suzuki. 1985. Morphological transformation of Chinese hamster cells by acylpeptides, inhibitors of cAMP phosphodiesterase, produced by *Bacillus subtilis*. Journal of Biological Chemistry 260(20): 11252–11255.

Huang, X., Z. Lu, H. Zhao, X. Bie, F. Lü and S. Yang. 2006. Antiviral activity of antimicrobial lipopeptide from *Bacillus subtilis* fmbj against pseudorabies virus, porcine parvovirus, newcastle disease virus and infectious bursal disease virus *in vitro*. International Journal of Peptide Research and Therapeutics 12(4): 373–377.

Huang, X., J. Suo and Y. Cui. 2011. Optimization of antimicrobial activity of surfactin and polylysine against *Salmonella enteritidis* in milk evaluated by a response surface methodology. Foodborne Pathogens and Disease 8(3): 439–443.

Igarashi, S., Y. Hattori and Y. Maitani. 2006. Biosurfactant MEL-A enhances cellular association and gene transfection by cationic liposome. Journal of Controlled Release 112(3): 362–368. https://doi.org/10.1016/j.jconrel.2006.03.003.

Irfan-Maqsood, M. and M. Seddiq-Shams. 2014. Rhamnolipids: Well-characterized glycolipids with potential broad applicability as biosurfactants. Industrial Biotechnology 10(4): 285–291. https://doi.org/10.1089/ind.2014.0003.

Ishigami, Y., M. Osman, H. Nakahara, Y. Sano, R. Ishiguro and M. Matsumoto. 1995. Significance of β-sheet formation for micellization and surface adsorption of surfactin. Colloids and Surfaces B: Biointerfaces 4(6): 341–348.

Kameda, Y., S. Ouhira, K. Matsui, S. Kanatomo, T. Hase and T. Atsusaka. 1974. Antitumor activity of *Bacillus natto*. V. Isolation and characterization of surfactin in the culture medium of *Bacillus natto* KMD 2311. Chemical and Pharmaceutical Bulletin 22(4): 938–944.

Kasture, M.B., P. Patel, A.A. Prabhune, C.V. Ramana, A.A. Kulkarni and B.L.V. Prasad. 2008. Synthesis of silver nanoparticles by sophorolipids: Effect of temperature and sophorolipid structure on the size of particles. Journal of Chemical Sciences 120(6): 515–520. https://doi.org/10.1007/s12039-008-0080-6.

Khalid, H.F., B. Tehseen, Y. Sarwar S.Z. Hussain, W.S. Khan, Z.A. Raza, S.Z. Bajwa, A.G. Kanaras, I. Hussain and A. Rehman. 2019. Biosurfactant coated silver and iron oxide nanoparticles with enhanced anti-biofilm and anti-adhesive properties. Journal of Hazardous Materials 364(October 2018): 441–448. https://doi.org/10.1016/j.jhazmat.2018.10.049.

Khan, Mohd., B. Singh and S. Cameotra. 2014. Biological Applications of Biosurfactants and Strategies to Potentiate Commercial Production (pp. 269–294). https://doi.org/10.1201/b17599-16.

Kiran, G.S., A. Sabu and J. Selvin. 2010. Synthesis of silver nanoparticles by glycolipid biosurfactant produced from marine *Brevibacterium casei* MSA19. Journal of Biotechnology 148(4): 221–225. https://doi.org/10.1016/j.jbiotec.2010.06.012.

Kitamoto, D., H. Yanagishita, T. Shinbo T. Nakane, C. Kamisawa and T. Nakahara. 1993. Surface active properties and antimicrobial activities of mannosylerythritol lipids as biosurfactants produced by *Candida antarctica*. Journal of Biotechnology 29(1-2): 91–96.

Kitamoto, D., T. Morita, T. Fukuoka, M. Konishi and T. Imura. 2009. Self-assembling properties of glycolipid biosurfactants and their potential applications. Current Opinion in Colloid and Interface Science 14(5): 315–328. https://doi.org/10.1016/j.cocis.2009.05.009.

Landman, D., C. Georgescu, D.A. Martin and J. Quale. 2008. Polymyxins revisited. Clinical Microbiology Reviews 21(3): 449–465.

Lin, S.C., M.A. Minton, M.M. Sharma and G. Georgiou. 1994. Structural and immunological characterization of a biosurfactant produced by *Bacillus licheniformis* JF-2. Applied and Environmental Microbiology 60(1): 31–38.

Lu, J.R., X. Zhao and M. Yaseen. 2007. Biomimetic amphiphiles: biosurfactants. Current Opinion in Colloid & Interface Science 12(2): 60–67.

Lydon, H.L., N. Baccile, B. Callaghan, Marchant, C.A. Mitchell and I.M. Banat. 2017. Adjuvant antibiotic activity of acidic sophorolipids with potential for facilitating wound healing. Antimicrobial Agents and Chemotherapy 61(5). https://doi.org/10.1128/AAC.02547-16.

Magalhães, L. and M. Nitschke. 2013. Antimicrobial activity of rhamnolipids against *Listeria monocytogenes* and their synergistic interaction with nisin. Food Control 29(1): 138–142. https://doi.org/10.1016/j.foodcont.2012.06.009.

Maget-Dana, R. and M. Ptak. 1995. Interactions of surfactin with membrane models. Biophysical Journal 68(5): 1937–1943.

Maier, R.M. and G. Soberón-Chávez. 2000. *Pseudomonas aeruginosa* rhamnolipids: Biosynthesis and potential applications. Applied Microbiology and Biotechnology 54(5): 625–633. https://doi.org/10.1007/s002530000443.

Mani, P., G. Dineshkumar, T. Jayaseelan, K. Deepalakshmi, C.G. Kumar and S.S. Balan. 2016. Antimicrobial activities of a promising glycolipid biosurfactant from a novel marine *Staphylococcus saprophyticus* SBPS 15. 3 Biotech 6(2): 163.

Mazaheri Assadi, M. and M.S. Tabatabaee. 2010. Biosurfactants and their use in upgrading petroleum vacuum distillation residue: a review. International Journal of Environmental Research 4(4): 549–572. https://doi.org/10.22059/ijer.2010.242.

McGarrity, G.J., V. Vanaman and J. Sarama. 1978. Methods of prevention, control and elimination of mycoplasma infection. pp. 213–241. In Mycoplasma Infection of Cell Cultures. Springer, Boston, MA.

Merghni, A., I. Dallel, E. Noumi, Y. Kadmi, H. Hentati, S. Tobji, A. Ben Amor and M. Mastouri. 2017. Antioxidant and antiproliferative potential of biosurfactants isolated from *Lactobacillus casei* and their anti-biofilm effect in oral *Staphylococcus aureus* strains. Microbial Pathogenesis 104: 84–89. https://doi.org/10.1016/j.micpath.2017.01.017.

Mireles, J.R., A. Toguchi and R.M. Harshey. 2001. Salmonella enterica serovar Typhimurium swarming mutants with altered biofilm-forming abilities: surfactin inhibits biofilm formation. Journal of Bacteriology 183(20): 5848–5854.

Mnif, I. and D. Ghribi. 2016. Glycolipid biosurfactants: main properties and potential applications in agriculture and food industry. Journal of the Science of Food and Agriculture 96(13): 4310–4320.

Moutinho, L.F., F.R. Moura, R.C. Silvestre and A.S. Romão-Dumaresq. 2020. Microbial biosurfactants: A broad analysis of properties, applications, biosynthesis, and techno-economical assessment of rhamnolipid production. Biotechnology Progress, October, 1–14. https://doi.org/10.1002/btpr.3093.

Moyne, A.L., R. Shelby, T.E. Cleveland and S. Tuzun. 2001. Bacillomycin D: an iturin with antifungal activity against *Aspergillus flavus*. Journal of Applied Microbiology 90(4): 622–629.

Naruse, N., O. Tenmyo, S. Kobaru, H. Kamei, T. Miyaki, M. Konishi and T. Oki. 1990. Pumilacidin, a complex of new antiviral antibiotics. The Journal of Antibiotics 43(3): 267–280.

Naughton, P.J., R. Marchant, V. Naughton and I.M. Banat. 2019. Microbial biosurfactants: Current trends and applications in agricultural and biomedical industries. Journal of Applied Microbiology 127(1): 12–28. https://doi.org/10.1111/jam.14243.

Neu, T.R., T. Härtner and K. Poralla. 1990. Surface active properties of viscosin: a peptidolipid antibiotic. Applied Microbiology and Biotechnology 32(5): 518–520.

Nguyen, T.T.L., A. Edelen, B. Neighbors and D.A. Sabatini. 2010. Biocompatible lecithin-based microemulsions with rhamnolipid and sophorolipid biosurfactants: Formulation and potential applications. Journal of Colloid and Interface Science 348(2): 498–504. https://doi.org/10.1016/j.jcis.2010.04.053.

Nielsen, T.H., C. Christophersen, U. Anthoni and J. Sørensen. 1999. Viscosinamide, a new cyclic depsipeptide with surfactant and antifungal properties produced by *Pseudomonas fluorescens* DR54. Journal of Applied Microbiology 87(1): 80–90.

Nurfarahin, A.H., M.S. Mohamed and L.Y. Phang. 2018. Culture medium development for microbial-derived surfactants production—an overview. Molecules: A Journal of Synthetic Chemistry and Natural Product Chemistry 23(5). https://doi.org/10.3390/molecules23051049.

Olasanmi, I. and R. Thring. 2018. The role of biosurfactants in the continued drive for environmental sustainability. Sustainability 10(12): 4817. https://doi.org/10.3390/su10124817.

Pacwa-Płociniczak, M., G.A. Płaza, Z. Piotrowska-Seget and S.S. Cameotra. 2011. Environmental applications of biosurfactants: recent advances. International Journal of Molecular Sciences 12(1): 633–654. https://doi.org/10.3390/ijms12010633.

Palanisamy, P. and A.M. Raichur. 2009. Synthesis of spherical NiO nanoparticles through a novel biosurfactant mediated emulsion technique. Materials Science and Engineering C 29(1): 199–204. https://doi.org/10.1016/j.msec.2008.06.008.

Pardhi, D., R. Panchal and K. Rajput. 2018. Biosurfactants: An Overview. Vol. No, 8.

Ramisse, F., C. Van Delden, S. Gidenne, J.D. Cavallo and E. Hernandez. 2000. Decreased virulence of a strain of *Pseudomonas aeruginosa* O12 overexpressing a chromosomal type 1 β-lactamase could be due to reduced expression of cell-to-cell signaling dependent virulence factors. FEMS Immunology and Medical Microbiology 28(3): 241–245. https://doi.org/10.1016/S0928-8244(00)00162-0.

Razin, S. 1978. The mycoplasmas. Microbiological Reviews 42(2): 414.

Reddy, A.S., C.Y. Chen, S.C. Baker, C.C. Chen, J.S. Jean, C.W. Fan, H.R. Chen and J.C. Wang. 2009. Synthesis of silver nanoparticles using surfactin: A biosurfactant as stabilizing agent. Materials Letters 63(15): 1227–1230. https://doi.org/10.1016/j.matlet.2009.02.028.

Reid, G. 2000. *In vitro* testing of *Lactobacillus acidophilus* NCFMTM as a possible probiotic for the urogenital tract. International Dairy Journal 10(5-6): 415–419.

Remichkova, M., D. Galabova, I. Roeva, E. Karpenko, A. Shulga and A.S. Galabov. 2008. Anti-herpesvirus activities of *Pseudomonas* sp. S-17 rhamnolipid and its complex with alginate. Zeitschrift für Naturforschung C 63(1–2): 75–81.

Resende, A.H.M., J.M. Farias, D.D.B. Silva, R.D. Rufino, J.M. Luna, T.C.M. Stamford and L.A. Sarubbo. 2019. Application of biosurfactants and chitosan in toothpaste formulation. Colloids and Surfaces B: Biointerfaces 181(May): 77–84. https://doi.org/10.1016/j.colsurfb.2019.05.032.

Ribeiro, B.G., J.C.M. Guerra and L.A. Sarubbo. 2020a. Potential food application of a biosurfactant produced by *Saccharomyces cerevisiae* URM 6670. Frontiers in Bioengineering and Biotechnology 8: 434. https://doi.org/10.3389/fbioe.2020.00434.

Ribeiro, B.G., J.C.M. Guerra and L.A. Sarubbo. 2020b. Potential Food Application of a Biosurfactant Produced by *Saccharomyces cerevisiae* URM 6670. Frontiers in Bioengineering and Biotechnology 8: 434. https://doi.org/10.3389/fbioe.2020.00434.

Rodrigues, L.R. and J.A. Teixeira. 2010. Biomedical and therapeutic applications of biosurfactants. pp. 75–87. In Biosurfactants. Springer, New York, NY.

Rodrigues, L., I.M. Banat, J. Teixeira and R. Oliveira. 2006. Biosurfactants: Potential applications in medicine. Journal of Antimicrobial Chemotherapy 57(4): 609–618. https://doi.org/10.1093/jac/dkl024.

Rodrigues, Lígia R. and J.A. Teixeira. 2010. Biomedical and therapeutic applications of biosurfactants. pp. 75–87. *In*: Sen, R. (ed.). Biosurfactants (Vol. 672). Springer New York. https://doi.org/10.1007/978-1-4419-5979-9_6.

Rodrigues, R. Ligia. 2015. Microbial surfactants: Fundamentals and applicability in the formulation of nano-sized drug delivery vectors. Journal of Colloid and Interface Science 449: 304–316. https://doi.org/10.1016/j.jcis.2015.01.022.

Rodríguez-López, L., M. Rincón-Fontán, X. Vecino, J.M. Cruz and A.B. Moldes. 2019. Preservative and irritant capacity of biosurfactants from different sources: a comparative study. Journal of Pharmaceutical Sciences 108(7): 2296–2304. https://doi.org/10.1016/j.xphs.2019.02.010.

Rottem, S. 1980. Membrane lipids of mycoplasmas. Biochimica et Biophysica Acta (BBA)-Biomembranes 604(1): 65–90.

Roy, A. 2018. A review on the biosurfactants: properties, types and its applications. Journal of Fundamentals of Renewable Energy and Applications 08(01). https://doi.org/10.4172/2090-4541.1000248.

Saimmai, A., W. Riansa-ngawong, S. Maneerat and P. Dikit. 2020. Application of biosurfactants in the medical field. Walailak Journal of Science and Technology (WJST) 17(2): 154–166.

Saini, H.S., B.E. Barragán-Huerta, A. Lebrón-Paler, J.E. Pemberton, R.R. Vázquez, A.M. Burns and R.M. Maier. 2008. Efficient purification of the biosurfactant viscosin from *Pseudomonas libanensis* strain M9-3 and its physicochemical and biological properties. Journal of Natural Products 71(6): 1011–1015.

Sałek, K. and S.R. Euston. 2019. Sustainable microbial biosurfactants and bioemulsifiers for commercial exploitation. Process Biochemistry 85(July): 143–155. https://doi.org/10.1016/j. procbio.2019.06.027.

Sana, S., A. Mazumder, S. Datta and D. Biswas. 2017. Towards the development of an effective *in vivo* wound healing agent from *Bacillus* sp. Derived biosurfactant using Catla catla fish fat. RSC Advances 7(22): 13668–13677. https://doi.org/10.1039/c6ra26904d.

Sana, S., S. Datta, D. Biswas, B. Auddy, M. Gupta and H. Chattopadhyay. 2018. Excision wound healing activity of a common biosurfactant produced by *Pseudomonas* sp. Wound Medicine 23: 47–52. https://doi.org/10.1016/j.wndm.2018.09.006.

Santos, D., R. Rufino, J. Luna, V. Santos and L. Sarubbo. 2016. Biosurfactants: Multifunctional biomolecules of the 21st century. International Journal of Molecular Sciences 17(3): 401. https:// doi.org/10.3390/ijms17030401.

Schmidt, J. and V. Erfle. 1984. Elimination of mycoplasmas from cell cultures and establishment of mycoplasma-free cell lines. Experimental Cell Research 152(2): 565–570.

Sen, S., S.N. Borah, R. Kandimalla, A. Bora and S. Deka. 2019. Efficacy of a rhamnolipid biosurfactant to inhibit *Trichophyton rubrum in vitro* and in a mice model of dermatophytosis. Experimental Dermatology 28(5): 601–608. https://doi.org/10.1111/exd.13921.

Sengupta, S., D. Eavarone, I. Capila, G. Zhao, N. Watson, T. Kiziltepe and R. Sasisekharan. 2005. Temporal targeting of tumour cells and neovasculature with a nanoscale delivery system. Nature 436(7050): 568–572. https://doi.org/10.1038/nature03794.

Seydlová, G. and J. Svobodová. 2008. Review of surfactin chemical properties and the potential biomedical applications. Open Medicine 3(2): 123–133.

Shah, V., G.F. Doncel, T. Seyoum, K.M. Eaton, I. Zalenskaya, R. Hagver and R. Gross. 2005. Sophorolipids, microbial glycolipids with anti-human immunodeficiency virus and sperm-immobilizing activities. Antimicrobial Agents and Chemotherapy 49(10): 4093–4100.

Shah, V., D. Badia and P. Ratsep. 2007. Sophorolipids having enhanced antibacterial activity. Antimicrobial Agents and Chemotherapy 51(1): 397–400.

Sheppard, J.D., C. Jumarie, D.G. Cooper and R. Laprade. 1991. Ionic channels induced by surfactin in planar lipid bilayer membranes. BBA - Biomembranes 1064(1): 13–23. https://doi.org/10.1016/0005-2736(91)90406-X.

Singh, P. and S.S. Cameotra. 2004. Potential applications of microbial surfactants in biomedical sciences. Trends in Biotechnology 22(3): 142–146.

Singh, S.K., N.D. Desai, G. Chikazawa, H. Tsuneyoshi, J. Vincent, B.M. Zagorski, F. Pen, F. Moussa, G.N. Cohen, G.T. Christakis and S.E. Fremes. 2010. The graft imaging to improve patency (GRIIP) clinical trial results. Journal of Thoracic and Cardiovascular Surgery 139(2): 294–301.e1. https:// doi.org/10.1016/j.jtcvs.2009.09.048.

Singh, S., P. Patel, S. Jaiswal, A.A. Prabhune, C.V. Ramana and B.L.V. Prasad. 2009. A direct method for the preparation of glycolipid-metal nanoparticle conjugates: Sophorolipids as reducing and capping agents for the synthesis of water re-dispersible silver nanoparticles and their antibacterial activity. New Journal of Chemistry 33(3): 646–652. https://doi.org/10.1039/b811829a.

Smith, M.L., S. Gandolfi, P.M. Coshall and P.K.S.M. Rahman. 2020. Biosurfactants: A Covid-19 perspective. Frontiers in Microbiology 11: 1–8. https://doi.org/10.3389/fmicb.2020.01341.

Stanbridge, E.J. and C.J. Doersen. 1978. Some effects that mycoplasmas have upon their infected host. pp. 119–134. In Mycoplasma Infection of Cell Cultures. Springer, Boston, MA.

Stipcevic, T., T. Piljac, J. Piljac, T. Dujmic and T. Piljac. 2004. Use of Rhamnolipids in Wound Healing, Treatment and Prevention of Gum Disease and Periodontal Regeneration.

Stipcevic, T., A. Piljac and G. Piljac. 2006. Enhanced healing of full-thickness burn wounds using di-rhamnolipid. Burns 32(1): 24–34. https://doi.org/10.1016/j.burns.2005.07.004.

Stipcevic, T., C.P. Knight and T.E. Kippin. 2013. Stimulation of adult neural stem cells with a novel glycolipid biosurfactant. Acta Neurologica Belgica 113(4): 501–506. https://doi.org/10.1007/ s13760-013-0232-4.

Sun, L., Z. Lu, X. Bie, F. Lu and S. Yang. 2006. Isolation and characterization of a co-producer of fengycins and surfactins, endophytic *Bacillus amyloliquefaciens* ES-2, from *Scutellaria baicalensis* Georgi.

World Journal of Microbiology and Biotechnology 22(12): 1259–1266. https://doi.org/10.1007/s11274-006-9170-0.

Tanaka, Y., T. Tojo, K. Uchida, J. Uno, Y. Uchida and O. Shida. 1997. Method of producing iturin A and antifungal agent for profound mycosis. Biotechnology Advances 15(1): 234–235.

Thimon, L., F. Peypoux, J. Wallach and G. Michel. 1995. Effect of the lipopeptide antibiotic, iturin A, on morphology and membrane ultrastructure of yeast cells. FEMS Microbiology Letters 128(2): 101–106.

Toraya, T., T. Maoka, H. Tsuji and M. Kobayashi. 1995. Purification and structural determination of an inhibitor of starfish oocyte maturation from a *Bacillus* species. Applied and Environmental Microbiology 61(5): 1799–1804.

Tsukagoshi, N., G. Tamura and K. Arima. 1970. A novel protoplast-bursting factor (surfactin) obtained from *Bacillus subtilis* IAM 1213: II. The interaction of surfactin with bacterial membranes and lipids. Biochimica et Biophysica Acta (BBA)-Biomembranes 196(2): 211–214.

Ueno, Y., N. Hirashima, Y. Inoh, T. Furuno and M. Nakanishi. 2007a. Characterization of biosurfactant-containing liposomes and their efficiency for gene transfection. Biological and Pharmaceutical Bulletin 30(1): 169–72. doi: 10.1248/bpb.30.169. PMID: 17202680.

Ueno, Y., Y. Inoh, T. Furuno, N. Hirashima, D. Kitamoto and M. Nakanishi. 2007b. NBD-conjugated biosurfactant (MEL-A) shows a new pathway for transfection. Journal of Controlled Release: Official Journal of the Controlled Release Society. 2007 Nov; 123(3): 247–253. DOI: 10.1016/j.jconrel.2007.08.012.

Uphoff, C.C., S.M. Gignac and H.G. Drexler. 1992. Mycoplasma contamination in human leukemia cell lines: II. Elimination with various antibiotics. Journal of Immunological Methods 149(1): 55–62.

Varvaresou, A. and K. Iakovou. 2015. Biosurfactants in cosmetics and biopharmaceuticals. Letters in Applied Microbiology 61(3): 214–223. https://doi.org/10.1111/lam.12440.

Van Bogaert, I.N., K. Saerens, C. De Muynck, D. Develter, W. Soetaert and E.J. Vandamme. 2007. Microbial production and application of sophorolipids. Applied Microbiology and Biotechnology 76(1): 23–34.

Vatsa, P., L. Sanchez, C. Clement, F. Baillieul and S. Dorey. 2010. Rhamnolipid biosurfactants as new players in animal and plant defense against microbes. International Journal of Molecular Sciences 11(12): 5095–5108.

Vecino, X., J.M. Cruz, A.B. Moldes and L.R. Rodrigues. 2017. Biosurfactants in cosmetic formulations: Trends and challenges. Critical Reviews in Biotechnology 37(7): 911–923. https://doi.org/10.1080/07388551.2016.1269053.

Vecino, X., L. Rodríguez-López, D. Ferreira, J.M. Cruz, A.B. Moldes and L.R. Rodrigues. 2018. Bioactivity of glycolipopeptide cell-bound biosurfactants against skin pathogens. International Journal of Biological Macromolecules 109: 971–979. https://doi.org/10.1016/j.ijbiomac.2017.11.088.

Velraeds, M.M., H.C. van der Mei, G. Reid and H.J. Busscher. 1996. Physicochemical and biochemical characterization of biosurfactants released by *Lactobacillus* strains. Colloids and Surfaces B: Biointerfaces 8(1-2): 51–61.

Velraeds, M.M., B. Van de Belt-Gritter, H.C. Van der Mei, G. Reid and H.J. Busscher. 1998. Interference in initial adhesion of uropathogenic bacteria and yeasts to silicone rubber by a *Lactobacillus acidophilus* biosurfactant. Journal of Medical Microbiology 47(12): 1081–1085.

Vollenbroich, D., G. Pauli, M. Ozel and J. Vater. 1997a. Antimycoplasma properties and application in cell culture of surfactin, a lipopeptide antibiotic from *Bacillus subtilis*. Applied and Environmental Microbiology 63(1): 44–49.

Vollenbroich, D., M. Özel, J. Vater, R.M. Kamp and G. Pauli. 1997b. Mechanism of inactivation of enveloped viruses by the biosurfactant surfactin from *Bacillus subtilis*. Biologicals 25(3): 289–297.

World Health Organization. 1993. Report of the WHO Meeting on the Development of Vaginal Microbicides for the Prevention of Heterosexual Transmission of HIV. World Health Organization, Geneva, Switzerland.

Wu, Y.-S., S.-C. Ngai, B.-H. Goh, K.-G. Chan, L.-H. Lee and L.H. Chuah. 2017. Anticancer activities of surfactin and potential application of nanotechnology assisted surfactin delivery. Frontiers in Pharmacology, 8. https://doi.org/10.3389/fphar.2017.00761.

Xie, Y.W., Y. Li and R.Q. Ye. 2005. Effect of alcohols on the phase behavior of microemulsions formed by a biosurfactant—Rhamnolipid. Journal of Dispersion Science and Technology 26(4): 455–461. https://doi.org/10.1081/DIS-200054576.

Zouari, R., D. Moalla-Rekik, Z. Sahnoun, T. Rebai, S. Ellouze-Chaabouni and D. Ghribi-Aydi. 2016. Evaluation of dermal wound healing and *in vitro* antioxidant efficiency of *Bacillus subtilis* SPB1 biosurfactant. Biomedicine and Pharmacotherapy 84: 878–891. https://doi.org/10.1016/j. biopha.2016.09.084.

Zuber, P., M.M. Nakano and M.A. Marahiel. 1993. Peptide antibiotics. *Bacillus subtilis* and Other Gram-Positive Bacteria: Biochemistry, Physiology, and Molecular Genetics 897–916.

5

Biosurfactants: Fermentation, Recovery, and Formulation Process and Constraints

Arnoldo Wong-Villarreal[1],* and *Gustavo Yañez-Ocampo*[2],*

1. Introduction

Biosurfactants (BS) are compounds of low molecular weight, amphiphilic molecules, the polar or hydrophilic side is a rhamnose, trehalose, mannose and the non-polar or hydrophobic side is an hydrocarbon chain saturated and/or unsaturated. The BS most known are glycolipids and lipopeptides (Fig. 1) (Banat et al. 2000, Banat et al. 2010, Abdel-Mawgoud et al. 2010, Müller et al. 2012).

The BS are produced mainly by fungi and bacteria, they have the ecological function of enhancing bioavailability and biodegradability of non-soluble organic and inorganic compounds (Chrzanowski et al. 2012). These molecules are emulsifiers, reduce water surface tension from 72 to 25 mN/m (Supaphol et al. 2011). Therefore, can be applied in the bioremediation processes of soil and water that have been polluted with pesticides, heavy metals and hydrocarbons (Banat et al. 2000, Yañez-Ocampo et al. 2009, Yañez Ocampo et al. 2011). Further, BS have applications in alimentary, pharmaceutics and cosmetics (Singh et al. 2007, Sajna et al. 2013).

Another application of biosurfactants is in the agricultural area, as a biocontrol agent for bacteria and fungi that affect crops, acting as a cytolytic agent, mainly cyclic lipopeptides, which cause the destruction of membranes of bacteria, fungi, oomycetes and viruses (D'aes et al. 2010). Thrane et al. 2000, reported that Viscosinamide a cyclic lipopeptide produced by *Pseudomonas fluorescens*, controls

[1] División Agroalimentaria, Universidad Tecnológica de la Selva, Ocosingo, Chiapas, México, C.P. 29950.
[2] Laboratorio de Edafología y Medio Ambiente, Facultad de Ciencias, Universidad Autónoma del Estado de México, México. C.P 50000.
* Corresponding authors: gyanezo@uaemex.mx, wova79@hotmail.com

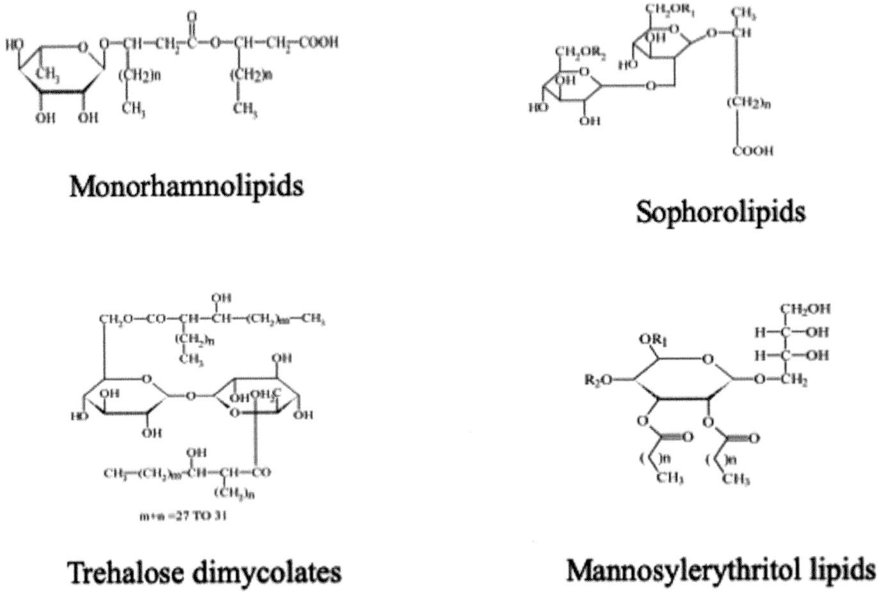

Monorhamnolipids

Sophorolipids

Trehalose dimycolates

Mannosylerythritol lipids

Figure 1. Chemical structures of the best studied microbial biosurfactants (Banat et al. 2010).

the phytopathogens *Pythium ultimum* and *Rhizoctonia solani* in the rhizosphere of sugar beet. While Nielsen et al. 2006, reported the effect of the biosurfactants ramonolipids and saponins for the biocontrol of *Phytophthora capsici* that causes root rot. Other phytopathogens that are controlled by biosurfactants are *Aspergillus*, *Cercospora*, *Colletotrichum* and *Fusarium* (Texeira et al. 2020). One of the last applications that has been reported of biosurfactants is the activity in larvae of *Aedes aegypti*, a vector that causes the transmission of diseases such as dengue, Zika, and Chikungunya, which are public health problems worldwide (Texeira et al. 2020). Due to the diversity of applications of biosurfactants, the production through different substrates is very important, as well as the recovery processes of these compounds.

2. Biotechnological Route for Microbial Biosurfactants Production

2.1 Upstream Processes

The isolation of microbial strains producers of BS from polluted or natural environments is the first step. The microbial samples collected from soil, water or sediment, are processed in the laboratory by several microbiological techniques using culture media. The methods for evidencing BS presence are shown in Fig. 2.

Biosurfactants are grouped as indirect and direct methods of screening and isolation for the microbial strains biosurfactant producers, they also report main fermentation conditions. According to Cassidy and Hudak (2001) and Kitamoto et al. (2002), in order to ensure the screening of microbial strains BS producers, assays, which are described below must be carried out.

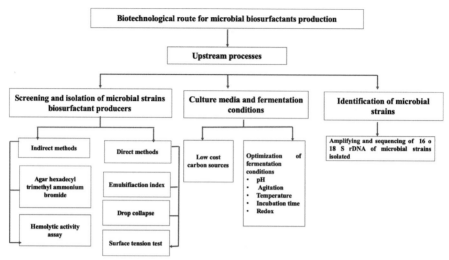

Figure 2. Flow diagram of general upstream processes considered for researching of microbial strains BS producers.

2.2 Screening and Isolation of Microbial Strains Biosurfactant Producers

Agar Hexadecyl Trimethy Lammonium Bromide (CTAB) Assay. This assay is used as an indirect method for anionic biosurfactant (rhamnolipids) detection. The culture medium is based on mineral salts agar, with a low soluble carbon source (*v. gr.* glycerol, vegetable oil, hydrocarbon), blue methylene and CTAB. The cationic salt CTAB reacts with rhamnolipid and blue methylene, this complex is visualized by the presence of a translucid halo around bacterial colony growth on the agar plates (Chandankere et al. 2013, Youssef et al. 2004).

2.2.1 Hemolytic Activity Assay

In this assay, biosurfactants presence are evidenced indirectly by streaking bacterial strain on a blood agar plate (5% v/v), halos around bacterial colonies indicates BS presence with hemolytic activity (Hassanshahian 2014). However, not all biosurfactants have hemolytic activity, for this reason it is necessary to confirm BS production by other tests (Youssef et al. 2004, Zhang et al. 2012).

2.2.2 Emulsification Index

Due to BS being extracellular molecules, in this assay, it is necessary to centrifuge the culture broth with biomass to obtain a supernatant cell-free with the extract crude of BS. The supernatant cell-free is mixed 1:1 v/v with different hydrophobic phases such as motor oil, corn oil, n-hexadecane or kerosene (Dubey et al. 2012). After homogenizing for two minutes, the mixture is left for 24 h in order to see a stable emulsion layer (Ayed et al. 2015). The emulsification index is calculated by dividing the emulsion layer height by the total height of the mixture and multiplying by 100% (Hassanshahian 2014, Luna et al. 2015). A representative image for this test is presented, in which positive controls using synthetic surfactants and distilled water as control negative have been considered. All glass material must be washed twice and rinsed with distilled water in order to prevent a positive false result caused by detergent residues.

2.2.3 Drop Collapse Test

In this qualitative assay, the supernatant cell-free with the extract crude of BS is added in form of a drop to a set of hydrophobic phases (v. gr. mineral oil, soja oil, motor oil). The positive result for this test is that the oil drop collapsed due to presence of BS in the supernatant cell-free. It suggests to run several positive controls, using synthetic surfactants as tween 80 (Tugrul and Cansuna 2005, Abdel-Mawgoud et al. 2011).

2.2.4 Oil Displacement Test

In this assay 50 mL of distilled water are placed in a Petri dish, 100 μL of oil and 30 μL of supernatant cell-free with the extract crude of BS (Affandi et al. 2014, Hassanshahian 2014). The diameter of the oil displaced after addition of extract crude of BS can be measured. This test indirectly provides information about tensoactive activity.

2.2.5 Surface Tension Test

The best way to quantitatively measure BS in supernatant cell-free is by a tensiometer with a Du Noüy ring or Wilhelmy plate (Xiao et al. 2013, Ayed et al. 2015). The biosurfactants breaks surface tension (ST) of aqueous solutions, it is well known that distilled water has a ST 68 mN/m, however by the addition of BS, the ST is 30 mN/m (Abdel-Mawgoud et al. 2011). According to Burgos-Díaz et al. (2013) this assay is highly sensitive therefore must be carried out carefully, all glass material must be washed and rinsed twice in order to ensure deleting detergent residues. Also they recommend running out the assay using synthetic surfactants as positive controls to compare with biosurfactants.

A parameter very important in the BS study is the Critical Micellar Concentration (CMC), defined as the concentration of biosurfactant that allows micelles formation. A good BS must have low CMC value, in order to perform a maximum surface activity with the minimum quantity of BS (Mao et al. 2015). In Table 1 CMC values for biosurfactants ordered from the minimum to maximum CMC are reported.

2.3 Culture Media and Fermentation Conditions

The selection of low-cost carbon sources is the key step for producing BS to an industrial scale because it represents approximately 60% of total production cost (Makkar et al. 2011, Reis et al. 2013).

Table 2 reports several carbon sources used for BS production, such as soja oil, corn oil used, glycerol, lacto serum, coffee waste, and hydrophobic compounds (n-decane, n-tetradecane, paraffin), the yield on BS produced vary for each one carbon sources and growth rate of the microbial strain (Abbasi et al. 2012, Abbasi et al. 2013).

Culture conditions in fermentation liquid medium must be optimized, by studying broad pH, agitation speed, temperature and incubation time. In Fig. 3 presents a scheme with methodological procedures key, in order to validate the stability of tensoactive properties of microbial biosurfactants, by response surface test.

Table 1. CMC and surface tension values from microbial biosurfactants.

Microorganism	Biosurfactant	CMC (g/L)	Surface tension (mN/m)	Reference
Bacillus subtilis	Surfactin	0.017	27.2	Sen and Swaminathan 2005
Pseudomonas fluorescens BD5	Lipopeptide	0.072	31.5	Janek et al. 2010
Pseudomonas aeruginosa SP4	Mono rhamnolipid	0.120	28.3	Pornsunthorntawee et al. 2008
Rhodococcus spp. MTCC 2574	N/I	0.120	30.8	Mutalik et al. 2008
Sphingobacterium sp.	Phospholipid/ Lipopeptide	6.3	22	Burgos-Díaz et al. 2011
Streptococcus thermophilus	Glycolipid	20	36	Rodrigues et al. 2006

N/I = not identified, CMC = Critical Micellar Concentration

Table 2. Carbon sources and BS yield for specific types of microbial biosurfactants (Abbasi et al. 2012, Abbasi et al. 2013).

Biosurfactant	Microorganisms	BS yield (g l^{-1})	Source carbon
Mannosyl erythritol lipids			
MEL-A, -B and -C	*Candida antarcitica* T-34	140	n-Octadecane
		47	Soybean oil,
MEL-SY16	*C. Antarctica* KCTC 7804	41	glycerol and oleic acid
RL-1 y-2	*P. aeruginosa* DSM 7107	112	Soybean oil
RL-1 y-2	*P. aeruginosa* UI 29791	46	Corn oil
RL-A y-B	*P. aeruginosa* BOP 101	14	n-Paraffin
Lipid trehalose			
Trehalose-tetraester	*R. erythropolis* DSM 43215	32	*n*-Decane
Lipid sophorose			
LS mixture	*C. bombicola* ATCC 22214	422	Lactoserum
LS mixture (SL-1: 73%)	*C. bombicola* ATCC 22214	160	Canola oil

N/A = data not available

By this test it is possible to combine several abiotic parameters settled and assess their effect on surface tension of BS as response variable.

2.3.1 *Molecular Identification*

The microorganism's candidate for BS synthesis must have the following characteristics (Konishi et al. 2015).

a) Non-pathogenic strain

b) Culturable at *in vitro* conditions

c) No loss of tensoactive properties (emulsifying, foaming, dispersant, decrease of surface tension) at different environmental conditions.

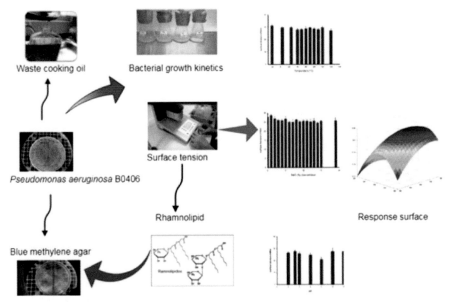

Figure 3. Schematic representation of procedures to assess stability of tensoactive properties of microbial BS.

2.4 Downstream Processes

The BS are extracellular products, therefore techniques used to separate and recover them depend of their biochemical nature (Sen and Swaminathan 2005). The procedures more widely used are extraction with cheap solvents with low environmental impact, acid precipitation, precipitation with ammonium sulfate, crystallization, centrifugation, or the use of columns of exchange ionic. In Fig. 4 represents the main methodological strategies for BS recovery considered in the downstream processes.

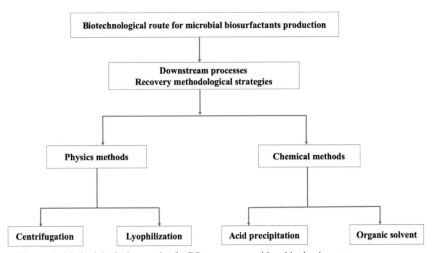

Figure 4. Methodological strategies for BS recovery considered in the downstream processes.

2.5 Recovery Methodological Strategies

Centrifugation. Through centrifuge force, it is possible to separate the biomass from the culture medium to obtain an extract crude of BS cell-free supernatant. Centrifugation speed values settled down in order to separate the biomass from the culture broth and obtain the supernatant with the total biosurfactants (Table 3).

Table 3. Speed centrifugation to extract biosurfactants from biomass in the culture medium of fermentation.

Speed G	Centrifugation time (min)	Temperature (°C)	Type of biosurfactant	Reference
5000	20	4	Lipopeptide	Rufino et al. 2014
7,656	10	N/A	Trehalolipid	Bajaj et al. 2014
8,000	15	4	Lipopeptide and Glycolipids	Burgos-Díaz et al. 2011
8,643	20	N/A	Lipopeptide	Sarafin et al. 2014
9,690	30	4	Rhamnolipids	Aparna et al. 2012
10,000	15	4	Rhamnolipids	Abbasi et al. 2012
10,800	20	4	Ochrosin	Kumar et al. 2014
10,956	10	N/A	Surfactin	Liu et al. 2015
11,952	20	4	Lipopeptide	Xia et al. 2014
11,952	10	4	Rhamnolipids	Bharali et al. 2013
11,952	15	4	Rhamnolipids	Zou et al. 2014
14,000	5	N/A	Rhamnolipids	Costa et al. 2011
20,217	20	4	Glycolipopeptid	Jain et al. 2013

2.5.1 Acid Precipitation

This methodological step is used for recovering mixture of biosurfactants. The cell-free supernatant is acidified with 2 N HCl (pH 2.0) and kept overnight at 4°C. The precipitate obtained is recovered by centrifugation, the pellet collected is washed with acidic water (pH 2.0 with HCl) after which it is washed with alkaline water (pH 11.0 with NaOH) to achieve a final pH 7.0, finally the precipitate is dried (Salleh et al. 2011). The Table 4 reports extraction of biosurfactants by acid precipitation and organic solvent extraction.

2.5.2 Organic Solvent

This technique is combined with acid precipitation to obtain higher yields of extraction. The cell-free supernatant is acidified in a separator funnel with HCl to pH 2.0, equal volume of solvent is added, both liquids are agitated and then allowed to stand until two separate phases are observed, the organic phase is recovered and placed in a rotary evaporator until an extract of brown colour is observe. The organic solvent extraction after acid precipitation is the most applied method to recover approximately 90% of BS (Gusmão et al. 2010, Salleh et al. 2011). The solvents

Table 4. Extraction methods to recover biosurfactants by acid precipitation and organic solvents.

Acid precipitation conditions	Solvent used	Microorganism producer of biosurfactant	Yield (g/L)	Reference
H_2SO_4 1 M, pH 2.0	Chloroform/ Methanol (2:1v/v)	*Pseudomonas* sp. 2B	4.97	Aparna et al. 2012
HCl 6 M, pH 2.0 Overnight at 4°C	Methanol	*Bacillus subtilis* BS-37	0.585	Liu et al. 2015
HCl 2 N, pH 2	Ethyl acetat	*Ochrobactrum* sp. BS-206 (MTCC 5720)	0.28	Kumar et al. 20
HCl 2 N, pH 2.0 Overnight at 4°C	Ethyl acetate	*Rhodococcus* sp. IITR03	N/A	Bajaj et al. 2014
HCl 6 M, pH 2.0 Overnight at 4 °C	Methanol	*Candida sphaerica* UCP0995	9	Luna et al. 2015
HCl 6 N, pH 2.	Chloroform/ Methanol (65:15 v/v)	*Kocuria marina* BS-15	0.00197	Sarafin et al. 2014
HCl 6 N, pH 2.0 Overnight at 4°C	Ethyl acetate	*Pseudomonas aeruginosa* MA01	12	Abbasi et al. 2012
HCl 6 N, pH 2.0 Overnight at 4°C	Ethyl acetate	*Pseudomonas aeruginosa* OBP1	N/A	Bharali et al. 2013
HCl 6 N, pH 2.0 Overnight at 4°C	Chloroform/ methanol (2:1 v/v)	*Pseudomonas* sp.	5.2	Silva et al. 2014
N/A	Chloroform	*Candida lipolytica* UCP 0988	8	Rufino et al. 2014
N/A	Ethyl acetate/ methanol (8:1 v/v)	*Sphingobacterium detergens*	0.466	Burgos-Díaz et al. 2013

N/A = data not available

used in the extraction can be recovered and reused, this represents an eco-friendly laboratory practice. Rufino et al. (2014) reported BS recovering from the culture of *Candida lipolytica* UCP 0988, using the mixture of chloroform/culture broth (1:1 v/v), with a yield of 8.0 g/L of tensoactive. It has also reported BS extraction with a mixture of ethyl acetate/methanol (8:1 v/v) and chloroform/methanol (2:1) v/v recovering 10 g/L BS crude (Burgos-Díaz et al. 2011, Affandi et al. 2014).

2.5.3 Lyophilization

The cell-free supernatant collected by centrifugation, is precipitated to pH 2.0 with chlorhydric acid overnight at 4°C. The pellet precipitated is recovered newly by centrifugation and it is resuspended in distilled alkaline water (pH 8.0) and then lyophilized to obtain a brown powder. Through this method, Al-Bahry et al. (2013) recovered 2.29 g/L of BS from a culture of *Bacillus subtilis* B20, also Xiao et al. (2013) in their research produced BS with *Klebsiella* sp., 10.1, 5.1, 3.25, 3.1, 2.75 and 2.62 g/L of BS produced with starch, sucrose, xylose, galactose, glucose and fructose respectively were recovered.

3. Constraints

According to the above considerations, this chapter shows that the main constraints around the fermentation stage for BS production, is the selection of a low-cost carbon source that are preferably non-soluble organic wastes, a microbial strain well adapted to *in vitro* conditions, with no lost of surface activity properties and culture conditions optimization for scale up industrially. For BS recovery main constraints are minimal, because BS are extracellular, they are possible to recover with simple and ecofriendly procedures (precipitation and extraction). However, for pharmaceutic, medic, cosmetic or alimentary applications, the BS recovery needs more procedures in order to control purity. Finally, BS formulation are commercialized as powder, or in liquid medium, in the first case half-life time is more prolonged than in liquid medium.

4. Conclusion

In this chapter, an analysis is made of the upstream processes that allow the identification and characterization of bacterial strains that produce biosurfactants, and carbon and nitrogen sources for the production of biosurfactants. The downstream processes for the recovery of biosurfactants are also addressed, where the selection depends on the biochemical characteristics of these molecules. The set of these techniques will make it possible to propose more economical biotechnological processes for the production of biosurfactants, using agro-industrial waste as raw material and low-cost recovery methods with little environmental impact. This will allow biosurfactants to be more widely used in applications in the environmental, food, agricultural, cosmetic and pharmaceutical fields.

References

Abbasi, H., M.M. Hamedia, T.B. Lotfabad, H.S. Zahirib, H. Sharafi, F. Masoomi, A.A. Moosavi-Movahedi, A. Ortiz, M. Amanlou and K.A. Noghabi. 2012. Biosurfactant-producing bacterium *Pseudomonas aeruginosa* MA01 isolated from spoiled apples: Physicochemical and structural characteristics of isolated biosurfactant. Biosci. Bioeng. 113: 211–219. http://dx.doi.org/10.1016/j.jbiosc.2011.10.002.

Abbasi, H., K.A. Noghabi, M.M. Hamedia, H.S. Zahirib, A.A. Moosavi-Movahedi, M. Amanlou, J.A. Teruele and A. Ortiz. 2013. Physicochemical characterization of a monorhamnolipid secreted by *Pseudomonas aeruginosa* MA01 in aqueous media. An experimental and molecular dynamics study. Colloids SurfB. Biointerfaces 101: 256–265. http://dx.doi.org/10.1016/j.colsurfb.2012.06.035.

Abdel-Mawgoud, A.M., F. Lépine and E. Déziel. 2010. Rhamnolipids: diversity of structures, microbial origins and roles. Appl. Microbiol. Biotechnol. 86: 1323–1336. http://dx.doi.org/10.1007/s00253-010-2498-2.

Abdel-Mawgoud, A.M., R. Hausmann, F. Lépine, M. Muller and E. Déziel. 2011. Rhamnolipids: detection, analysis, biosynthesis, genetic regulation, and bioengineering of production. pp. 13–56. *In*: Soberón-Chávez, G. (eds.). Microbiology Monographs 20. Springer, Heidelberg, Berlin. http://dx.doi.org/10.1007/978-3-642-14490-5_2.

Affandi, I.E., N.H. Suratman, S. Abdullah, W. Ahmad and Z.A. Zakaria. 2014. Degradation of oil and grease from high-strength industrial effluents using locally isolated aerobic biosurfactant-producing bacteria. Int. Biodeterior. Biodegradation 95: 33–40. http://dx.doi.org/10.1016/j.ibiod.2014.04.009.

Al-Bahry, S.N., Y.M. Al-Wahaibi, A.E. Elshafie, A.S. Al-Bemani, S.J. Joshi, H.S. Almakhmari and H.S. Al-Sulaimani. 2013. Biosurfactant production by *Bacillus subtilis* B20 using date molasses and its possible application in enhanced oil recovery. Int. Biodeterior. Biodegradation 81: 141–146. http://dx.doi.org/10.1016/j.ibiod.2012.01.006.

Aparna, A., G. Srinikethan and H. Smith. 2012. Production and characterization of biosurfactant produced by a novel *Pseudomonas* sp. 2B. Colloids SurfB. Biointerfaces 95: 23–29. http://dx.doi.org/10.1016/j.colsurfb.2012.01.043.

Arutchelvi, J., C.H. Joseph and M. Doble. 2011. Process optimization for the production of rhamnolipid and formation of biofilm by *Pseudomonas aeruginosa* CPCL on polypropylene. Biochem. Eng. J. 56: 37–45. http://dx.doi.org/10.1016/j.bej.2011.05.004.

Ayed, H.B., N. Hmideta, M. Béchet, M. Chollet, G. Chataigné, V. Leclére, P. Jacques and M. Nasria. 2014. Identification and biochemical characteristics of lipopeptides from *Bacillus mojavensis* A21. Process Biochem. 48: 1699–1707. http://dx.doi.org/10.1016/j.procbio.2014.07.001.

Ayed, H.B., N. Jemil, H. Maalej, A. Bayoudh, N. Hmider and M. Nasri. 2015. Enhancement of solubilization and biodegradation of diesel oil by biosurfactant from *Bacillus amyloliquefaciens* An6. Int. Biodeterior. Biodegradation 98: 8–14. http://dx.doi.org/10.1016/j.ibiod.2014.12.009.

Bajaj, A., S. Mayilrah, M.K.R. Mudiam, D.K. Patel and N. Manickam. 2014. Isolation and functional analysis of a glycolipid producing *Rhodocous* sp. strain IITR03 with potential for degradation of 1,1,1-tricloro-2,2-bis(4-chlorophenyl)ethane (DDT). Bioresour. Technol. 167: 398–406. http://dx.doi.org/10.1016/j.biortech.2014.06.007.

Banat, I., R. Makkar and S. Cameotra. 2000. Potential commercial applications of microbial surfactants. Appl. Microbiol. Biotechnol. 53: 495–508. http://dx.doi.org/10.1007/s002530051648.

Banat, I.A., A. Franzetti, I. Gandolfi, G. Bestetti, M.G. Martinotti, L. Fracchia, T.J. Smyth and R. Marchant. 2010. Microbial biosurfactants production, applications and future potential. Appl. Microbiol. Biotechnol. 87: 427–444. https://doi.org/10.1007/s00253-010-2589-0.

Bharali, P., J.P. Saikia, A. Ray and B.K. Konwar. 2013. Rhamnolipid (RL) from *Pseudomonas aeruginosa* OBP1: A novel chemotaxis and antibacterial agent. Colloids SurfB. Biointerfaces 103: 502–509. http://dx.doi.org/10.1016/j.colsurfb.2012.10.064.

Burgos-Díaz, C., R. Pons, M.J. Espuny, F.J. Aranda, J.A. Teruel, A. Manresa, A. Ortiz and A.A. Marqués. 2011. Isolation and partial characterization of a biosurfactant mixture produced by *Sphingobacterium* sp. isolated from soil. J. Colloid Interface Sci. 361: 195–204. http://dx.doi.org/10.1016/j.jcis.2011.05.054.

Burgos-Díaz, C., R. Pons, M.J. Espuny, F.J. Aranda, J.A. Teruel, A. Manresa, A. Ortiz and A.A. Marqués. 2013. The production and physicochemical properties of a biosurfactant mixture obtained from *Sphingobacterium* detergents. J. Colloid Interface Sci. 394: 368–379. http://dx.doi.org/10.1016/j.jcis.2012.12.017.

Cassidy, D.P. and A.J. Hudak. 2001. Microorganism selection and biosurfactant production in a continuously and periodically operated bioslurry reactor. J. Hazard. Mater. 84: 253–264. http://dx.doi.org/10.1016/S0304-3894(01)00242-4.

Chadwick, D.T., K.P. McDonnell, L.P. Brennan, C.C. Fagan and C.D. Everard. 2014. Evaluation of infrared techniques for the assessment of biomass and biofuel quality parameters and conversion technology processes: A review. Renew. Sust. Energ. Rev. 30: 672–81. http://dx.doi.org/10.1016/j.rser.2013.11.006.

Chandankere, R., J. Yao, M.M.F. Choi, K. Masakorala and Y. Chan. 2013. An efficient biosurfactant-producing and crude-oil emulsifying bacterium *Bacillus methylotrophicus* USTBa isolated from petroleum reservoir. Biochem. Eng. J. 74: 46–53. http://dx.doi.org/10.1016/j.bej.2013.02.018.

Chrzanowski, Ł., M. Dzdiadas, Ł. Ławniczar, P. Cyplik, W. Bialas, A. Szulc, P. Lisiecki and H. Jelen. 2012. Biodegradation of rhamnolipids in liquid cultures: Effect of biosurfactant dissipation on diesel fuel/B20 blend biodegradation efficiency and bacterial community composition. Bioresour. Technol. 111: 328–335. http://dx.doi.org/10.1016/j.biortech.2012.01.181.

Compaoré, C.S., D.S. Nielsen, L.I.I. Ouoba, T.S. Berner, K.F. Nielsen, H. Sawadogo-Linganni, B. Diawara, G.A. Ouédraogo, M. Jakobsen and L. Thorsen. 2013. Co-production of surfactin and a nonvel bacteriocin by *Bacillus subtilis* subsp. *subtilis* H4 isolated from Bikalga, an African alkaline Hibiscus sabariffa seedn fermented condiment. Int. J. Food Microbiol. 162: 297–307. http://dx.doi.org/10.1016/j.ijfoodmicro.2013.01.013.

Cortés-Sánchez, A.J., H. Hernández-Sánchez and M.E. Jaramillo-Flores. 2013. Biological activity of glycolipids produced by microorganisms: New trends and possible therapeutic alternatives. Microbiol. Res. 168: 22–32. http://dx.doi.org/10.1016/j.micres.2012.07.002.

Costa, S.G.V.A.O., E. Déziel and F. Léine. 2011. Characterization of rhamnolipid production by *Burkholderia glumae*. Appl. Microbiol. Biotechnol. 53: 620–627. http://dx.doi.org/10.1111/j.1472-765X.2011.03154.x.

D'aes, J., K. De Maeyer, E. Pauwelyn and M. Höfte. 2010. Biosurfactants in plant-*Pseudomonas* interactions and their importance to biocontrol. Environ. Microbiol. Rep. 2(3): 359–72. doi: 10.1111/j.1758–2229.2009.00104.x.

de Andrade Teixeira Fernandes, N., A.C. de Souza, L.A. Simões, G.M. Ferreira Dos Reis, K.T. Souza, R.F. Schwan and D.R. Dias. 2020. Eco-friendly biosurfactant from *Wickerhamomyces anomalus* CCMA 0358 as larvicidal and antimicrobial. Microbiol. Res. 2020. doi: 10.1016/j.micres.2020.126571.

Dubey, K.V., P.N. Chard, S.U. Meshram, L.P. Shendre, V.S. Dubey and A.A. Juwarkar. 2012. Surface-active potential of biosurfactants produced in curd whey by *Pseudomonas aeruginosa* strain-PP2 and *Kocuria turfanesis* strain-J at extreme environmental conditions. Bioresour. Technol. 126: 368–374. http://dx.doi.org/10.1016/j.biortech.2012.05.024.

Gudiña, E.J., A.I. Rodrigues, E. Alves, M.R. Domingues, J.A. Teixeira and L.R. Rodrigues. 2015. Bioconversion of agro-industrial by-products in rhamnolipids toward applications in enhanced oil recovery and bioremediation. Bioresour. Technol. 177: 87–93. http://dx.doi.org/10.1016/j.biortech.2014.11.069.

Gusmão, C.A.B., R.D. Rufino and L.A. Sarubbo. 2010. Laboratory production and characterization of a new biosurfactant from *Candida glabrata* UCP1002 cultivated in vegetable fat waste applied to the removal of hydrophobic contaminant. World J. Microbiol. Biotechnol. 26: 1683–1692. http://dx.doi.org/10.1007/s11274-010-0346-2.

Hassanshahian, M. 2014. Isolation and characterization of biosurfactant producing bacteria from Persian Gulf (Bushehr provenance). Mar. Pollut. Bull. 86: 361–366. http://dx.doi.org/10.1016/j.marpolbul.2014.06.043.

Hoskova, M., O. Schreiberova, R. Jezdik, J. Chudoba, J. Mask, K. Sigler and T. Rezanka. 2013. Characterization of rhamnolipids produced by non-pathogenic *Acinetobacter* and *Enterobacter* bacteria. Bioresour. Technol. 130: 510–516. http://dx.doi.org/10.1016/j.biortech.2012.12.085.

Ibrahim, M.L., U.J.J. Ijah, S.B. Manga, L.S. Bilbis and S. Umar. 2013. Production and partial characterization of biosurfactant produced by crude oil degrading bacteria International. Int. Biodeterior. Biodegradation 81: 28–34. http://dx.doi.org/10.1016/j.ibiod.2012.11.012.

Ismail, W., I.S. Al-Rowaihi, A.A. Al-Humam, R.Y. Hamza, A.M. El-Nayal and M. Bououdina. 2013. Characterization of a lipopeptide biosurfactant produced by a crude-oil-emulsifying *Bacillus* sp. I-15. Int. Biodeterior. Biodegradation 84: 168–178. http://dx.doi.org/10.1016/j.ibiod.2012.04.017.

Jain, R.M., K. Mody, A. Mishra and B. Jha. 2012. Physicochemical characterization of biosurfactant and its potential to remove oil from soil and cotton cloth. Carbohydr. Polym. 89: 1110–1116. http://dx.doi.org/10.1016/j.carbpol.2012.03.077.

Jain, R.M., K. Mody, N. Joshi, A. Mishra and B. Jha. 2013. Effect of unconventional carbon sources on biosurfactant production and its application in biorremediation International. Int. J. Biol. Macromol. 62: 52–58. http://dx.doi.org/10.1016/j.ijbiomac.2013.08.030.

Jain, R.M., K. Mody, N. Joshi, A. Mishra and B. Jha. 2013. Production and structural characterization of biosurfactant produced by an alkaphilic bacterium, *Klebsiella* sp.: Evaluation of different carbon sources. Colloids SurfB. Biointerfaces 108: 199–204. http://dx.doi.org/10.1016/j.colsurfb.2013.03.002.

Janek, T., M. Łukaszewicz, T. Rezanca and A. Krasowska. 2010. Isolation and characterization of two new lipopeptide biosurfactants produced by *Pseudomonas fluorescens* BD5 isolated from water from the Artic Archipielago of Svalbard. Bioresour. Technol. 101: 118–123. http://dx.doi.org/10.1016/j.biortech.2010.02.109.

Janek, T., M. Łukaszewicz and A. Krasowska. 2013. Identification and characterization of biosurfactants produced by the Arctic bacterium *Pseudomonas putida* BD2. Colloids SurfB. Biointerfaces 110: 379–386. http://dx.doi.org/10.1016/j.colsurfb.2013.05.008.

Jara, A.M.A.T., R.F.S. Andrade and G.M. Campos-Takaki. 2013. Physicochemical characterization of tensio-active produced by *Geobacillus stearothermophilus* isolated from petroleum-contaminated soil. Colloids SurfB. Biointerfaces 201: 315–318. http://dx.doi.org/10.1016/j.colsurfb.2012.05.021.

Kitamoto, D., H. Isoda and T. Nakahara. 2002. Functions and potential applications of glycolipid biosurfactants from energy saving materials to gene delivery carriers. J. Biosci. Bioeng. 94: 187–201. http://dx.doi.org/10.1016/S1389-1723(02)80149-9.

Konishi, M., Y. Yoshida and J. Horiuchi. 2015. Efficient production of sophorolipids by *Starmerella bombicola* using a corncob hydrolysate médium. J. Biosci. Bioeng. 119: 317–322. http://dx.doi.org/10.1016/j.jbiosc.2014.08.007.

Kryachko, Y., S. Nathoo, P. Lai, J. Voordouw, E.J. Prenner and G. Voordouw. 2013. Prospects for using native and recombinant rhamnolipid producers for microbially enhanced oil recovery. Int. Biodeterior. Biodegradation 81: 133–140. http://dx.doi.org/10.1016/j.ibiod.2012.09.012.

Kumar, C.G., P. Sujithaa, S.K. Mamidyalab, P. Usharanic, B. Dasd and C.R. Reddy. 2014. Ochrosin, a new biosurfactant produced by halophilic *Ochrobactrum* sp. strain BS-206 (MTCC 5720): Purification, characterization and its biological evaluation. Process Biochem. 49: 1708–1717. http://dx.doi.org/10.1016/j.procbio.2014.07.004.

Kuyukina, M.S., I.B. Ivshina, T.N. Kamenskikh, M.V. Bulicheva and G.I. Stukova. 2013. Survival of cryogel-immobilized *Rhodococcus* strains in crude oil-contaminated soil and their impact on biodegradation efficiency. Int. Biodeterior. Biodegradation 84: 118–25. http://dx.doi.org/10.1016/j.ibiod.2012.05.035.

Lee, S., S. Lee, S.H. Kim, I. Park, Y.S. Lee, S.Y. Chung and Y.L. Choi. 2008. Characterization of new biosurfactant produced by *Klebsiella* sp. Y6-1 isolated from waste soybean oil. Bioresour. Technol. 99: 2288–2292. http://doi:10.1016/j.biortech.2007.05.020.

Li, X., X. Dai, J. Takahashi, N. Li, J. Jin, L. Dai and B. Dong. 2014. New insight into chemical changes of dissolved organic matter during anaerobic digestion of dewatered sewage sludge using EEM-PARAFAC and two-dimensional FTIR correlation spectroscopy. Bioresour. Technol. 159: 412–420. http://doi:10.1016/j.biortech.2014.02.085.

Liu, Q., J. Lin, W. Wang, H. Huanga and S. Li. 2015. Production of surfactin isoforms by *Bacillus subtilis* BS-37 and its applicability to enhanced oil recovery under laboratory conditions. Biochem. Eng. J. 93: 31–37. http://doi:10.1016/j.bej.2014.08.023.

Luna, J.M., R.D. Rufino, L.A. Sarubbo and G.M. Campos-Takaki. 2015. Characterization, surface properties and biological activity of a biosurfactant produced from industrial waste by *Candida sphaerica* UCP0995 for application in the petroleum industry. Colloids SurfB. Biointerfaces 102: 202–209. http://doi:10.1016/j.colsurfb.2012.08.008.

Madsen, M., J.B. Holm-Nielsen and K.H. Esbensen. 2011. Monitoring of anaerobic digestion processes: A review perspective. Renew. Sust. Energ. Rev. 15: 3141–3155. http://doi:10.1016/j.rser.2011.04.026.

Makkar, R.S., S.S. Cameotra and I.M. Banat. 2011. Advances in utilization of renewable substrates for biosurfactant production. AMB Express. 1: 1–19. doi: 10.1186/2191-0855-1-5.

Mao, X., R. Jiang, W. Xiao and J. Yu. 2015. Use of biosurfactant for the remediation of contaminated soils: A review. J. Hazard. Mater. 285: 419–435. http://doi:10.1016/j.jhazmat.2014.12.009.

Müller, M.M., J.H. Kügler, M. Henkel, M. Gerlitzki, B. Hörmanna, M. Pöhnleina, CH. Syldatka and R. Hausmann. 2012. Rhamnolipids—Next generation surfactants? J. Biotechnol. 162: 366–380. http://doi:10.1016/j.jbiotec.2012.05.022.

Mutalik, S.R., B.K. Vaidya, R.M. Joshi, K.M. Desai and S. Nene. 2008. Use of response optimization for the production of biosurfactant from *Rhodococus* spp. MTCC 2574. Bioresour. Technol. 99: 875–880. http://doi:10.1016/j.biortech.2008.02.027.

Nalini, S. and R. Parthasarathi. 2014. Production and characterization of rhamnolipids produced by *Serratia rubidaea* SNAU02 under solid state fermentation and its application as biocontrol agent. Bioresour. Technol. 173: 231–238. http://doi:10.1016/j.biortech.2014.09.051.

Nielsen, C.J., D.M. Ferrin and M.E. Stanghellini. 2006. Efficacy of biosurfactants in the management of *Phytophthora capsici* on pepper in recirculating hydroponic systems. Can. J. Plant Pathol. 28(3): 450–460. doi: 10.1080/07060660609507319.

Pantazaki, A.A., CH.P. Papaneophytou and D.A. Lambropoulou. 2011. Simultaneous polyhydroxyalkanoates and rhamnolipids production by *Thermus thermophilus* HB8. AMB Express. 1: 1–13. doi: 10.1186/2191-0855-1-17.

Pedetta, A., K. Pouyte, S.M.K. Herrera, P.A. Babay, M. Espinosa, M. Costagliola, C.A. Studdert and S.R. Peressutti. 2013. Phenanthrene degradation and strategies to improve its bioavailability to microorganisms isolated from brackish sediments. Int. Biodeter. Biodegradation 84: 161–167. http://doi:10.1016/j.ibiod.2012.04.018.

Pornsunthorntawee, O., N. Arttaweeporn, S. Paisanjit, P. Somboonthanate, M. Abe, R. Rujiravanit and S. Chavadej. 2008. Isolation and comparison of biosurfactant produced by *Bacillus subtilis* PT2 and *Pseudomonas aeroginosa* SP4 for microbial surfactant-enhanced oil recovery. Biochem. Eng. J. 42: 172–179. http://doi:10.1016/j.bej.2008.06.016.

Reis, R.S., G.J. Pacheco, A.G. Pereira and G. Freire. 2013. Biosurfactants: Production and applications. pp. 31–61. *In*: Chamy, R. and F. Rosenkranz (eds.). Biodegradation—Life of Science. IntechOpen, London, United, Kingdom. http://dx.doi.org/10.5772/56144.

Ribeiro, I.A., M.R. Bronze, M.F. Castro and M.H.L. Ribeiro. 2012. Optimization and correlation of HPLC-ELSD and HPLC-MS/MS methods for identification and characterization of sophorolipids. ChromatogrB. Analyt. Technol. Biomed. Life Sci. 899: 72–80. doi: 10.1016/j.jchromb.2012.04.037.

Rikalović, M.G., G. Gojgić-Cvijović, M.M. Vrvić and I. Karadžić. 2012. Production and characterization of rhamnolipids from *Pseudomonas aeruginosa* san-ai. J. Serb. Chem. Soc. 77: 27–42. http://doi:10.2298/JSC110211156R.

Rodrigues, L.R., J.A. Teixeira, H.C. Van Der Mei and R. Oliveira. 2006. Isolation and partial characterization of a biosurfactant produced by *Streptococus thermophilus* A. Colloids SurfB. Biointerfaces 53: 105–112. http://doi:10.1016/j.colsurfb.2006.08.009.

Rufino, R.D., J. Moura de Luna, G.M. Takaki and L.A. Sarubbo. 2014. Characterization and properties of the biosurfactant produced by *Candida lipolytica* UCP 0988. Electron. J. Biotechnol. 17: 34–38. http://doi:10.1016/j.ejbt.2013.12.006.

Sajna, K.V., R.K. Sukumaran, H. Jayamurthy, K.K. Reddy, S. Kanjilal, R.B.N. Prasad and A. Pandey. 2013. Studies on biosurfactants from *Pseudozyma* sp. NII 08165 and their potential application as laundry detergent additives. Biochem. Eng. J. 78: 85–92. http://doi:10.1016/j.bej.2012.12.014.

Salleh, S.M., N.A.M. Noh and A.R.M. Yahya. 2011. Comparative study: Different recovery techniques of rhamnolipid produced by *Pseudomonas aeruginosa* USMAR-2. Int. Conf. Biotechnol. Environ. Manag. IPCBEE 18: 432–562.

Sarafin, Y., M.B.S. Donio, S. Velmurugan, M. Michaelbabu and T. Citarasu. 2014. *Kocuria marina* BS-15 a biosurfactant producing halophilic bacteria isolated from solar salt works in India. Saudi J. Biol. Sci. 21: 511–519. http://doi:10.1016/j.sjbs.2014.01.001.

Sen, R. and T. Swaminathan. 2005. Characterization of concentration and purification parameters and operating conditions for the small-scale recovery of surfactin. Process Biochem. 40: 2953–2958. http://doi:10.1016/j.procbio.2005.01.014.

Sharma, D., B.S. Saharan, N. Chauhan, S. Procha and S. Lal. 2015. Isolation and functional characterization of novel biosurfactant produced by *Enterococcus faecium*. SpringerPlus 4: 1–14. doi: 10.1186/2193-1801-4-4.

Silva, E., N.M.P. Rocha, E. Silva, R.D. Rufino, J.M. Luna, R.O. Silva and L.A. Sarubbo. 2014. Characterization of a biosurfactant produced by *Pseudomonas cepacia* CCT6659 in the presence of industrial wastes and its application in the biodegradation of hydrophobic compounds in soil. Collois SurfB. Biointerfaces 117: 36–41. http://doi:10.1016/j.colsurfb.2014.02.012.

Singh, A., J. Van Hamme and O. Ward. 2007. Surfactants in microbiology and biotechnology: Part 2. Application aspects. Biotechnol. Adv. 25: 99–121. http://doi:10.1016/j.biotechadv.2006.10.004.

Smyth, T., A. Perfumo, R. Marchant and I. Banat. 2010. Isolation and analysis of low molecular weight microbial glycolipids. pp. 3705–3723. *In*: Timmis, K.N. (ed.). Handbook of Hydrocarbon and Lipid Microbiology. Springer, Berlin.

Soberón-Chávez, G., F. Lépine and E. Déziel. 2005. Production of rhamnolipids by *Pseudomonas aeruginosa*. Appl. Microbiol. Biotechnol. 68: 718–725. http://doi:10.1007/s00253-005-0150-3.

Souza, E.C., T.CH. Vessoni-Penna and R. PInheiro De Souza Oliveira. 2014. Biosurfactant-enhanced hydrocarbon bioremediation: An overview. Int. Biodeterior. Biodegradation 89: 88–94. http://dx.doi.org/10.1016/j.ibiod.2014.01.007.

Supaphol, S., S.N. Jenkins, P. Intomo, I.S. Waite and A.G. O'donnell. 2011. Microbial community dynamics in mesophilic anaerobic codigestion of mixed waste. Bioresour. Technol. 102: 4021–4027. http://doi:10.1016/j.biortech.2010.11.124.

Swaathy, S., V. Kavitha, A.S. Pravin, A.B. Mandal and A. Gnanamani. 2014. Microbial surfactant mediated degradation of anthracene in aqueous phase by marine *Bacillus licheniformis* MTCC 5514. Biotechnol. Rep. 4: 161–170. http://doi:10.1016/j.btre.2014.10.004.

Thavasi, R., V. Subramanyam, N. Jayalakshmi, T. Balasubramanian and I. Banat. 2011. Biosurfactant production by *Pseudomonas aeruginosa* from renewable resources. Indian J. Microbiol. 51: 30–36. http://doi:10.1007/s12088-011-0076-7.

Thrane, C., T.H., Nielsen, M.N. Nielsen, J. Sorensen and S. Olsson. 2000. Viscosinamide-producing *Pseudomonas fluorescens* DR54 exerts a biocontrol effect on *Pythium ultimum* in sugar beet rhizosphere. FEMS Microbiol. Ecol. 33: 139–146. doi: 10.1111/j.1574-6941.2000.tb00736.x.

Tugrul, T. and E. Cansunar. 2005. Detecting surfactant-producing microorganisms by the drop-collapse test. World J. Microbiol. Biotechnol. 21: 851–853. http://doi:10.1007/s11274-004-5958-y.

Vecino, X., R. Devesa-Rey, J.M. Cruz and A.B. Moldes. 2013. Evaluation of biosurfactant obtained from *Lactobacillus pentosus* as foaming agent in froth flotation. J. Environ. Manage. 128: 655–660. http://doi:10.1016/j.jenvman.2013.06.011.

Xia, W., Z. Du, Q. Cui, H. Dong, F. Wang, P. He and Y. Tang. 2014. Biosurfactant produced by novel *Pseudomonas* sp. WJ6 with biodegradation of n-alkanes and polycyclic aromatic hydrocarbons. J. Hazard. Mater. 276: 489–498. http://doi:10.1016/j.jhazmat.2014.05.062.

Xiao, M., Z. Zhang, J. Wang, G. Zhang, Y. Luo, Z. Song and J. Zhang. 2013. Bacterial community diversity in a low-permeability oil reservoir and its potential for enhancing oil recovery. Bioresour. Technol. 147: 110–116. http://doi:10.1016/j.biortech.2013.08.031.

Yañez-Ocampo, G., E. Sanchez-Salinas, G., Jimenez-Tobon, M. Penninckx and M.L. Ortíz-Hernández. 2009. Removal of two organophosphate pesticides by a bacterial consortium immobilized in alginate or tezontle. J. Hazard. Mat. 168: 1554–1561. http://doi:10.1016/j.jhazmat.2009.03.047.

Yañez-Ocampo, G., E. Sanchez-Salinas and M.L. Ortíz-Hernandez. 2011. Removal of methyl parathion and tetrachlorvinphos by a bacterial consortium immobilized on tezontle-packed up-flow reactor. Biodeg. 22: 1203–1213. http://10.1007/s10532-011-9475-z.

Youssef, N.H., K.E. Duncan, D.P. Nagle, K.N. Savage, R.M. Knapp and M.J. Mcinerney. 2004. Comparison of methods to detect biosurfactant production by diverse microorganisms. J. Microbiol. Methods. 56: 339–347. http://doi:10.1016/j.mimet.2003.11.001.

Zhang, X., D. Xu, C. Zhu, T. Lundaa and K.E. Scherr. 2012. Isolation and identification of biosurfactant producing and crude oil degrading *Pseudomonas aeruginosa* strains. Chem. Eng. J. 209: 138–146. http://doi:10.1016/j.cej.2012.07.110.

Zou, C.H., M. Wang, Y. Xing, G. Lan, T. Ge, X. Yan and T. Gu. 2014. Characterization and optimization of biosurfactants produced by *Acinetobacter baylvi* ZJ2 isolated from crude oil-contaminated soil sample toward microbial enhanced oil recovery applications. Biochem. Eng. J. 90: 49–58. http://doi:10.1016/j.bej.2014.05.007.

6

Biosurfactant from the Marine Microorganisms
Potentials and Future Prospects

Nisha S Nayak,[1] *Mamta S Purohit,*[1] *Ranjan R Pradhan,*[2]
Devayani R Tipre[1,*] and *Shailesh R Dave*[3]

1. Introduction

The marine environment holds the extensive majority of the earth's surface and is an abundant source of dynamic and active compounds. It is possible to derive a vast and enormous supply of natural compounds from the marine ecosystem (Kim et al. 2013). However, due to the enormity of the marine ecosystem, most of the marine microbial world remains unexplored. Only < 0.1% of marine isolates have been explored. The universal property of marine microorganisms involves its low nutritional requirements and resistance to high salt concentrations with the capacity to stabilize the osmotic pressure of the environment (Ventosa et al. 1998). They are able to thrive in severe environments through specific metabolic and physiological capacities and thus develop new metabolites that cannot be found elsewhere (De Carvalho and Fernandes 2010, Satpute et al. 2010). Many novel extracellular and intracellular by-products like antibiotics, enzymes, biopolymers, pigments, toxins and microbial biosurfactants (BS) are well known derivatives from marine microbes. Their structural and functional diversity and industrial applications, are of great significance (Banat 1995a, Banat 1995b, Banat et al. 1991, Rodrigues et al. 2006).

Biosurfactants are amphiphilic surface active compounds which have both hydrophilic and hydrophobic moieties as shown in Fig. 1 (Fracchia et al. 2012).

---
[1] Department of Microbiology and Biotechnology, School of Sciences, Gujarat University, Ahmedabad 380009. India.
[2] School of Engineering, University of Guelph, Guelph, Ontario, Canada N1G 2W1.
[3] Xavier's Research Foundation, St. Xavier's College Campus, Navrangpura, Ahmedabad 380009. India.
* Corresponding author: devayanitipre@hotmail.com

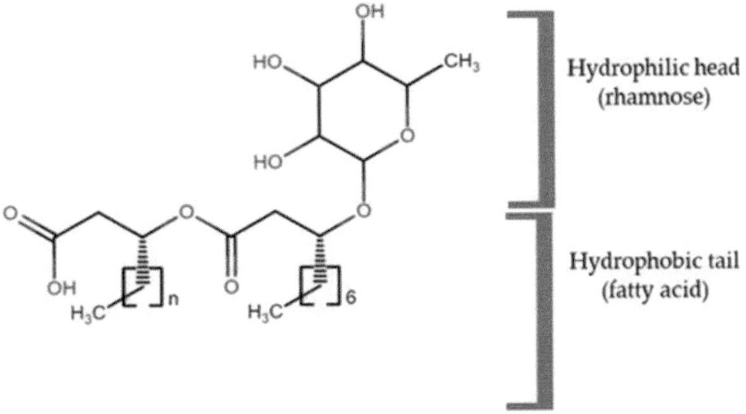

Figure 1. Structure of rhamnolipid biosurfactant molecule.

Broadly, these biomolecules can be classified into two groups—low and high molecular weight biosurfactants. The low molecular weight biosurfactants (molecular weight 1–2 kDa) are generally glycolipids or lipopeptides and are potential interfacial and surface tension reducers. The high molecular weight biosurfactants (molecular weight > 1 MDa) involves amphipathic polysaccharides, lipoproteins, proteins and lipopolysaccharides which are good stabilizers of oil-in-water emulsions. The amphipathic nature of biosurfactants exhibits a wide range of surface activities, which allows for their application in different fields related to foaming, detergent, emulsification, solubilisation, wetting and dispersion of hydrophobic compounds (Marchant and Banat 2012, Banat et al. 2000).

More than 2000 different biosurfactant structures are reported which include chemically different families of compounds, but also groups of congeners, that is, minor structural variations with structurally closely related compounds (Hausmann and Syldatk 2014). This structural diversity of biosurfactants indicates a large variety of biological and physicochemical properties. Biosurfactants have a wide range of properties which includes low critical micelle concentrations (CMC), strong surface tension reduction, metal ion complexation, prominent bioactivities, and low eco-toxicity. A low CMC indicates higher function at much lower concentrations of biosurfactants than many chemical surfactants (Bhadoriya and Madoriya 2013). In addition to their higher biodegradability and effectiveness at extreme temperatures, salinity and pH values, many biosurfactants have been documented to have comparable or better performance than synthetic surfactants. These properties make them a green substitute in various applications to their chemical counterparts, including agriculture, cosmetics, food, and petroleum industries, as well as bioremediation (Bhadoriya and Madoriya 2013, Poremba et al. 1991). To isolate BS-producing microbes, combination of various screening methodologies have been studied (Maneerat and Phetrong 2007, Satpute et al. 2008). Marine microbial species of genera like *Pseudomonas, Enterobacter, Azotobacter, Halomonas, Acinetobacter, Arthrobacter, Bacillus, Rhodococcus, Corynebacterium, Lactobacillus* and yeast have been reported to produce biosurfactants (Schulz et al. 1991, Passeri et al. 1992, Abraham et al. 1998, Maneerat et al. 2006, Thavasi et al. 2007, Das et al. 2008b,

Thavasi et al. 2009, Mukherjee et al. 2009, Perfumo et al. 2010a, Thavasi et al. 2011a). In addition, some biological functions of biosurfactants such as antibacterial, antifungal, antiviral or anti-tumor activities make them plausible alternatives to traditional medicinal agents in many biomedical applications (Gudiña et al. 2013, Fracchia et al. 2012).

2. Marine Microbial Biosurfactants

Microorganisms reported for biosurfactant production till date are mostly from terrestrial sources. The marine environment, holding three-fourth part of the earth's surface, is a robust reservoir of diverse microorganisms including biosurfactant producers. Few reports on biosurfactants from marine microbes are described later within the text. Most of the biosurfactants of marine origin have been evaluated for their environmental remediation application potentials, but their therapeutic potentials have not been exploited thoroughly. Therefore the scientific community is constantly searching for marine compounds with therapeutic applications by considering the great diversity of structures obtained from this source.

In addition, surface active metabolites producing microorganisms are ubiquitous and occupy both water (fresh water, ground water and sea water) and land (sediment, mud and sludge) in extreme conditions (e.g., hypersaline locations, oil reservoirs) and expand actively in a wide variety of temperatures, pH and salinity values (Margesin and Schinner 2001, Olivera et al. 2003, Floodgate 1978), however, this fact may not be always true. Few reports state that microbes produce biosurfactant on water-soluble substrates (Gunther et al. 2005, Turkovskaya et al. 2001). It has been proposed that the surface active molecules on the microbial cell surface increases the hydrophobicity of the cell which helps it to survive in hydrophobic environments (Abraham et al. 1998, Perfumo et al. 2010b). Microbes produce biosurfactant either extracellularly into the environment or it is localized on surfaces, i.e., to become associated with the cell membrane. When biosurfactants are associated with the cell, the organism itself behaves as a biosurfactants in controlling the adherence property to water-insoluble substrates (Maneerat and Dikit 2007).

2.1 Classification of Marine Biosurfactants

Biosurfactants produced by different microorganisms manifests a wide range of diverse chemical structures. The amphiphilic character of biosurfactants is usually formed by both hydrophobic and hydrophilic components. The hydrophobic part is constituted of saturated or unsaturated fatty acids, hydroxy fatty acids, or fatty alcohols with a chain length between 8 and 18 carbon atoms. The hydrophilic components are made up of small hydroxyl, phosphate or carboxyl groups or by carbohydrate (such as mono-, oligo-, or polysaccharides) or (poly-) peptide moieties. Biosurfactants are mainly anionic and non-ionic compounds. Most biosurfactants are anionic or neutral in nature, whereas those that contain amine groups are cationic in nature. The molar mass of biosurfactants generally ranges from 500 to 1500 Da (Bognolo 1999). Biosurfactants are generally categorized by their microbial origin and chemical composition as shown in Fig. 2 (Banat et al. 2010, Vijayakumar and Saravanan 2015, Rahman and Gakpe 2008).

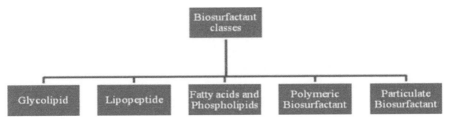

Figure 2. Classes of biosurfactant.

2.1.1 *Glycolipids*

Glycolipids are biosurfactant which have carbohydrates in combination with long chain aliphatic acid or hydroxyaliphatic acid. Rhamnolipids, sophorolipids and trehalolipids are the known glycolipids. Pathogenic strains of *Pseudomonas aeruginosa* are found to be the producers of rhamnolipid biosurfactant. It is a combination of α-L-rhamnopyranosyl-α-L-rhamnopyranosyl-β-hydroxydecanoyl-β-hydroxydecanoate (Rha-Rha-C_{10}-C_{10}) and α-L-rhamnopyranosyl-α-L-rhamnopyranosyl-β-hydroxydecanoate (Rha-Rha-C_{10}) along with their mono-rhamnolipid analogue (Rha-C_{10}-C_{10} and Rha-C_{10}) (Abdel-Mawgoud et al. 2010). A marine strain, *Plantococcus maritinus* SAMP MCC 3013 showed the therapeutic potential of glycolipid biosurfactant produced by it (Waghmode et al. 2020). Different concentrations of rhamnolipid of species of *Pseudomonas* and other bacterial strains have been discovered by sensitive analytical techniques (Maneerat and Phetrong 2007). For example, rhamnolipid producer *Pseudomonas aeruginosa* ZS1 is isolated from petroleum sludge. A majority reports on rhamnolipids focus mainly on evaluating the biodegradation potential of petroleum hydrocarbons (Passeri et al. 1992, Abraham et al. 1998). Different classes of biosurfactants produced by marine microbes are enlisted in Table 1.

2.1.2 *Sophorolipids*

Sophorolipids are biosurfactants composed of a hydroxylated fatty acid and the glucose disachharide sophorose. They are produced by yeasts that belong to the genus *Candida* (Thavasi et al. 2011c). *Candida bombicola* and *Starmerella bombicola* produce biosurfactant that offer a green alternative for traditional surfactants. Mannosylerythriol lipids are yeast glycolipids which have the most efficient biosurfactant properties and are produced by *Pseudozyma* sp. (Faria et al. 2014) from vegetable oils, sugars, glycerol and hydrocarbons as a substrate.

Trehalose is mainly produced by several microorganisms belonging to *Mycobacterium, Nocardia, Rhodococcus, Arthrobacter, Gordonia* and *Corynebacterium* (Christova and Stoineva 2014, Kügler et al. 2015). Trehalose lipids are made up of disaccharide trehalose, which is acylated with long-chained, which can be a straight or a branched 3-hydroxy fatty acids named mycolic acids.

2.1.3 *Fatty Acids and Phospholipids*

Fatty acids and phospholipids derivatives may act as surface-active substances. Strong reduction of surface and interfacial tensions were, for instance, observed for branched fatty acids known as corynomycolic acids with chain lengths of C_{12}–C_{14} (Fujii et al.

Table 1. Biosurfactant/Bioemulsifiers produced by marine microbes.

Biosurfactant type	Source organism	Compound	Reference
Fatty acids	*Cobetia* sp. strain MM1IDA2H-1	3-Hydroxy fatty acids	Ibacache-Quiroga et al. 2013
	Serratia rubidaea	Rubiwettin R1	Matsuyama et al. 1990
Lipoamino acids	*Alcanivorax dieselolei*	Proline lipid	Qiao et al. 2010
	Brevibacterium luteolum	Proline lipid	Unás et al. 2018
Glycolipids	*Pseudomonas aeruginosa*	Rhamnolipid	Cheng et al. 2017
	Pseudomonas aeruginosa	Rhamnolipid	Chakraborty and Das 2016
	Pseudomonas aeruginosa	Rhamnolipid	Du et al. 2019
	Alcanivorax borkumensis	Glucose lipid	Yakimov et al. 1998
	Paracoccus	Sophorolipid	Antoniou et al. 2015
	Arthrobacter sp. EK 1	Trehalose lipid tetraester	Passeri et al. 1991
	Rhodococcus sp. strain PML026	Trehalose lipid	White et al. 2013
	Cyberlindnera saturnus	Cybersan (galactose lipid)	Balan et al. 2019
Lipopeptides	*Brevibacterium luteolum*	Thr-Pro-Pro-Leu/Ile-Leu/Ile-Ala-Phe	Unás et al. 2018
	Bacillus pumilus	Pumilacidin	Saggese et al. 2018
	Aneurinibacillus aneurinilyticus	Aneurinifactin	Balan et al. 2017
	Bacillus sp. WSM1	Iturin W	Zhou et al. 2020
	Bacillus amyloliquefaciens SH-B74	Plipastatin A1	Ma et al. 2018
	Tistrella mobilis	Didemnin B	Xu et al. 2012
	Staphylococcus lentus	Threose diester	Hamza et al. 2017
	Pontibacter korlensis	Pontifactin	Balan et al. 2016

1999, Käppeli and Finnerty 1980, Kretschmer et al. 1982). Various bacteria and yeasts produce huge amounts of fatty acids and phospholipids surfactants. Phosphotidyl ethanolamine rich vesicles are produced by *Acinetobacter* sp., which forms optically clear microemulsions of alkane in water (Santos et al. 2016). These biosurfactants are useful in medical applications. Linear 3-hydroxy fatty acids were produced by *Cobetia* sp. (Ibacache-Quiroga et al. 2013); *Serratia rubidaea* excreted hydroxy fatty acid dimers designated as rubiwettin R1 (Matsuyama et al. 1990), which is just like the marine bacterium described in 1944 as *Serratia marinorubra*. Furthermore, *Alcanivorax dieselolei* and the *Brevibacterium luteolum* produced identical proline lipids with long chain fatty acids (Qiao and Shao 2010, Unás et al. 2018).

2.1.4 Lipopeptides

Lipopeptides are a very well-known group of biosurfactants derived from amino acids. Several species of marine microorganisms have been reported for the production these types of biosurfactant which include the species of *Bacillus*, *Lactobacillus*, *Streptomyces*, *Pseudomonas* and *Myroides*. For instance, two marine strains MK90e85 and MK91CC8 which were identified as *Pseudomonas* sp. produced cyclic depsipeptides and viscosin (Das et al. 2010). Many lipopeptides shows reductions in surface tension and significant bioactivities (Nayak et al. 2020b, Baltz et al. 2005, Inès and Dhouha 2015). *Streptomyces roseosporus* produces antibiotics daptomycin and *Paenibacillus polymyxa* produces polymyxin B which shows prominent bioactivity (Tally and DeBruin 2000, Trimble et al. 2016). Surfactin produced by different species of the genus *Bacillus* is known to be the most efficient biosurfactants among all. It reduces the air/water surface tension from 72.5 mN/m down to 27 mN/m (Yeh et al. 2005). Iturin W produced by *Bacillus* sp. WSM1 showed potential antifungal activity (Zhou et al. 2020). A pontifactin lipopeptide was described as an antibiotic which was produced by the *Pontibacter korlensis*, isolated from petroleum-contaminated seawater (Balan et al. 2016).

2.1.5 Polymeric Biosurfactants

Emulsan, lipomanan, alasan and liposan are the best-studied polymeric biosurfactants. They are made up of polysaccharides, proteins, lipopolysaccharides, lipoproteins, or complex mixtures of these compounds that are referred to as lipoheteropolysaccharides. *Acinetobacter calcoaceticus* RAG-1, an isolate from the Mediterranean Sea produces the prototypical high molecular weight bioemulsifier emulsan (Rosenberg and Ron 1997, Sar and Rosenberg 1983). It has a heteropolysaccharide backbone with a repeating tri-saccharide unit. This repeating unit is probably made up of N-acetyl-d-galactosamine, N-acetylgalactosamine uronic acid and a di-amino-6-deoxy-d-glucose with fatty acids covalently linked to the polysaccharide through ester linkages (Zuckerberg et al. 1979).

2.1.6 Particulate Biosurfactants

Particulate biosurfactants partition the extracellular membrane vesicles to form a micro-emulsion that exerts an influence on the alkane uptake in microbial cells. The *Acinetobacter* sp. has vesicles having 20 to 50 nm in diameter and 1.158 g^3/cm of a buoyant density, which are a combination of proteins, phospholipids and lipopolysaccharides (Vijayakumar and Saravanan 2015).

3. Biosynthesis Mechanism of Biosurfactant Production

The insufficient knowledge of fundamental biosynthetic pathways concurs with lack of molecular structures leading to the production of many marine biosurfactants. Most of the reports that explored known pathways are of non-marine strains. Microorganisms need water soluble substrates to grow and produce hydrophilic moiety of biosurfactant, while hydrophobic moieties are produced from hydrophobic substrates like oils and fats (Desai and Banat 1997, Weber et al. 1992). Numbers of metabolic pathways are involved in production of precursors for biosurfactant production. One of the major factors is the carbon source that can be found in the culture medium which is illustrated in Fig. 3.

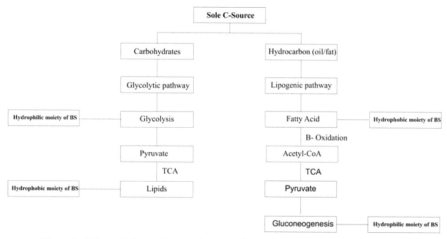

Figure 3. Pathway followed by microbes according to the carbon source in the medium.

Glycolysis

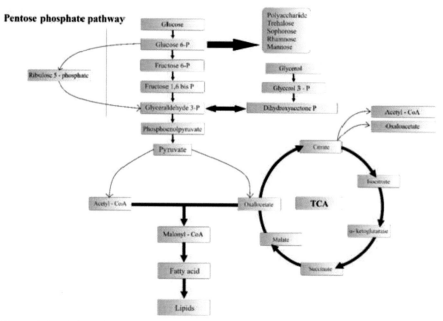

Figure 4. Metabolic pathways involved in the synthesis of biosurfactant using water soluble substrates (Korla and Mitra 2014, Giraud and Naismith 2000).

For instance, carbon flux will be regulated by two major pathways namely, (i) Glycolytic pathway (hydrophilic moiety generation), (ii) Lipogenic pathway (lipid generation) as shown in Fig. 4 and Fig. 5. Both of these pathways are restrained by microbial metabolism (Hommel and Huse 1993).

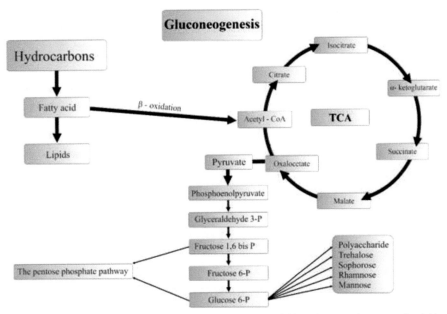

Figure 5. Biosynthetic pathways involved in the synthesis of biosurfactant using water insoluble (hydrocarbon) substrate (Santos et al. 2016, Papagianni 2012).

In the glycolytic pathway (Fig. 4), a water soluble substrate like glucose is broken down into the intermediate product glucose-6-phosphate (G6P). This G6P is one of the major precursors of carbohydrates, which are found in the hydrophilic part of the biosurfactant. A series of enzymes are involved in the catalysis of G6P to form different hydrophilic moieties of trehalose, sophorose, rhamnose, mannose and polysaccharide.

For the generation of hydrophobic moiety, glucose is oxidized to pyruvate. Then pyruvate is converted to acetyl Co-A which synthesizes malonyl-Co-A. Both malonyl Co-A and oxaloacetate are then converted into fatty acid which acts as a precursor for lipid production (Hommel and Huse 1993). Although, in some microorganisms like *Lactobacillus delbrueckii* N2, *L. cellobiosus* TM1, and *L. plantarum* G88, biosurfactants have high lipid content when they are produced using glycerol as a substrate as compared to sugarcane molasses (Mouafo et al. 2018). This might be because of the fact that these microorganisms are directed to the lipogenic pathway and gluconeogenesis pathway. When hydrophobic substrates like hydrocarbons, oils and fats are used as a substrate, the metabolic mechanism might switch into the lipolytic pathway and gluconeogenesis which directs the formation of hydrophobic moieties of the biosurfactant. In this case, the hydrophilic moieties will be produced *de novo* through gluconeogenesis. Gluconeogenesis starts with the formation of acetyl Co-A by oxidation of fatty acids. Transformation of pyruvate into G6P involved series of enzymes similar as glycolysis.

Figure 5 demonstrates the main reactions involved in the biosynthesis of hydrophobic and hydrophilic moieties of biosurfactant using hydrocarbon as a substrate.

Additionally, some biosurfactant production requires multienzyme complexes to finish biosurfactant production. Till date, very scanty reports exist that shows the molecular biosynthetic regulation for biosurfactant. Rhamnolipids by *Pseudomonas aeruginosa* and surfactin synthesized by *Bacillus subtilis* were the earliest to be decoded for pathway. Rhamnolipid production is executed by two different glycosyltransferase reactions. Both of them are named as rhamnosyl transferase I and rhamnosy ltransferase II. The gene RhlA and RhlB are designated as rhamnosyl transferase I which is responsible for mono-rhamnolipid synthesis while RhlC gene is designated as rhamnosyl transferase II which produces di-rhamnolipids. The surfactin production represents the best example of non-ribosomal protein synthesis (NRPS). The large multienzyme peptide complex is involved in the surfactin production. There are multifunctional proteins encoded in SrfA operon, which comprises SrfA-A, SrfA-B and SrfA-C.

4. Potential Applications of Marine Derived Biosurfactants

The increasing breakthrough in marine biotechnology has drawn the attention of many researchers towards marine microbial surface active agents and exopolysaccharides. Till date, many marine microbes of various bioactive compounds have been reported for production. The marine biosurfactants have proved their efficiency in bioremediation of environmental pollution. Studies showed that these biosurfactants can be used for clean-up of the environments polluted with crude oil or polyaromatic hydrocarbons. These molecules have been found to be potential for environmental clean-up, industrial emulsification and stabilization processes.

Different industrial sectors are continually searching for novel bioactive products for their use in different fields like textile, pharmaceutical, cosmetics, food, metal mining, and oil recovery. Due to novelty in the structural and functional activity, these compounds have numerous applications in various fields. Tensiometric and emulsification properties of biosurfactants are important for application in multiple fields (Baird et al. 1983, Sutherland 1998).

4.1 Environmental Applications of Marine Biosurfactant

4.1.1 Hydrocarbon Degradation Potential of Marine Biosurfactant

Oil pollution may be a common phenomenon that causes significant ecological and social problems in terrestrial and aquatic environments. The utilization of traditional available treatment processes to decontaminate polluted areas is limited in their application (Perfumo et al. 2010a). Physical collection methods like booms, skimmers, and adsorbents typically recover maximum 10%–15% of the spilled oil, and therefore the use of synthetic surfactants for remediation is not preferred thanks to their toxic effects on the prevailing biota within the polluted area. Hence, despite decades of research, successful bioremediation of oil contaminated environment remains a challenge (Perfumo et al. 2010b). Oil pollution in marine environments encourages obligate hydrocarbonoclastic bacteria to flourish which becomes the bulk of the entire microbial population (Yakimov et al. 2007). Oil degrading microorganisms have adopted different strategies to reinforce the bioavailability and gain access to hydrophobic compounds, one among them includes BS-mediated solubilization. The unique structure of BS resides within the coexistence of the hydrophilic (a sugar or peptide) and therefore the hydrophobic (fatty acid chain) domains within the same molecule which allows them to occupy the interfaces of mixed-phase systems (e.g., oil/water, air/water, and oil/solid/water) and accordingly altering the forces governing the equilibrium conditions. It's a prerequisite for the existence of a broad range of surface activities including emulsification, dispersion, dissolution, wetting and foaming (Banat et al. 2000, Desai and Banat 1997). Many scientists reported such advantages by investigating the effect of BS and fertilizer on biodegradation of petroleum by four marine oil-degrading bacteria namely *B. megaterium*, *Corynebacterium kutscheri*, *P. aeruginosa*, and *Lactobacillus delbrueckii* (Thavasi et al. 2011a, Thavasi et al. 2011b, Thavasi et al. 2007). It has been reported that the degradation process increased from 19% to 37.7% into the lab-scale biodegradation system as compared to the experiments where no BS and fertilizer was added. Lipopeptide biosurfactant produced by *Bacillus licheniformis* LRK1 also has been reported for engine oil degradation potential which degraded 24.23% of engine oil after 21 days of incubation (Nayak et al. 2020a). Marine microorganisms like *Halomonas*, *Marinobacter*, *Myroides* and *Yarrowia lipolytica* may play a key role in removing hydrocarbons from polluted areas by forming bioemulsifiers. The lipopeptide produced by *Alcaligens aquatilis* YGD 2906 potentially degraded petroleum (Yalaoui-Guellal et al. 2020). *Halomonas* sp. produces the surface active emulsifiers to participate within the oil spill removal. Glycolipid molecules produced by *Halomonas* sp. enhances the solubility of hydrocarbon on their cell surface and increases their bioavailability for degradation (Pepi et al. 2005, Dhasayan et al. 2014).

Marine microbial BSs also exhibited higher stability under various conditions to market the degradation process (Thaniyavarn et al. 2006). The researchers isolated a marine *P. aeruginosa* A41 strain which produced rhamnolipid BS having good stability and activity at wide range of temperatures (40°C–121°C), pH (2–12), and NaCl concentrations (0%–5%). Higher stability at various conditions and broad-spectrum emulsification activities in marine BSs showed their potential broad-spectrum application against various hydrocarbons and in several environments.

4.1.2 Microbially Enhanced Oil Recovery (MEOR) by Marine Biosurfactant

Biosurfactants are utilized in the formulations of emulsifying/demulsifying agents, anti-corrosives, biocides, and other innovative applications within the petroleum industry (De Almeida et al. 2016). Biosurfactants have proved their usefulness in reside recovery by solubilizing trapped oil in rock formations, which may be a prerequisite for enhanced oil recovery (EOR). BSs have extensive potential application within the petroleum industry like emulsifiers, demulsifiers, and oil recovery agents. MEOR may be a technique that either uses a crude preparation of BS or sterilized BS containing culture broth to liberate petroleum from a binding substrate (Marchant and Banat 2012). For instance, Banat et al. (1991) administered a petroleum sludge tank clean-up and oil recovery process during which BS-containing sterilized culture broth was used to pack up oil sludge from an oil tank. After 5 days of treatment involving energy addition and circulation to reinforce the method of emulsification followed by a de-emulsification, 91% of the petroleum present within the oil tank was recovered. Hydrocarbon analysis of the recovered crude showed a

100% hydrocarbon content. This result indicated that MEOR process doesn't require live microorganisms or pure BSs, sterilized BS-containing broth is sufficient to mobilize and recover a significant amount of oil from oil sludge deposits.

4.1.3 Heavy Metal Remediation Potential of Marine Biosurfactant

Microbial BSs are known for its metal-complexing activities that are reported to be beneficial within the bioremediation of heavy metal-contaminated environments (Mulligan et al. 2001, Singh and Cameotra 2004). Das et al. (2009a) studied the *B. circulans* bacterial cells and BS-mediated cadmium and lead metal binding and suggested that metal binding was not cell-mediated, but it became so thanks to BS concentration. An increase in metal binding was observed with a rise in BS concentration.

Gnanamani et al. (2010) reported that the marine *Bacillus* sp. MTCC 5514 reduces chromium and trivalent chromium tolerance through its extracellular enzyme reductase and BS production. Biosurfactant molecules entrap the trivalent chromium within the micelle of biosurfactants, which prevents microbial cells from exposure to trivalent chromium. It has been found that extracellular chromium reductase and BS mediate the remediation process and keep the cells active and supply tolerance and resistance toward high concentration of hexavalent chromium and trivalent chromium.

4.2 Commercial Applications of Marine Biosurfactant

4.2.1 Food and Cosmetic Application of Marine Biosurfactant

Food industries are seeking to identify additives that can enhance food value and features due to higher use of emulsifying and thickening agents in food products. Lecithin, xanthan gum and gum acacia are commonly used emulsifiers of hydrocolloids that emulsify and stabilize oil-in-water emulsions effectively. Dried fungus culture from *Fusarium venenatum* is used as a binder in Quorn goods alongside egg albumen. Whereas in vegan food products, egg albumen is added to potato protein *in situ*. Global climate change, though, and weather variability such as drought can generate defoliation and impact plant-based gum assembly. There is an increased interest in discovering new additives to reduce the focus on conventional emulsifiers and plant-based emulsifiers. Marine BS derivatives are non-toxic and/or less toxic, with high heat, pH and salinity safety, as opposed to chemical surfactants. To date, not many studies have been published on the use of marine-derived biosurfactant (BS) in food and cosmetic formulations. Emulsan derived from *A. calcoaceticus* is the sole commercially available microbial biosurfactant, which can enhance the consistency, appearance and solubilization of fat globules and the fragrance of food products due to its emulsion-stabilizing properties. Incorporation of BS has been found to improve dough rheology, increase fat quantity and emulsification, and thus find useful applications in the industries of bakery and meat processing. A glycoprotein emulsion stabilizer developed by marine *Antarctobacter* sp. TG22, is able to produce stable oil-in-water emulsions with commercial food-grade oils (Gutiérrez et al. 2007a). Similarly, in accordance with commercial emulsifiers with stable operation under acidic conditions and elevated temperatures, two other glycoprotein emulsifiers of the marine *Halomonas* species TG39 and TG67 indicate superior emulsifying properties (Gutiérrez et al. 2007b). The lipopeptide from the marine species *Nesterenkonia* MSA31 is a powerful emulsifier, with significant antioxidant activity and increased softness and retained food texture exhibited by its inclusion to muffins. 90% of the biofilm developed by *Staphylococcus aureus* was effectively reduced by the lipopeptide MSA31. Marine-derived biosurfactants are commonly used as emulsifiers, stabilizing agents, antimicrobial and antiadhesive/ antibiofilm agents in the food industry. Skin irritations and reactions are also caused by ingredients found in cosmetic formulations. There is considerable interest among customers in natural cosmetic products. Replacement of surface active compounds (SACs) in such products with biosurfactant can significantly decrease such adverse effects. The surface active properties are essential to work out the sort and amount of BS in detergents, cosmetic, pharmaceuticals and various other industries. To figure out the nature and quantity of BS in detergents, cosmetics, pharmaceuticals and various other industries, the surface active properties are important. The type of BS compound to be introduced into the formulations is often chosen to help their emulsifying capacity and/or surface activity, such as hydrophilic lipophilic balance (HLB) and significant concentration of micelle (CMC), respectively. The HLB determines the polarity of biosurfactant, which in many processes gives an indication of its solubility. Biosurfactant with a high HLB value means that it is extremely hydrophilic, whereas a high lipophilic character is seen by a low HLB value. A biosurfactant can be an emulsifier, antifoaming agent and wetter assisted by the HLB properties, which are ideal properties for cosmetic products.

4.3 Biological Applications of Marine Biosurfactants

4.3.1 Antimicrobial, Anti-adhesive and Biofilm Disruption Activity of Marine Biosurfactant

Marine sponges and other marine invertebrates are predominant sources of novel bioactive compounds, including antimicrobial, anti-adhesive and anti-biofilm agents. These bioactive compounds play a pivotal role in their defence against predators, infectious agents and biofilm-forming microorganisms (Dusane et al. 2011, Kiran et al. 2009). Earlier, biosurfactants were used as antibiotics due to their antimicrobial activities. Even till date many biosurfactant/bioemulsifier with antimicrobial activities are available and exploited. *Bacillus* species showed highest examples of antimicrobial activity of biosurfactant. Lipopeptide biosurfactant from *B. circulans* features a powerful antimicrobial activity, hence these sort of surface active agents have applications in antimicrobial chemotherapy (Das et al. 2008a). Mukherjee et al. (2009) purified a biosurfactant from marine *B. circulans* in a glucose mineral salts medium which showed enhanced surface and antimicrobial activities. Thus, the authors proposed that these sorts of biosurfactants are often used as new potential drugs in antimicrobial chemotherapy.

Antibiotic therapies are successful in the treatment of most human bacterial infections. Nevertheless, in recent years, it has been observed that there's a big increase within the emergence of pathogenic microorganisms immune to the available antimicrobials including multi-drug resistant (MDR) pathogens. As a result, it has become difficult to treat infectious diseases which are found to be the source of great public ill health (Coates et al. 2011). Antimicrobial, anti-adhesive or anti-biofilm activities against different pathogenic and opportunistic microorganisms by marine biosurfactants are listed in Table 2.

Many of those biosurfactants are powerful against a broad spectrum of human pathogens, including Gram-negative and Gram-positive bacteria, also against the yeast *Candida albicans*. Moreover, in some cases, they are also effective against

Table 2. Antimicrobial, antiadhesive and antibiofilm activity by biosurfactant producing marine microbes.

Biosurfactant type	Marine organism	Activity	Name of human pathogen	Reference
Glycolipid	*Brevibacterium casei* MSA19	Antimicrobial activity	*Escherichia coli* *Klebsiella pneumoniae* *Proteus mirabilis* *Pseudomonas aeruginosa* haemolytic *Streptococcus* *Vibrio parahaemolyticus* *Vibrio vulnificus*	Kiran et al. 2010
	Serratia marcescens	Antimicrobial, anti-adhesive anti-biofilm	*Candida albicans* *P. aeruginosa*	Dusane et al. 2011
	Streptomyces sp. MAB36	Antimicrobial activity	*Aspergillus niger* *Bacillus cereus* *C. albicans* *Enterococcus faecalis* *Shigella boydii* *Shigella dysenteriae* *S. aureus*	Manivasagan et al. 2014
Lipopeptide	*Bacillus circulans*	Anti-adhesive anti-biofilm	*C. freundii* *E. coli* *P. vulgaris* *Salmonella typhimurium* *S. marcescens*	Das et al. 2009b
	Bacillus licheniformis NIOT-AMKV06	Antimicrobial activity	*E. faecalis* *K. pneumoniae* *M. luteus* *P. mirabilis* *Salmonella typhi* *Shigella flexineri* *S. aureus* *Vibrio cholera*	Lawrance et al. 2014
	Nocardiopsis alba MSA10	Antimicrobial activity	*C. albicans* *E. faecalis* *K. pneumoniae* *M. luteus* *P. mirabilis* *S. aureus* *S. epidermidis*	Gandhimathi et al. 2009

MDR clinical isolates. Therefore, they will be a substitute for the prevailing drugs to treat infections caused by pathogens. Additionally, many of those biosurfactants exhibited a big anti-adhesive and anti-biofilm activity. The biosurfactant produced by *B. circulans* showed anti-adhesive activity at concentrations between 0.1 and 10 mg/mL. At the very best concentration, microbial adhesion was inhibited between 84% and 89%, and pre-formed biofilms were removed (with efficiencies between 59% and 94%) for all the pathogenic microorganisms tested (Das et al. 2009b).

A partially purified glycolipid biosurfactant produced by *Brevibacterium casei* MSA19 removed pre-formed biofilms of all the pathogenic microorganisms tested at 30 g/mL (Kiran et al. 2010). Strain S*treptomyces* sp. ISP2-49E, isolated from marine sediment samples obtained from Galveston Bay (Texas) produced the rhamnolipid biosurfactant L-rhamnosyl-3-hydroxydecanoyl-3-hydroxydecanoate (Rha-Rha-C_{10}-C_{10}), which possess a broad spectrum of antimicrobial and anti-adhesive activities (Haba et al. 2003). The antagonistic activities exhibited by these biosurfactants against human pathogens (including MDR pathogens) make them candidates to be used as an alternate to traditional antibiotics.

4.3.2 Anticancer Activity of Biosurfactant

Worldwide, many people, face a health risk caused by cancer (Siegel et al. 2015); therefore, any development resulting in increased survival of such patients may be a global priority. Several strategies have followed over the years like checking out new biomarkers, drugs or treatments. Although, a successful targeted selective and non-toxic therapy remains to be developed. In traditional cancer chemotherapy, highly cytotoxic drugs used are non-specifically targeting any dividing cells that end in a modest improvement in patient survival. Thus, the overall prognosis of most patients remains dismal and treatment is non-specific, non-selective and toxic. Therefore, the search and development of latest anti-cancer drugs remains an excellent challenge. Currently, many anticancer drugs utilized in clinical practice are natural products or derivatives (Bolhassani 2015, Cragg and Newman 2005). So, it is possible that systematic exploration of natural sources, like the marine microbiota, will lead to the discovery of compounds with interesting biological activities, including anticancer activity (Janakiram et al. 2015, Sawadogo et al. 2015). Biosurfactants, especially lipopeptides and glycolipids, are being focused on for potential use as anticancer agents interfering with cancer progression processes (Gudiña et al. 2013).

These bioactive compounds are involved in many intercellular molecular recognition steps, including signal transduction, cell differentiation and cell immune response (Rodrigues et al. 2006). Additionally, they manifest low toxicity, easy biodegradability and high efficacy, which are appropriate features in any anticancer agent. Different mechanisms underlying the anti-cancer activity of biosurfactants are proposed including the delay of cell cycle progression; inhibition of crucial signalling pathways like Akt (Protein kinase B), extracellular signal-regulated kinase/c-Jun N-terminal kinase (ERK/JNK) and Janus kinase/signal transducer and activator of transcription (JAK/STAT); reduction of angiogenesis; activation of natural killer T (NKT) cells; and induction of apoptosis through death receptors in cancer cells. Additionally, the power of biosurfactants to disrupt cell membranes, resulting in a sequence of events that include lysis, increased membrane permeability and metabolite leakage, has also been pointed to as a probable mechanism of anticancer activity (Janek et al. 2013). Though, many reports have shown that LPs (lipopeptides) and glycolipids can selectively inhibit the proliferation of cancer cells and disrupt cell membranes, causing their lysis through apoptosis pathways (Gudina et al. 2013).

The lipopeptides (LPs) and sophorolipids (SLs) are the biosurfactants most studied in terms of anticancer potential. The LPs are composed of a peptide and a carboxylic acid chain and are shown to exhibit antitumor activity *in vitro* (Zhao et al. 2018). Reports on the *Bacillus* LPs, namely surfactin, iturin and fengycin, suggest that they possess antitumor activities. Iturin has been shown to inhibit the proliferation of MDa-MB-231 cancer cells (Dey et al. 2015). Fengycin can block non-small cell carcinoma cell 95D and inhibit the expansion of xenografted 95D cells in nude mice (Yin et al. 2013). Recently, Zhao et al. (2018) showed the *B. subtilis* LPs consisting of a majority of iturin exhibited promising potential in inhibiting chronic myelogenous leukaemia *in vitro* via simultaneously causing paraptosis, apoptosis and inhibition of autophagy. The anticancer mechanisms of *Bacillus* LPs are extensively studied and SUR has been found to display an anti-proliferative effect via apoptosis induction, cell cycle arrest and survival signalling suppression.

It is clear that the marine environment represents a promising source of novel added value compounds. Among these compounds, some new biosurfactant structures are reported, namely fellutamides (Lee et al. 2010, Shigemori et al. 1991) and rakicidin (Poulsen 2011, Takeuchi et al. 2011). A sponge-derived fungus named *Aspergillus versicolor* produced fellutamides C and F which exhibited anticancer activity against HCT-15 carcinoma, A549 carcinoma, SK-MEL-2 carcinoma and SK-OV-3 ovarian neoplastic cell lines (Shigemori et al. 1991, Poulsen 2011). Among these lipopeptides, rakicidin A, shows cytotoxicity against several neoplastic cell lines, such as HCT-8 and PANC-1 (Wilson and Hay 2011). Actions against oesophageal squamous carcinoma cells (EC109), cancer cell lines (A549 and 95D), gastric cancerous cells (SGC7901), cervix cancer cells (HeLa), and lung cancer cells (HepG2) have also been documented with another lipopeptide named rakicidin B (Xie et al. 2011). This rakicidin derivative mediated apoptosis by caspase-3, -7 and -9 activation and blocked the signalling pathways of MAPK and JNK/p38. Rakicidin derivatives C and D are found to be non-cytotoxic while Rakicidin D showed interference with the invasiveness of aggressive carcinoma cells (Poulsen 2011).

4.3.3 *Nanoparticle Based Therapeutics by using Marine Biosurfactant*

Nanoparticle-based therapeutics are accounted as the foremost promising platforms in drug delivery applications due to their capacity to extend drug accumulation in solid tumors by enhanced permeability retention (EPR) and MDR reversal through bypassing or inhibiting P-gp activity (Bao et al. 2016). Furthermore, it had been also reported that membrane bursting, oozing out of proteins and intracellular materials occurred in *C. albicans* necrobiosis mediated by SL-capped ZnO nanoparticle which caused necrobiosis (Basak et al. 2014). In addition to functioning as a cyclic lipopeptide, the biosurfactant, SUR, has been found to exhibit versatile bioactive features including adjuvant for immunization and antitumor properties. Supported by its unique amphipathic properties, surfactin has the potential for self-assembly (under certain conditions) into nanoparticles to function as a drug carrier for loading hydrophobic drugs. Combining the anticancer activity of SUR and therefore the characteristics of nanoparticles like EPR effects and MDR reversal might improve cancer chemotherapy by designing surfactin as a carrier to load anticancer drugs. In an investigation by (Huang et al. 2018), Surfactin (SUR) was assembled by a solvent-emulsion method to load the anticancer drug doxorubicin (DOX). The DOX@SUR assembly was shown to induce stronger cytotoxicity against DOX-resistant human carcinoma MCF-7/ADR cells compared to free DOX. The DOX@-SUR nanoparticles exhibited enhanced cellular uptake and decreased cellular efflux. Moreover, *in vivo* DOX@-SUR nanoparticles accumulated more efficiently in tumors than free DOX. The DOX@SUR showed stronger tumor inhibition activity and fewer side effects in MCF-7/ADR-bearing nude mice, suggesting that SUR based nanoparticles could be used as potential anticancer drug carriers to reverse MDR in cancer chemotherapy.

5. Metagenomics for the Invention of Novel Substances

The use of molecular techniques to examine the variety of marine ecosystems has recently revealed that the majority of marine microbes remain unexplored, particularly due to the problem of growing these microbes under laboratory conditions (Kennedy et al. 2010). Through the study of the genetic resources of otherwise inaccessible marine microorganisms and the discovery of previously unknown natural compounds with significant biological activities, culture-independent techniques (metagenomics) are promising methodologies that do away with the need for cultivation.

Metagenomics can be a series of molecular techniques that permit culture-independent studies of microbial communities from any environmental sample by specifically extracting and analysing their genetic material, providing access to the entire genetic diversity and biosynthetic potential of all the microorganisms in the community (Coughlan et al. 2015, Kennedy et al. 2010). It begins with the DNA extraction from the environmental sample under study, although it is also based on RNA. The DNA sample must represent all the microbial communities present within the population (both qualitative and quantitative). The very next stage is to build metagenomic libraries using appropriate cloning vectors. Various vectors are used, counting on the lengths of the DNA fragments retrieved like plasmids, fosmids, cosmids and artificial bacterial chromosomes. The metagenomic library is subsequently transferred to a suitable host strain, generally *E. coli*. Eventually, the individual clones of recombinants are screened. Sequences of DNA inserts cloned (sequence-based metagenomics) or inside the roles bestowed by such DNA inserts to the host (function-based metagenomics) are also provided by the screening process (Kennedy et al. 2010, Coughlan et al. 2015).

Sequence-based metagenomics has mostly been carried out by random sequencing of metagenomic libraries on a wide scale, which produces a high number of sequences. These sequences are then analysed using bioinformatics tools and are compared through homology-based investigations with sequences deposited within the databases. Novel biosurfactant producing genes with unknown roles may lead to a variety of these sequences. In recent years, the introduction of next-generation sequencing technology has also been attributed to the resulting substantial decrease in DNA sequencing costs, leading to significant advancement in sequence-based metagenomics, facilitating its widespread use. The gene similarity is provided by sequence-based metagenomics using Polymerase Chain Reaction (PCR) or DNA hybridization techniques. In this case, after the target genes (or proteins) have been selected, PCR primers or DNA probes are constructed in conjunction with consensus sequences unique to the most conserved regions of these genes (or proteins), and the metagenomic library will not be screened any further.

The key downside of this method is that it needs advanced knowledge of gene sequences, which, of course, prevents the invention of new functions or activities (Jackson et al. 2015, Felczykowska et al. 2012). In the particular case of biosurfactants, for example, these methods often do not search for new genes involved in their biosynthesis that endorse the sequences of genes that are already identified. As a consequence, it is predicted that the latest biosurfactants detected are much like those previously mentioned. In certain cases, though, these genes may display minor modifications that will allow biosurfactants with different properties and activities to be assembled.

Function-based metagenomics, on the other hand, consists of screening metagenomic libraries for the existence of behaviours or phenotypes that arise from the heterologous expression of the microbial community's genes. An even more advantage of this approach is that it is not directly linked to the previous knowledge of the DNA sequences; therefore, it is the easiest way to identify new genes and gene families encoding new biomolecules that in comparison with previously described genes would not be identified using the sequence-based approaches. Function-based metagenomics, however, entails several challenges and difficulties. Some examples of metagenomic library functional screening for brand spanking of new antimicrobial compounds using *E. coli* and other alternative hosts simultaneously (*Bacillus subtilis* or *Streptomyces lividans*) resulted in the antimicrobial activity identification inside the alternative hosts, but not in *E. coli* (Iqbal et al. 2014). For that reason, with the aim of expressing several of the genes present in the microbial population, there is an increasing trend to use multiple hosts concurrently. In this case, it is important to use broad-host vectors in order to create metagenomic libraries. In addition, if a feature is encoded by several genes clustered in one operon, it could be identified as long as the entire operon is cloned during a single DNA insert. As an example, the genes involved in lipopeptide biosurfactant biosynthesis (e.g., surfactin, lichenysin, fengycin, or iturin) are grouped into the operons of a size range from 25 kb to 40 kb (Sen 2010). This means that DNA inserts greater than 20 to 30 kb must be ready for the vectors to create metagenomic libraries to spot new biosurfactants. Another concern is that, while the underlying gene is correctly expressed, it cannot be identified during the screening processes if the resulting product is not excreted. These problems are not applicable to sequence-based metagenomics, where there is no need for the expression of DNA inserts.

In addition, function-based metagenomics needs appropriate screening methods to be responsive and accessible to thousands of transformants at the same time, as metagenomic libraries with ample community-wide coverage needs an especially large number of clones. Heterologous complementation of host strains or mutants by target genes is often performed to enable their growth under selective conditions or to enable the use of a chosen substrate, or to detect particular enzymatic activities using chromogenic or fluorescent substrates (Kennedy et al. 2010, Felczykowska et al. 2012). Thus, function-based metagenomics shows a beneficial approach to the study of genetic resources of inaccessible marine microbes, providing access to all microbes in the community's total genetic pool and biosynthetic potential, as well as enabling the discovery of novel biosurfactants.

6. Conclusion

The marine ecosystem is a rich source for a broad range of surface-active metabolites. Due to their structural and functional diversity interest in the research and discovery of marine organisms has increased in recent years. The detection of potential biosurfactants from marine microorganisms would boost their application in several environmental fields of bioremediation, metal removal and microbial enhanced oil recovery. In the biomedical field their prospective use is observed in antimicrobial, anti-biofilm, anticancer and nanoparticle synthesis.

References

Abdel-Mawgoud, A.M., F. Lépine and E. Déziel. 2010. Rhamnolipids: diversity of structures, microbial origins and roles. Applied Microbiology and Biotechnology 86(5): 1323–1336.

Abraham, W.R., H. Meyer and M. Yakimov. 1998. Novel glycine containing glucolipids from the alkane using bacterium *Alcanivorax borkumensis*. Biochimica et Biophysica Acta (BBA)-Lipids and Lipid Metabolism 1393(1): 57–62.

Antoniou, E., S. Fodelianakis, E. Korkakaki and N. Kalogerakis. 2015. Biosurfactant production from marine hydrocarbon-degrading consortia and pure bacterial strains using crude oil as carbon source. Frontiers in Microbiology 6: 274.

Baird, J.K., P.A. Sandford and I.W. Cottrell. 1983. Industrial applications of some new microbial polysaccharides. Biotechnology 1(9): 778–783.

Balan, S.S., C.G. Kumar and S. Jayalakshmi. 2016. Pontifactin, a new lipopeptide biosurfactant produced by a marine *Pontibacter korlensis* strain SBK-47: purification, characterization and its biological evaluation. Process Biochemistry 51(12): 2198–2207.

Balan, S.S., C.G. Kumar and S. Jayalakshmi. 2017. Aneurinifactin, a new lipopeptide biosurfactant produced by a marine *Aneurinibacillus aneurinilyticus* SBP-11 isolated from Gulf of Mannar: Purification, characterization and its biological evaluation. Microbiological Research 194: 1–9.

Balan, S.S., C.G. Kumar and S. Jayalakshmi. 2019. Physicochemical, structural and biological evaluation of Cybersan (trigalactomargarate), a new glycolipid biosurfactant produced by a marine yeast, *Cyberlindnera saturnus* strain SBPN-27. Process Biochemistry 80: 171–180.

Baltz, R.H., V. Miao and S.K. Wrigley. 2005. Natural products to drugs: daptomycin and related lipopeptide antibiotics. Natural Product Reports 22(6): 717–741.

Banat, I.M., N. Samarah, M. Murad, R. Horne and S. Banerjee. 1991. Biosurfactant production and use in oil tank clean-up. World Journal of Microbiology and Biotechnology 7(1): 80–88.

Banat, I.M. 1995a. Biosurfactants characterization and use in pollution removal: state of the art. A review. Acta Biotechnology 15: 251–67.

Banat, I.M. 1995b. Biosurfactants production and possible uses in microbial enhanced oil recovery and oil pollution remediation: a review. Bioresource Technology 51(1): 1–12.

Banat, I.M., R.S. Makkar and S.S. Cameotra. 2000. Potential commercial applications of microbial surfactants. Applied Microbiology and Biotechnology 53(5): 495–508.

Banat, I.M., A. Franzetti, I. Gandolfi, G. Bestetti, M.G. Martinotti, L. Fracchia and R. Marchant. 2010. Microbial biosurfactants production, applications and future potential. Applied Microbiology and Biotechnology 87(2): 427–444.

Bao, Y., M. Yin, X. Hu, X. Zhuang, Y. Sun, Y. Guo and Z. Zhang. 2016. A safe, simple and efficient doxorubicin prodrug hybrid micelle for overcoming tumor multidrug resistance and targeting delivery. Journal of Controlled Release 235: 182–194.

Basak, G., D. Das and N. Das. 2014. Dual role of acidic diacetate sophorolipid as biostabilizer for ZnO nanoparticle synthesis and biofunctionalizing agent against *Salmonella enterica* and *Candida albicans*. Journal of Microbiology and Biotechnology 24(1): 87–96.

Bhadoriya, S.S., N. Madoriya, K. Shukla and M.S. Parihar. 2013. Biosurfactants: A new pharmaceutical additive for solubility enhancement and pharmaceutical development. Biochemical Pharmacology 2: 113.

Bognolo, G. 1999. Biosurfactants as emulsifying agents for hydrocarbons. Colloids and Surfaces A: Physicochemical and Engineering Aspects 152(1-2): 41–52.

Bolhassani, A. 2015. Cancer chemoprevention by natural carotenoids as an efficient strategy. Anti-Cancer Agents in Medicinal Chemistry (Formerly Current Medicinal Chemistry-Anti-Cancer Agents) 15(8): 1026–1031.

Chakraborty, J. and S. Das. 2016. Characterization of the metabolic pathway and catabolic gene expression in biphenyl degrading marine bacterium *Pseudomonas aeruginosa* JP-11. Chemosphere 144: 1706–1714.

Cheng, T., J. Liang, J. He, X. Hu, Z. Ge and J. Liu. 2017. A novel rhamnolipid-producing *Pseudomonas aeruginosa* ZS1 isolate derived from petroleum sludge suitable for bioremediation. AMB Express 7(1): 120.

Christova, N. and I. Stoineva. 2014. Trehalose biosurfactants. pp. 197–216. *In*: Mulligan, C.N., S.K. Sharma and A. Mudhoo (eds.). Biosurfactants—Recent Trends and Applications. CRC Press.

Coates, A.R., G. Halls and Y. Hu. 2011. Novel classes of antibiotics or more of the same? British Journal of Pharmacology 163(1): 184–194.

Coughlan, L.M., P.D. Cotter, C. Hill and A. Alvarez-Ordóñez. 2015. Biotechnological applications of functional metagenomics in the food and pharmaceutical industries. Frontiers in Microbiology 6: 672.

Cragg, G.M. and D.J. Newman. 2005. Plants as a source of anti-cancer agents. Journal of Ethnopharmacology 100: 72–79.

Das, P., S. Mukherjee and R. Sen. 2008a. Improved bioavailability and biodegradation of a model polyaromatic hydrocarbon by a biosurfactant producing bacterium of marine origin. Chemosphere 72: 1229–1234.

Das, P., S. Mukherjee and R. Sen. 2008b. Antimicrobial potential of a lipopeptide biosurfactant derived from a marine *Bacillus circulans*. Journal of Applied Microbiology 104(6): 1675–1684.

Das, P., S. Mukherjee and R. Sen. 2009a. Biosurfactant of marine origin exhibiting heavy metal remediation properties. Bioresource Technology 100(20): 4887–4890.

Das, P., S. Mukherjee and R. Sen. 2009b. Antiadhesive action of a marine microbial surfactant. Colloids and Surfaces B: Biointerfaces 71(2): 183–186.

Das, P., S. Mukherjee, C. Sivapathasekaran and R. Sen. 2010. Microbial surfactants of marine origin: potentials and prospects. pp. 88–101. *In*: Biosurfactants. Springer, New York, NY.

De Almeida, D., R.C.F. Soares De Silva, J.M. Luna, R.D. Rufino, V.A. Santos, I.M. Banat and L.A. Sarubbo. 2016. Biosurfactants: promising molecules for petroleum biotechnology advances. Frontiers in Microbiology 7: 1718

De Carvalho, C.C. and P. Fernandes. 2010. Production of metabolites as bacterial responses to the marine environment. Marine Drugs 8(3): 705–727.

Desai, J.D. and I.M. Banat. 1997. Microbial production of surfactants and their commercial potential. Microbiology and Molecular Biology Reviews 61(1): 47–64.

Dey, G., R. Bharti, R. Sen and M. Mandal. 2015. Microbial amphiphiles: a class of promising new-generation anticancer agents. Drug Discovery Today 20(1): 136–146.

Dhasayan, A., G.S. Kiran and J. Selvin. 2014. Production and characterisation of glycolipid biosurfactant by *Halomonas* sp. MB-30 for potential application in enhanced oil recovery. Applied Biochemistry and Biotechnology 174(7): 2571–2584.

Du, J., A. Zhang, X. Zhang, X. Si and J. Cao. 2019. Comparative analysis of rhamnolipid congener synthesis in neotype *Pseudomonas aeruginosa* ATCC 10145 and two marine isolates. Bioresource Technology 286: 121380.

Dusane, D.H., V.S. Pawar, Y.V. Nancharaiah, V.P. Venugopalan, A.R. Kumar and S.S. Zinjarde. 2011. Anti-biofilm potential of a glycolipid surfactant produced by a tropical marine strain of *Serratia marcescens*. Biofouling 27(6): 645–654.

Faria, N.T., M. Santos, C. Ferreira, S. Marques, F.C. Ferreira and C. Fonseca. 2014. Conversion of cellulosic materials into glycolipid biosurfactants, mannosylerythritol lipids, by *Pseudozyma* spp. under SHF and SSF processes. Microbial Cell Factories 13(1): 155.

Felczykowska, A., S.K. Bloch, B. Nejman-Faleńczyk and S. Barańska. 2012. Metagenomic approach in the investigation of new bioactive compounds in the marine environment. Acta Biochimica Polonica 59(4).

Floodgate, G.D. 1978. The formation of oil emulsifying agents in hydrocarbonclastic bacteria. Microbial Ecology 82–85.

Fracchia, L., M. Cavallo, M.G. Martinotti and I.M. Banat. 2012. Biosurfactants and bioemulsifiers biomedical and related applications—present status and future potentials. Biomedical Science, Engineering and Technology 14: 326–335.

Fujii, T., R. Yuasa and T. Kawase. 1999. Biodetergent IV. Monolayers of corynomycolic acids at the air-water interface. Colloid and Polymer Science 277(4): 334–339.

Gandhimathi, R., G.S. Kiran, T.A. Hema, J. Selvin, T.R. Raviji and S. Shanmughapriya. 2009. Production and characterization of lipopeptide biosurfactant by a sponge-associated marine actinomycetes *Nocardiopsis alba* MSA10. Bioprocess and Biosystems Engineering 32(6): 825–835.

Giraud, M.F. and J.H. Naismith. 2000. The rhamnose pathway. Current Opinion in Structural Biology 10(6): 687–696.

Gnanamani, A., V. Kavitha, N. Radhakrishnan, G.S. Rajakumar, G. Sekaran and A.B. Mandal. 2010. Microbial products (biosurfactant and extracellular chromate reductase) of marine microorganism are the potential agents reduce the oxidative stress induced by toxic heavy metals. Colloids and Surfaces B: Biointerfaces 79(2): 334–339.

Gudiña, E.J., V. Rangarajan, R. Sen and L.R. Rodrigues. 2013. Potential therapeutic applications of biosurfactants. Trends in Pharmacological Sciences 34(12): 667–675.

Gunther, N.W., A. Nunez, W. Fett and D.K. Solaiman. 2005. Production of rhamnolipids by *Pseudomonas chlororaphis*, a nonpathogenic bacterium. Applied and Environmental Microbiology 71(5): 2288–2293.

Gutiérrez, T., B. Mulloy, C. Bavington, K. Black and D.H. Green. 2007a. Partial purification and chemical characterization of a glycoprotein (putative hydrocolloid) emulsifier produced by a marine bacterium *Antarctobacter*. Applied Microbiology and Biotechnology 76(5): 1017.

Gutiérrez, T., B. Mulloy, C. Bavington, K. Black and D.H. Green. 2007b. Glycoprotein emulsifiers from two marine *Halomonas* species: chemical and physical characterization. Journal of Applied Microbiology 103(5): 1716–1727.

Haba, E., A. Pinazo, O. Jauregui, M.J. Espuny, M.R. Infante and A. Manresa. 2003. Physicochemical characterization and antimicrobial properties of rhamnolipids produced by *Pseudomonas aeruginosa* 47T2 NCBIM 40044. Biotechnology and Bioengineering 81(3): 316–322.

Hamza, F., S. Satpute, A. Banpurkar, A.R. Kumar and S. Zinjarde. 2017. Biosurfactant from a marine bacterium disrupts biofilms of pathogenic bacteria in a tropical aquaculture system. FEMS Microbiology Ecology 93(11): fix140.

Hausmann, R. and C. Syldatk. 2014. Types and classification of microbial surfactants. pp. 3–18. *In*: Kosaric, N. and F. Varder-Sukan (eds.). Biosurfactants: Production and Utilization—Processes, Technologies, and Economics. CRC Press Taylor & Francis Group: Boca Raton, FL, USA.

Hommel, R.K. and K. Huse. 1993. Regulation of sophorose lipid production by *Candida (Torulopsis) apicola*. Biotechnology Letters 15(8): 853–858.

Huang, W., Y. Lang, A. Hakeem, Y. Lei, L. Gan and X. Yang. 2018. Surfactin-based nanoparticles loaded with doxorubicin to overcome multidrug resistance in cancers. International Journal of Nanomedicine 13: 1723.

Ibacache-Quiroga, C., J. Ojeda, G. Espinoza-Vergara, P. Olivero, M. Cuellar and M.A. Dinamarca. 2013. The hydrocarbon-degrading marine bacterium *Cobetia* sp. strain MM1IDA2H-1 produces a biosurfactant that interferes with quorum sensing of fish pathogens by signal hijacking. Microbial Biotechnology 6(4): 394–405.

Inès, M. and G. Dhouha. 2015. Lipopeptide surfactants: production, recovery and pore forming capacity. Peptides 71: 100–112.

Iqbal, H.A., J.W. Craig and S.F. Brady. 2014. Antibacterial enzymes from the functional screening of metagenomic libraries hosted in *Ralstonia metallidurans*. FEMS Microbiology Letters 354(1): 19–26.

Jackson, S.A., E. Borchert, F. O'Gara and A.D. Dobson. 2015. Metagenomics for the discovery of novel biosurfactants of environmental interest from marine ecosystems. Current Opinion in Biotechnology 33: 176–182.

Janakiram, N.B., A. Mohammed and C.V. Rao. 2015. Sea cucumbers metabolites as potent anti-cancer agents. Marine Drugs 13(5): 2909–2923.

Janek, T., A. Krasowska, A. Radwańska and M. Łukaszewicz. 2013. Lipopeptide biosurfactant pseudofactin II induced apoptosis of melanoma A 375 cells by specific interaction with the plasma membrane. PloS One 8(3): e57991.

Käppeli, O. and W.R. Finnerty. 1980. Characteristics of hexadecane partition by the growth medium of *Acinetobacter* sp. Biotechnology and Bioengineering 22(3): 495–503.

Kennedy, J., B. Flemer, S.A. Jackson, D.P. Lejon, J.P. Morrissey, F. O'gara and A.D. Dobson. 2010. Marine metagenomics: new tools for the study and exploitation of marine microbial metabolism. Marine Drugs 8(3): 608–628.

Kim, S.K. (ed.). 2013. Marine Biomaterials: Characterization, Isolation and Applications. CRC press.

Kiran, G.S., T.A. Hema, R. Gandhimathi, J. Selvin, T.A. Thomas, T.R. Ravji and K. Natarajaseenivasan. 2009. Optimization and production of a biosurfactant from the sponge-associated marine fungus *Aspergillus ustus* MSF3. Colloids and Surfaces B: Biointerfaces 73(2): 250–256.

Kiran, G.S., B. Sabarathnam and J. Selvin. 2010. Biofilm disruption potential of a glycolipid biosurfactant from marine *Brevibacterium casei*. FEMS Immunology and Medical Microbiology 59(3): 432–438.

Korla, K. and C.K. Mitra. 2014. Modelling the Krebs cycle and oxidative phosphorylation. Journal of Biomolecular Structure and Dynamics 32(2): 242–256.

Kretschmer, A., H. Bock and F. Wagner. 1982. Chemical and physical characterization of interfacial-active lipids from *Rhodococcus erythropolis* grown on n-alkanes. Applied and Environmental Microbiology 44(4): 864–870.

Kügler, J.H., L. Roes-Hill, C. Syldatk and R. Hausmann. 2015. Surfactants tailored by the class *Actinobacteria*. Frontiers in Microbiology 6: 212.

Lawrance, A., M. Balakrishnan, T.C. Joseph, D.P. Sukumaran, V.N. Valsalan, D. Gopal and K. Ramalingam. 2014. Functional and molecular characterization of a lipopeptide surfactant from the marine sponge-associated eubacteria *Bacillus licheniformis* NIOT-AMKV06 of Andaman and Nicobar Islands, India. Marine Pollution Bulletin 82(1–2): 76–85.

Lee, Y.M., J.K. Hong, C.O. Lee, K.S. Bae, D.K. Kim and J.H Jung. 2010. A cytotoxic lipopeptide from the sponge-derived fungus *Aspergillus versicolor*. Bulletin of the Korean Chemical Society 31(1): 205–208.

Ma, Z. and J. Hu. 2018. Plipastatin A1 produced by a marine sediment-derived *Bacillus amyloliquefaciens* SH-B74 contributes to the control of gray mold disease in tomato. 3 Biotech 8(2): 125.

Maneerat, S., T. Bamba, K. Harada, A. Kobayashi, H. Yamada and F. Kawai. 2006. A novel crude oil emulsifier excreted in the culture supernatant of a marine bacterium, *Myroides* sp. strain SM1. Applied Microbiology and Biotechnology 70(2): 254–259.

Maneerat, S. and P. Dikit. 2007. Characterization of cell-associated bioemulsifier from *Myroides* sp. SM1, a marine bacterium. Cell 29(3): 770.

Maneerat, S. and K. Phetrong. 2007. Isolation of biosurfactant-producing marine bacteria and characteristics of selected biosurfactant. Songklanakarin Journal of Science and Technology 29(3): 781–791.

Manivasagan, P., P. Sivasankar, J. Venkatesan, K. Sivakumar and S.K. Kim. 2014. Optimization, production and characterization of glycolipid biosurfactant from the marine actinobacterium, *Streptomyces* sp. MAB36. Bioprocess and Biosystems Engineering 37(5): 783–797.

Marchant, R. and I.M. Banat. 2012. Microbial biosurfactants: challenges and opportunities for future exploitation. Trends in Biotechnology 30(11): 558–565.

Margesin, R. and F. Schinner. 2001. Bioremediation (natural attenuation and biostimulation) of diesel-oil-contaminated soil in an alpine glacier skiing area. Applied and Environmental Microbiology 67(7): 3127–3133.

Matsuyama, T.O.H.E.Y., K. Kaneda, I. Ishizuka, T. Toida and I. Yano. 1990. Surface-active novel glycolipid and linked 3-hydroxy fatty acids produced by *Serratia rubidaea*. Journal of Bacteriology 172(6): 3015–3022.

Mouafo, T.H., A. Mbawala and R. Ndjouenkeu. 2018. Effect of different carbon sources on biosurfactants' production by three strains of *Lactobacillus* spp. BioMed. Research International 2018: 5034783.

Mukherjee, S., P. Das, C. Sivapathasekaran and R. Sen. 2009. Antimicrobial biosurfactants from marine *Bacillus circulans*: extracellular synthesis and purification. Letters in Applied Microbiology 48(3): 281–288.

Mulligan, C.N., R.N. Yong and B.F. Gibbs. 2001. Heavy metal removal from sediments by biosurfactants. Journal of Hazardous Materials 85(1-2): 111–125.

Nakar, D. and D.L. Gutnick. 2001. Analysis of the wee gene cluster responsible for the biosynthesis of the polymeric bioemulsifier from the oil-degrading strain *Acinetobacter lwoffii* RAG-1. The GenBank/ EMBL accession number for the sequence analysis of the eight fragments determined in this work is AJ243431. Microbiology 147(7): 1937–1946.

Nayak, N.S., M.S. Purohit, D.R. Tipre and S.R. Dave. 2020a. Biosurfactant production and engine oil degradation by marine halotolerant *Bacillus licheniformis* LRK1. Biocatalysis and Agricultural Biotechnology 101808.

Nayak, N.S., S.C. Thacker, D.R. Tipre and S.R. Dave. 2020b. *Bacillus pumilus*—A marine bacteria: unexplored source for potential biosurfactant production. Bioscience Biotechnology Research Communications 13(1): 180–187.

Olivera, N.L., M.G. Commendatore, O. Delgado and J.L. Esteves. 2003. Microbial characterization and hydrocarbon biodegradation potential of natural bilge waste microflora. Journal of Industrial Microbiology and Biotechnology 30(9): 542–548.

Papagianni, M. 2012. Recent advances in engineering the central carbon metabolism of industrially important bacteria. Microbial Cell Factories 11(1): 1–13.

Passeri, A., S. Lang, F. Wagner and V. Wray. 1991. Marine biosurfactants, II. Production and characterization of an anionic trehalose tetraester from the marine bacterium *Arthrobacter* sp. EK 1. Zeitschrift für Naturforschung C 46(3–4): 204–209.

Passeri, A., M. Schmidt, T. Haffner, V. Wray, S. Lang and F. Wagner. 1992. Marine biosurfactants. IV. Production, characterization and biosynthesis of an anionic glucose lipid from the marine bacterial strain MM1. Applied Microbiology and Biotechnology 37(3): 281–286.

Pepi, M., A. Cesàro, G. Liut and F. Baldi. 2005. An Antarctic psychrotrophic bacterium *Halomonas* sp. ANT-3b, growing on n-hexadecane, produces a new emulsyfying glycolipid. FEMS Microbiology Ecology 53(1): 157–166.

Perfumo, A., T. Smyth, R. Marchant and I.M. Banat. 2010a. Production and roles of biosurfactants and bioemulsifiers in accessing hydrophobic substrates. pp. 1501–1512. *In*: Timmis, K.N. (eds.). Handbook of Hydrocarbon and Lipid Microbiology. Springer, Berlin, Heidelberg.

Perfumo, A., I. Rancich and I.M. Banat. 2010b. Possibilities and challenges for biosurfactants use in petroleum industry. pp. 135–145. *In*: Sen, R. (eds.). Biosurfactants. Advances in Experimental Medicine and Biology. vol 672. Springer, New York, NY.

Poremba, K., W. Gunkel, S. Lang and F. Wagner. 1991. Toxicity testing of synthetic and biogenic surfactants on marine microorganisms. Environmental Toxicology and Water Quality 6(2): 157–163.

Poulsen, T.B. 2011. A concise route to the macrocyclic core of the rakicidins. Chemical Communications 47(48): 12837–12839.

Qiao, N. and Z. Shao. 2010. Isolation and characterization of a novel biosurfactant produced by hydrocarbon-degrading bacterium *Alcanivorax dieselolei* B-5. Journal of Applied Microbiology 108(4): 1207–1216.

Rahman, P.K. and E. Gakpe. 2008. Production, characterisation and applications of biosurfactants— review. Biotechnology 7: 360–370.

Rodrigues, L., I.M. Banat, J. Teixeira and R. Oliveira. 2006. Biosurfactants: potential applications in medicine. Journal of Antimicrobial Chemotherapy 57(4): 609–618.

Rosenberg, E. and E.Z. Ron. 1997. Bioemulsans: microbial polymeric emulsifiers. Current Opinion in Biotechnology 8(3): 313–316.

Saggese, A., R. Culurciello, A. Casillo, M.M. Corsaro, E. Ricca and L. Baccigalupi. 2018. A marine isolate of *Bacillus pumilus* secretes a pumilacidin active against *Staphylococcus aureus*. Marine Drugs 16(6): 180.

Santos, D.K.F., R.D. Rufino, J.M. Luna, V.A. Santos and L.A. Sarubbo. 2016. Biosurfactants: multifunctional biomolecules of the 21st century. International Journal of Molecular Sciences 17(3): 401.

Sar, N. and E. Rosenberg. 1983. Emulsifier production by *Acinetobacter calcoaceticus* strains. Current Microbiology 9(6): 309–313.

Satpute, S.K., B.D. Bhawsar, P.K. Dhakephalkar and B.A. Chopade. 2008. Assessment of different screening methods for selecting biosurfactant producing marine bacteria. Indian Journal of Marine Sciences 37(3): 243–250.

Satpute, S.K., I.M. Banat, P.K. Dhakephalkar, A.G. Banpurkar and B.A. Chopade. 2010. Biosurfactants, bioemulsifiers and exopolysaccharides from marine microorganisms. Biotechnology Advances 28(4): 436–450.

Sawadogo, W.R., R. Boly, C. Cerella, M.H. Teiten, M. Dicato and M. Diederich. 2015. A survey of marine natural compounds and their derivatives with anti-cancer activity reported in 2012. Molecules 20(4): 7097–7142.

Schulz, D., A. Passeri, M. Schmidt, S. Lang, F. Wagner, V. Wray and W. Gunkel. 1991. Marine biosurfactants, I. Screening for biosurfactants among crude oil degrading marine microorganisms from the North Sea. Zeitschrift für Naturforschung C 46(3–4): 197–203.

Sen, R. 2010. Surfactin: biosynthesis, genetics and potential applications. pp. 316–323. *In*: Sen, R. (ed.). Biosurfactants. Advances in Experimental Medicine and Biology. vol 672. Springer, New York, NY.

Shigemori, H., S. Wakuri, K. Yazawa, T. Nakamura, T. Sasaki and J.I. Kobayashi. 1991. Fellutamides A and B, cytotoxic peptides from a marine fish-possessing fungus *Penicillium fellutanum*. Tetrahedron 47(40): 8529–8534.

Siegel, R.L., K.D. Miller and A. Jemal. 2015. CA: a cancer journal for clinicians. Cancer Statistics 65(1): 5.

Singh, P. and S.S. Cameotra. 2004. Enhancement of metal bioremediation by use of microbial surfactants. Biochemical and Biophysical Research Communications 319(2): 291–297.

Sutherland, I.W. 1998. Novel and established applications of microbial polysaccharides. Trends in Biotechnology 16(1): 41–46.

Takeuchi, M., E. Ashihara, Y. Yamazaki, S. Kimura, Y. Nakagawa, R. Tanaka and T. Maekawa. 2011. Rakicidin A effectively induces apoptosis in hypoxia adapted Bcr-Abl positive leukemic cells. Cancer Science 102(3): 591–596.

Tally, F.P. and M.F. DeBruin. 2000. Development of daptomycin for gram-positive infections. Journal of Antimicrobial Chemotherapy 46(4): 523–526.

Thaniyavarn, J., A. Chongchin, N. Wanitsuksombut, S. Thaniyavarn, P. Pinphanichakarn, N. Leepipatpiboon and S. Kanaya. 2006. Biosurfactant production by *Pseudomonas aeruginosa* A41 using palm oil as carbon source. The Journal of General and Applied Microbiology 52(4): 215–222.

Thavasi, R., S. Jayalakshmi, T. Balasubramanian and I.M. Banat. 2007. Biosurfactant production by *Corynebacterium kutscheri* from waste motor lubricant oil and peanut oil cake. Letters in Applied Microbiology 45(6): 686–691.

Thavasi, R., V.S. Nambaru, S. Jayalakshmi, T. Balasubramanian and I.M. Banat. 2009. Biosurfactant production by *Azotobacter chroococcum* isolated from the marine environment. Marine Biotechnology 11(5): 551.

Thavasi, R., S. Jayalakshmi and I.M. Banat. 2011a. Application of biosurfactant produced from peanut oil cake by *Lactobacillus delbrueckii* in biodegradation of crude oil. Bioresource Technology 102(3): 3366–3372.

Thavasi, R., S. Jayalakshmi and I.M. Banat. 2011b. Effect of biosurfactant and fertilizer on biodegradation of crude oil by marine isolates of *Bacillus megaterium*, *Corynebacterium kutscheri* and *Pseudomonas aeruginosa*. Bioresource Technology 102(2): 772–778.

Thavasi, R., S. Sharma and S. Jayalakshmi. 2011c. Evaluation of screening methods for the isolation of biosurfactant producing marine bacteria. Journal of Petroleum and Environmental Biotechnology 1(2).

Trimble, M.J., P. Mlynárčik, M. Kolář and R.E. Hancock. 2016. Polymyxin: alternative mechanisms of action and resistance. Cold Spring Harbor Perspectives in Medicine 6(10): a025288.

Turkovskaya, O.V., T.V. Dmitrieva and A.Y. Muratova. 2001. A biosurfactant-producing *Pseudomonas aeruginosa* strain. Applied Biochemistry and Microbiology 37(1): 71–75.

Unás, J.H., D. de Alexandria Santos, E.B. Azevedo and M. Nitschke. 2018. *Brevibacterium luteolum* biosurfactant: Production and structural characterization. Biocatalysis and Agricultural Biotechnology 13: 160–167.

Ventosa, A., J.J. Nieto and A. Oren. 1998. Biology of moderately halophilic aerobic bacteria. Microbiology and Molecular Biology Reviews 62(2): 504–544.

Vijayakumar, S. and V. Saravanan. 2015. Biosurfactants-types, sources and applications. Research Journal of Microbiology 10(5): 181–192.

Waghmode, S., S. Swami, D. Sarkar, M. Suryavanshi, S. Roachlani, P. Choudhari and S. Satpute. 2020. Exploring the pharmacological potentials of biosurfactant derived from *Planococcus maritimus* SAMP MCC 3013. Current Microbiology 77(3): 452–459.

Weber, L., C. Döge, G. Haufe, R. Hommel and H.P. Kleber. 1992. Oxygenation of hexadecane in the biosynthesis of cyclic glycolipids in *Torulopsis apicola*. Biocatalysis 5(4): 267–272.

White, D.A., L.C. Hird and S.T. Ali. 2013. Production and characterization of a trehalolipid biosurfactant produced by the novel marine bacterium *Rhodococcus* sp., strain PML026. Journal of Applied Microbiology 115(3): 744–755.

Wilson, W.R. and M.P. Hay. 2011 Targeting hypoxia in cancer therapy. Nature Reviews Cancer 11: 393–410.

Xie, J.J., F. Zhou, E.M. Li, H. Jiang, Z.P. Du, R. Lin and L.Y. Xu. 2011. FW523-3, a novel lipopeptide compound, induces apoptosis in cancer cells. Molecular Medicine Reports 4(4): 759–763.

Xu, Y., R.D. Kersten, S.J. Nam, L. Lu, A.M. Al-Suwailem, H. Zheng and P.Y. Qian. 2012. Bacterial biosynthesis and maturation of the didemnin anti-cancer agents. Journal of the American Chemical Society 134(20): 8625–8632.

Yakimov, M.M., P.N. Golyshin, S. Lang, E.R. Moore, W.R. Abraham, H. Lünsdorf and K.N. Timmis. 1998. *Alcanivorax borkumensis* gen. nov., sp. nov., a new, hydrocarbon-degrading and surfactant-producing marine bacterium. International Journal of Systematic and Evolutionary Microbiology 48(2): 339–348.

Yakimov, M.M., K.N. Timmis and P.N. Golyshin. 2007. Obligate oil-degrading marine bacteria. Current Opinion in Biotechnology 18(3): 257–266.

Yalaoui-Guellal, D., S. Fella-Temzi, S. Djafri-Dib, S.K. Sahu, V.U. Irorere, I.M. Banat and K. Madani. 2020. The petroleum-degrading bacteria *Alcaligenes aquatilis* strain YGD 2906 as a potential source of lipopeptide biosurfactant. Fuel 285: 119112.

Yeh, M.S., Y.H. Wei and J.S. Chang. 2005. Enhanced Production of Surfactin from *Bacillus subtilis* by addition of solid carriers. Biotechnology Progress 21(4): 1329–1334.

Yin, S.Y., W.C. Wei, F.Y. Jian and N.S. Yang. 2013. Therapeutic applications of herbal medicines for cancer patients. Evidence-Based Complementary and Alternative Medicine, Article ID 302426.

Zhao, H., L. Yan, X. Xu, C. Jiang, J. Shi, Y. Zhang and Q. Huang. 2018. Potential of *Bacillus subtilis* lipopeptides in anti-cancer I: induction of apoptosis and paraptosis and inhibition of autophagy in K562 cells. AMB Express 8(1): 1–16.

Zhou, S., G. Liu, R. Zheng, C. Sun and S. Wu. 2020. Structural and functional insights of iturin W, a novel lipopeptide produced by the deep-sea bacterium *Bacillus* sp. wsm-1. Applied and Environmental Microbiology 86(21): e01597–20.

Zuckerberg, A., A. Diver, Z. Peeri, D.L. Gutnick and E. Rosenberg. 1979. Emulsifier of *Arthrobacter* RAG-1: chemical and physical properties. Applied and Environmental Microbiology 37(3): 414–420.

7

Mannoprotein
A Biosurfactant Produced
by Yeast

Niyati Uniyal, Renitta Jobby and *Pamela Jha**

1. Introduction

Surfactants are substances that are widely utilized for cleaning in general, removing undesirable particles or dirtiness by a process called emulsification. Surfactants are amphipathic compounds, containing both hydrophilic (polar) and hydrophobic (non-polar) groups, capable of lowering the surface or inter-facial tension between two liquid phases such as oil/water, or air/liquid interfaces (Luna et al. 2011, 2013, Gudiña et al. 2016). This largely happens due to the increase in aqueous solubility of the Non-Aqueous Phase Liquids (NAPLS) (Yin et al. 2009). Surfactants are amphiphilic moieties, both lipophilic and hydrophilic parts. The amphiphiles that form micelles can be potentially used for surface chemical works are termed as surface active agents or surfactants (Jaysre et al. 2011, Kigsley and Pekdemir 2004). It is widely used in different fields like industrial, agricultural, food, cosmetic, pharmaceutical, etc. These compounds are chemically synthesized and create environmental and toxic problems (Schramm et al. 2003, Makkar and Rockne 2003). In the past few decades' surface-active molecules having microbial origin has gained interest in these fields. These surface-active metabolites are called biosurfactants (Desai and Banat 1997). Biosurfactant has better proficiency in compare to chemical surfactant in terms of low toxicity, low molecular weight with better degradation nature and, they do have the potency to reduce the inter-facial tension between the surface of two immiscible liquid or between a solid and a liquid (Gamil 2010, Joshi 2013). A widely accepted role of biosurfactants is to enhance the uptake of insoluble substrates. Three methods of uptake of hydrocarbons are mentioned in literature, i.e., the direct interfacial uptake

Amity Institute of Biotechnology, Amity University, Mumbai, Maharashtra, India.
* Corresponding author: pamelajha@gmail.com

of the hydrocarbon by hydrophobic cell membranes, interfacial uptake enhanced by emulsification by biosurfactants and solubilization (micelle transfer) of hydrocarbons by biosurfactants. Of these methods, two methods (emulsification and solubilization) are directly influenced by biosurfactants (Jitesh et al. 2015). Increasing awareness for the need to protect the environment and the tightening of environmental regulations has resulted in an increased interest in biosurfactants as alternatives to chemical surfactants (Banat et al. 2000). Biosurfactants can be grouped into two categories namely; low-molecular-mass molecules with lower surface and inter-facial tensions and high-molecular-mass polymers, which bind tightly to surfaces (Rosenberg and Ron 1999). High-molecular-weight biosurfactants are also called bio-emulsifiers (BE). They form and stabilize oil-in-water or water-in-oil emulsions (Dastgheib et al. 2008). If oil is in the dispersed phase, it is an oil-in-water emulsion and when water droplets are dispersed in oil, the resulting emulsion is called a water-in-oil emulsion.

1.1 Merits of Biosurfactants

Research has shown that biosurfactants exhibit many advantages over chemically synthesized surfactants. The following are some of the advantages of biosurfactants (Kosaric 1992, Mulligan and Wang 2006) (Fig. 1(A)).

1) **Availability of raw material:** Biosurfactants can be produced from cheap raw materials that are available in large quantities.

2) **Biodegradability:** Biosurfactants are easily degraded by bacteria and other microscopic organisms; hence they do not pose much of a threat to the environment.

3) **Use in environmental control:** Biosurfactants can be efficiently used in handling industrial emulsions, control of oil spills, biodegradation and detoxification of industrial effluents and bioremediation of contaminated soil.

4) **Generally low toxicity:** For instance, glycolipids from *Rhodococcus* sp. 413A were 50% less toxic than Tween 80 in naphthalene solubilization tests (Kanga et al. 1997).

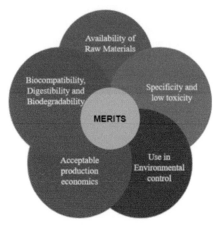

Figure 1(A). Merits of biosurfactant.

5) **Acceptable production economics:** Depending on its application, biosurfactants can also be produced from industrial wastes and by-products and this is of particular interest for their bulk production.

6) **Bio-compatibility and digestibility:** This ensures their application in cosmetic, pharmaceuticals and as functional food additives.

7) **Specificity:** Biosurfactants being complex organic molecules with specific functional groups are often specific in their action. This would be of particular interest in detoxification of specific pollutants, de-emulsification of industrial emulsions, specific cosmetic, pharmaceutical and food applications.

1.2 Demerits of Biosurfactants

Despite the numerous advantages that biosurfactants have been known to exhibit, they are also known to have the following associate demerits (Kosaric 1992) (Fig. 1(B)).

1) **Expensive:** Large scale production of biosurfactants may be expensive. However, this problem could be overcome by coupling the process to utilization of waste substrates, combating at the same time their polluting effects that balance the overall costs.

2) **Difficult purification process:** There is difficulty in obtaining pure substances (biosurfactants), which is of particular importance in pharmaceutical, food and cosmetic applications. This is because the downstream processing of diluted broths is involved and that may require multiple consecutive steps.

3) **Low productivity:** Over producing strains of bacteria are rare and those found generally display a low productivity. In addition, complex media need to be applied to the sample.

4) **Biosurfactant synthesis is difficult:** The regulation of biosurfactant synthesis is hardly understood, seemingly it represents secondary metabolite regulation.

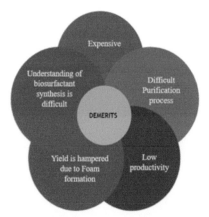

Figure 1(B). Demerits of biosurfactants.

Thus considering a batch culture, secondary metabolite production begins when the culture is stressed due to the depletion of a nutrient. This phenomenon is closely correlated with the transition phase-slow growth rate of culture and with the morphological changes that this phase implies. Among others O_2-limitation has been described as an essential parameter to govern biosurfactant production.

5) **Yield is hampered due to foam formation:** An improvement of the production yield is hampered by the strong foam formation. Consequently, diluted media must be applied and only immobilised systems provide an increased productivity of about 3 gl^{-1} h^{-1} (Fiechter 1992).

Biosurfactants are amphiphilic biological compounds produced extra cellularly or as part of the cell membranes by a variety of yeast, bacteria, and filamentous fungi (Anandaraj and Thivakaran 2010, Ghayyomi et al. 2012) from various substances including sugars, oils, and wastes. Depending on the organism there are numbers of biosurfactants available (Table 1).

Out of the biosurfactants mentioned in Table 1, the mannoprotein produced by yeast is discussed in this chapter.

Table 1. Classification of biosurfactants depending on the organism.

Organism	Biosurfactant	Species name
Bacteria	Rhamnolipids	*Pseudomonas aeruginosa, Burkholderia* sp.
	Mannosylerythritol Lipids (MEL)	*Arthrobacter* sp., *Rhodococcuserythropolis, Mycobacterium* sp.
	Surfactin, Iturin A	*Bacillus subtilis*
	Lichenysin	*Bacillus licheniformis*
	Viscosin	*Pseudomonas fluorescens*
	Serrawettin	*Serratia marcescens*
	Anthrofactin	*Arthrobacter* sp.
	Polymyxin	*Bacillus polymyxa*
	Emulsan	*Acinetobacter calcoaceticus* RAG-1, *Arthrobacter calcoaceticus*
	Biodispersan	*Acinetobacter calcoaceticus* A2
	Alasan	*Acinetobacter radioresistens* KA53
Fungi	Trehalolipids	*Torulopsisbombicola, T. apicola*
	Liposan	*Candida lipolytica*
	Mannoprotein	*Saccharomyces cerevisiae, Kluyveromycesmarxianus*
	Lipomanan	*Candida tropicalis*
	Sophorolipids	*Candida bombicola, Candida batistae, Trichosporanashii*
	Glycolipoprotein	*Aspergillus ustus*

* Data obtained from Vijayakumar et al. 2015.

2. Mannoprotein: Its Source and Composition

Yeast cell wall, which represents up to 20% of yeast cell dry weight, is mainly composed of β-glucans and mannoproteins. Mannoproteins are glycoproteins extracted from the cell walls of a few yeasts, mostly obtained from *Saccharomyces cerevisiae* and *Kluyveromyces marxianus*. These mannoproteins are highly glycosylated (~ 90% sugars, mainly mannose), located in the outermost layer of the yeast cells acting as structural components (Quiros et al. 2012) and released from the cell wall of yeast using pressurized heat treatments.

These molecules are classified as structural and enzymatic mannoproteins depending on their chemical compositions and specific functions in living systems. Structural mannoproteins are the most abundant and are composed of a small protein portion linked to a greater carbohydrate portion (mannopyranosyl) while enzymatic mannoproteins have more protein moieties in their structures. These molecules are not only effective emulsifiers but have been associated with stimulation of host immunity by activating immune cells and proteins as well as triggering the production of antibodies (Casanova et al. 1992, Oliveira et al. 2009). A mannoprotein bioemulsifier from *Kluyveromyces marxianus* has been reported to form a 3-month-old stable emulsion in corn oil (Lukondeh et al. 2003).

The two main classes of mannoproteins are as follows:

A) **Structural mannoproteins:** These contain approximately 90% mannose and 4–6% protein. The components are considered necessary for their action as emulsifiers, which are interspersed within a network of glucan to form the outer layer in the yeast cell wall.

B) **Mannan enzymes:** These contain approximately 30–50% protein and the remainder is carbohydrate. They are located mainly in the periplasmic space between the plasma membrane and the cell wall.

One of the examples of mannoprotein, a biosurfactant produced from *Saccharomyces cerevisiae 2031* is reported to consist of 77% carbohydrate and 23% protein (Alcantara et al. 2014).

Mannoprotein bioemulsifier is a glycoprotein with a molecular weight of about 14 to 15 KDa present within the cellular wall of *Saccharomyces* spp. and *Kluyveromyces marxianus* of yeast (Lukondeh et al. 2003). Mannoprotein is stable in pH range of 3–11 (Shepherd et al. 1995, Nitschke et al. 2007). Currently there are four kinds of mannoproteins, delineated by their structure, location, and function.

A) In the cell wall, intermeshed with the glucan network, is a structural mannoprotein containing up to 90% mannose (Peat et al. 1961).

B) In the periplasm of the cell are the mannoprotein enzymes, such as invertase and acid phosphatase, that contain about 50% mannose (Lampen 1968).

C) Inside the cell, localized predominantly in the vacuole, are hydrolytic enzymes, such as carboxypeptidase, with about 15% mannose (Hayashi 1976).

D) The fourth type of mannoprotein, represented by the sexual agglutinin carried on the surface of *Hansenulawingei* (Crandall and Brock 1968) and other strongly agglutinative yeasts (Burke et al. 1980), has about 85% mannose, most or all of which is linked to serine and threonine residues in the protein (Yen and Ballou 1974b).

In the first three types, most of the carbohydrate is attached to the protein by way of Di-N-acetylchitobiose units linked to asparagine (Nakajima and Ballou 1974b). About 10% of the mannose in the periplasmic and cell wall mannoproteins mentioned above is linked to protein in a similar fashion (Nakajima and Ballou 1974a).

2.1 *Mannoprotein: Production, Extraction and Purification*

Mannoprotein emulsifier is usually extracted in a high yield from whole cells of fresh bakers' yeast by two methods:

A) **Autolysis:** Mannoproteins are released from the yeast cell wall during yeast autolysis. Autolysis is a phenomenon that begins with the disorganization of membranous systems (cytoplasmic membrane and other organelle membranes) caused by cell's death. During autolysis enzymes glucanase and proteinase degrade the cell wall and, as result, the cell wall becomes porous and different compounds such as mannoproteins are released into the surrounding medium (Alexandre and Guilloux-Benatier 2006). Heat-extracted emulsifier was purified by ultra-filtration and contained approximately 44% carbohydrate (mannose) and 17% protein.

B) **Enzymatic digestion (β-1 and 3-glucanases):** Endo-1,4-β-D-glucanase (cellulase) (EC 3.2.1.4) promotes the hydrolysis of β-1,4 bonds in the amorphous regions of cellulose molecules, decreasing the degree of polymerization, exposing the micro-fibrils to other enzymatic attacks. Currently, fungal cellulases is used in industrial processes, emphasizing the hydrolysis of lignocellulosic biomass (Wang et al. 2012, Zhao et al. 2012).

Kerosene-in-water emulsions were stabilized over a broad range of conditions, from pH 2 to 11, with up to 5% sodium chloride or up to 50% ethanol in the aqueous phase. In the presence of a low concentration of various solutes, emulsions were stable to three cycles of freezing and thawing. An emulsifying agent was extracted from each species or strain of yeast tested, including 13 species of genera other than *Saccharomyces*. Spent yeast from the manufacture of beer and wine was demonstrated to be a possible source for the large-scale production of this bioemulsifier (Cameron et al. 1988).

Many mannoproteins extracted from yeasts' wall, have been reported to have high emulsification properties due to the presence of hydrophilic mannose polymers covalently attached to the protein backbone providing the amphiphilic structure common to surface-active agents.

There are 5 stages in the Production of Mannoprotein

Inoculum Development

↓

Biosurfactant Production

↓

Screening of biosurfactant

↓

Purification of biosurfactant

↓

Determination of properties of biosurfactant

Figure 2. Schematic representation for the production of biosurfactant by *Saccharomyces.*

Stage 1: Inoculum Development: The inoculum development was performed with cells grown in YEGP broth (Composition: 10 g/l yeast extract, 20 g/l Glucose and 20 g/l Peptone) at stationary growth phase, which was achieved after 48 h of incubation.

Step 1	Step 2	Step 3	Step 4
Strains of *S. cerevisiae* (Dry baker yeast) were collected	Yeasts were grown in 600 mL of Sabouraud-Dextrose broth under agitation at 25°C	Growth was monitored by measuring absorbance at 600 nm and by plate count method in PDA using Thoma counting chamber	For cultivation, fresh culture was inoculated in 25 mL of YEGP broth and incubated at 30°C for 24 hrs

Figure 3. Inoculum development for the production of biosurfactant (Martínez et al. 2016, Mahmood 2018).

Stage 2: Biosurfactant production: Biosurfactant production was performed according to Farahnejad et al. 2004 with slight modification.

Step 1	Step 2	Step 3	Step 4
Prepare 1L YEGP broth and divide it into 2 erlenmeyer flask	After cooling the medium to 50°C, medium was inoculated with 5% seed culture	Incubate the flask at 28-30°C for 4 days under constant agitation	Biomass was separated by centrifugation at 8,000 rpm for 15 min; Supernatant was collected for screening

Figure 4. Biosurfactant production (Farahnejad et al. 2004).

Stage 3: Screening of biosurfactant (Shoeb et al. 2015)

A) **Oil displacement test:** The oil displacement test is a reliable, rapid, and easy qualitative test carried out for the screening of biosurfactants, and requires no specialized equipment and just a small volume of samples. This test is used to detect biosurfactant production by diverse microorganisms based on the ability of the biosurfactants present in the supernatant to isolate solutions capable of spreading the oil and producing a clear zone. The diameter of this clearing zone on the oil surface correlates to surfactant activity, also called oil displacement activity (Walter et al. 2010).

Step 1	Step 2	Step 3
Oil was layered over water in a petri plate	Drop of supernatant was added to the surface of oil; water drop was used as negative control	The diameter of the clear zone on the oil was measured for each isolate **ODA=22/7 x (Radius)^2**

Figure 5(A). Screening by oil displacement method.

B) **Measurement of emulsification activity:** Emulsification activity was measured using the method described by Cooper and Goldenberg (1986). A vernier caliper was used to take measurements after 24 h. The emulsification index (E_{24}) was obtained by dividing the height of the emulsion layer by the total height and multiplied by 100. The E_{24} is used to characterize the biosurfactant/bioemulsifier in emulsifying the hydrophobic phase in the hydrophilic phase (Methods in Biotechnology 2009).

A negative control using the production medium without fermentation was carried out. All the tests were done in triplicate and the data presented was the mean and standard deviation of three independent samples.

Step 1	Step 2	Step 3
1.5 mL of Kerosene was added to 1.5mL of supernatant in a test tube	Vortexed at high speed for 2 mins and allowed to stand for 24 hrs	Percentage of the E_{24} was calculated using the following eq: **E_{24}= (Height of emulsion formed x 100)/Total height of solution**

Figure 5(B). Measurement of emulsification activity.

Stage 4: Purification of Biosurfactants: Biosurfactants have been purified from *S. cerevisiae* by two purification steps included Ion exchange chromatography using DEAE-Cellulose followed by Gel filtration chromatography using Sepharose-6B. Then these characteristics of the purification compound were studied.

A) **Ion Exchange Chromatography:** The method depends on the surface molecule charge, protein, and the buffer conditions. This is the most practical method for protein purification, the protein will have net positive or negative charge (Segel 1976).

The analysis by DEAE ion exchange chromatography can provide qualitative information about biosurfactants. Many advantages have been found in DEAE-Cellulose chromatography including high resolution power, easy handling, high capacity, good separation, and ability of reactivation for using many times besides the simplicity of separation principle which depend on charge differences (Karlesson et al. 1998).

B) **Gel Filtration Chromatography:** Additional purification carried out by a gel filtration using Sepharose-6B. Protein fractions from DEAE-cellulose were pooled and passed through gel filtration columns.

Step 1: Ion Exchange Chromatography	Step 2: Gel Filtration Chromatography
1. Use DEAE-Cellulose column of 1.5x8.5cm to apply the supernatant	1. Concentrated material from the prior purification step was applied to a column of Sepharose-6B of 1.5 x 60 cm, equilibrated with PBS.
2. Equilibrate the column with PBS buffer	
3. Washed column with PBS & eluted with 0.25,0.5,1.0M of NaCl at a flow rate of 1mL/min	2. The column was eluted with PBS buffer at a flow rate of 1mL/min
4. 5mL of sample were assayed for absorbance at 280 nm & 490 nm for measurement of protein and carbohydrate concentration respectively.	3. 5ml of sample were assayed for absorbance at 280 nm & 490 nm to determine the protein and carbohydrate concentration respectively
5. The biosurfactant peak fractions were collected and concentrated with sucrose at 4°C	4. The biosurfactant peak fractions were pooled and concentrated with sucrose at 4°C

Figure 6. Steps involved in purification processes.

2.2 *Mannoprotein: Properties, Stability and Applications*

A) **Molecular weight:** The method of gel filtration on a column Sepharos-6B was followed to estimate the molecular weight of the Mannoprotein, using a standard protein by drawing the relationship between the logarithm of a standard protein molecular weight and the size of recovery size of Void (Ve/Vo), molecular weight was calculated as shown in the following steps:

Step 1: Determination of the void volume of the column	Step 2: Determination of elution volume for the standard protein
1. Void volume was determined by using blue dextran to recovered parts of the PBS buffer.	1. To determine elution volume for the standard protein, the relationship between elution volume percentages was blotted for each standard protein to the elution volume of blue dextran (Ve/Vo) against molecular weight logarithm.
2. Measured absorbance in separate parts (5mL) at a wavelength of 600 nm	
	2. In this way, it helps to measure enzymatic molecular weight.

Figure 7. Determination of molecular weight process.

B) **Stability Studies:** Role of biosurfactants in industrial fields for various applications is dependent on its stability at different physico-chemical conditions. By considering the industrial importance of biosurfactants its stability was assessed.

i) **pH:** To determine pH stability cell free broth was adjusted to different pH using 1 N NaOH and 1 N HCl. The experimental control was prepared without a dry emulsifier as described below.

ii) **Temperature:** To determine its thermal stability cell free broth was maintained at a temperature range of 4–70°C for 15 minutes and then it allowed to reach at room temperature.

iii) **Salt concentration:** Salt tolerance capacity of the biosurfactant was studied in different concentrations of NaCl (2–10%).

C) **Applications:** Mannoprotein emulsifiers produced by *Saccharomyces cerevisiae* (Cameron et al. 1988) are classified as high-molecular weight biosurfactants. Since *S. cerevisiae* is generally regarded as safe (GRAS), it may be used in the food and pharmaceutical (Barth and Gaillard 1997), beverage and cosmetic industries (Cameron et al. 1988). According to researchers, bioemulsifiers can be used for producing mayonnaise along with carboxymethyl cellulose (CMC), instead of using expensive ingredients such as ginseng for mayonnaise formulation. Baker's yeast (*Saccharomyces cerevisiae*) is an affordable, inexpensive and non-toxic source used for producing this bioemulsifier. Emulsification properties of the biosurfactant revealed that it is not affected by a wide range of environmental conditions such as pH (3–11), temperature (4–70°C) and salt concentration (2–10%), thus making it useful for various applications. Due to the emulsification activity, mannoprotein emulsifiers were capable of forming stable emulsions with different hydrocarbons, organic solvents and waste oils, suggesting their applications as cleaning agents.

2.3 Comparison of Mannoprotein with known Bioemulsifiers from Yeasts

The mannoprotein emulsifier of *S. cerevisiae* has a chemical composition similar to that of the bioemulsifiers produced by alkane-grown yeasts. Purified liposan emulsifier from *C. lipolytica* is a glycoprotein which contains 83% carbohydrate and 17% protein (Cirigliano et al. 1985). Also the emulsifiers from *C. petrophilum* and *E. lipolytica* also contain carbohydrate and protein (Iguchi et al. 1969, Roy et al. 1979). Many yeast species have a cell wall similar to that of *S. cerevisiae* and contain glycoproteins with a structural role similar to that of the mannoprotein in *S. cerevisiae* (Ballou et al. 1974, Kaneko et al. 1973). As an emulsifying agent, mannoprotein from *S. cerevisiae* may present certain advantages. The difficulty of removing residual hydrocarbons from bioemulsifiers from alkane-grown yeasts would preclude their use in certain applications. Since *S. cerevisiae* is edible and is used in the manufacture of food and beverage products, it is expected that a mannoprotein bioemulsifier would be non-toxic. The yield of mannoprotein emulsifier was far greater than those of previously known bioemulsifiers from yeasts.

The yield of mannoprotein emulsifier was approximately 8% of the wet weight of yeast biomass. The yield of liposan from a 300 ml broth culture of *C. lipolytica* was 50 mg (Cirigliano et al. 1985); from an equivalent culture (containing 15 g [wet weight] of *S. cerevisiae* cells), the yield of mannoprotein emulsifier would be approximately 1.2 g. The yields of bioemulsifiers from *C. petrophilum,* and *E. lipolytica* were also a very small proportion of the yeast biomass (Iguchi et al. 1969, Kappeli et al. 1978, Roy et al. 1979). Mannoprotein can be extracted in high yield from yeast cells and processed to a relatively pure dry form by simple procedures. The spent yeast produced as a by-product in the brewing and wine industries could provide a source of raw material for the mass production of mannoprotein emulsifiers. This would eliminate the need to grow the yeast specifically for the production of emulsifiers, as is the case with the bioemulsifiers from other yeasts. Currently, spent yeast has low value and is dried for use as a protein supplement in animal feed or is treated as waste with a high biological oxygen demand. In addition, it may be possible to extract glycoprotein bioemulsifier as a high-value product from fodder yeasts that are used subsequently for single-cell protein (Cameron et al. 1988).

References

Alexandre, H. and M. Guilloux-Benatier. 2006. Yeast autolysis in sparkling wine—a review. Aust. J. Grape Wine Res. 12: 119–127. doi: 10.1111/j.1755–0238. 2006.tb00051.x.

Anandaraj, B. and P. Thivakaran. 2010. Isolation and production of biosurfactant producing organism from oil spilled soil. J. Biosci. Technol. 1(3): 120–126.

Anwar, M.I., F. Muhammad, M.M. Awais and M. Akhtar. 2017. A review of β-glucans as a growth promoter and antibiotic alternative against enteric pathogens in poultry. World's Poultry Science Journal 73(3): 651–661.

Ballou, C.E. and W.C. Raschke. 1974. Polymorphism of the somatic antigen of yeast. Science 184: 127–134.

Banat, I.M., R.S. Makkar and S.S. Cameotra. 2000. Potential commercial applications of microbial surfactants. Appl. Microbiol. Biotechnol. (53): 495–508.

Barriga, J.A., D.G. Cooper, E.S. Idziak and D.R. Cameron. 1999. Components of the bioemulsifier from *S. cerevisiae*. Enzyme Microb. Tech. 25(1): 96–102.

Barth, G. and C. Gaillard. 1997. Physiology and genetics of the dimorphic fungus Yarrowialipolytica. FEMS Microbiol. Rev. 19: 219–237.

Burke, D., L.M. Previato and C.E. Ballou. 1980. Cell-cell recognition in yeast. Isolation of the 21-cell sexual agglutination factor from Hansenulawingei and comparison of the factors from three genera. Proc. Natl. Acad. Sci. USA 77: 318–322.

Cameron, D.R., D.G. Cooper and R.J. Neufeld. 1988. The mannoprotein of *Saccharomyces cerevisiae* is an effective bioemulsifier. Appl. Environ. Microbiol. 54: 1420–1425.

Casanova, M., J.L.L. Ribot, J.P. Martínez and R. Sentandreu. 1992. Characterization of cell wall proteins from yeast and mycelial cells of *Candida albicans* by labelling with biotin: comparison with other techniques. Infect Immun. 60: 4898–4906.

Cirigliano, M.C. and G.M. Carman. 1985. Purification and characterization of liposan, a bioemulsifier from Candida lipolytica. Appl. Environ. Microbiol. 50: 846–850.

Crandall, M.A. and T.D. Brock. 1968. Molecular basis of mating in the yeast Hansenulawingei. Bacteriol. Rev. 32: 139–163.

Dastgheib, S.M., M.A. Amoozegar, E. Elahi, S. Asad and I.M. Banat. 2008. Bioemulsifier production by a halothermophilic *Bacillus* strain with potential applications in microbial enhanced oil recovery. Biotechnol. Lett. 30: 263–270.

Desai, J.D. and I.M. Banat. 1997. Microbial production of surfactants and their commercial potential. Microbiol. Mol. Biol. Rev. 61: 47–64.

Farahnejad, Z.J., M. Rasaee, H. Yadegari and M.F. Moghadam. 2004. Purification and characterization of cell wall mannoproteins of *Candida albicans* using intact cell method. Medical Journal of the Islamic Republic of Iran 18(2): 167–172.

Fiechter, A. 1992. Integrated systems for biosurfactant synthesis. Pure Applied Chem. 64: 1739–1743.

Gamil Amin, A. 2010. Potent biosurfactant producing bacterial strain for application in enhanced oil recovery. Journal of Petroleum & Environmental Biotechnology 1(2): 1–104.

Ghayyomi, J.M., F. Forghani and D.- H. Oh. 2012. Biosurfactant production by *Bacillus* sp. isolated from petroleum contaminated soil of Sirri Island. Ame. J. Appl. Sci. 9(1).

Gudiña, E.J., J.A. Teixeira and L.R. Rodrigues. 2016. Biosurfactants produced by marine microorganisms with therapeutic applications. Mar Drugs 14: 1–15, doi: 10.3390/md14020038.

Hayashi, R. 1976. Carboxypeptidase Y. pp. 568–587. *In*: Lorand, L. (ed.). Methods in Enzymology, vol XLVB. Academic Press, London New York.

Iguchi, I., I. Takeda and M. Ohsana. 1969. Emulsifying factor of hydrocarbon produced by a hydrocarbon-assimilating yeast. Agric. Biol. Chem. 33: 1657–1658.

Jaysree, R.C., S. Basu, P.P. Singh, T. Ghosal, P.A. Patra, Y. Keerthi and N. Rajendran. 2011. Isolation of biosurfactant producing bacteria from environmental samples. Pharmacology online, 3: 1427–1433.

Joshi, S.J., H. Suthar, A.K. Yadav, K. Hingurao and A. Nerurkar. 2013. Occurrence of biosurfactant producing *Bacillus* spp. in diverse habitats. International Scholarly Research Notices, 2013.

Kaneko, T., K. Kitamura and Y. Yamamoto. 1973. Susceptibilities of yeasts of yeast wall lytic enzyme of *Arthrobacter luteus*. Agric. Biol. Chem. 37: 2295–2302.

Kanga, S.H., J.S. Bonner, C.A. Page, M.A. Mills and R.L. Autenrieth. 1997. Solubilization of naphthalene from crude oil using biosurfactants. Environ. Sci. Technol. 31: 556–561.

Kappeli, O. and A. Fiechter. 1978. Chemical and structural alterations at the cell surface of *Candida tropicalis*, induced by hydrocarbon substrate. J. Bacteriol. 133: 952–958.

Karlesson, E., L. Rydnen and J. Brewer. 1998. Ion exchange chromatography. *In*: Protein Purification (ed. Wiley. Liss). Ajohn Wily and Sons, INS-publication.

Kigsley, U. and T. Pekdemir. 2004. Evaluation of biosurfactants for crude oil contaminated soil washing. Chemosphere 57: 1139–1150.

Kosaric, N. 1992. Biosurfactants in industry. Pure Applied Chem. 64: 1731–1737.

Lampen, J.O. 1968. External enzymes of yeast: their nature and formation. Antonie von Leeuwenhoek J. Microbiol. Serol. 34: 1–18.

Lukondeh, T., N.J. Ashbolt and P.L. Rogers. 2003. Evaluation of *Kluyveromyces marxianus* fii 510700 grown on a lactose-based medium as a source of a natural bioemulsifier. J. Ind. Microbiol. Biotechnol. 30(12): 715–20.

Luna, J.M., R.D. Rufifino, C.D.C. Albuquerque, L.A. Sarubbo and G.M. Campos Takaki. 2011. Economic optimized medium for tensio-active agent production by Candida sphaerica UCP0995 and application in the removal of hydrophobic contaminant from sand. Int. J. Mol. Sci. 12: 2463–2476, doi: 10.3390/ijms12042463.

Luna, J.M., R.D. Rufifino, L.A. Sarubbo and G.M. Campos Takaki. 2013. Characterization, surface properties and biological activity of a biosurfactant produced from industrial waste by Candida sphaerica UCP0995 for application in the petroleum industry. Colloids Surf B Biointerfaces 102: 202–209, doi:10.1016/j.colsurfb.2012.08.008.

Mahmood, N.N. 2018. Effect of biosurfactants purified from Saccharomyces cerevisiae against Corynebacteriumurealyticum. Journal of Pharmaceutical Sciences and Research 10(3): 481–486.

Makkar, R.S. and K.J. Rockne. 2003. Comparison of synthetic surfactants and biosurfactants in enhancing biodegradation of polycyclic aromatic hydrocarbons. Environ. Toxicol. Chem. 22: 2280–2292.

Martínez, J.M., G. Cebrián, I. Álvarez and J. Raso. 2016. Release of Mannoproteins during *Saccharomyces cerevisiae* autolysis induced by pulsed electric field. Front. Microbiol. 7: 1435. doi: 10.3389/fmicb.2016.014352012;1–6.

Methods in biotechnology. 2009. Surfactant properties: How to quantitatively measure. The alternative resource to find the laboratory methods for biotechnology based on published journals and reference books. http://biotechmethods.blogspot.com/2009/02/references.html.

Mulligan, C.N. and S. Wang. 2006. Remediation of a heavy metal-contaminated soil by a rhamnolipid foam. Eng. Geol. 85: 75–81.

Nakajima, T. and C.E. Ballou. 1974a. Characterization of the carbohydrate fragments obtained from *Saccharomyces cerevisiae* mannan by alkaline degradation. J. Biol. Chem. 249: 7679–7684.

Nakajima, T. and C.E. Ballou. 1974b. Structure of the linkage region between the polysaccharide and protein parts of *Saccharomyces cerevisiae* mannan. J. Biol. Chem. 249: 7685–7694.

Nitschke, M. and S. Costa. 2007. Biosurfactants in food industry. Trends Food Sci. Tech. 18(5): 252–9.

Oliveira, M.C., D.F. Figueiredo-Lima, D.E. FariaFilho, R.H. Marques and V.M.B. Moraes. 2009. Effect of mannan oligosaccharides and/or enzymes on antibody titers against infectious bursal and Newcastle disease viruses. Arq. Bras. Med. Vet. 61: 6–11 10.1590/S0102-09352009000100002.

Peat, S., W.J. Whelan and T.E. Edwards. 1961. Polysaccharides of baker's yeast. Part IV. Mannan. Biochem. J. 29–34.

Quiros, M., R. Gonzalez and P. Morales. 2012. A simple method for total quantification of mannoprotein content in real wine samples. Food Chem. 134: 1205–1210. doi: 10.1016/j.foodchem.2012.02.168.

Rosenberg, E. and E.Z. Ron. 1999. High- and low-molecular-mass microbial surfactants. Applied Microbiol. Biotechnol. 52: 154–162.

Roy, P.K., H.D. Singh, S.D. Bhagat and J.N. Baruah. 1979. Characterization of hydrocarbon emulsification and solubilization occurring during the growth of Endomycopsislipolytica on hydrocarbons. Biotechnol. Bioeng. 21: 955–974.

Schramm, L.L., E.N. Stasiuk and D.G. Marangoni. 2003. Surfactants and their applications. Ann. Rep. Program Chem. Sec. 99: 3–48.

Seagel, I.H. 1976. Biochemical Calculations, 2nd ed. John and Sons. Inc. New York.

Shepherd, R., J. Rockey, I.W. Sutherland and S. Roller. 1995. Novel bioemulsifiers from microorganisms for use in foods. J. Biotechnol. 40(3): 207–17.

Shoeb, E., N. Ahmed, J. Akhater, U. Badar, K. Siddiqui, F.A. Ansari, M. Wqar, S. Imtiaz, N. Akhatar, Q.A. Shaikh, R. Baig, S. Butt, S. Khan, S. Husain, B. Ahmed and M.A. Ansari. 2015. Turk J. Biol. 39: 210–216.

Sony, J., S.K. Arora, A. Sharma and M. Taneja. 2015. Production and characterization of biosurfactant from *Pseudomonas* spp. Int. J. Curr. Microbiol. App. Sci. 4(1): 245–253.

Vijayakumar, S. and V. Saravanan. 2015. Biosurfactants - types, sources and applications. Research Journal of Microbiology, 10(5): 181–192.

Walter, V., C. Syldatk and R. Hausmann. 2010. Screening Concepts for the Isolation of Biosurfactant Producing Microorganisms. *In*: Sen, R. (ed.). Biosurfactants. Advances in Experimental Medicine and Biology, vol 672. Springer, New York, NY. https://doi.org/10.1007/978-1-4419-5979-9_1.

Wang, M., Z. Li, X. Fang, L. Wang and Y. Qu. 2012. Cellulolytic enzyme production and enzymatic hydrolysis for second-generation bioethanol production. Biotechnology in China III: Biofuels and Bioenergy, 1–24.

Yen, P.H. and C.E. Ballou. 1974b. Partial characterization of the sexual agglutination factor from Hansenulawingei Y -2340 type 5 cells. Biochemistry 13: 2428–2437.

Yin, H., J. Qiang, Y. Jia, J. Ye, H. Peng, H. Qin, N. Zhang and B. He. 2009. Characteristics of biosurfactant produced by *Pseudomonas aeruginosa* S6 isolated from oil–containing wastewater. Process Biochem. 44: 302–308.

Zhao, X.Q., L.H. Zi, F.W. Bai, H.L. Lin, X.M. Hao, G.J. Yue and N.W. Ho. 2011. Bioethanol from lignocellulosic biomass. Biotechnology in China III: Biofuels and Bioenergy, 25–51.

8

Biomedical Application of Biosurfactants

Shrikant S Sonawane,[1] *Shreyas V Kumbhare*[2] and
Nitinkumar P Patil[2,*]

1. Introduction

It is clear, that oil and water are irreconcilable. Emulsion is caused due to its mixture. Formation of oil droplets occurs when an emulsion stays still for some time. With this regard, emulsifiers are used to stop this process, emulsifiers are used to prevent the emulsion from breaking. Examples of emulsions currently used in the industry include milk, butter, mayonnaise and ice cream. Emulsifiers are principally categories into surface active agents or surfactant. The word surfactant is used for molecules that migrate to the surface between phases.

Bioemulsifiers are higher in molecular weight, they are complex mixtures of heteropolysaccharides, lipopolysaccharides, lipoproteins and proteins. Emulsifiers have double lipophilic and hydrophilic properties. Emulsions are generally displacement of oil in water or water in oil. In oil emulsions, small droplets of oil form the dispersed phase and are discrete in water, while in water emulsions, they are distributed as small droplets of water in oil. Addition of an emulsifier to an un-mixable compound, reduces surface tension between the two phases and prevents

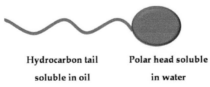

Hydrocarbon tail   Polar head soluble

soluble in oil   in water

Figure 1. Surfactant molecule with apolar (hydrophobic) and polar (hydrophilic) moieties.

[1] Smt. Chandibai Himathmal Mansukhani College, Ulhasnagar, Thane, Maharashtra-421003.
[2] National Centre for Cell Science, Pune. Current affiliation - University of Manitoba, Winnipeg, Canada.
* Corresponding author: nitinkumarpatil1@gmail.com

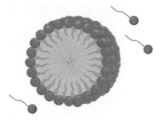

Figure 2. Schematic illustration of tensioactive agent and micelle formation.

it from separating. Allocating the two liquids to form an emulsion. Since an emulsion consists of water-soluble and oil-soluble fragments, an emulsifier is placed on the surface of the area where the two liquids (water and oil) are connected. The water-soluble fragment ambulates towards the water fragment and the fat-soluble fragment places near the oil.

Surfactants are either derived synthetically or biologically. Naturally derived surfactants are denominated biosurfactants since they are produced from biological entities, especially microorganisms. Biosurfactants of a diverse variety of molecular structures are reported to be produced by different species and strains of Fungi, bacteria, and yeast. Bacterial domains such as *Pseudomonas*, *Bacillus*, and *Acinetobacter* are found to be excellent biosurfactant producers. Among these genera of excellent biosurfactant producer's some species to highly explore are *Pseudomonas aeruginosa*, *Bacillus subtilis*, and *Acinetobacter calcoaceticus*, apart from other species. Based on their biochemical nature biosurfactants are classified as low molecular weight (LMW) and high molecular weight (HMW). The former competently lowers surface and interfacial tensions while the latter is slightly more of an emulsion-stabilizing agent. On the basis of chemical composition, biosurfactants are grouped into different types as lipopeptides (serrettin, lichenysin, iturin, surfactin, fengycin), glycolipids (trehalolipids rhamnolipids, sophorolipids, mannosylerithritol lipids), fatty acids/neutral lipids/phospholipids (spiculisporic

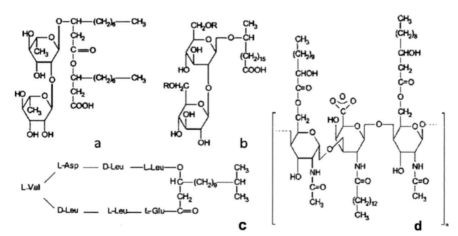

Figure 3. Chemical structure of most studied microbial surface-active compounds. (a) Rhamnolipid; (b) Sophorolipid; (c) Surfactin and (d) Emulsan.

acid and phosphatidylethanolamine), particulate biosurfactants (whole-cell, vesicles) and polymeric biosurfactants (biodispesan, alasan, emulsan, liposan). Among the different types Glycolipids, Lipopeptides, and phospholipids belong to the LMW biosurfactants while the HMW biosurfactants consists of particulate and polymeric biosurfactants.

To serve purposes of preservation, stabilization, colouring and sweetness, synthetic food additives such namely carboxymethylcellulose (CMC) and Polysorbate–80 (P80) are added to food and drinks. The use of food additives has increased dramatically since they were intentionally used for food preservation. Today, it is nearly impossible to avoid processed foods, a modern dietary pattern that is characterised by the low intake of fruit, legume and vegetable fibre and high intake of meat, dairy, eggs and refined grains, saturated fat, sugar and salt. Some processed foods can form part of a healthy, balanced diet (e.g., wholemeal bread; low-sugar, high-fibre breakfast cereals), whilst others may be considered more detrimental for health (e.g., processed meats, high-fat dairy and bakery products, confectionery, foodstuffs containing hydrogenated oils and high fructose corn syrups). Presently, there are almost more than 2500 permitted additives that are included in foods to enhance appearance, smell, texture and taste, and/or to extend shelf-life.

Different microbial and synthetic surfactants are employed in varied industries, including the cosmetics, food, healthcare and the pharmaceutical sector. Surfactants have a versatile phase character and diversity of colloidal structures, thus finding applications in many industrial processes, because of its easy modification in interface activity or stability of the colloidal systems. They are used in manufacturing shampoos and cosmetics. So, surfactants can be used in the petroleum industry, health, pharmaceuticals, agriculture, detergents, cosmetics, bioprocessing, environmental remediation, textiles, paint, leader, papermaking, and other industries and activities where water serves as an interactive medium.

The application of biosurfactants in certain industrial sectors that are condemnatory for sustainable economic development, especially in developing countries, remains a focus for every microbiologist and biotechnologist. This chapter offers an overview of the multifunctional properties of biosurfactants that influence their applications in current and diverse industrial sectors. The number of industries in which bioemulsifiers have found its applications has proved that they have the potential to substitute synthetic surfactants in the future.

There are reports of associations between 'ultra-processed' foods and adverse health outcomes in populations around the world, including allergic and autoimmune disorders, some types of cancer, cardiovascular diseases and metabolic disorders, such as type 2 diabetes and obesity. 'Ultra-processed' foods can be defined as foodstuffs made from processed substances extracted or refined from whole food, are typically energy dense; have a high glycaemic load; are low in dietary fiber, micronutrients, and phytochemicals; and are high in unhealthy types of dietary fat, free sugars, and sodium. Furthermore, that 'intense palatability' achieved by a high content of fat, sugar, salt and cosmetics and other additives encourages over consumption of such food. 'Ultra-processed' products include those used to imitate or enhance the sensory qualities of foods or to disguise unpalatable aspects of the final product. These additives include dyes and other colours, colour stabilizers; flavour enhancers,

Table 1. Functional classes and examples of additives in foods. Adapted from European Parliament.

Functional class	Description	Example additive (E number)*
Acidity regulators	Alter or control the acidity or alkalinity pH of a foodstuff.	E325 Sodium lactate
Acids	Rise in the acidity of a foodstuff and/or impart a sour taste to it.	E507 Hydrochloric acid
Anti-caking agents	Decrease the tendency of individual particles of a foodstuff to adhere to one another.	E341 Calcium phosphate
Anti-foaming agents	Stop or decrease foaming.	E905a Mineral oil
Antioxidants	Protect foods against deterioration caused by oxidation and colour changes which results in increase in shelf life of food.	E300 Ascorbic Acid
Bulking agents	Effect on volume of foodstuff without any effect on its available energy value.	E336 Potassium tartrates
Carriers	Dissolve, dilute, disperse or otherwise physically modify a food additive or a flavoring, food enzyme, nutrient and/or other substance added for nutritional or physiological purposes to a food without any changing its function.	E1200 Polydextrose
Colours	Add or restore colour in a food, and include natural constituents of foods and natural sources, which are normally not consumed as foods as such and not normally used as characteristic ingredients of food.	E100 Curcumin
Emulsifiers	Useful to maintain homogenous mixture of oil and water in the food sample.	E322 Lecithin
Salts Emulsifier	Help full to convert protein present in cheese so that it will disperse and resulting in uniform distribution of fat with other components of cheese.	E325 Sodium lactate
Stiffening agents	It provides a firm compactness to the tissue of fruit and vegetables. It imparts uniform strengthening to the food.	E333 Calcium citrates
Flavour enhancers	Help to increase the taste and odour of food material.	E620 Glutamic acid
Flour treatment agents	Mainly use to enhance baking quality of four or dough.	E927b Carbamide
Foaming agents	In liquid or solid food materials this foaming agent help for uniform homogenous dispersion.	E999 Quillaia extract
Gelling agents	Formation of gel provide uniform texture to food material.	E441 Gelatine
Glazing agents	Helpful for formation of shiny appearance and protective coating on to the external surface of a food materials.	E901 Bees wax
Humectants	Mainly avoids drying of foodstuffs by counteracting the effect of an atmosphere having a low degree of humidity, or enhance the dissolution of a powder in an aqueous medium.	E965 Maltitol

Table 1 Contd. ...

...Table 1 Contd.

Functional class	Description	Example additive (E number)*
Modified starches	Mainly derived from one step or multiple step chemical treatments of edible starches, which may have undergone in various treatment like enzymatic or a physical treatment and may be treatment of acid or alkali.	E1404 Oxidized starch
Packaging gases	Different types of gases other than air are introduced into a container every time before, during or after the placing of a food material in it.	E938 Argon
Preservatives	Protect food material from deterioration caused by microorganisms which results in increased in the shelf-life of foodstuffs. Also, growth of pathogenic micro-organisms is prevented.	E200 Sorbic acid
Propellants	Gases other than air that expel a food material from a container.	E942 Nitrous oxide
Raising agents	The volume of dough or batter is increased by liberation of gas from it using substances or combinations of substances.	E500 Sodium carbonate
Sequestrants	Develop a chemical complex with different metallic ions.	E385 Calcium disodium ethylene diamine tetraacetate
Stabilizers	Help to maintain the physico-chemical state of a food materials.	E415 Xanthan gum
Sweeteners	Impart a sweet taste to foods or in table-top sweeteners.	E955 Sucralose
Thickeners	Increase the viscosity of a foodstuff.	E1400 Dextrin

* E stands for Europe, are codes for substances used as food additives for use within the European Union (EU) and European Free Trade Association (EFTA)

non-sugar sweeteners; and processing aids such as carbonating, firming, bulking and anti-bulking, de-foaming, anti-caking and glazing agents, emulsifiers, sequestrants and humectants. Regulatory bodies ensure that food additives are rigorously tested for safety and additives continue to undergo long-term monitoring for their effects on health conditions. Food additives that pass these safety tests are given an 'E' number which are be listed on packaging.

2. Widespread Use of Emulsifiers

Consumption of some food additives (e.g., artificial sweeteners such as sucralose) may be much more difficult to avoid ingestion of emulsifiers (also known as surfactants or detergents) because they are commonly added to a wide variety of foods. Regulatory bodies can define limits on amounts that can be added to food products, information regarding actual content within foods usually lacks on food labels, limiting our knowledge of levels consumed and our ability to avoid consumption of a diverse array of surfactant compounds used in foods. The term 'emulsifier' is commonly used for surfactants that are used in both the food and pharmaceutical industries, while the term 'detergent' is more commonly used to refer to specific surfactants used in household and cleaning products (e.g., washing liquids, shampoos, toothpastes). A wide range of surfactants is available, both those that are synthetic (e.g., polysorbates, derived from polyethoxylated sorbitan and oleic acid, also known as Tween) and biological (e.g., lecithin), many of which can also be modified. Surfactants have the common property of being amphophilic [i.e., with a molecular structure that includes both a hydrophilic (water-loving, polar) and a lipophilic (fat-loving) component]. Lipophilic components tend to be similar, but hydrophilic components vary and form the basis for the classification of surfactants as non-ionic, anionic, cationic and amphoteric. Within the food industry, synthetic polysorbates were initially incorporated into margarines and then used extensively in the baking industry as preservatives to prevent staling, and enhance firmness and volume of bakery goods. Polysorbates, and other synthetic emulsifiers, are frequently incorporated into dietary products, either singularly or in combination. The same synthetic emulsifiers are used in pharmaceutical products as absorption enhancers. Natural emulsifiers such as lecithin (phosphatidylcholine) are broken down to choline-rich nutrients on passage through the small intestine by intestinal lipases and then acted upon by bacteria to produce triethylamine. There is a greater resistance to breakdown by digestion of synthetic emulsifiers, such as the polysorbate series of surfactants, as seen for polysorbate 80 where the fatty acid moieties are effectively metabolised but the sorbitol part of the molecule is seen to be highly resistant to digestion in the intestine. Likewise, carboxymethylcellulose is a non-digestible polysaccharide polymer, hence its common use as a thickening agent and stabilizer in food emulsions. Citric acid esters of mono- and diglycerides used to stabilise emulsions in food are completely hydrolysed in the gut into constituent free fatty acids, glycerol and citric acid. Evidences suggest that the ester bond between citric acid and glycerol is likely not fully hydrolysable.

3. Potential Concerns

Permitted dietary emulsifiers impact on gut health through impairing intestinal barrier function, thus increasing antigen exposure, or by modulating the microbiota, thus potentially increasing the incidence of inflammatory bowel disease (IBD) and metabolic syndrome. Key food stabilizers and additives, including maltodextrin, have been associated with increased early life intestinal stress, damage and inflammation in animal studies. Exposure to carboxymethylcellulose is 2–3 times higher than the estimated mean daily exposure seen in the population consuming processed foodstuffs. Potential effects of food additives on the gut microbiome have generally been overlooked; however, emerging evidence, mainly from animal studies, suggests that several common food additives, not just emulsifiers, can induce microbiota-mediated adverse effects. Taken together, the emerging effects on intestinal inflammation and gut microbiota are consistent with those observed in IBD. Food exclusion diets for Crohn's disease, which encourage the avoidance of additive-rich 'processed foods', have been observed to induce remission, although lots of other dietary factors may be involved.

4. Antiadhesive Agents

A biofilm is described as a collection of bacteria that paperwork a colony or a film on a ground. The biofilm no longer best consists of microorganisms, but it is also all the extracellular material produced at the surface and any material trapped in the normal matrix. Bacterial biofilms that are present inside the food industry surfaces are ability property of contamination that may cause food spoilage and transmission of illness (Hood and Zottola 1995). Therefore, controlling the adherence of microorganisms to meals-contact surfaces is a critical step in presenting cozy and great merchandise to clients. Surfactants produced by *Streptococcus thermophilus* has been mainly used for fouling manipulate of heat-exchanger plates in pasteurizers because it decelerates the colonization of different thermophilic microorganisms due to which it is utilized as fats stabilizers and anti-spattering strains of *Streptococcus* which might be chargeable for fouling. Biosurfactant obtained from *Pseudomonas fluorescens* inhibits the attachment of *L. monocytogenes* by the treatment of stainless-steel surfaces with a biosurfactant. The bio conditioning of surfaces via the usage of microbial surfactants has been cautioned as an ultra-modern strategy to lessen adhesion.

5. Therapeutic and Biomedical Applications and Antimicrobial Activity

The antibiotic activity of bioemulsifiers is primarily due to the ability of lipopeptide molecules to self-companion and forming a pore bearing channel or micellular combination at the inner lipid membrane. Surfactins are capable of penetrating into the membrane via hydrophobic interactions, and as a consequence influencing the order of hydrocarbon chains and varying membrane thickness. The bioactive surfactants produced by the *Bacillus circulans* has an antimicrobial motion in the direction of not only both gram-positive and gram-negative pathogenic microorganisms but also towards semi pathogenic microorganism along with *Bacillus pumilis, E. coli, Serratia marcescens* and *Proteus vulgaris*. Several biosurfactants have tested antimicrobial movement in competition to diverse micro organism, algae, fungi, and viruses. The lipopeptide iturin from *B. subtilis* confirmed strong antifungal activity (Kitamoto et al. 1993). The cellobiose lipid flocculosin synthesized with the aid of *Pseudomonas flocculosa* has tested antifungal activity towards pathogenic yeast which includes *Cryptococcus neoformans, Trichosporonashii* and *Candida albicans* (Ahmed and Hussan et al. 2013).

5.1 Anticancer Activity

The organic activity for the seven different microbial extracellular glycolipids, mannosylerythritol lipids-a, mannosylerythritol lipids-b, sophorose lipid, polyol lipid, rhamnolipid are studied, it is reported that maximum of these glycolipids, except for rhamnolipid, are able to spark off cell differentiation instead of cellular proliferation within the human promyelocytic HL60 leukaemia cell line. Particularly extended common place differentiation characteristics in granulocytes and monocytes was observed for MEL and STL respectively. The condensation of chromatin, DNA fragmentation and sub-G1 arrest is due to publicity of B16 cells to MEL.

5.2 Antitumor Activity

In several intercellular molecular recognitions such as cellular differentiation, signal transduction and mobile immune responses significant contribution of bio emulsifiers are taken into consideration. MCF-7 cells via a ROS/JNK-mediated mitochondrial/caspase pathway in human breast cancers is a result of apoptosis mediated by surfactins, visconsin is another surface lively compound that is a cyclic lipopeptidfe recovered from M9-3 inhibited the migration of metastatic prostate cancers without any toxic results (Fracchia et al. 2012).

It has been observed that microbial products must be taken into consideration to take part in numerous intercellular molecular recognitions together with signal transduction, cellular differentiation, cell immune reaction and so forth. In human breast cancers MCF-7 cells via a ROS/JNK mediated mitochondrial/caspase pathway apoptosis induced by surfactins.

Viscosin, isolated from *Pseudomonas libanensis* M9-3 is a powerful surface lively cyclic lipopeptide, without visible toxicity outcomes inhibited the migration of the metastatic prostate cancer mobile line, pc-3M. Now a days, lipopeptides specifically surfactins and fengycins which are derived from a marine *Bacillus circulans* exhibited interesting cytotoxic interest in competition to most cancers cell traces. Large outcomes in the direction of every tumor mobile traces in comparison to non-tumor cellular line representing the specific inhibitory activity of these molecules. A Serratamolide AT514 which is the group of serrawettins cyclic d epsipeptide isolated from *Serratia marcescens* reported to be a strong inducer of apoptosis of diverse cellular strains derived from numerous human tumors and B lymphocytic leukemia cells. It is mainly associated with meddling with AKT/NF-kB survival indicators and the mitochondria-mediated apoptotic pathway and (Escobar-díaz et al. 2005).

5.3 Biosurfactants/Bioemulsifiers Ability in Drug Transport

Detergency, emulsification, foaming, dispersion are such properties of bioemulsifiers the motive them to appropriate for use in drug delivery. Rhamnolipids and sophorolipids have been combined with lecithin's to put together biocompatible micro emulsions wherein phase behavior became unaffected by modifications in temperature and electrolyte cognizance making them appropriate for drug delivery. Rhamnolipids liposomes are used as microcapsules for capsules, proteins, nucleic acids and dyes. The lipopeptide, fengycin and surfactins act as enhancers for transdermal penetration and pores and pores and skin accumulation of acyclovir (Fracchia et al. 2012).

5.4 Anti-inflammatory Activity

Bioemulsifier exhibit anti-inflammatory activity. Surfactin is able to down regulating LPS brought on nitric oxide, it moreover down regulates number one macrophages inhibiting NF-κβ activation which indicates a brilliant ability as bacterium derived anti-inflammatory agent. *Bacillus subtilis* secrete surfactin that are said to inhibit phospholipase, concerned in the pathophysiology anti-inflammatory bowel illnesses. Lipopeptides have a strong inhibitory property on production of nitric oxide. Surfactins isomers derived from *Bacillus* spp. (Banat et al. 2010).

Surfactin mediate down regulation of LPS-precipitated nitric oxide production in RAW264.7 cells and also number one macrophages through inhibiting NF-κb activation, signifying an excellent ability as an agent derived from bacterium. The *B. subtilis* PB6, a connate probiotic, in colon mucosal infection have impact on plasma cytokine levels and anti-inflammatory bowel ailment.

5.5 *Nanoparticles*

Surfactants are used for the synthesis of metallic nanoparticles that are used significantly within the field which include catalysis, mechano and electric powered utility and biomolecules. Surfactin mediated gold nanoparticles are utilized in location of drug transport, gene therapy, focused therapy (Yakimov et al. 1995, Satpute et al. 2010, Ghosh et al. 2012).

5.6 *Antiviral Activity*

Surfactins and its analogues have proven antiviral activity, they show physic-chemical interactions with virus envelope. Surfactins have tested effective inactivation of enveloped virus inclusive of retrovirus and herpes virus. Rhamnolipids a few different surface-active compounds have shown antiviral activity in opposition of herpes virus (Cirigliano and Carman 1985, Fracchia et al. 2012).

In vitro studies demonstrated that effective deactivation of cell-loose virus stocks of porcine bursal illness virus, new castle sickness virus and parvovirus, pseudiorabies virus mediated by the surfactin and fengycin formed by *B. subtiliis* which will competently prevent infections and replication of these viruses (Huang et al. 2006).

5.7 *Antifungal Activity*

Significant antifungal activity has been reported against pathogenic yeast including *Cryptococcus neoformans*, *Trichosporon asahii* and *Candida albicans* by cellobiose lipid flocculosin synthesized by *Pseudomonas flocculosa*. Antifungal activity of biosurfactants procrastinate despite the fact that their movement towards human pathogenic fungi has been hardly ever defined. The cellobiose lipid flocculosin extracted from *Pseudozyma flocculosa*, observed to give significant in vitro antifungal activity against several pathogenic yeasts, associated with human fungal infections. The byproducts undoubtedly suppressed all the pathogenic lines examined beneath acidic conditions and confirmed synergistic activity with amphotericin B. Different antifungal activity of biosurfactants in opposition to phytopathogenic fungi has additionally reported. It has been currently proven that surfactin, iturin and fenfyin which contain glycolipids along with cellobiose lipids rhamnolipids and cyclic lipopeptides can all have varying levels of antimicrobial activities (Mimee et al. 2009).

5.8 Anti-adhesion Interest of Biosurfactants/Bioemulsifiers

Formation of microbial biofilm on medical and technical device's motives risky impact such that the bacteria inner biofilm emerges as exceptional evidence towards antibiotics and create negative environment demanding situations. Biomedical tools along with urinary catheters, heart valves, venous catheters are prone to biofilm formation on its surface. Surfactins are capable of inhibiting biofilm formation of *E. coli, Salmonella typhimurium* and *Proteus mirabilis* in polyvinyl chloride partitions, similarly to vinyl urethral catheters. Rhamnolipids are likewise able to inhibit adhesion of microorganisms on silicon rubber. This indicates biosurfactants can be used as better coating sellers for medical intentional substances that may cause cut price in massive variety of health facility infections without the want for use of artificial capsules and chemical compounds (Ron and Rosenberg 2002, Fracchia et al. 2012).

Microbial biofilm formation on clinical and technical gadgets is a crucially and common risk prevalence. Strict hygienic practices through healthcare personnel inclusive of hand washing and ordinary disinfection of tools and surroundings becomes of grave importance.

5.9 Anti-biofilm Activity

There are numerous microbes which include *Pseudomonas* which might be capable of forming colonization that effects into formation of a sticky layer refereed as a biofilm. Biofilm are located on surfaces of urinary catheters, venous catheters, etc. which forms a coat on the floor, but treatment of devices with biosurfactants produced by using *S. typhimurium* exhibit inhibition of biofilm formation (Banat et al. 1995).

The infections which are device associated are often known as having a biofilm aetiology and biofilm formation can every now and then be facilitated with the useful resource of the host anti-inflammatory reaction molecules that may make adhesion to the floor of the tool much less difficult. Biomedical gadgets aren't the exception, biofilms are frequently located at the surface of big venous catheters, urinary catheters, voice prostheses, hip prostheses and intrauterine devices.

Since the nosocomial infections remains as significant problem even for hospitals with severe infection control programmes, infection control programs remain particularly sought. Development of a success generation primarily based at the biofilm formation and control is required and is expected to be a main advance in scientific exercise and protective remedy.

In conclusion, the anti-adhesive activity of biosurfactants in competition to numerous pathogens suggest their capability utility as coating entrepreneurs for clinical insertion substances that could bring about a discount in a massive variety of medical institution infections without the need to use synthetic drugs and chemicals.

5.10 Anti-mycoplasma Activity

Mycoplasma contamination in cell culture has been a regular issue because of biomedical research especially inside the case of mammalian cell traces. But it has been proved that remedy of the mammalian cell traces with the biosurfactants which include surfactin kill mycoplasma without adverse cellular strains. It is essential to perform the cytotoxicity findings that permits to avoid the same.

Mycoplasma infections in cellular lifestyle are an often-happening crucial catch situation to biomedical research, especially at the same time as it impacts the irreplaceable cell lines which in the end eventually gets destroyed. There are apparent caution signs and symptoms of constrained exclusions, found to be significant in the decontamination using surfactin. Studies suggest surfactin ability to remove mycoplasma cells separately of the aim cell, over the mode of movement of conventional antibiotics. It has furthermore been cited that surfactin showed a synergistic effect in combination with enrofloxacin, and let to mycoplasma killing activity in orders of magnitude greater than at the same time as the molecules had been used one by one (Kumar et al. 2007).

5.11 Anti-human Immunodeficiency Virus and Sperm Immobilizing Activity

The urgent need for a controlled, effective and secure vaginal topical microbicide reported with the increase diagnosis and prevalence of human immunodeficiency virus (HIV)/AIDS. To overcome this, sophorolipid synthesized by means of manner of *C. bombicola* and its different structural analogues had been effect on his or her spermicidal, anti-HIV and cytotoxic activities. Among different agents the series of sophorolipids maximum strong spermicidal and virucidal activity was reported for sophorolipid diacetate ethyl ester. The virucidal activity in opposition to HIV and sperm-immobilizing activity in competition to human semen are found to be just like those of nonoxynol. Although, it is furthermore observed that induction of vaginal mobile toxicity to develop more concerns about its applicability for long term microbicidal contraception (Kitamoto et al. 1993).

5.12 Agents for Respiratory Failure

In premature born babies, the failure of breathing detected due to a deficiency of pulmonary surfactant that could be a phospholipid protein complex. Advancement of molecular biology techniques make it feasible for isolation of the important genes for protein molecules of this surfactant and its successive cloning in bacteria which is an important prerequisite in fermentative manufacture for scientific applications (Gautam and Tyagi 2006).

5.13 Agents for the Stimulation of Pores and Skin Fibroblast Metabolism

The usage of sophorolipids in lactone form includes a primary a part of diacetyl lactones as marketers for stimulating pores and skin dermal fibroblast cellular metabolism and especially, as outlets for the stimulization of collagen neosynthesis. The purified lactone sophorolipid spinoff is of significance within the formulation of dermis antiaging products because of its impact at the stimulation of cells of the epidermis. By way of encouraging the producing of recent collagen fibres, purified lactone sophorolipids may be used both as a preventive measure in competition to developing older of the pores and pores and skin and applied in lotions for the frame, and in the frame milks, creams and gels that are used for the pores and skin (Borzeix 2003).

5.14 *Anti-adhesive Dealers in Surgicals*

Pre-treatment of silicone rubber with surfactant produced through *S. thermophilus* inhibited with the aid of the usage of 85% the adhesion of *C. albicans* while surfactants acquired from *L. fermentum* and *L. acidophilus* adsorbed on glass, reduced by the use of 77% the amount of adhering uropathogenic cells of *Enterococcus faecalis*. The biosurfactants obtained from *L. fermentum* inhibited *S. aureus* infection and adhered to surgical implants. Surfactin reduced the amount of biofilm formation with the useful resource of *Proteus mirabilis, S. enterica, Salmonella typhimurium* and, *Escherichia coli* in plates and vinyl urethral catheters (Mireles et al. 2001).

5.15 *Probiotics Surfactant Activity*

Probiotics can be considered as beneficial microorganisms which are when provided in required quantity can confer a fitness gain to the host. Now a days, global development of antimicrobial resistance has provided understanding of importance of activity of probiotics. Evidence shows that probiotic organisms can have important role in lowering the prevalence or the period of antibiotic-associated diarrhea and treatment of recurrent lower urinary tract infections, vaginal candidiasis and bacterial vaginosis.

Furthermore, they stimulate progressed immunological defense responses and may reduce the use of several toxic antimetabolites. Though probiotics mechanisms of movement show variation still few are regarded to supply numerous antimicrobial sellers at the side of natural acids, diacetyl, carbon peroxide, hydrogen peroxide, low molecular weight antimicrobial substances and bacteriocins. The producing of antimicrobial lipopeptides by means of *Bacillus,* probiotics produce as the number one mechanism for inhibition of the growth of pathogenic microorganisms within the gastrointestinal tract. Also oppose other organisms for adhering to epithelial cells in addition to biosurfactants manufacturing are widely known mechanisms of probiotics from *lactobacillus* to interfere with vaginal pathogens (Cribby et al. 2008, Barrons and Tassone 2001, Falagas et al. 2007).

Bacterial pathogens adhesion to silicone rubber with a layer of adsorbed biosurfactants has been prevented by biosurfactants acquired from the probiotic bacterium *Lactococcus lactis.* Adhesion of yeasts turned into significantly reduced in the presence of biosurfactant, but to a lesser quantity. Substantially reduction of microbial numbers on voice prostheses and induced a lower within the airflow resistance of voice prostheses after biofilm formation, which can also prolong the lifestyles of indwelling silicone rubber voice prostheses was reported by use of an artificial model, biosurfactants received from probiotic traces. In a greater ultra-modern painting, it became observed biosurfactant produced through the stress *Streptococcus thermophilus* that the treatment of silicon rubber with reduced adhering bacterial pathogen through as a whole lot as 97% and adhering *Candida* spp. via up to 70%. Significant adhesion inhibition of the pathogenic enteric bacteria through a biosurfactant produced via way of *Lactobacillus* pressure and it has been further observed the an vital dose-associated inhibition of the initial deposition of *E. coli* induced by biosurfactant and also one kind microorganism adherent on every hydrophobic and hydrophilic substrata (Velraeds et al. 1997). Recently probiotics used in stopping oral infections showed that *Streptococcus mitis* biosurfactants efficiently repressed adhesion of *Streptococcus sobrinus* HG 1025 and *Streptococcus mutans* ATCC 25175 to naked enamel, on the same time as *S. mitis* biosurfactant end up able to inhibit salivary pellicles adhesion of *S. sobrinus* HG 1025 to (Van Hoogmoed et al. 2006).

A completely unique xylolipid biosurfactant from *Lactococcus lactis* showed an awesome antibacterial hobby closer to medical pathogens of *E. coli* and MRSA strains. Xylolipid found to be safe and also non-pathogenic for different applications like oral intake and dermal packages signifying that it could be precisely used as a recovery agent and also as an effective preservative in food or splendor products. Thus, probiotic organisms may also represent a safe and powerful involvement because of their significance for human fitness and their safety. Thus, probiotics themselves or their biosurfactants might be used with an affected persons care system, along with tubes or catheters, with the reduction colonization of those through nosocomial pathogens and block a vital step within the pathogenesis of nosocomial infections (Falagas and Makris 2009).

6. Conclusion and Perspectives

The call for logo spanking new region of information surfactants inside pharmaceutical, the agriculture, food, beauty, and environmental industries is gradually developing and biosurfactants serving as powerful and environmentally well-matched compounds, perfectly to meet this demand. The complexity and excessive cost of production are important considerations for the cost-effective use of biosurfactants which can be taken care of as important and constrained in their use on a large-scale application. The detected antimicrobial, anti-adhesive, immune-modulating activity of biosurfactants, gene remedy, and clinical insertion protection propose that it is definitely worth persisting at this discipline. Furthermore, the increase rate of production in pharmaceutical and biomedical sectors can be compensated by way of the small amount of derivative. Biosurfactants used as pharmaceutical retailers are wished for only in very low concentrations. Situations for making biosurfactant production extra valuable and economically feasible encompass by optimized increase/production conditions with novel and inexperienced multi-step downstream processing techniques in addition to using recombinant types of microorganisms or decided on hyper generating mutants, that increases the possibilities of use of large variation of renewable substrates (Muthusamy et al. 2008).

Recent developments in the biomedical effectiveness are likely getting importance because of a higher functionality economic return. Furthermore, various newer applications of biosurfactants can be studied for their substantial applications in nanotechnology can be explored. Studies in their herbal roles in microbial damaging interaction, pathogenesis, motility and biofilm formation, cell to cell communication and preservation needed to propose advanced and interesting future perspectives.

References

Banat, I.M., A. Franzetti, I. Gandolfi, G. Bestetti, M.G. Martinotti, L. Fracchia, T.J. Smyth, R. Barrons and D. Tassone. 2008. Use of *Lactobacillus* probiotics for bacterial genitourinary infections in women: a review. Clin. Ther. 30: 453–468.

Barrons, R. and D. Tassore. 2002. Use of Lactobacillus probiotics for bacterial genitourinary infections in women: a review. Clinical. Ther. (30): 453–468.

Borzeix. 2003. Mapping of patents on bioemulsifiers and biosurfactants : A review. Journal of Scientific and Industrial Research (65): 91–115.

Cirigliano, M.C. and G.M. Carman. 1985. Purification and characterization of liposan, a bioemulsifier from *Candida lipolytica*. Appl. Environ. Microbiol. 50: 846–850.

Cribby, S., M. Taylor and G. Reid. 2008. Vaginal microbiota and the use of probiotics. Interdisciplinary Perspectives on Infectious Diseases. Article ID. 256490: 01–09.

Danyelle Khadydja, F. Santos, Raquel O. Rufino, Juliana M. Luna, Valdemir A. Santos and Leorie A. Sarubbo. 2016. Biosurfactants : Multifuntional biomolecules of 21st century. International Journal of Molecular Sciences 17(401): 1–31.

Escobar Diaz et al. 2005. Biosurfactant production by Serratiamarcescens SS-1 and its isogenic strain SMR defective in SpnR, a quorum sensing LuxR family protein. Biotechnology Letters 26(10): 799–802.

Falagas, M.E., G.I. Betsi and S. Athanasiou. 2007. Probiotics for the treatment of women with bacterial vaginosis. Clin. Microbiol. Infect. 13: 657–664.

Falagas, M.E. and G.C. Makris. 2009. Probiotic bacteria and biosurfactants for nosocomial infection control: a hypothesis. J. Hosp. Infect. 71(4): 301–306.

Fracchia, L., M. Cavallo, G. Allegrone and M.G. Martinotti. 2012. A *Lactobacillus*-derived biosurfactant inhibits biofilm formation of human pathogenic *Candida albicans* biofilm producers. *In*: Current Research, Technology and Education Topics in Applied Microbiology and Applied Biotechnology. 827–837.

Gautam, K.K. and V.K. Tiagi. 2006. Microbial surfactants: A review. J. Oleo Sci. 55: 155–166.

Ghosh, S., M. Ahire, A. Jabgunde, M. Bhat-Dusane, B.N. Joshi, K. Pardesi, S. Jachak, D.D. Dhavale and B.A. Chopade. 2011. Biosurfactants, bioemulsifiers and exopolysaccharides from marine microorganisms. Biotechnology Advances 28(4): 436–450.

Ghosh, Sougata, Mehul Ahire, Sumersing Patil, Amit Jabgunbde and Meenakshi Dusane. 2012. Antidiabetic activity of Cnidiaglauca and Dioscoreabulbifera: Potent Amylase and Glucosidase Inhibitors. Hindawi Publishing Corporation, Evidenvces based complementary and Alternative medicine 10.1155(2012): 929–951.

Hood, S. and E.A. Zottola. 1995. Biofilms in food processing. Food Control 6: 8–18.

Huang, X., Z. Lu, H. Zhao, X. Bie, F.X. Lü and S. Yang. 2006. Antiviral activity of antimicrobial investigating biosurfactants and bioemulsifiers: a review. Crit. Rev. Biotechnol. 30: 127–144.

Kitamoto, D., H. Yanagishita, T. Shinbo, T. Nakane, C. Kamisawa and T. Nakahara. 1993. Surface active properties and antimicrobial activities of mannosylerythritol lipids as biosurfactants produced by *Candida antartica*. Journal of Biotechnology 30: 161–163.

Kumar, A., A. Ali and L.K. Yerneni. 2007. Effectiveness of a mycoplasma elimination reagent on lipopeptide from *Bacillus subtilis* fmbj against pseudorabies virus, porcine lipopeptide surfactant produced by thermotolerant and halotolerant subsurface *Bacillus licheniformis* BAS 50. Appl. Environ. Microbiol. 61: 1706–1713.

Marchant, R. 2010. Microbial biosurfactants production, applications and future potential. Appl. Microbiol. Biotechnol. 87: 427–44.

Mimee, B., R. Pelletier and R.R. Bélanger. 2009. *In vitro* antibacterial activity and antifungal mode of action of flocculosin, a membrane-active cellobiose lipid. J. Appl. Microbiol. 107: 989–996.

Mireles, J.R. A. Toguchi and R.M. Harshey. 2001. Salmonella enteric serovartyphimurium swarming mutants with altered biofilm forming abilities. Surfactin inhibits biofilm formation. J. Bacteriology (183): 5848–5854.

Muthusamy, K., S. Gopalakrishnan, T.K. Ravi and P. Sivachidambaram. 2008. Biosurfactants: Properties, commercial production and application. Curr. Sci. 94: 736–747.

Ron, E.Z. and E. Rosenberg. 2002. Biosurfactants and oil bioremediation. Else Sci. Biotechnology 13: 249–252.

Satpute, S.K., A.G. Banpurkar, P.K. Dhakephalkar, I.M. Banat and B.A. Chopade. 2010. Methods for investigating biosurfactant and bioemulsifiers: a review. Crit. Rev. Biotchnol. 30: 127–144.

Van Hoogmoed, C.G., R.J.B. Dijkstra, H.C. van der Mei and H.J. Busscher. 2006. Influence of biosurfactant on interactive forces between mutans *Streptococci* and enamel measured by atomic force microscopy. J. Dent. Res. 85: 54–58.

Velraeds, M.M.C., H.C. Van der Mei, G. Reid and H.J. Busscher. 1997. Inhibition of initial adhesion of uropathogenic *Enterococcus faecalis* to solid substrata by an adsorbed biosurfactant layer from *Lactobacillus acidophilus*. Urology 49(5): 790–794.

Yakimov, M.M., K.N. Timmis, V. Wray and H.L. Fredrickson. 1995. Characterization of a new lipopetide surfactant produced by thermotolerant and halotolerant subsurface *Bacillus licheniformis* BAS50. Appl. Environmental Microbiology 61(5): 1706–1713.

9

Challenges and Potential Applications of Plant and Microbial-Based Biosurfactant in Cosmetic Formulations

Nur Izyan Wan Azelee,[1,2,]* *Nor Hasmaliana Abdul Manas,*[1,2]
Rohaida Che Man,[3] *Mohd Akmali Mokhter,*[4]
Nurrulhidayah Salamun,[4] *Shalyda Md Shaarani,*[3]
Fuad Mohamad[4] and *Zehra Edis*[5,6]

1. Introduction

Surfactants are chemical compounds that reduce the surface tension of a liquid. These compounds are amphiphilic consisting of two main parts: hydrophilic (water-loving) and hydrophobic (water-repelling)—thus giving surfactants the ability to reduce surface tension and promote emulsion in immiscible liquids (Costa et al. 2018, Fenibo et al. 2019). Surfactants have been employed in many applications such as in cosmetics, pharmaceuticals, food, textile, and painting to name a few where they could function as stabilizers, wetting agents, emulsifiers, and antimicrobial agents.

---

[1] Institute of Bioproduct Development (IBD), Universiti Teknologi Malaysia, 81310, UTM Skudai, Johor, Malaysia.
[2] School of Chemical and Energy Engineering, Faculty of Engineering, Universiti Teknologi Malaysia, 81310 UTM Skudai, Johor, Malaysia; Email: hasmaliana@utm.my
[3] Department of Chemical Engineering, College of Engineering, Universiti Malaysia Pahang, Lebuhraya Tun Razak, 26300 Gambang, Pahang, Malaysia; Emails: rohaida@ump.edu.my; shalyda@ump.edu.my
[4] Department of Chemistry, Faculty of Science, Universiti Teknologi Malaysia, 81310, UTM Skudai, Johor, Malaysia; Emails: mohdakmali@utm.my; nurrulhidayah@utm.my; m.fuad@utm.my
[5] Department of Pharmaceutical Sciences, College of Pharmacy and Health Sciences, Ajman University, Ajman, United Arab Emirates; Email: shalyda@ump.edu.my
[6] Center of Medical and Bio-allied Health Sciences Research, Ajman University, Ajman, United Arab Emirates; Email: z.edis@ajman.ac.ae
* Corresponding author: nur.izyan@utm.my

Surfactants could be produced synthetically in the laboratory or extracted from natural sources. For natural surfactants, they are produced by microorganisms or derived from plants. These surfactants are referred to as biosurfactants. The first biosurfactant reported in the literature is *surfactin* which is produced by *Bacillus subtilis*. It was first characterized and purified by Arima et al. (1968). Nowadays, surfactin is among other biosurfactants that are commercially available in the market. Examples of plant-based biosurfactants include saponins, lecithin, and soy proteins. Although they have been reported to have excellent emulsification properties, they are expensive when it comes to larger-scale production and also other challenges that will be covered in the rest of the chapter (Sekhon Randhawa and Rahman 2014). Advantages of biosurfactants over chemically produced surfactants include lower toxicity, improved biocompatibility, higher selectivity and specific activity in adverse conditions (temperature, salinity, and pH) and importantly the ability to be produced from renewable raw materials (microorganisms and plants) (Akbari et al. 2018, Costa et al. 2018, Fenibo et al. 2019). On top of that, biosurfactants are usually more biodegradable and eco-friendlier than synthetic surfactants (Vecino et al. 2017).

Biosurfactants are categorized based on their biochemical natures, chemical composition, and their microbial origins. On the basis of biochemical natures, biosurfactants are divided into low molecular weight (LMW) and high molecular weight (HMW) biosurfactants (Costa et al. 2018, Fenibo et al. 2019). LMW biosurfactants are more efficient at lowering the surface tension while HMW biosurfactants are good emulsion-stabilizing agents (Fenibo et al. 2019). For chemical compositions, biosurfactants are categorized into glycolipids, lipopeptides, phospholipids (or fatty acids), polymeric biosurfactants, and particulate biosurfactants—where amongst those glycolipids are the most exploited microbial biosurfactants (Costa et al. 2018, Fenibo et al. 2019, Sekhon Randhawa and Rahman 2014, Vecino et al. 2017). On the basis of the ionic natures of biosurfactants (based on the polarity of head groups of biosurfactants), there are four types of biosurfactants, namely anionic, cationic, non-ionic, and amphoteric (zwitterion) (Fenibo et al. 2019, Vecino et al. 2017). Surfactin from *Bacillus subtilis* is an example of zwitterion which exhibits both anionic and cationic behavior. For microbial origins, they could be either from fungi, bacteria, and yeast (Costa et al. 2018, Fenibo et al. 2019).

Worldwide, different countries have different regulations towards cosmetic production; however, the main goal is the same—to provide safe cosmetics for everyone. In Malaysia, cosmetic productions are regulated under the Control of Drugs and Cosmetic Regulations (CDCR) 1984 (CDCR 2020). The cosmetic industry shall follow the guidelines if they intend to supply safe cosmetic products to meet consumers' needs. Synthetically surfactants in cosmetics have been reported to cause skin irritations and allergies. Thus, an alternative approach to reduce these harmful effects is by using biosurfactants (Vecino et al. 2017). Several biosurfactants have been used in cosmetic formulations such as rhamnolipids and sophorolipids which belong to the glycolipids group (Sekhon Randhawa and Rahman 2014). However, the same guidelines as above should be followed if selected biosurfactants are intended to be used in cosmetic formulation.

Several parameters that determine the function of biosurfactants in cosmetic formulations are the critical micelle concentration, hydrophilic-lipophilic balance

(HLB) and ionic performance. The critical micelle concentration is the lowest concentration of biosurfactants that could produce the maximum reduction in surface tension (Vecino et al. 2017). HLB predicts the emulsifying ability of biosurfactants. It acts as an indicator if a biosurfactant forms water in oil or *vice versa,* which will be based on the HLB values (Costa et al. 2018). In general, hydrophilic biosurfactants will have a high value of HLB, whereas low HLB values are for lipophilic biosurfactants. The ionic performance of biosurfactants depends on the ionic nature of biosurfactants (Fenibo et al. 2019, Vecino et al. 2017). Anionic biosurfactants are known to have excellent foaming, wetting, and emulsifying properties compared to the other biosurfactants. Cationic biosurfactants are good at antimicrobial properties (Vecino et al. 2017).

2. Natural Surfactants

Natural surfactants may be produced from natural sources such as plants and microbes (bacteria and fungi). Different techniques have been applied for the extraction of surfactants from these natural sources such as extraction, filtration, precipitation, and distillation (Pradhan and Bhattacharyya 2017). Natural surfactants have been applied in many applications such as in the cosmetics, food, and pharmaceutical industries. Surfactants are usually used to reduce the surface tension, which resulted from the intermolecular force imbalance at a liquid-vapor or liquid-solid interface (Pradhan and Bhattacharyya 2017). A good surface reduction indicates good detergency and surface activity.

2.1 Plant-based Biosurfactants

2.1.1 Saponins

Plant-based surfactants are gaining more attention due to the biodegradability, biocompatibility, less toxic, acid balance, renewable, low cost, and poses less of a threat to the environment. Their structural biodiversities are based on physicochemical and biological properties such as emulsification, solubilization, foaming effect, antimicrobial, insecticide, and many others (Vincken et al. 2007). A significant variation in terms of the structure and properties of biosurfactants was found in diverse plant species (Vincken et al. 2007). Generally, the behavior of bulk solutions of surfactants is being studied, which include surface tension, density, viscosity, conductivity, and others (Zdziennicka et al. 2012). The most popular, well-known and established plant-based biosurfactant is saponin, a non-ionic biosurfactant. It is also the secondary metabolites in most of the plant species. However, it has been reported that only a few plant species are suitable to be extracted for commercial use due to the amount of available saponin inside the species (Tmáková et al. 2016). A toxicology test on saponin has been performed by (Menghao et al. 2015) on rats and rabbits, and the results show non-toxic and non-dermal irritant effects. The capability of saponins has been compared with other types of natural biosurfactants and was found that only a low level of saponins are required to form fine oil droplets and remain stable over a varied environmental conditions (Zhang et al. 2016). However, there is some distinction in the surface activity and molecular structure between different sources of botanical plants used (Zhu et al. 2019). A study on Pyagi Phool saponin and Ritha saponin by Pradhan and Bhattacharyya show that both have moderate to good detergency properties. Ritha shows a better performance in surface reduction and formed a very stable emulsion compared to Pyagi Phool. Saponin also helps to generate large surface areas which are an excellent factor for foam formation and foam stability where a good surfactant solubility produces stable foam in the cleansing agent (Mousli and Tazerouti 2007).

A study by Tmáková et al. (2016) has investigated the performance of the surface and antioxidant activities of 5 different plant-derived surfactants. Saponin extracts were obtained from *Sapindus mukorossi, Verbascum densiflorum, Equisetum arvense, Betula pendula,* and *Bellis perennis.* Interestingly, all of the plant extracts showed lower critical micelle concentration (CMC) than the synthetic surfactant, sodium lauryl sulfate (SLS) with two of the extracts, *E. arvense* and *B. perennis* showed even better CMC properties compared to the Tween 80 (a better surfactant than SLS). Moreover, *B. perennis* achieved the minimum surface tension of 36.8 mN m^{-1}. Critical micelle concentration is amongst the critical characteristics as it is a measurement of the solubilization strength, viscosity, osmotic pressure, density, and polarity of a surfactant. High surface activity and high formation of micelles in water were observed for all the plant extracts. Besides, other parameters were also tested, and were found that *S. mukorossi* and *B. pendula* extracts showed the best foaming properties, *S. mukorossi, V. densiflorum* and *E. arvense* extracts showed good emulsification properties, while *B. pendula* extract showed the highest antioxidant activity (Tmáková et al. 2016). The commercialization of a surfactant is usually based on the CMC values (Essa Mahmood 2013).

Nevertheless, a comparison study between natural and synthetic surfactants has been conducted by Zhu et al. (2019) for their interfacial and emulsification properties. Measurements of the interfacial tension at an oil-water interface were taken to study the characteristic behavior of tea saponin, quillaja saponin, and Tween 80. Saponin extracted from tea leaves was also tested for its forming and stabilizing nanoemulsions. Interfacial characteristics of tea saponin show that it has both a hydrophobic (aglycone) and a hydrophilic (carbohydrate) part, which indicates it to be highly surface-active. The high surface-active of saponin helps decrease the interfacial tension of the oil-water interface due to the absorption of the saponin molecules to the interface while screening the thermodynamically unfavorable molecular interactions between the two phases. From the result obtained, tea saponins showed a better performance compared to quillaja saponins and Tween 80 (Zhu et al. 2019) with the capability of producing nano-scaled droplets at low surfactant-to-oil ratios. However, one factor that may still hinder the application of tea saponin in commercial products is the saponin stabilized-emulsion has worse stability in high ionic strength (high salt level). A study by Mølgaard et al. (2000) has proven the biodegradability property of biosurfactants. From the study, saponin from *Phytolacca dodecandra* L'Herit was degraded in 10 days in an aquatic environment.

2.1.2 *Lecithin*

Lecithin is another alternative of plant-based surfactants that can be used rather than saponins. Lecithin consists of a mixture of phospholipids and is an amphoteric type of surfactant. Besides egg yolk, plant-based lecithin is typically obtained from soybean oil as well as several other seeds such as cottonseed, rapeseed, and sunflower oil. As lecithin is part of the cellular membrane constituents, they demonstrate excellent biocompatibility and emulsifying properties (Vater et al. 2019). From a study by Vater et al. (2019), it was also found that emulsifiers containing amphoteric lecithin were excellent in terms of skin permeation compared to non-ionic surfactants. In another study by Nguyen et al. (2010), they have formulated a mixture of biosurfactants containing lecithin, rhamnolipid, and sophorolipid. From the study conducted, they have found that mixed biosurfactant shows more ability and robustness in forming microemulsions for different types of oils. Kang et al. (2004) have studied the formulation of nano-sized particles using a different type of lecithin, oils, and non-ionic surfactants by controlling the phase inversion temperature. The performance in terms of moisturizing effects and cell toxicity of the lecithin-based microemulsions were compared with the conventional non-ionic emulsifying emulsions. The cell activity obtained from the lecithin-based nanoemulsion was recorded to be four times superior compared to other emulsions tested. Table 1 summarises the different plant-based biosurfactants and its properties.

2.2 *Microbial Based Biosurfactants*

Biosurfactants from microorganisms have gained much attention as they are potential substitutes of chemically-synthesized industrial surfactants. Microbial biosurfactants are highly degradable and less toxic, making them valuable and sustainable compounds for industrial applications. Microbial biosurfactants can be produced by a fermentation process. Several microorganisms from the class of fungi, yeast, and bacteria have been identified to produce biosurfactants. These second metabolites are produced by microorganisms as a response to the environment for cell defense, mobility, and nutrient access (Ahmadi-Ashtiani et al. 2020). Biosurfactants can be classified as low molecular weight biosurfactants such as glycolipids, lipopeptides, fatty acids, and high molecular weight polymeric biosurfactants such as polysaccharides, proteins, liposaccharides, and lipoproteins (Renard et al. 2019, Vecino et al. 2017).

Glycolipids are the most-characterized biosurfactants, which can be divided according to the carbohydrate group that is linked to the lipid tail of the molecule. For example, rhamnolipids, sophorolipids, trehalolipids, cellobiolipids, mannosylerythritol lipids, and others (Ahmadi-Ashtiani et al. 2020). Rhamnolipids are produced by *Pseudomonas* strain such as *P. aeruginosa* (Araújo et al. 2018, Ben Belgacem et al. 2015, Diab et al. 2020, Shen et al. 2016). Sophorolipids are commonly produced by yeasts such as *Starmerella bombicola* and *Candida kuoi* as reported by (Zerhusen et al. 2020) and (Bollmann et al. 2019). Actinobacteria produce Trehalolipids in the *Rhodococcus* genus (Kuyukina et al. 2020, Wang et al. 2019). Cellobiolipids are produced by *Cryptococcus humicola* (Imura et al. 2012), *Trichosporon porosum* (Kulakovskaya et al. 2010). Mannosylerythritol lipids are mostly synthesized by *Pseudozyma* yeasts (Saika et al. 2017, Shu et al. 2019) and fungi *Ceriporia lacerate* and *Ustilago scitaminea* (Niu et al. 2019).

Lipopeptides and lipoproteins are also well-characterized biosurfactants. Examples of lipopeptides are surfactin, iturin, fengycin, gramicidin, polymyxin, and

Table 1. Types and properties of plant-based biosurfactants.

Plant/Parts	Scientific name	Biosurfactant	Properties	Reference
Pyagi Phool/ Pink Rain Lily	*Zephyranthes carinata* Herbert.	Saponin	• Acid balanced. • Reduces surface tension to 40.76 mN/m. • Possesses high viscosity. • Shows good dirt dispersion. • A natural cleansing agent. • Better emulsification at higher concentrations as compared to synthetic surfactant.	Pradhan and Bhattacharyya 2017
Ritha/ Soapnut	*Sapindus mukorossi* Gaerth.	Saponin	• Acid balanced. • Exhibits a prominent surface tension reduction to 35.30 mN/m. • High foaming, wetting, and cleaning properties. • Better emulsification at higher concentrations as compared to synthetic surfactant.	Pradhan and Bhattacharyya 2017
Raw ritha	*Sapindus mukorossi*	Saponin	• Glycosidic • Compounds containing either a triterpenoid or an alkaloid steroid • As a hydrophobic nucleus (aglycone).	Rupeshkumar et al. 2011
Endod plant (berries)	*Phytolacca dodecandra* L'Herit	Saponin	• Fully degraded in 10 days.	Mølgaard et al. 2000
Pericarp	*Sapindus mukorossi*	Saponin	• Extract yield: 8.7g (43.5%). • Soluble in water. • CMC: 0.243 g L^{-1}	Tmáková et al. 2016
Flowers	*Verbascum densiflorum*	Saponin	• Extract yield: 6.9 g (34.5%). • Soluble in water. • CMC: 0.355 g L^{-1}.	Tmáková et al. 2016
Haulm	*Equisetum arvense*	Saponin	• Extract yield: 1.9 g (9.5%). • CMC: 0.033 g L^{-1}. • Non-soluble in water. • Minimum surface tension: 37.9 mN m^{-1}.	Tmáková et al. 2016
Leaf	*Betula pendula*	Saponin	• Extract yield: 5.2 g (26.0%). • Non-soluble in water. • CMC: 0.240 g L^{-1}.	Tmáková et al. 2016
Flowers	*Bellis perennis*	Saponin	• Extract yield: 4.9 g (24.5%). • CMC: 0.076 g L^{-1}. • Soluble in water. • Minimum surface tension: 36.8 mN m^{-1}.	Tmáková et al. 2016

Table 1 Contd. ...

...Table 1 Contd.

Plant/Parts	Scientific name	Biosurfactant	Properties	Reference
Tea leaves	*Camellia Lutchuensis*	Saponin	• Stabilized oil-in-water emulsions. Tea saponins behaved almost similar to commercial quillaja saponins	Zhu et al. 2019
Molina tree	*Quillaja Saponaria*	Saponin	• Could be used to solubilize hydrophobic • Carotenoids in aqueous solutions.	Tippel et al. 2016
Soybean	*Glycine max*	Lecithin	• Formulations containing lecithin resulted in lower mean drug • Fluxes compared to anionic surfactants. • Showed much lower cytotoxicity than anionic and non-ionic emulsifiers.	Vater et al. 2019
Sunflower	*Helianthus annuus*	Lecithin	• Faster sedimentation kinetics at 0.1% concentration. • 0.1% lecithins were more stable against coalescence.	Pan et al. 2002
Soybean	*Glycine max*	Lecithin	• The behaviour of the mixed microemulsion phase of lecithin/rhamnolipid/sophorolipid biosurfactant remain stable changing temperature and electrolyte concentration.	Nguyen et al. 2010
Lipoid-S75-3N (commercial)	N/A	Lecithin	• Particle size of the emulsion depends on phosphatidyl choline of lecithin as the primary emulsifier. • Unsaturated lecithin emulsified better than the saturated lecithin.	Kang et al. 2004

*CMC: critical micelle concentration; N/A: data not available

lichenysin (Nelson et al. 2020, Wu et al. 2019, Wu et al. 2017) that are produced by *Bacillus* species. Renard et al. (2019) identified viscosin and massetolide E produced by *Pseudomonas* species and syringafactins by *Pseudomonas* species and *Xanthomonas campestris* from cloud water. In addition, rare biosurfactants, glycolipopeptide that consists of sugar, lipid, and amino acid groups are also reported by several studies, produced by *Lactobacillus* spp. (Morais et al. 2017, Satpute et al. 2019).

The important properties of biosurfactants to be used in cosmetics are the critical micelle concentration (CMC), hydrophilic–lipophilic balance (HLB), ionic performance, surface tension, and solubility (Ahmadi-Ashtiani et al. 2020, Renard et al. 2019). CMC value determines the efficiency of biosurfactants. A low CMC value indicates an efficient biosurfactant to reduce the water surface tension. Table 2 summarises the CMC values for commonly characterized microbial biosurfactants.

Table 2. Properties of microbial biosurfactants.

Biosurfactant	Source	Critical micelle concentration (mg/mL) (surface tension in mN/m)	Functional properties	Reference
Rhamnolipid	*P. aeruginosa* SR17	110 (26.5)	Antimicrobial effect Anti-fibrotic function (scar therapy) Anti-aging	Diab et al. 2020, Rikalovic et al. 2015, Shen et al. 2016
	P. aeruginosa DN1	50 (25.88)		
	P. aeruginosa	50 (30)		
	P. aeruginosa SWP-4	27 (24.1)		
	P. aeruginosa ATCC 9027	22 (26)		
Sophorolipid	*S. bombicola* and *C. kuoi*	5.4 (35.9)	N/A	Bollmann et al. 2019
Mannosylerythritol lipids	*C. lacerate* CHZJU	3.0×10^6 M (31.11)	Antibacterial and sporicidal activity Anti-aging	Niu et al. 2019, Shu et al. 2019, Takahashi et al. 2012
	Pseudozyma aphidis ZJUDM34	20 (30.63)		
Viscosin	*Pseudomonas* sp. PDD-14b-2	21.6 (25)	Anticancer	Mnif and Ghribi 2015, Renard et al. 2019, Saini et al. 2008
	P. libanensis M9 while	54 (27.5)		
	P. libanensis M9-3	54 (28)		
Surfactin	*B. clausii* BS02	45 (30)	Anticancer, Insecticidal activity, Negligible cytotoxic effect against the mammalian cells HEK293	Wu et al. 2017
	Genetically modified *B. subtilis* BS37	20 (~ 30)		
	B. subtilis 573	160 (30.7)		
	B. stratosphericus FLU5	50 (28)		
	B. nealsonii	40 (34.15)		
Syringafactin	*X. campestris* PDD-32b-52 and *P. syringae* PDD-32b-74	1200 (25)	N/A	Renard et al. 2019
Glycoprotein biosurfactant	*L. paracasei*	2500 (41.8)	Antimicrobial activity and antiadhesive property against pathogens	Madhu and Prapulla 2014, Takahashi et al. 2012
	L. plantarum CFR 2194	6000 (44.3)		
	L. agilis CCUG31450	7500 (42.5)		

Table 2 Contd. ...

...Table 2 Contd.

Biosurfactant	Source	Critical micelle concentration (mg/mL) (surface tension in mN/m)	Functional properties	Reference
Glycolipoprotein	*L. acidophilus* NCIM 2903	23600 (26)	Antimicrobial activity	Morais et al. 2017, Satpute et al. 2019
	L. jensenii P6A *L. gasseri* P65.	7100 (42.5) 8580 (42.5)		
Lipoprotein	*Pediococcus dextrinicus* SHU1593	2700 (39.01)	Antimicrobial and anti-adhesive	Ghasemi et al. 2019

Smaller biosurfactants such as glycolipids and lipopeptides exhibit lower CMC values compared to bigger biosurfactants. HLB value indicates the emulsifying degree of a biosurfactant. Biosurfactants with low HLB values show lipophilic character, which is suitable as water in oil emulsifiers. While, biosurfactants with high HLB values show hydrophilic character which is suitable to be used as oil in water emulsifiers (Vecino et al. 2017). Generally, low molecular weight biosurfactants reduce the surface tension at oil/water interface and vice versa (Banat et al. 2010). The ionic behavior of biosurfactants determines their properties, with anionic biosurfactants giving good foaming, emulsifying, and wetting properties (Ahmadi-Ashtiani et al. 2020).

Microbial biosurfactants are also reported to exhibit antimicrobial activities against a wide range of microorganisms. Cellobiose lipids of *T. porosum* exhibited antifungal activities against *Candida albicans* and *Filobasidiella neoformans* (Imura et al. 2012, Kulakovskaya et al. 2010). MELs exhibited antibacterial and sporicidal activities against vegetative cells and spores of *Bacillus cereus* (Shu et al. 2019). Rhamnolipid also demonstrated an antimicrobial effect (Diab et al. 2020). Glycoprotein, glycolipoprotein, and lipoprotein also exhibited antimicrobial activity and antiadhesive property against various pathogens (Ghasemi et al. 2019, Madhu and Prapulla 2014, Morais et al. 2017, Satpute et al. 2019). These antimicrobial activities of biosurfactants give an advantage in cosmetic application as it could serve as a bio preservative which would potentially replace chemical preservatives used in cosmetic products. Besides, microbial biosurfactants also exert anti-ageing effect (Rikalovic et al. 2015, Takahashi et al. 2012), anti-fibrotic function for scar therapy (Shen et al. 2016), and anticancer activity (Mnif and Ghribi 2015, Saini et al. 2008, Wu et al. 2017). Di-rhamnolipid from *P. aeruginosa* exhibited anti-fibrotic function to be used for scar therapy (Shen et al. 2016). Surfactin also showed anticancer properties (Wu et al. 2017). These functional properties of biosurfactants add value to the cosmetic products by exhibiting various healing functions to the skin.

Microbial biosurfactants have the potential to break into the market of synthetic surfactants owing to its outstanding properties and easy of scaling-up production by microorganism through the fermentation process. Furthermore, the production is economically advantageous as it can be produced from renewable raw materials as carbon source. The widely diverged source of biosurfactants also provides a broader opportunity for usage in various field of application.

3. Advantages of Biosurfactant in Cosmetics

Biosurfactants have gained a special interest in the cosmetic industry due to the potential use as wetting agents, detergents, foaming agents, and emulsifiers as well as solubilizers (Ram et al. 2019). The usage of biosurfactants in the cosmetic industry is pervasive because they are critical components used in producing products for example cleansers, shower gel shampoo, soap, hair conditioners, toothpaste, creams, and moisturizers (Chakraborty et al. 2015). Biosurfactant have many advantages compared to chemical surfactants. Table 3 summarises the benefits of using biosurfactants.

The use of chemical surfactants in the cosmetic formulation is one of the most challenging problems because of the potential risk of irritation, and skin allergies resulting from direct interaction of surfactants with skin keratinocytes (Bujak et al. 2015, Perez-Ameneiro et al. 2015). However, the outstanding characteristics

Table 3. Benefits of biosurfactants in cosmetics.

Advantages	Description	Reference
Biodegradability	- Biosurfactant is easily degraded by the microorganism.	Jahan et al. 2020
Less toxicity	- Biosurfactant shows lower toxicity than the chemical surfactant. - Biosurfactant depicted higher *EC 50 values than chemical surfactant.	Jimoh and Lin 2019
Sustainability	- Biosurfactant can be produced from inexpensive raw materials that can be obtained in the large quantities	Jimoh and Lin 2019
Effectiveness in extreme environments	- Many biosurfactants can be operated effectively at extreme environmental conditions such as temperature, pH, and ionic strength tolerances. - Lichenysin produced by *Bacillus licheniformis* can be used at temperatures up to 50°C and pH at 4.5–9.0.	Anjum et al. 2016
Surface and interface activity	- A good surfactant can reduce the surface tension of water. - The value of surface tension of the water can decrease from 75 to 35 mN/m. - Surfactin can reduce the surface tension of the water to 25 m N/m.	Mulligan 2005, Muthusamy et al. 2008
Low critical micellization concentration (CMC)	- make the biosurfactant more suitable to be used in certain commercial applications.	Anjum et al. 2016
Other advantages	- Extensive foaming activities, environmentally friendliness, and long storage time. - Antimicrobial and anti-adhesive properties, heavy metal binding, aggregation, quorum sensing, biofilm formation, and solubility in hydrophobic compounds.	Anjum et al. 2016, Roming et al. 2016

* EC 50 values show the effective concentration of the test population to decrease up to 50%.

of biosurfactants make them an excellent green ingredient used in cosmetic formulations (Ferreira et al. 2017). Recently, biosurfactants have been valuable for skin moisturizing which are similar to ceramides (Akbari et al. 2018). Ceramides are used for skin barrier and dryness. The depletion of ceramides in the stratum corneum layer (outermost surface layer of the skin) can cause chronic skin diseases such as psoriasis, atopic dermatitis, and aged skin due to water loss and barrier dysfunction in the epidermis (Tessema et al. 2017). Stratum corneum forms a barrier between the internal body and the external environment (Thakur et al. 2009). This layer is used to avoid the loss of water from the skin and sustain the skin barrier function.

A proper selection of surfactants for cosmetic formulations is essential to minimize irritation, skin allergy, and inflammation (Otto et al. 2009, Sałek and Euston 2019). Some of the surfactants can lead to skin dehydration by depleting the ceramides from the stratum corneum. Skin irritation is associated with the contact time between the surfactant and skin at concentrations below the CMC. The irritation happens due to the high concentration of surfactant monomers that can lead to the denaturation of the keratin protein in the stratum corneum (Seweryn 2018). Thus, it is generally favored to use the mixtures of surfactant that are characterized by lower CMC and resulted in limitation of the irritation process.

Currently, the cosmetic industry is seeking natural ingredients as alternatives to the commonly used chemicals (Vecino et al. 2017). In this case, biosurfactants could be the most suitable natural ingredients that can be used in the cosmetic industry. However, the costs involved in biosurfactant production can limit their application in this industry. Nevertheless, the biosurfactants still can be used as the natural ingredients in the cosmetics industry as this industry presents very high profits that can overcome the costs involved.

4. Challenges of Biosurfactants in Cosmetic Formulations

Biosurfactant is considered as a better possible alternative to existing commercially available synthetic/chemical surfactants due to its higher biodegradability, low toxicity, and higher selectivity and specificity in optimum conditions (Varvaresou and Iakovou 2015). Because of this, there is enormous awareness amongst consumers these days towards environmental sustainability and health. Therefore, the cosmetic industry is seeking natural-based ingredients that exhibit equal and additional benefits as compared to chemical-based ingredients to their cosmetic products. The demand for natural, safe, and bio-based products is increasing by 5 day, and more environmental-friendly solutions are sought for every process.

Despite their promising potential and commercial interest, their pilot-scale production is currently difficult to achieve due to high raw-material and processing costs, but low manufacturing output. The cost to produce the same amount of biosurfactants compared to chemical or synthetic surfactants is about 10-folds more (Banat et al. 2014). It has been reported that about 10 to 30% of the total production cost is due to the cost of raw materials (Chong and Li 2017, Drakontis and Amin 2020). This happens because hydrophobic substrates are more expensive as compared to sugar-based ones, and hydrophobic substrates are usually the ones that had been used to achieve high product yield. For this reason, the current research challenges are to increase the production yield but at the same time, reduce the cost of raw materials by using renewable agro-industrial wastes as substrates. Agro-industrial waste contains a high amount of carbohydrates, lipids and thus, can be utilized as a rich carbon source for microbial growth. Examples of low-cost substrates reported from previous studies are corn oil, soybean oil, frying oil, and many more (Mukherjee et al. 2006). Other than the low prices of the raw materials, several other factors have to be considered, for example, the local availability of each low-cost renewable substrates and the average composition. The availability of these low-cost renewable substrates is usually locally confined, thus lacks the capacities to withstand a commercial-scale production of biosurfactants.

Furthermore, these substrates usually compose of variable components which, in return, contribute to significant variability in different batches of these potential substrates. Thorough monitoring of the applied substrates with a tight process control will prevent this occurrence (Henkel et al. 2012). The renewable substrate with average composition will also lead to a high cost for the concentration and purification processes due to highly diluted raw materials. Apart from that, higher transportation costs are also needed for these highly diluted materials. Biosurfactants are usually produced in batch fermentation; therefore, the cost is also mainly associated with the cost of fermentation and downstream processing. Submerged fermentation is one of the most common methods for biosurfactant production. However, recently solid-state fermentation has been discovered as an alternate method due to its less energy consumption for cultivation, no agitation unit, and utilization of a small amount of solvent for extraction of the product (Nalini and Parthasarathi 2018). In solid-state fermentation, for optimum function, all the essential nutrients from the solid substrates were supplied to the microorganisms. Therefore, it is vital for an appropriate selection of the solid substrate. For instance, an improvement was observed for the fermentation of rhamnolipids by mutated strains of *Pseudomonas* with a maximum yield of 120 g L^{-1} as compared when using corn oil as the substrate with only 35 to 70 g L^{-1} of yield (Solaiman et al. 2005). This indicated that choosing a competent and suitable strain for fermentation is also a critical element for production's cost reduction.

Downstream processing costs account for approximately 60% of the total manufacturing costs. In recent years, with improvement in the upstream fermentation technology to obtain higher product yields, downstream processing turns out to be the economic bottleneck of manufacturing. The major challenges faced in downstream purification of biosurfactants include risk of contamination with undesired multiple components in low concentration, unknown or insufficient thermodynamic data, and high water content (Invally et al. 2019, Weber et al. 2012). Therefore, extensive purification steps are needed to obtain a pure product, and this will not only time-consuming but also increases expenses on additional operation and reagent. Over the years, there is still a limited number of economically downstream technology that has been developed. A simple-wise method such as precipitation and organic extraction has difficulty in producing a final product with sufficient purity; whereas, the use of HPLC separation would dramatically increase the final price.

Other than cost consuming, several environmental factors and growth conditions affect the biosurfactant production, such as changes in pH, temperature, aeration, or/and agitation speed. Cell growth and production of secondary metabolites for numerous organisms are much dependent on the changes in pH. Some researchers reported the optimum pH for a higher production yield is within the alkaline condition range which is also the natural pH of seawater (pH $\geq$ 8.0) (Jimoh and Lin 2019, Patil et al. 2014). In contrast, the optimum temperature for improved production of biosurfactant yields occurs in the range of 25 to 300°C (Desai and Banat 1997). Other factors that affect the production of biosurfactants are agitation speed and aeration. Both factors act to facilitate the exchange of oxygen from the gaseous to the aqueous phase. Kronemberger et al. (2008) reported that the production of rhamnolipid by *Pseudomonas aeruginosa* is dependent on a specific number of oxygen uptake. It is always necessary to optimize the agitation speed to maximize the efficiency of the mass transfer between medium components with oxygen molecules.

5. Application of Biosurfactants in Cosmetic Formulations

As can be seen, the utilization of biosurfactants is becoming a new trend in the recent formulations of cosmetic products that can be found in markets, where the existence of these biological or so-called 'natural' products in the ingredients is playing an important role and leading in this industry. The reason why it is becoming a new trend now is that many industries nowadays want to shift towards greener materials in developing their products in accordance with the high demand from customers that would prefer more biocompatible products, especially cosmetics. In addition, as has been recorded, synthetic surfactants are creating some harmful effects on both ecological and biological systems, thus making them more unpopular nowadays.

Usually, a surfactant is added in the formulation of cosmetics like body washes, lotions, and shampoos as an organic compound or surface-active agent which is amphiphilic and objectively to reduce the surface tension of a solution where it is dissolved. Biosurfactants can be classified as a group of surface-active molecules that have been synthesized by microorganisms (Desai and Banat 1997). Microbial and plant-based biosurfactants are the most common biosurfactants used in the cosmetics formulation due to their advantages on low toxicity and biodegradability besides can be easily obtained from renewable energy resources (Makkar and Cameotra 2002). Most of the cosmetic and pharmaceutical products require antibacterial and antifungal agents in the formulations to make the commercialized products safe for users or customers (Haque et al. 2016). Moreover, from a biological point of view, these biosurfactants produced by bacteria possess many advantages like antifungal and antiviral properties besides other biomedical and therapeutic benefits (Naughton et al. 2019). Various applications were seen to use biosurfactants as an active agent in their formulation, for example, in detergency. There is one group of researchers who have added *Bacillus subtilis* biosurfactants in laundry detergent formulations. In this remedy, Bouassida et al. have mixed the *Bacillus subtilis* with sodium tripolyphosphate and sodium sulfate as a builder and as a filler, respectively. This formulation has shown a promising capacity in removing oil with optimal washing parameters (Bouassida et al. 2018).

Moreover, Khushboo Bhange and his team have optimized the production of feather meal, rapeseed cake, and potato peel-based biosurfactant (from industrial waste) for detergent additive by evaluating its biochemical properties. This biosurfactant was extracted using chloroform and based on the outcome obtained, and it was seen that it is stable at different pH and temperature while the reduction in the deterging activities was observed at temperatures beyond 70°C (Bhange et al. 2016). Jahanbani Veshareh and Ayatollahi (2020) have isolated and screened *Bacillus subtilis*-based surfactant, and the wetting effect on carbonate surfaces was also investigated in this study. They have discovered that this lipopeptide-like surfactin biosurfactant showed a good wettability result where at zero salinity surfactin, the original oil-waste state can be changed to a water-wet state. The emulsifying capacity of biosurfactant from the yeast *Candida albicans* and *Candida glabrata* were studied by Gaur and his team. It is found that the emulsifying capacity for the *Candida albicans* and *Candida glabrata* were recorded to be 51% and 53% respectively for the assays against castor oil. Besides, the capacity remains stable for pH vary between 2 to 10 and temperature between 4 to 120°C (Gaur et al. 2019).

The use of biosurfactants is also applied as solubilization agent as done by Soderlind et al., where sugar-based biosurfactants were compared to the commercialized polyexythylene-based surfactants. Both of them were tested on the solubilization capacity of felodipine and haemolytic activity. However, the polyexythylene-based surfactants were observed to be more stable to act as solubilizing agents compared to sugar-based surfactants (Söderlind et al. 2003). Besides that, a natural lung surfactant was also used as a dispersing agent of single-walled carbon nanotubes in a biological medium. The test showed that this natural surfactant had successfully produced well-dispersed carbon nanotubes without having any fibroge and cytotoxic effect (Wang et al. 2010). Then, synergistic foaming and surface properties of the mixture of soy glycinin and biosurfactant stevioside were studied by Wan et al. (2014). The results from this study showed some good surface properties through the foams produced, thus exhibited an excellent foaming capacity and stability (Wan et al. 2014). A few researchers have added antimicrobial ingredients in the cosmetic product as a preservative booster since they act as components of self-preserving cosmetic products. For example, glycolipopeptide cell-bound biosurfactants from *Lactobacillus pentosus* and *Lactobacillus paracasei* were tested on their bioactivity against skin pathogens like *Pseudomonas aeruginosa, Staphylococcus aureus, Escherichia coli, Streptococcus pyogenes,* and *Candida albicans*. Both biosurfactants showed a good result on antimicrobial activity with 67% to 100% of capacity (Vecino et al. 2018).

An eco-friendly glycolipid biosurfactant capacity was also studied on the adsorption of hair care products. Based on the results obtained, there were specific physicochemical properties of those surfactants that play an essential role in the adsorption process control. The properties include the number of sugar rings available in the structure of surfactant, charge, and length of the hydrocarbon (Fernández-Peña et al. 2020). Besides, many personal care products like shampoo, creams, toothpaste, and makeup have started to replace petroleum-based surfactants with new so-called green surfactants in their formulations. Ferreira et al. (2017) have shown that the biosurfactants from *Lactobacillus paracasei* was used as a stabilizing agent in the emulsions of oil-water. The emulsion volume percentage was obtained 100% with the presence of this biosurfactant, and the result is comparable with the one obtained from the test using sodium dodecyl sulfate (SDS). Thus, it is very promising for the cosmetic and healthcare products to move towards green ingredient formulations. In another example, the addition of eco-friendly biosurfactant extract from corn stream in the formulation of sunscreen lotion has successfully increased the sunscreen protection factor (SPF) up to 200% compared to its formulation without the presence of this biosurfactant (Rincón-Fontán et al. 2018). A study done by Lee et al. has confirmed that phospholipid biosurfactants from rapeseed oil and rapeseed acid are promising materials to be added in the cosmetic and household products formulation. This is due to their excellent detergency test results, which are comparable to the conventional surfactants, excellent mildness, and good environmental compatibility (Lee et al. 2018). Recently, it was also revealed that a biosurfactant extracted from corn could be a potential stabilizing agent for the formulations of antidandruff on Zn Pyrithione powder. The promising results were obtained based on the study on the particle size, the solubility of Zn pyrithione, and stability beyond the period of 30 days (Myriam Rincón-Fontán et al. 2020).

Besides, vegetable oil-based nano-structured biosurfactants have also been tested for their interfacial properties like static surface tension, wetting property, and foam property. This study showed an interesting outcome where the new formulation of cosmetic products by adding these vegetable oil-based biosurfactants lead to superior interfacial properties and have an excellent mildness (Yea et al. 2019). Sorpholipids are one of the biosurfactants that are commonly added in the cosmetic products since it has a multifunctional behavior like a foaming agent, emulsifier and solubilizer and according to Lourith and Kanlayavattanakul, this biosurfactant is also present in the lip cream, eye shadow and compressed powder makeup (Lourith and Kanlayavattanakul 2009). Furthermore, Peng et al. have discovered the Sorpholipid's exciting properties in the application of drug delivery system by producing sorpholipid coated curcumin nanoparticles which can enhance the curcumin bioavailability (Peng et al. 2018). In addition, mannosylerythritol lipid, which is derived from natural products like soy oil and olive oil is also an excellent candidate to replace conventional surfactants as it is safe to be consumed by the human body. Kitagawa et al. confirmed that this lipid most probably can make the water-resistant pigments available to overcome the problem of hydrophobicity of typical surfactants, thus facilitate the production of cosmetic pigments that are stable and has good skin adhesion (Masaru et al. 2015).

6. Conclusion and Future Perspectives of Biosurfactants

Biosurfactants are emerging as ideal alternatives for petrochemical and synthetic surfactants in various applications. They possess several interesting features, including excellent physicochemical characteristics (surface/interfacial activity), renewability, and bioavailability (Sałek and Euston 2019). Biosurfactants demonstrate superior biodegradation properties and are environmentally friendly, hence becoming a popular choice in the development of new formulations for industrial and consumer use (Bhadani et al. 2020). The heightened awareness among consumers to seek natural products in recent years has encouraged the use of biosurfactants in key industries linked to human health such as food, cosmetics, personal care, and pharmaceuticals. The origin of a biosurfactant, either plant or microbial-based, determines its structure, composition and subsequent physicochemical properties (Jahan et al. 2020). In this chapter, we discuss the types, production and functional properties of the plant and microbial-based biosurfactants to compare and distinguish their useful characteristics. We highlight the advantages of biosurfactant application in an upcoming industry–cosmetics. We also shed light on the challenges of biosurfactants in cosmetic formulations, detailing the availability of feedstocks, production as well as factors affecting biosurfactant production. The vast applications of biosurfactants in cosmetic products are also extensively discussed. These should be useful to those seeking to apply biosurfactants in designing innovative cosmetic products.

Although biosurfactants have shown practical use, their commercial success is still hampered by the high cost and low production yield resulting from optimization complexities, purification costs, and variation in product features (Liu et al. 2020, Sałek and Euston 2019). Therefore, multidisciplinary researches involving the adoption of novel and useful multistep downstream processing techniques, optimization of production conditions, and development of more inexpensive renewable raw materials are necessary to ameliorate the challenges (Liu et al. 2020). For instance, fermentation and purification costs could be alleviated by carefully choosing the appropriate strain, its composition, and fermentation conditions. Substrates used could be replaced by wastes and renewable sources based on the most effective option (Drakontis and Amin 2020). Genetic engineering is beneficial in enhancing the production yields and reproducibility of the surface-active compounds (Sałek and Euston 2019). In a nutshell, global awareness on utilizing bio-friendly ingredients and the obligation for sustainable development drives the advancement in biosurfactants. Given the fast development in chemical/biotechnological engineering, biosurfactants might replace their chemical counterpart, specifically in cosmetic formulations soon. Figure 1 summarizes the advantages and potential applications of biosurfactants in cosmeceutical products as well as the different type of biosurfactants that can be extracted from the plant and microbes.

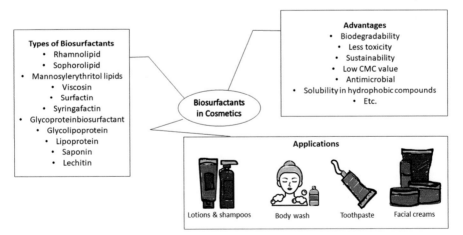

Figure 1. Summary of biosurfactants in cosmetics.

References

Ahmadi-Ashtiani, H.-R., A. Baldisserotto, E. Cesa, S. Manfredini, H. Sedghi Zadeh, M. Ghafori Gorab, M. Khanahmadi, S. Zakizadeh, P. Buso and S. Vertuani. 2020. Microbial biosurfactants as key multifunctional ingredients for sustainable cosmetics. Cosmetics 7(2): 1–34. doi:https://doi.org/10.3390/COSMETICS7020046.

Akbari, S., N.H. Abdurahman, R.M. Yunus, F. Fayaz and O.R. Alara. 2018. Biosurfactants—a new frontier for social and environmental safety: a mini review. Biotechnology Research and Innovation 2(1): 81–90. doi:https://doi.org/10.1016/j.biori.2018.09.001.

Anjum, F., G. Gautam, G. Edgard and S. Negi. 2016. Biosurfactant production through *Bacillus* sp. MTCC 5877 and its multifarious applications in food industry. Bioresource Technology 213: 262–269. doi:https://doi.org/10.1016/j.biortech.2016.02.091.

Araújo, J., J. Rocha, M. Oliveira Filho, S. Matias, S. Júnior, C. Padilha and E. Santos. 2018. Rhamno lipids biosurfactants from *Pseudomonas aeruginosa*—A review. Biosciences Biotechnology Research Asia 15: 767–781. doi:https://doi.org/10.13005/bbra/2685.

Arima, K., A. Kakinuma and G. Tamura. 1968. Surfactin, a crystalline peptidelipid surfactant produced by *Bacillus subtilis*: isolation, characterization and its inhibition of fibrin clot formation. Biochem. Biophys. Res. Commun. 31(3): 488–494. doi:https://doi.org/10.1016/0006-291x(68)90503-2.

Banat, I.M., A. Franzetti, I. Gandolfi, G. Bestetti, M.G. Martinotti, L. Fracchia, T.J. Smyth and R. Marchant. 2010. Microbial biosurfactants production, applications and future potential. Appl. Microbiol. Biotechnol. 87(2): 427–444. doi:https://doi.org/10.1007/s00253-010-2589-0.

Banat, I.M., S.K. Satpute, S.S. Cameotra, R. Patil and N.V. Nyayanit. 2014. Cost effective technologies and renewable substrates for biosurfactants' production. Frontiers in Microbiology 5(697). doi:https://doi.org/10.3389/fmicb.2014.00697.

Ben Belgacem, Z., S. Bijttebier, S. Verreth, S. Voorspoels, I. Van de Voorde, G. Aerts, K.A. Willems, H. Jacquemyn, S. Ruyters and B. Lievens. 2015. Biosurfactant production by *Pseudomonas* strains isolated from floral nectar. J. Appl. Microbiol. 118(6): 1370–1384. doi:https://doi.org/10.1111/jam.12799.

Bhadani, A., A. Kafle, T. Ogura, M. Akamatsu, K. Sakai, H. Sakai and M. Abe. 2020. Current perspective of sustainable surfactants based on renewable building blocks. Current Opinion in Colloid & Interface Science 45: 124–135. doi:https://doi.org/10.1016/j.cocis.2020.01.002.

Bhange, K., V. Chaturvedi and R. Bhatt. 2016. Simultaneous production of detergent stable keratinolytic protease, amylase and biosurfactant by *Bacillus subtilis* PF1 using agro industrial waste. Biotechnology Reports 10: 94–104. doi:https://doi.org/10.1016/j.btre.2016.03.007.

Bollmann, T., C. Zerhusen, B. Glüsen and U. Schörken. 2019. Structures and properties of sophorolipids in dependence of microbial strain, lipid substrate and post-modification. Tenside Surfactants Detergents 56: 367–377. doi:https://doi.org/10.3139/113.110640.

Bouassida, M., N. Fourati, I. Ghazala, S. Ellouze-Chaabouni and D. Ghribi. 2018. Potential application of *Bacillus subtilis* SPB1 biosurfactants in laundry detergent formulations: Compatibility study with detergent ingredients and washing performance. Engineering in Life Sciences 18(1): 70–77. doi:https://doi.org/10.1002/elsc.201700152.

Bujak, T., T. Wasilewski and Z. Nizioł-Łukaszewska. 2015. Role of macromolecules in the safety of use of body wash cosmetics. Colloids Surf. B Biointerfaces 135: 497–503. doi:https://doi.org/10.1016/j.colsurfb.2015.07.051.

Chakraborty, S., M. Ghosh, S. Chakraborti, S. Jana, K.K. Sen, C. Kokare and L. Zhang. 2015. Biosurfactant produced from *Actinomycetes nocardiopsis* A17: Characterization and its biological evaluation. International Journal of Biological Macromolecules 79: 405–412. doi:https://doi.org/10.1016/j.ijbiomac.2015.04.068.

Chong, H. and Q. Li. 2017. Microbial production of rhamnolipids: opportunities, challenges and strategies. Microbial Cell Factories 16(1): 137. doi:https://doi.org/10.1186/s12934-017-0753-2.

Costa, J.A.V., H. Treichel, L.O. Santos and V.G. Martins. 2018. Chapter 16—Solid-state fermentation for the production of biosurfactants and their applications. pp. 357–372. *In*: Pandey, A., C. Larroche and C.R. Soccol (eds.). Current Developments in Biotechnology and Bioengineering. Elsevier.

Desai, J.D. and I.M. Banat. 1997. Microbial production of surfactants and their commercial potential. Microbiology and Molecular Biology Reviews: MMBR 61(1): 47–64. doi:https://doi.org/10.1128/.61.1.47-64.1997.

Diab, A., S. Ibrahim and H. Abdulla. 2020. Safe application and preservation efficacy of low-toxic rhamnolipids produced from *Ps. aeruginosa* for cosmetics and personal care formulation. doi:https://doi.org/10.21608/ejm.2020.19131.1124.

Drakontis, C.E. and S. Amin. 2020. Biosurfactants: Formulations, properties, and applications. Current Opinion in Colloid & Interface Science 48: 77–90. doi:https://doi.org/10.1016/j.cocis.2020.03.013.

Control of Drugs and Cosmetics Regulations (2020) Drug Registration Guidance Document (DRGD). Available at: https://www.npra.gov.my/index.php/en/component/sppagebuilder/925-drug-registration-guidance-document-drgd.html (Accessed: 17 March 2021).

Essa Mahmood, D.A.F.A.-K.M. 2013. Effect of temperature changes on critical micelle concentration for tween series surfactant. Global Journal of Science Frontier Research 13(4-B).

Fenibo, E.O., G.N. Ijoma, R. Selvarajan and C.B. Chikere. 2019. Microbial surfactants: The next generation multifunctional biomolecules for applications in the petroleum industry and its associated environmental remediation. Microorganisms 7(11): 581. doi:https://doi.org/10.3390/microorganisms7110581.

Fernández-Peña, L., E. Guzmán, F. Leonforte, A. Serrano-Pueyo, K. Regulski, L. Tournier, Couturier, F. Ortega, R.G. Rubio and G.S. Luengo. 2020. Effect of molecular structure of eco-friendly glycolipid biosurfactants on the adsorption for hair-care conditioning polymers. Colloids and Surfaces B: Biointerfaces 185: 110578. doi:https://doi.org/10.1016/j.colsurfb.2019.110578.

Ferreira, A., X. Vecino, D. Ferreira, J.M. Cruz, A.B. Moldes and L.R. Rodrigues. 2017. Novel cosmetic formulations containing a biosurfactant from *Lactobacillus paracasei*. Colloids Surf. B Biointerfaces 155: 522–529. doi:https://doi.org/10.1016/j.colsurfb.2017.04.026.

Gaur, V.K., R. Regar, N. Dhiman, K. Gautam, J.K. Srivastava, S. Patnaik, M. Kamthan and N. Manickam. 2019. Biosynthesis and characterization of sophorolipid biosurfactant by *Candida* spp.: Application as food emulsifier and antibacterial agent. Bioresour. Technol. 285: 121314. doi:https://doi.org/10.1016/j.biortech.2019.121314.

Ghasemi, A., M. Moosavi-Nasab, P. Setoodeh, G. Mesbahi and G. Yousefi. 2019. Biosurfactant production by lactic acid bacterium *Pediococcus dextrinicus* SHU1593 grown on different carbon sources: strain screening followed by product characterization. Scientific Reports 9(1): 5287. doi:https://doi.org/10.1038/s41598-019-41589-0.

Haque, F., M. Alfatah, K. Ganesan and M.S. Bhattacharyya. 2016. Inhibitory effect of sophorolipid on *Candida albicans* biofilm formation and hyphal growth. Scientific Reports 6(1): 23575. doi:https://doi.org/10.1038/srep23575.

Henkel, M., M.M. Müller, J.H. Kügler, R.B. Lovaglio, J. Contiero, C. Syldatk and R. Hausmann. 2012. Rhamnolipids as biosurfactants from renewable resources: Concepts for next-generation rhamnolipid production. Process Biochemistry 47(8): 1207–1219. doi:https://doi.org/10.1016/j.procbio.2012.04.018.

Imura, T., D. Kawamura, Y. Ishibashi, T. Morita, S. Sato, T. Fukuoka, Y. Kikkawa and D. Kitamoto. 2012. Low molecular weight gelators based on biosurfactants, cellobiose lipids by *Cryptococcus humicola*. J. Oleo Sci. 61(11): 659–664. doi:https://doi.org/10.5650/jos.61.659.

Invally, K., A. Sancheti and L.K. Ju. 2019. A new approach for downstream purification of rhamnolipid biosurfactants. Food and Bioproducts Processing 114: 122–131. doi:https://doi.org/10.1016/j.fbp.2018.12.003.

Jahan, R., A.M. Bodratti, M. Tsianou and P. Alexandridis. 2020. Biosurfactants, natural alternatives to synthetic surfactants: Physicochemical properties and applications. Advances in Colloid and Interface Science 275: 102061. doi:https://doi.org/10.1016/j.cis.2019.102061.

Jahanbani Veshareh, M. and S. Ayatollahi. 2020. Microorganisms' effect on the wettability of carbonate oil-wet surfaces: implications for MEOR, smart water injection and reservoir souring mitigation strategies. Journal of Petroleum Exploration and Production Technology 10(4): 1539–1550. doi:10.1007/s13202-019-00775-6.

Jimoh, A.A. and J. Lin. 2019. Biosurfactant: A new frontier for greener technology and environmental sustainability. Ecotoxicology and Environmental Safety 184: 109607. doi:https://doi.org/10.1016/j.ecoenv.2019.109607.

Kang, K.C., H.B. Pyo, J.D. Lee and N. Jeong. 2004. Properties of nanoemulsions containing lecithin in cosmetics. Journal of Industrial and Engineering Chemistry 10: 564–568.

Kronemberger, F.d.A., L.M.M. Santa Anna, A.C.L.B. Fernandes, R.R.d. Menezes, C.P. Borges and D.M.G. Freire. 2008. Oxygen-controlled biosurfactant production in a bench scale bioreactor. Appl. Biochem. Biotechnol. 147(1): 33–45. doi:https://doi.org/10.1007/s12010-007-8057-3.

Kulakovskaya, T.V., W.I. Golubev, M.A. Tomashevskaya, E.V. Kulakovskaya, A.S. Shashkov, A.A. Grachev, A.S. Chizhov and N.E. Nifantiev. 2010. Production of antifungal cellobiose lipids by *Trichosporon porosum*. Mycopathologia 169(2): 117–123. doi:https://doi.org/10.1007/s11046-009-9236-2.

Kuyukina, M.S., O.A. Kochina, S.V. Gein, I.B. Ivshina and V.A. Chereshnev. 2020. Mechanisms of immunomodulatory and membranotropic activity of trehalolipid biosurfactants (a review). Applied Biochemistry and Microbiology 56(3): 245–255. doi:https://doi.org/10.1134/S0003683820030072.

Lee, S., J. Lee, H. Yu and J. Lim. 2018. Synthesis of environment friendly biosurfactants and characterization of interfacial properties for cosmetic and household products formulations. Colloids and Surfaces A: Physicochemical and Engineering Aspects 536: 224–233. doi:https://doi.org/10.1016/j.colsurfa.2017.05.001.

Liu, K., Y. Sun, M. Cao, J. Wang, J.R. Lu and H. Xu. 2020. Rational design, properties, and applications of biosurfactants: a short review of recent advances. Current Opinion in Colloid & Interface Science 45: 57–67. doi:https://doi.org/10.1016/j.cocis.2019.12.005.

Lourith, N. and M. Kanlayavattanakul. 2009. Natural surfactants used in cosmetics: glycolipids. Int. J. Cosmet. Sci. 31(4): 255–261. doi:https://doi.org/10.1111/j.1468-2494.2009.00493.x.

Madhu, A.N. and S.G. Prapulla. 2014. Evaluation and functional characterization of a biosurfactant produced by *Lactobacillus plantarum* CFR 2194. Appl. Biochem. Biotechnol. 172(4): 1777–1789. doi:https://doi.org/10.1007/s12010-013-0649-5.

Makkar, R.S. and S.S. Cameotra. 2002. An update on the use of unconventional substrates for biosurfactant production and their new applications. Appl. Microbiol. Biotechnol. 58(4): 428–434. doi:https://doi.org/10.1007/s00253-001-0924-1.

Masaru, K., N. Kenji and T. Takumi. 2015. United States Patent No. United States Patent O.-s. J. Toyobo Co. Ltd. & O.-s. J. Daito Kaseikogyo Co. Ltd.

Menghao, D., H. Sumei, Z. Jinping, W. Jingwen, H. Lisong and J. Jingmin. 2015. Toxicolological test of saponins from *Sapindus mukorossi* Gaerth. Open Journal of Forestry 5(7): 749–753. doi:https://doi.org/10.4236/ojf.2015.57067.

Mnif, I. and D. Ghribi. 2015. Lipopeptides biosurfactants, main classes and new insights for industrial; biomedical and environmental applications. Biopolymers 104. doi:https://doi.org/10.1002/bip.22630.

Mølgaard, P., A. Chihaka, E. Lemmich, P. Furu, C. Windberg, F. Ingerslev and B. Halling-Sørensen. 2000. Biodegradability of the molluscicidal saponins of *Phytolacca dodecandra*. Regulatory Toxicology and Pharmacology 32(3): 248–255. doi:https://doi.org/10.1006/rtph.2000.1390.

Morais, I.M.C., A.L. Cordeiro, G.S. Teixeira, V.S. Domingues, R.M.D. Nardi, A.S. Monteiro, R.J. Alves, E.P. Siqueira and V.L. Santos. 2017. Biological and physicochemical properties of biosurfactants produced by *Lactobacillus jensenii* P(6A) and *Lactobacillus gasseri* P(65). Microbial Cell Factories 16(1): 155–155. doi:https://doi.org/10.1186/s12934-017-0769-7.

Mousli, R. and A. Tazerouti. 2007. Direct method of preparation of dodecanesulfonamide derivatives and some surface properties. Journal of Surfactants and Detergents 10(4): 279–285. doi:https://doi.org/10.1007/s11743-007-1043-5.

Mukherjee, S., P. Das and R. Sen. 2006. Towards commercial production of microbial surfactants. Trends in Biotechnology 24(11): 509–515. doi:https://doi.org/10.1016/j.tibtech.2006.09.005.

Mulligan, C.N. 2005. Environmental applications for biosurfactants. Environmental Pollution 133(2): 183–198. doi:https://doi.org/10.1016/j.envpol.2004.06.009.

Muthusamy, K., S. Gopalakrishnan, T.K. Ravi and P. Sivachidambaram. 2008. Biosurfactants: Properties, commercial production and application. Current Science 94: 736–747. doi:https://www.jstor.org/stable/24100627.

Nalini, S. and R. Parthasarathi. 2018. Optimization of rhamnolipid biosurfactant production from *Serratia rubidaea* SNAU02 under solid-state fermentation and its biocontrol efficacy against Fusarium wilt of eggplant. Annals of Agrarian Science 16(2): 108–115. doi:https://doi.org/10.1016/j.aasci.2017.11.002.

Naughton, P.J., R. Marchant, V. Naughton and I.M. Banat. 2019. Microbial biosurfactants: current trends and applications in agricultural and biomedical industries. J. Appl. Microbiol. 127(1): 12–28. doi:https://doi.org/10.1111/jam.14243.

Nelson, J., A.O. El-Gendy, M.S. Mansy, M.A. Ramadan and R.K. Aziz. 2020. The biosurfactants iturin, lichenysin and surfactin, from vaginally isolated lactobacilli, prevent biofilm formation by pathogenic Candida. FEMS Microbiology Letters 367(15). doi:https://doi.org/10.1093/femsle/fnaa126.

Nguyen, T.T.L., A. Edelen, B. Neighbors and D.A. Sabatini. 2010. Biocompatible lecithin-based microemulsions with rhamnolipid and sophorolipid biosurfactants: Formulation and potential applications. Journal of Colloid and Interface Science 348(2): 498–504. doi:https://doi.org/10.1016/j.jcis.2010.04.053.

Niu, Y., J. Wu, W. Wang and Q. Chen. 2019. Production and characterization of a new glycolipid, mannosylerythritol lipid, from waste cooking oil biotransformation by *Pseudozyma aphidis* ZJUDM34. Food Science & Nutrition 7(3): 937–948. doi:https://doi.org/10.1002/fsn3.880.

Otto, A., J. du Plessis and J.W. Wiechers. 2009. Formulation effects of topical emulsions on transdermal and dermal delivery. Int. J. Cosmet. Sci. 31(1): 1–19. doi:https://doi.org/10.1111/j.1468-2494.2008.00467.x.

Pan, L.G., M.C. Tomás and M.C. Añón. 2002. Effect of sunflower lecithins on the stability of water-in-oil and oil-in-water emulsions. Journal of Surfactants and Detergents 5(2): 135–143. doi:https://doi.org/10.1007/s11743-002-0213-1.

Patil, S., A. Pendse and K. Aruna. 2014. Studies on optimization of biosurfactant production by *Pseudomonas aeruginosa* f23 isolated from oil contaminated soil sample. International Journal of Current Biotechnology 2: 20–30.

Peng, S., Z. Li, L. Zou, W. Liu, C. Liu and D.J. McClements. 2018. Enhancement of curcumin bioavailability by encapsulation in sophorolipid-coated nanoparticles: an *in vitro* and *in vivo* study. Journal of Agricultural and Food Chemistry 66(6): 1488–1497. doi:https://doi.org/10.1021/acs.jafc.7b05478.

Perez-Ameneiro, M., X. Vecino, J.M. Cruz and A.B. Moldes. 2015. Wastewater treatment enhancement by applying a lipopeptide biosurfactant to a lignocellulosic biocomposite. Carbohydrate Polymers 131: 186–196. doi:https://doi.org/10.1016/j.carbpol.2015.05.075.

Pradhan, A. and A. Bhattacharyya. 2017. Quest for an eco-friendly alternative surfactant: Surface and foam characteristics of natural surfactants. Journal of Cleaner Production 150: 127–134. doi:https://doi.org/10.1016/j.jclepro.2017.03.013.

Ram, H., A. Kumar Sahu, M.S. Said, A.G. Banpurkar, J.M. Gajbhiye and S.G. Dastager. 2019. A novel fatty alkene from marine bacteria: A thermo stable biosurfactant and its applications. Journal of Hazardous Materials 380: 120868. doi:https://doi.org/10.1016/j.jhazmat.2019.120868.

Renard, P., I. Canet, M. Sancelme, M. Matulova, I. Uhliarikova, B. Eyheraguibel, L. Nauton, J. Devemy, M. Traïkia, P. Malfreyt and A.M. Delort. 2019. Cloud microorganisms, an interesting source of biosurfactants. In IntechOpen 1–17. doi:https://doi.org/10.1016/j.colsurfa.2011.12.014.

Rikalovic, M., M. Vrvic and I. Karadzic. 2015. Rhamnolipid biosurfactant from *Pseudomonas aeruginosa*: From discovery to application in contemporary technology. Journal of the Serbian Chemical Society 80: 279–304. doi:https://doi.org/10.2298/JSC140627096R.

Rincón-Fontán, M., L. Rodríguez-López, X. Vecino, J.M. Cruz and A.B. Moldes. 2018. Design and characterization of greener sunscreen formulations based on mica powder and a biosurfactant extract. Powder Technology 327: 442–448. doi:https://doi.org/10.1016/j.powtec.2017.12.093.

Rincón-Fontán, M., L. Rodríguez-López, X. Vecino, J.M. Cruz and A.B. Moldes. 2020. Novel multifunctional biosurfactant obtained from corn as a stabilizing agent for antidandruff formulations based on Zn pyrithione powder. ACS Omega 5(11): 5704–5712. doi:https://doi.org/10.1021/acsomega.9b03679.

Roming, F.I., O.A. Sicuia, C. Voaide, M. Zamfir and S. Grosu-Tudor. 2016. Biosurfactant production by *Lactobacillus* spp. strains isolated from Romanian traditional fermented food products.

Rupeshkumar, G., K.S. Surekha, A.C. Balu and G.B. Arun. 2011. Study of functional properties of *Sapindus mukorossi* as a potential bio-surfactant. Indian Journal of Science and Technology 4(5): 530–533. doi:https://doi.org/10.17485/ijst/2011/v4i5.8.

Saika, A., H. Koike, S. Yamamoto, T. Kishimoto and T. Morita. 2017. Enhanced production of a diastereomer type of mannosylerythritol lipid-B by the basidiomycetous yeast Pseudozyma tsukubaensis expressing lipase genes from Pseudozyma antarctica. Appl. Microbiol. Biotechnol. 101(23): 8345–8352. doi:https://doi.org/10.1007/s00253-017-8589-6.

Saini, H.S., B.E. Barragán-Huerta, A. Lebrón-Paler, J.E. Pemberton, R.R. Vázquez, A.M. Burns, M.T. Marron, C.J. Seliga, A.A. Gunatilaka and R.M. Maier. 2008. Efficient purification of the biosurfactant viscosin from *Pseudomonas libanensis* strain M9-3 and its physicochemical and biological properties. J. Nat. Prod. 71(6): 1011–1015. doi:https://doi.org/10.1021/np800069u.

Sałek, K. and S.R. Euston. 2019. Sustainable microbial biosurfactants and bioemulsifiers for commercial exploitation. Process Biochemistry 85: 143–155. doi:https://doi.org/10.1016/j.procbio.2019.06.027.

Satpute, S.K., N.S. Mone, P. Das, I.M. Banat and A.G. Banpurkar. 2019. Inhibition of pathogenic bacterial biofilms on PDMS based implants by *L. acidophilus* derived biosurfactant. BMC Microbiology 19(1): 39. doi:https://doi.org/10.1186/s12866-019-1412-z.

Sekhon Randhawa, K.K. and P.K.S.M. Rahman. 2014. Rhamnolipid biosurfactants—past, present, and future scenario of global market. Frontiers in Microbiology 5: 454–454. doi:https://doi.org/10.3389/fmicb.2014.00454.

Seweryn, A. 2018. Interactions between surfactants and the skin—Theory and practice. Advances in Colloid and Interface Science 256: 242–255. doi:https://doi.org/10.1016/j.cis.2018.04.002.

Shen, C., L. Jiang, H. Shao, C. You, G. Zhang, S. Ding, T. Bian, C. Han and Q. Meng. 2016. Targeted killing of myofibroblasts by biosurfactant di-rhamnolipid suggests a therapy against scar formation. Scientific Reports 6(1): 37553. doi:https://doi.org/10.1038/srep37553.

Shu, Q., Y. Niu, W. Zhao and Q. Chen. 2019. Antibacterial activity and mannosylerythritol lipids against vegetative cells and spores of *Bacillus cereus*. Food Control 106: 106711. doi:https://doi.org/10.1016/j.foodcont.2019.106711.

Söderlind, E., M. Wollbratt and C. von Corswant. 2003. The usefulness of sugar surfactants as solubilizing agents in parenteral formulations. International Journal of Pharmaceutics 252(1–2): 61–71. doi:https://doi.org/10.1016/s0378-5173(02)00599-9.

Solaiman, D., R.D. Ashby and T.A. Foglia. 2005. Production of biosurfactants by fermentation of fats, oils and their coproducts. Boca Raton, FL: Taylor & Francis.

Takahashi, M., T. Morita, T. Fukuoka, T. Imura and D. Kitamoto. 2012. Glycolipid biosurfactants, mannosylerythritol lipids, show antioxidant and protective effects against H(2)O(2)-induced oxidative stress in cultured human skin fibroblasts. J. Oleo Sci. 61(8): 457–464. doi:https://doi.org/10.5650/jos.61.457.

Tessema, E.N., T. Gebre-Mariam, S. Lange, B. Dobner and R.H.H. Neubert. 2017. Potential application of oat-derived ceramides in improving skin barrier function: Part 1. Isolation and structural characterization. Journal of Chromatography B 1065–1066: 87–95. doi:https://doi.org/10.1016/j.jchromb.2017.09.029.

Thakur, R., P. Batheja, D. Kaushik and B. Michniak. 2009. Chapter 4—Structural and biochemical changes in aging skin and their impact on skin permeability barrier. pp. 55–90. *In*: Dayan, N. (ed.). Skin Aging Handbook. Norwich, NY: William Andrew Publishing.

Tippel, J., M. Lehmann, R. von Klitzing and S. Drusch. 2016. Interfacial properties of Quillaja saponins and its use for micellisation of lutein esters. Food Chemistry 212: 35–42. doi:https://doi.org/10.1016/j.foodchem.2016.05.130.

Tmáková, L., S. Sekretár and Š. Schmidt. 2016. Plant-derived surfactants as an alternative to synthetic surfactants: surface and antioxidant activities. Chemical Papers 70(2): 188–196. doi:https://doi.org/10.1515/chempap-2015-0200.

Varvaresou, A. and K. Iakovou. 2015. Biosurfactants in cosmetics and biopharmaceuticals. Lett. Appl. Microbiol. 61(3): 214–223. doi:https://doi.org/10.1111/lam.12440.

Vater, C., A. Adamovic, L. Ruttensteiner, K. Steiner, P. Tajpara, V. Klang, M. Elbe-Bürger, M. Wirth and C. Valenta. 2019. Cytotoxicity of lecithin-based nanoemulsions on human skin cells and *ex vivo* skin permeation: Comparison to conventional surfactant types. International Journal of Pharmaceutics 566: 383–390. doi:https://doi.org/10.1016/j.ijpharm.2019.05.078.

Vecino, X., J.M. Cruz, A.B. Moldes and L.R. Rodrigues. 2017. Biosurfactants in cosmetic formulations: trends and challenges. Critical Reviews in Biotechnology 37(7): 911–923. doi:https://doi.org/10.1080/07388551.2016.1269053.

Vecino, X., L. Rodríguez-López, D. Ferreira, J.M. Cruz, A.B. Moldes and L.R. Rodrigues. 2018. Bioactivity of glycolipopeptide cell-bound biosurfactants against skin pathogens. International Journal of Biological Macromolecules 109: 971–979. doi:https://doi.org/10.1016/j.ijbiomac.2017.11.088.

Vincken, J.-P., L. Heng, A. de Groot and H. Gruppen. 2007. Saponins, classification and occurrence in the plant kingdom. Phytochemistry 68(3): 275–297. doi:https://doi.org/10.1016/j.phytochem.2006.10.008.

Wan, Z.-L., L.Y. Wang, J.M. Wang, Y. Yuan and X.Q. Yang. 2014. Synergistic foaming and surface properties of a weakly interacting mixture of soy glycinin and biosurfactant stevioside. Journal of Agricultural and Food Chemistry 62(28): 6834–6843. doi:https://doi.org/10.1021/jf502027u.

Wang, L., V. Castranova, A. Mishra, B. Chen, R.R. Mercer, D. Schwegler-Berry and Y. Rojanasakul. 2010. Dispersion of single-walled carbon nanotubes by a natural lung surfactant for pulmonary *in vitro* and *in vivo* toxicity studies. Part Fibre Toxicol. 7: 31. doi:https://doi.org/10.1186/1743-8977-7-31.

Wang, Y., M. Nie, Z. Diwu, Y. Lei, H. Li and X. Bai. 2019. Characterization of trehalose lipids produced by a unique environmental isolate bacterium *Rhodococcus qingshengii* strain FF. J. Appl. Microbiol. 127(5): 1442–1453. doi:https://doi.org/10.1111/jam.14390.

Weber, A., A. May, T. Zeiner and A. Górak. 2012. Downstream Processing of Biosurfactants (Vol. 27).

Wu, Q., Y. Zhi and Y. Xu. 2019. Systematically engineering the biosynthesis of a green biosurfactant surfactin by *Bacillus subtilis* 168. Metabolic Engineering 52: 87–97. doi:https://doi.org/10.1016/j.ymben.2018.11.004.

Wu, Y.-S., S.C. Ngai, B.H. Goh, K.G. Chan, L.H. Lee and L.H. Chuah. 2017. Anticancer activities of surfactin and potential application of nanotechnology assisted surfactin delivery. Frontiers in Pharmacology 8: 761–761. doi:https://doi.org/10.3389/fphar.2017.00761.

Yea, D., S. Jo and J. Lim. 2019. Synthesis of eco-friendly nano-structured biosurfactants from vegetable oil sources and characterization of their interfacial properties for cosmetic applications. MRS Advances 4(7): 377–384. doi:https://doi.org/10.1557/adv.2018.619.

Zdziennicka, A., K. Szymczyk, J. Krawczyk and B. Jańczuk. 2012. Critical micelle concentration of some surfactants and thermodynamic parameters of their micellization. Fluid Phase Equilibria 322-323: 126–134. doi:https://doi.org/10.1016/j.fluid.2012.03.018.

Zerhusen, C., T. Bollmann, A. Gödderz, P. Fleischer, B. Glüsen and U. Schörken. 2020. Microbial synthesis of nonionic long-chain sophorolipid emulsifiers obtained from fatty alcohol and mixed lipid feeding. European Journal of Lipid Science and Technology 122(1): 1900110. doi:https://doi.org/10.1002/ejlt.201900110.

Zhang, J., L. Bing and G.A. Reineccius. 2016. Comparison of modified starch and Quillaja saponins in the formation and stabilization of flavor nanoemulsions. Food Chemistry 192: 53–59. doi:https://doi.org/10.1016/j.foodchem.2015.06.078.

Zhu, Z., Y. Wen, J. Yi, Y. Cao, F. Liu and D.J. McClements. 2019. Comparison of natural and synthetic surfactants at forming and stabilizing nanoemulsions: Tea saponin, Quillaja saponin, and Tween 80. Journal of Colloid and Interface Science 536: 80–87. doi:https://doi.org/10.1016/j.jcis.2018.10.024.

10

Yeast Biosurfactants Biosynthesis, Production and Application

Daniel Joe Dailin,[1,2,]* *Roslinda Abd Malek,*[1,2]
Shanmugaprakasham Selvamani,[1,2] *Nurul Zahidah Nordin,*[1,2]
Li Ting Tan,[1,2] *R Z Sayyed,*[3] *Adibah Mohd Hisham,*[1]
Racha Wehbe[4] and *Hesham El Enshasy*[1,2,5]

1. Introduction

Biosurfactants production was first reported in 1941 by Bushnell and Haas (1941). Microorganisms such as bacteria, fungi, and yeast produced secondary metabolites having surface-active properties which are defined as biosurfactants (Biniarz et al. 2017). The attention on using yeast as a biofactory in biosurfactants production is because most of the species are generally recognized as safe (GRAS), which helps the applicability in the food, pharmaceutical and cosmetic industries (Sałek and Euston 2019, Ribeiro 2020a). Due to this reason, the investigation of biosurfactant production produced by yeast has increased and several non-pathogenic yeast strains have been reported as producers (Souza et al. 2018, Eldin et al. 2019). Biosurfactants exhibit an amphiphilic nature and can be produced extracellularly or intracellularly. Their hydrophilic and a hydrophobic fraction allows them to decrease interfacial and

[1] Institute of Bioproduct Development (IBD), Universiti Teknologi Malaysia (UTM), Skudai, Johor, Malaysia.
[2] School of Chemical and Energy Engineering, Faculty of Engineering, Unviersiti Teknologi Malaysia (UTM), Skudai, Johor, Malaysia.
[3] Department of Microbiology, PSGVP Mandal's Arts, Science, and Commerce College, Shahada 425409, Maharashtra, India.
[4] Laboratory of Cancer Biology and Molecular Immunology, Faculty of Science I, Lebanese University, Hadath, Lebanon.
[5] City of Scientific Research and Technology Applications, New Burg Al Arab, Alexandria, Egypt.
* Corresponding author: jddaniel@utm.my

surface tension (Kapadia and Yagnik 2013). The family of biosurfactants comprises fatty acids, neutral lipids, phospholipids, glycolipids, polymeric, lipopeptides and particulate chemicals (Desai et al. 1997).

In the last decade, environmentally friendly surfactants having good biodegradability are in great demand. Various biosurfactants that match both environmental and technical requirements are frequently produced by microorganisms. Surfactants generated using microorganism as a biofactory act as natural wetting, dispersing, foaming, emulsifying and surface adjustment reagents (Liu et al. 2020). In addition, biosurfactants also have high surface activity and a low critical micelle concentration and are therefore promising replacements for synthetic surfactants. Despite the high cost of production of the natural biomolecules, biosurfactants are considered to have some advantages over synthetic surfactants, including high biodegradability, low toxicity, compatibility with skin, irritability and effectiveness at multiple temperatures, salinities, and pH levels (Vecino et al. 2017, Perfumo et al. 2018). Moreover, a number of researchers have also reported that biosurfactants are promising for use in cosmetics and personal hygiene products due to their anti-adherence, antifungal, antiviral and antibacterial activities against several pathogens (Ongena et al. 2008). Biosurfactants molecules are classically branded as emulsifiers of hydrocarbon compounds and are extensively applied in several fields including pharmaceutics, food processing, biodegradation, bioremediation, cosmetics, and pest control management (Sajid et al. 2020). Surfactants are widely used in industry due to their wide range of applications in the field of personal care products, such as detergents, lubricants, dyes, pomades and drug-delivery systems.

2. Biosurfactant Classification and Characteristics

Generally, microorganisms (bacteria, fungus, yeast, actinomycetes, etc.) have the ability to produce molecules with surface activity. Two main types of surface-active compounds are produced by microorganisms: biosurfactants and bioemulsifiers. Biosurfactants are produced by a number of yeasts, either extracellularly or attached to parts of the cell, predominantly during their growth on water-immiscible substrates. However, some yeasts may produce biosurfactants in the presence of different types of substrates, such as carbohydrates. The use of different carbon sources changes the structure of the biosurfactant produced and, consequently, its properties. A number of studies in biosurfactant production involving the optimization of their physicochemical properties were done for various industrial applications.

There are four major types of biosurfactant which are classified as glycolipids (rhamnolipids, trehalose lipids, sophorolipids), lipopeptides, phospholipids, and polymeric. Biosurfactants have their own unique characteristics which are generally categorized by their microbial origin and chemical composition as shown in Table 1. The properties of biosurfactants when compared to their chemically synthesized counterparts and broad substrate availability have made them suitable for commercial applications.

Biosurfactants are surface-active compounds derived from biological sources, usually extracellular, produced by bacteria, yeast or fungi. A growing number of aspects related to the production of biosurfactants from yeasts have been the topic

Table 1. Classifications of biosurfactant and its characteristics.

Types of biosurfactants	Characteristics	References

1) Glycolipids	• Comprised of sugars with long-chain of aliphatic acids or hydroxyaliphatic acids by an ester group.
	• The best known glycolipids are rhamnolipids, sophorolipids and trehalolipids.
	a) Rhamnolipids
	   Glycolipids in which one or two molecules of rhamnose are linked to one or two molecules of hydroxydecanoic acid. It is the widely studied biosurfactant which are the principal glycolipids produced by *P. aeruginosa*.
	b) Sophorolipids
	   Glycolipids which are produced by yeasts and consist of a dimeric carbohydrate sophorose linked to a long-chain hydroxyl fatty acid by glycosidic linkage.
	c) Trehalolipids
	   Glycolipids in which disaccharide trehalose connected at C-6 and C-6 to mycolic corrosive is connected with most types of *Mycobacterium*, *Corynebacterium* and *Nocardia*.
		Roy (2017)

2) Lipopeptides and lipoproteins	• Consist of a lipid attached to a polypeptide chain which involves an extraordinary number of cyclic lipopetides, including decapeptide anti-toxins (gramicidins) and lipopeptide anti-toxins (polymyxins) are delivered. • Examples: surfactin, subtilisin, viscosin, serrawettin, fengycin, arthrofactine, ornithine, etc.	Roy (2017)
3) Phospholipid	• A few microbes and yeast create huge amounts of unsaturated fats and phospholipid surfactants amid development on n-alkanes such as *Acinetobacter* spp.	Vijayakumar and Saravanan (2015)
4) Polymeric	• Polymeric biosurfactants such as emulsan, liposan, alasan, lipomanan and other polysaccharide-protein complexes are the best studied. • Emulsan is an effective emulsifying agent for hydrocarbons in water whereas liposan is an extracellular water-soluble emulsifier synthesized by *Candida lipolytica* which composed of 83% carbohydrate and 17% protein.	Vijayakumar and Saravanan (2015)

of research during the last decade as shown in Table 2. The study of biosurfactant production by yeasts has been growing in importance, with production being reported mainly by *Candida* sp., *Pseudozyma* sp. and *Yarrowia* sp. The great advantage of using yeasts in biosurfactant production is the GRAS (generally regarded as safe) status that most of these species present, for example *Yarrowia lipolytica*, *Saccharomyces cerevisiae* and *Kluyveromyces lactis*. Organisms with GRAS status are not toxic or pathogenic, allowing for the application of their products in the food and pharmaceutical industries.

Table 2. Main biosurfactant producing yeast species.

Biosurfactant	Producing microorganisms	References
1. Glycolipids	*Candida bombicola* ATCC 22214	Elshafie et al. (2015)
	Starmerella bombicola	Konishi et al. (2016)
	Candida floricola	
	Candida antarctica	Accorsini et al. (2012)
	Candida albicans	
	Trichosporon ashii	Chandran and Das (2010)
	Pichia anomala	Thaniyavarn et al. (2008)
	Ustilago maydis	Alejandro et al. (2011)
	Saccharomyces cerevisiae	Ribeiro et al. (2020a)
2. Lipopeptides and lipoproteins	*Yarrowia lipolytica*	Amaral et al. (2006)
	Saccharomyces cerevisae 2031	Alcantara et al. (2010)
3. Phospholipid	*Candida* sp. SY16	Kim et al. (2006)
4. Polymeric	*Candida lipolytica*	Sarubbo et al. (2007)
	Saccharomyces cerevisae	Saha and Rao (2017)
	Candida tropicalis	Saha and Rao (2017)

3. Biosynthesis of Biosurfactants

Biosurfactants are molecules with both hydrophobic and hydrophilic properties. Therefore, they belong to glycolipid, lipopeptide, lipoprotein or phospholipids classes. Several biosurfactants have been reported in various yeast species. The most common type of yeast biosurfactant are the sophorolipids, mannosylerythritol lipids, and carbohydrate-protein-lipid complex such as cellulobiose lipids and trehalose lipids (Sharma et al. 2019). Production of biosurfactants in yeast starts with nitrogen depletion in the cultivation medium. The nitrogen limitation is the key regulator for biosynthetic pathways for various glycolipids in yeast. These biosurfactants display many interesting properties such as lower toxicity levels, biodegradability, and higher efficiency than their chemically synthesized counterparts (Jezierska et al. 2018, Jezierska et al. 2019).

Sophorolipids are the pioneer group of biosurfactant reported in non-pathogenic yeast species. In 1961, Gorin et al. reported the synthesis of sophorolipids in *C. bombicola*. This group of biosurfactant is mainly composed of a disaccharide covalently linked to a long chain of fatty acid, known as sophorose (Sharma et al. 2019). In nature, the sophorolipids are found in a mixture with differences in acetylation and lactonization. Other minor variations reported were the number of carbon atoms, saturation degree and the position of hydroxyl groups. Recently, Price et al. (2012) had demonstrated new types sophorolipids, called bola-sophorolipids in *Candida* spp. This novel molecule is composed of a fatty acid with a carbohydrate head on both sides. The unique feature of a long hydrophobic spacer with hydrophilic groups at both ends, renders this biosurfactant to be more water soluble (Price et al. 2012). Microorganism and culture medium conditions plays a vital role in the structure of the sophorolipids. Biosurfactants from the yeast strains of Stramerella sp. were reported as nearly identical. Differences observed only on the ratio of lactonic to acidic forms, number of carbon atoms and types of hydroxylation (Ciesielska et al. 2014).

In a biosynthetic pathway of sophorolipid-type biosurfactant synthesis in yeast, a cytochrome P450 monooxygenase enzyme in yeast hydroxylate a fatty acid (Van Bogaert et al. 2009). Then, a glucosyltransferase enzyme adds a molecule of glucose to the fatty acid structure. A similar enzyme also linked the second glucose molecule to the structure, producing a non-acetylated sophorolipid. An acetytransferace enzyme would mediate the conversion to acetylation. Both the acetylated and non-acetylated sophorolipids will be recognized by a specific membrane transporter to export them into extracellular space. Diffused molecules will be further lactonized through esterification by a cell wall-bound lactone esterase (Jezierska et al. 2018, Jezierska et al. 2019, Sharma et al. 2019).

Another promising group of yeast biosurfactant could be mannosylerythritol lipid. The mannosylerythritol lipid biosurfactants were mostly reported in *Pseudozyma* sp. (Fukuoka et al. 2007a); *Ustilago maydis* and *Geotrichum* sp. (Kurz et al. 2003). This group of biosurfactant is made up of a mannose molecule linked to an erythritol residue at position 1, forming mannosylerythritol. The hydrophilic head could be acylated with a short chain. Mono- and di-acylated mannosylerythritol lipids are the most abundantly reported (Fukuoka et al. 2007b). The variations reported were the presence of an extra acylation at terminal hydroxyl group of erythritol, disaccharides, and some have no acetyl group. Similar to sophorolipids, the mannosylerythritol lipid biosurfactants are also species specific (Jezierska et al. 2018, Jezierska et al. 2019).

In the biosynthetic pathway of mannosylerythritol lipid production in yeast species, the cellular beta oxidation of fatty acids in yeast cytoplasm generates acetyl-conzyme A, which then generates the precursors for mannosylerythritol lipid production, a GDP-mannose and a erythritol. The biosynthesis of surfactant lipid begins with mannosylation of the erythritol by a glycosyltransferase enzyme. Similar to the synthesis of sophorolipids, these glycosyl transferases are strongly induced under nitrogen limited conditions (Nwaguma et al. 2019a). The disaccharide mannosylerythritol formed is acylated with fatty acids. However, study by Hewald et al. (2006) had reported that acetylation is not essential for glycolipid synthesis in yeast, as the presence of non-acetylated mannosylerythritol biosurfactants. Synthesis of mannosylerythritol lipid biosurfactants is greatly affected by presence of a hydrophobic substrate. Morita et al. (2014) demonstrated that the type of carbon source does not affect the mannosylerythritol lipid biosynthetic genes in *P. antarctica*. In contrast, addition of soybean oil had strongly induced the gene cluster with glucose as sole carbon source (Morita et al. 2014). This indicates that the fatty acid metabolism in yeast plays vital role in biosynthesis of mannosylerythritol lipid biosurfactants.

Another type of carbohydrate-fatty acid linked yeast biosurfactant would be the cellobiose lipids. The most simplest structure will consist of disaccharide cellobiose glycosidically linked to the terminal hydroxyl group of tri-hydroxyl palmitic acids. Acetylation leads to formation of more complex structures of cellobiose lipids. Yeast *U. maydis* was reported as the first cellobiose lipids producer, which was named as ustilagic acids. Mimee et al. (2005) reported the production of a rare cellobiose lipids in *P. flocculosa* yeast, the flocculosin. Both ustilagic acid and flocculosin type cellobiose lipids, are made up of a short-chain fatty acid which hydroxylated at beta-position. In contrast, Kulakovskaya et al. (2004) had reported production of cellobiose lipids without additional hydroxyl groups, in *Sympodiomycopsis paphiopedili*.

The biosynthesis of cellobiose lipids biosurfactant begins with terminal hydroxylation of palmitic acid. This reaction is catalysed by a cytochrome P450 monooxygenase, where the process subsequently attaches a cellobiose residue to the hydroxylated fatty acid (Van Bogaert et al. 2009). Then a glucosyltransferace catalyse the glycosylation process which add two glucose molecules to the structure. In the final step, the structure is acylated and further hydroxylated. The fully assembled cellobiose lipid was then recognized by a membrane transporter and exported. As mentioned in other biosynthetic pathways of yeast biosurfactants, nitrogen limited condition and types of carbon, controls the synthesis of cellobiose lipids (Nwaguma et al. 2019b). A study by Hammami et al. (2008) reported that nitrogen sources in the medium enhanced yeast biomass production. This leads to a nitrogen depleted condition and the higher cell mass produced have induced biosurfactant synthesis in *P. flocculosa*.

4. Production Process

Biosurfactants can be produced in large scale by using yeast in submerged fermentation. The two crucial factors effecting the production of biosurfactants are culture medium compositions and processing conditions. These two factors are important and can be manipulated to increased biosurfactants production.

4.1 Effect of Culture Medium Compositions

Biosurfactants are produced extracellularly by a variety of microbes or attached to the parts of cells (Baki et al. 2017, Sekhon et al. 2012) in the presence of different types of substrates, for example carbon and nitrogen sources. Mostly, the carbon source used in the biosurfactant production is hydrophobic, nonetheless, renewable and low-cost hydrophilic carbon sources are gaining attention in the recent years (Maneerat and Phetrong 2007, Qiao and Shao 2010, Batool et al. 2017, Fenibo et al. 2019a). Different carbon sources used will affect the structure of the biosurfactant produced and therefore its properties. Besides carbon sources, nitrogen sources and the presence of sulphur, iron, phosphorous, manganese and magnesium are also reported to have significant impact on the characteristics and composition of biosurfactants.

In an optimization study, different physiochemical and nutrimental factors which influence the biosurfactant production are usually evaluated by the researchers (Fontes et al. 2010). The physicochemical factors normally encompass pH, temperature, agitation and aeration whilst the nutritional factors include the effect of concentration, types and the ratios of carbon and nitrogen sources (Heryani and Putra 2017, Hu et al. 2015). Generally, yeasts are inoculated into broth media with appropriate nutrient compositions such as carbon and nitrogen sources under controlled environment to produce specific biosurfactants (Fenibo et al. 2019a). Since nitrogen and carbon sources are important for biosurfactant synthesis, thus, selecting appropriate nitrogen and carbon sources is utmost vital for the efficiency production of the biosurfactant (Bertrand et al. 2018). The influence of medium compositions on the biosurfactant production is discussed below.

4.1.1 Carbon Source

Carbon source is believed to be one of the main limiting factors in biosurfactant production (Sen 1997) and is required mostly in cell growth, metabolism and reproduction. Apart from that, carbon sources are also used as building blocks for biosurfactant synthesis. For example, substrates such as soybean oil and coconut oil are required in the biosynthesis of lipid tails of lipopeptides and glycolipids (George and Jayachandran 2013). The types of carbon substrate used for production affect both the quantity and quality of biosurfactant (Abouseoud et al. 2008, Panilaitis et al. 2007).

Commercial carbon sources such as glucose, sucrose, glycerol, diesel and crude oil are considered as good carbon sources and are generally used for biosurfactant production (Fakruddin 2012). Different carbon sources are required by different yeast strains to produce biosurfactants. For instance, Dhivya et al. (2014) reported that the *Saccharomyces cerevisiae* was proficient in producing biosurfactant using groundnut oil as a carbon substrate. Another study by Sen et al. (2017) showed that the production medium for *Rhodotorula babjevae* to produce biosurfactants contained glucose (10% w/v) as the sole carbon source. Besides, the study by Silva et al. (2020) demonstrated that *Candida bombicola* produced biosurfactants in a culture medium containing 5% residual soybean oil, 5% sugar cane molasses and 3% corn steep liquor while Punrat et al. (2020) reported that *Wickerhamomyces anomalus* MUE24 secreted biosurfactant in a production medium containing glucose and soybean oil.

Other than that, the optimization study by Luepongpattana et al. (2014) has reported that the highest biosurfactant activity of yeast was attained at the production medium containing 2.5% (w/v) glucose supplement with 2.5% (v/v) glycerol as carbon source. Moreover, in the optimization study by Sharma et al. (2019), three types of carbon sources were used which were glucose, olive oil and diesel. The result depicted that the maximum oil displacement by yeast *Meyerozyma guilliermondii* YK32 equaling 7.5 cm was achieved with olive oil at 8% (v/v) as carbon source. This research study indicated that although glucose is a simple sugar and readily utilized by any microbial cells, nevertheless, it gets exhausted relatively fast, particularly at low concentration (2–6%). Thus, it cannot support *M. guilliermondii* YK32 to further increase the biosurfactant production up to the fifth day of incubation. At higher concentrations, however, increased the biosurfactant production up to the fifth day of incubation, afterwards, the increase was found to be statistically insignificant. Diesel on the other hand, being a complex hydrocarbon did not support the biosurfactant production as it cannot be easily utilized by yeast.

In spite of the numerous advantages of biosurfactants, however, they are still incapable to compete economically with commercially available surfactants in the market due to the low yields generated and high-cost of production which are the key obstacles for large-scale production and application (Alcantara et al. 2012). Generally, the economy of production is the major bottleneck in biosurfactant production as the inefficient recovery method and the type of raw material used for its production can contribute significantly to the cost (Neboh et al. 2016, Silva et al. 2020). The raw material usually used are non-renewable and expensive (Terán-Hilares et al. 2019). Thus, to make the production of biosurfactants economically feasible, it is necessary to cut down the cost of the fermentation media which represents approximately 50% of the final cost of the product (Jimoh and Lin 2020, Lima et al. 2017, Santos et al. 2016). These costs can be reduced remarkably by using alternative sources of nutrients such as several inexpensive and renewable raw materials from agro-industrial wastes (Camargo et al. 2018, Neboh et al. 2016, Silva et al. 2020). This will improve the feasibility of large-scale production of biosurfactants as well as their competitiveness (Lima et al. 2020, Sarubbo et al. 2015). Besides being used as a cheap raw material for biosurfactants production, bioconversion of these wastes are also crucial for the near future due to its relative ease of operation, low energy cost, eco-friendly nature and most importantly in reducing environmental pollution (Camargo et al. 2018, Neboh et al. 2016, Ribeiro et al. 2020a).

There are several studies using agro-waste as a carbon source to produce biosurfactants. Waste frying oil was used by Batista et al. (2010) as carbon source to synthesize biosurfactant in the cultivation medium of *Candida tropicalis*. The authors demonstrated that the biosurfactant produced in the medium formulated with 2% waste frying oil can remove nearly 78 to 97% of the motor oil and petroleum adsorbed in sand samples. Moreover, in the research study by Marcelino et al. (2019), sugarcane bagasse hemicellulosic hydrolysate was used as carbon source by yeasts in biosurfactant production. The results acquired in this study illustrated that the yeasts have successfully produced biosurfactants in the media containing sustainable hemicellulosic hydrolysate and the feasibility of these products for biorefineries. In addition, another study by Camargo et al. (2018) used residual soybean oil as a low-cost carbon source for the production of biosurfactants from yeasts under acidic conditions. The results showed that the application of 2% of biosurfactants produced by yeast *Meyerozyma guilliermondii* with pH adjusted to 2 was capable to solubilize 15.9% of cadmium in the sewage sludge.

Furthermore, a study conducted by Ribeiro et al. (2020b) demonstrated that the yeast *Saccharomyces cerevisiae* URM 6670 can produce biosurfactant with glycolipid structure in the medium containing 1% soybean oil as carbon source which has the potential application in the food industry to replace egg yolk in a cookie formulation. Besides, Silva et al. (2020) have conducted a study to produce the biosurfactant from the yeast *Candida bombicola* URM 3718 in the medium containing 5% residual soybean oil and 5% sugar cane molasses as a cheap carbon source. The results depicted that the yield of biosurfactant produced was 251.02 g/L and has the potential for application in the food industry as an emulsifier for flour dessert. Research study by Sarubbo et al. (2016) showed that the yeast *Candida guilliermondii* UCP0992 can produce biosurfactant in a low-cost medium formulated with 4.0% molasses, 2.5% corn steep liquor and 2.5% waste frying oil. The biosurfactant produced by *Candida guilliermondii* UCP0992 can effectively recovery 50% of motor oil from seawater. Besides, Lima et al. (2017) indicated that the yeast *Candida glabrata* UCP 1556 can produce an anionic lipopeptide biosurfactant in a low-cost medium containing 20%

corn steep liquor and 40% waste whey with appropriate conditions for large scale production.

4.1.2 Nitrogen Source

Nitrogen is a limiting factor and it is important in the medium formulation for biosurfactant production as it is essential for protein and enzyme synthesis which are critical for microbial growth (Saharan et al. 2011). Numerous sources of nitrogen have been used for biosurfactants production and can be classified into organic (peptone, urea, corn steep liquor, yeast extract, malt extract and meat extract) and inorganic (sulphate nitrate, ammonium sulphate, ammonium nitrate and sodium nitrate). Generally, organic nitrogen sources are much cheaper than inorganic nitrogen sources which mostly originate from nitrate or ammonium salts (Fenibo et al. 2019a, Saharan et al. 2011). With reference to the nitrogen sources, yeast extract is the most widely used for biosurfactant production, nevertheless, its required concentration depends on the culture medium and the nature of the microorganism. The biosurfactant production often happens during the stationary phase of cell growth, when the nitrogen source in the medium is depleted (Saharan et al. 2011).

In the previous study by Batista et al. (2010), the best culture medium for *Candida tropicalis* to produce biosurfactant according to the factorial design was the medium containing 0.067% NH_4Cl as nitrogen sources compare to yeast extract. Ribeiro et al. (2020b), in their research, formulated the production medium with the combination of two types of nitrogen sources that are 0.01% NH_4NO_3 and yeast extract for the yeast *Candida utilis* UFPEDA1009 in order to produce biosurfactant. Similar to Ribeiro et al. (2019), the work by Souza et al. (2018) indicated that the optimized medium for *Wickerhamomyces anomalus* CCMA 0368 to increase the biosurfactant production contained two types of nitrogen sources, including 4.22 g/L ammonium sulphate and 4.64 g/L yeast extract. The biosurfactant production in this medium was validated in both bioreactor and shake flasks level in which the surface tension was reduced from 49.0 mN/m to 29.3 mN/m and 31.4 mN/m respectively. The highest biosurfactant production in both cases was attained after 24 h of growth. In another study by Nwaguma et al. (2019a), they also used two types of nitrogen sources ($NaNO_3$ and yeast extract) as well in the production medium for yeast to produce biosurfactant. Furthermore, the results from the experiment conducted by Punrata et al. (2020) showed that the yeast *Wickerhamomyces anomalus* MUE24 produced sophorolipid biosurfactant in a medium containing yeast extract and $NaNO_3$ as nitrogen sources. This strain released biosurfactant into culture medium at 0.55 g/L after 7 days of cultivation.

Since inorganic nitrogen sources are more expensive as compared to organic nitrogen sources, the later are more preferred in most of the studies in order to produce sustainable biosurfactants. Bueno et al. (2019) used a production medium containing yeast extract as the sole nitrogen source supplemented with glycerol. The biosurfactant was released during the end of the exponential phase and at the beginning of the stationary growth phase. At the end of the fermentation process, the Emulsification Index (E.I.) of 30% was obtained. Besides yeast extract, corn steep liquor is also frequently used by researchers as a renewable organic substance for biosurfactant production (Lima et al. 2017, Ribeiro et al. 2020b, Silva et al. 2020, Sarubbo et al. 2016).

In optimization studies, there are several researchers using various nitrogen sources for different strain of yeasts in order to achieve optimal biosurfactant production. Lira and colleagues (2020) reported that the maximum biosurfactant production by *Candida guilliermodii* was achieved with a yield of 21 g/L when using corn steep liquor as nitrogen source. In the optimization study by Nwaguma et al. (2019a), the authors used different combinations of nitrogen sources that included yeast extract + peptone, yeast extract + NH_4SO_4, yeast extract + urea, yeast extract + NH_4SO_4 and $NaNO_3$ + yeast extract to improve biosurfactant production. Among the combination, yeast extract and $NaNO_3$ more favoured biosurfactant production and yeast's growth with E_{24} value of $6.17 \pm 3.53\%$ and OD value of 2.286 ± 0.01 respectively. Dejwatthanakomol et al. (2016) have reported that $NaNO_3$ supported the biosurfactant production and the growth of *Wickerhamomyces anomalus* PY189. However, the other two nitrogen sources which are acidic, for instance $(NH_4)_2SO_4$ and NH_4NO_3, diminish the culture pH and yield low biosurfactant production.

4.2 *Process Conditions Affecting the Biosurfactant Production*

Environmental factors and growth conditions are very important because they affect the characteristics and yield of biosurfactant production. To attain huge output of biosurfactant, it is necessary to optimize the bioprocess conditions as the biosurfactant production is influenced greatly by variables, for example temperature, pH, agitation speed and aeration (Fenibo et al. 2019a, Saharan et al. 2011).

4.2.1 *pH*

Typically, yeast prefers an acidic pH environment as showed by the maximum biosurfactant production by *Pichia anomala* PY1 at an initial pH of 5.5 (Thaniyavarn et al. 2008) and at pH 6 for *Candida bombicola* (Deshpande and Daniels 1995). Besides, several studies also showed that in the acidic environment, yeast produces the maximum amount of biosurfactants. Ali and Ali (2019) found that the optimum pH for *Saccharomyces cerevisiae* to produce biosurfactant is 5. An increase in pH led to the reduction in the productivity of biosurfactant by *S. cerevisiae* and no productivity was observed at pH 7 and 8. The effect of pH on biosurfactant production by the yeast *Candida* spp. have been studied by Alwaely et al. (2019). In this research, the media pH was adjusted to achieve the pH range of 2, 3, 4, 5, 6, 7, 8 and 9. Among the pH values, the best pH for yeast to produce biosurfactant at its highest level was reported at pH 4 with an activity of 2.185 and began to drop with the increase in the pH value toward neutral and alkaline. The activity level also reduced when the pH level was decreased to a value lower than pH 4 which indicated a reduction in the production quantity.

Furthermore, yeast *Candida* spp. isolated from the sap of *Elaeis guineensis* has been reported to achieve its optimal biosurfactant production at pH 2 and decrease sharply when the pH increased (Nwaguma et al. 2019a). Two *Candida* species, *Candida apicola* and *Candida antarctica* studied by Bednarski et al. (2004) have been reported to has maximum glycolipids production at pH 5.5. Without controlling pH level, the synthesis of the decrease of biosurfactant demonstrated the importance of maintaining the pH during fermentation process. According to Sharma et al. (2019), the biosurfactant production by *Meyerozyma guilliermondii* YK32 was highly sensitive toward pH. The E_{24} index and oil displacement were found to be increased to 67.3% and 10.4 cm respectively when the pH was increased from 5.5 to 6.0. Since most of the studies demonstrated that yeast thrives best in acidic conditions, but there are some exceptions. For example, Zinjarde and Pant (2002) reported that *Yarrowia lipolytica* experience its optimal growth and biosurfactant production at pH 8.

4.2.2 Temperature

Temperature is one of the key factors that affects biosurfactant production. Various microbial processes are temperature dependant. A small temperature change has a great impact on the microbial metabolism processes (Neboh et al. 2016). Most of the biosurfactant productions reported have been performed in a range of temperatures of 25–30°C (Bednarski et al. 2004, Bhardwaj et al. 2013a). Based on the study by Sharma et al. (2019), *Meyerozyma guilliermondii* YK32 produces biosurfactant at its maximum at the temperature of 30°C with oil displacement of 8.8 cm. Any shift in the temperature, either to 25°C or 35°C decreased the biosurfactant production remarkably as revealed by the reduction in oil displacement to 7.6 cm and 5.2 cm respectively. The biosurfactant production is decreased further when the temperature reached 40°C with an oil displacement of 2.5 cm. Ali and Ali (2019) also examined the effect of temperature on the biosurfactant production by *Saccharomyces cerevisiae* and observed that at temperature of 25°C, the biosurfactant production reached its maximal with a recorded 27 mm in diameter of clear zone by OSM.

Moreover, in the study by Alwaely et al. (2019), *Candida* spp. reached its highest emulsification activity (2.285) at the temperature of 30°C and began to drop with the increase in temperature, reaching the activity of 1.007 at 40°C. The optimal temperature for *Candida bombicola* to produce Sophorolipids was 27°C while for its maximum growth, the optimal temperature was at 30°C (Deshpande and Daniels 1995). Another study depicted that the production of sophorolipids from this yeast was similar at both 30°C and 25°C (Casas and García-Ochoa 1999). The maximum biosurfactant production by both *Candida rugosa* and *Rhodotorula muciliginosa* was reported at 35°C (Chandran and Das 2011) whereas 30°C was found optimum for *Pichia anomala* (Thaniyavarn et al. 2008). Furthermore, Lima and Alegre (2009) found that the *Saccharomyces lipolytica* CCT-0913 was able to produce biosurfactant at 32°C. Other than that, *Pseudozyma siamensis* CBS 9960 achieved maximum biosurfactant production at 25°C with a yield of 7 g/L and reduced to approximately 4 g/L and 6 g/L when the temperature shifted to 30°C or 20°C respectively (Morita et al. 2008). Studies conducted by Nwaguma et al. (2019b) report that the isolated *Candida* spp. was capable of producing biosurfactant at the optimum temperature of 20°C with an E_{24} value of 54.7 ± 0.282%.

4.2.3 Aeration and Agitation

Aeration and agitation rates are crucial as they affect the biosurfactant production since they enhance the mass transfer characteristics with respect to products/by-products, oxygen and substrate. This also may be associated to the physiological function of microbial emulsifiers. The production of biosurfactants can facilitate transportation of nutrient to microorganisms as it increases the solubilization of hydrophobic substances. It has been suggested that high shear stress may induce biosurfactant production because the contact between microorganisms and organic droplets dispersed in water becomes more difficult. Nevertheless, an opposite condition may occur with other microorganisms as the increase in agitation speed can result in a reduction in biosurfactant production due to the increase in cells' mechanical stress (Fontes et al. 2010, Roy 2017). Normally, the most common agitation speed in microbial studies is between 120 rpm to 200 rpm (Singh and Tiwary 2016, Yeh et al. 2005).

Adamczak and odzimierz Bednarski (2000) examined the effect of aeration on the biosurfactant synthesis by *Candida antarctica* and observed that the improved yield value of biosurfactant (45.5 g/L) was achieved when the airflow rate was 1 vvm and the dissolved oxygen concentration was sustained at 50% saturation. However, an increase in the airflow rate to 2 vvm causes foam formation and decreased 84% of the biosurfactant production. The formation of foam is not appropriate for the production of biosurfactants and must be prevented as it can remove some lipids, biosurfactants and biomass. Efforts have been taken by separating the biosurfactant from the culture using aqueous two-phase cultivation and foam separation in order to diminish the end product inhibition during the biosurfactant production (Besson and Michel 1992, Sandrin et al. 1990). Further, Guilmanov et al. (2002) have conducted the study in agitated flasks to examine the effect of aeration rate on the biosurfactant production by *Candida bombicola*. The result illustrated that *C. bombicola* achieved the maximum biosurfactant production at the aeration rate between 50 and 80 mM of O_2L/h. On the other hand, in their review paper, Desai and Banat (1997) reported some opposite outcomes regarding the bacteria *Acinetobacter calcoaceticus* and *Norcadia erythropolis* which produce less biosurfactant in the increasing shear stress condition. Nonetheless, the authors mention that in general, the biosurfactant produced by yeasts increases with aeration rates and agitation.

5. Applications of Biosurfactants

Due to their unique characteristics biosurfactants have many potential applications which include those in the food, pharmaceutical, cosmeceutical, petroleum and for bioremediation.

5.1 Food Industry

For several decades, surfactants were found beneficial and used extensively in the food processing industry. As biocompatible compounds, biodegradable and nontoxic, biosurfactants display many useful food industry assets, mainly as food additives, emulsifiers, foaming, wetting, solubilizes, antiadhesives, thickener, lubricating agent and antimicrobial action (Khan et al. 2014, Campos et al. 2014). The application of biosurfactant from yeast as food additives is consistent with the growing awareness of using natural ingredients, organic, edible foods by end-users, encouraging the production of biomolecules that can minimize or avoid the utilization of synthetic surfactants (Ribeiro et al. 2020, Jahan et al. 2020). Biosurfactants also encourage vegetable oils to solubilize, stabilize fats during food preparation, and enhance the organoleptic effects of bread (Ribeiro et al. 2020, Rawat et al. 2020, Roy 2017).

Literature reports show that biosurfactants can boost the stability of salad emulsions (usually used as food dressings) (Campos et al. 2019, Amaral et al. 2010), the texture profile of various food types, such as muffins (baking powder and eggs as ingredients) (Silva et al. 2020) or cookies (for synthetic additive substitution) (Ribeiro et al. 2020, Ribeiro et al. 2020b). Biosurfactants can also be used in baking products and ice cream to monitor consistency, decelerate staling and solubilize flavoring oils (Ribeiro et al. 2020, Pessoa et al. 2019, Vijayakumar and Saravanan 2015).

Saccharomyces URM 6670 obtained promising biosurfactants (glycolipid) that are a substitute for egg yolk with possible applications in cookie formulation. After baking, the biosurfactant showed no affect on the physical or physicochemical effects of the food in the cookie dough. Furthermore, a basic assessment of the scent, taste, color, and texture did not show any major variations between the biosurfactant formulations and the regular formula. Concerning the physical chemistry of the dough, moisture dramatically changed for the cookies formulation using egg yolk (2%) and 2% biosurfactant and formulation with 4% biosurfactant. The lipid contents, however, increased dramatically as biosurfactants concentrated, affecting their energy value in proportion (Ribeiro et al. 2020).

The objective of the current Ribeiro et al. 2020 research was therefore to characterize and apply a *Candida* species that can have generated biosurfactant. Using *Candida* UFPEDA1009 in the formulation of a cookie as a substitute for animal fat and to determine its effect on the physical and organoleptic properties of end-product quality. The biosurfactant development in *Candida utilis* showed thermal strength and antioxidant action. In addition to the positive contribution to cookie texture analysis, the study showed biosurfactant contains high lipid content which contributes to softer and spongier cookies. When the yolk is substituted, it was low, with a moisture content of 0.02% to 0.07%. Thus, this biosurfactant is provided in flour-based food formulations as a possible ingredient as an antioxidant and absence of cytotoxicity (Ribeiro et al. 2020b).

Campos et al. (2015) established that a biosurfactant developed by *Candida utilis* is useful in severe toxicity examination and can be applied in the formulation of mayonnaise in a medium that contains waste canola frying oil. The two substances in combination with the biological agent have provided the emulsion for thirty days' stability. The biosurfactant developed by *Candida utilis* can therefore be manufactured and used as a suggested innovative component in food production at an industrial scale. The biosurfactant use has shown surface stability and emulsification in extreme conditions against vegetable oils of alkaline pH, which has favored high emulsifying activities and emulsifying stability. It was marked by the greatest stability within the formulation at 0.7 percent as a carbohydrate lipid-protein complex. The biosurfactant and guar gum (w/v) were not detected as microbial contamination (Campos et al. 2019). Amaral et al. (2010) successfully used a *Candida utilis* emulsifier, demonstrating its possible use for salad dressing, and successful results have been reported with mayonnaise formulation mannoproteins derived from the cerevisiae cell wall. The mannoproteins molecules can familiarize and fixed at pH 3 to 11 with a molecular weight of 14,000 to 15,800 dalton and are capable to alleviate oil in water emulsions (Alizadeh-Sani et al. 2018).

This review explains the use of the *Candida bombicola* URM 3718 biosurfactant as a cupcake food additive. The biosurfactant was used in the cupcake dessert formulation and in the regular formulation it replaces 50%, 75%, and 100% of vegetable fat. Thermal analysis has shown that while cooking these cupcakes (180°C) have stable biosurfactants. The biosurfactant was promising in foodstuffs with low antioxidants and did not demonstrate a cytotoxic ability in the cell lines examined. After baking, biosurfactant cupcakes in their dough displayed no major deviations from the standardized formulation in physical and physical-chemical characteristics (Silva et al. 2020).

5.2 Petroleum Industry

Petroleum has been one of the primary sources of energy and chemical raw material Industries. The world relies on oil and the fuel use of oil has helped to grow the economy intensively. Though oil refineries and petrochemical plants are useful for society, they produce a significant quantity of hazardous waste. Oil spills have also created substantial environmental issues during mining, transportation, and refining (Osman et al. 2019, Silva et al. 2014, Souza et al. 2014). The petroleum industry, where these compounds can benefit in washing up oil spills, extracting oil waste from storage tanks, microbial oil recovery, or bioremediation process of soil and water, is presently the main market for biosurfactants. In the oil industry, biological agents for heavy oil exploration were successfully applied, providing benefits over their synthetic counterparts over the entire oil chain (crude oil transportation, cleaning of storage tanks for crude oil, crude oil waste treatment) (Bhardwaj et al. 2013b, Fenibo et al. 2019). Biosurfactants are used to extract microbial oil, clean polluted vessels, and to make the transport of heavy crude oil by pipeline easier (Osman et al. 2019, Silva et al. 2014).

The developed biosurfactant can be used to remediate oil-contaminated water. It should be noted that the best results have been achieved with the crude biosurfactant, which means that the costs for manufacturing this compound have been drastically reduced and the chances of actual industrial use increased. The biosurfactant developed by *Candida bombicola* has demonstrated high dispersive effects for car engine lubrication oil that can make oily spots on the sea easier to target. This raw or crude biosurfactant succeeded in a level of dispersion of 80% of the initial oil diameter, whereas the achieved rate of dispersion with biosurfactant isolated was 50%. Effects were found when low levels were used in the transport and solubilization of oil spots in the aquatic marine environment. The ability of these compounds was demonstrated (Luna et al. 2016a).

The *Wickerhamomyces anomalous* CCMA 0358 biosurfactant had excellent active surface properties, as shown by low surface tension values with high productivity of about 108.3 mg/L/h. The biosurfactants produced remain stable between 6 to 12 pH, can adapt to high salinity (300 g/L NaCl), and can withstand high temperature when exposed at 121°C. The crude biosurfactant has allowed 20% of crude oil to be recovered from polluted sand. The results obtained was used as an alternative to established chemical surfactants as promising candidates for application in bioremediation or the petroleum industry (Souza et al. 2018).

Candida sphaerica's biosurfactant may be used in improving oil recovery activities was archived also by Chaprão et al. (2015). Under kinetic conditions (70–90%), crude and isolated biosurfactants demonstrated excellent efficacy for removal from polluted sand, while synthetic surfactants extracted among 55% and 80% of the oil. The raw and the isolated *C. sphaerica* biosurfactant was able to eliminate from packaged columns large concentrations of upto 90% of motor oil. Cell-free broth with biosurfactants can also be used immediately without purification, reducing extra costs of biosurfactants output. *Candida bombicola* ATCC22214 and the research model has been investigated and has significant application for promoting oil recovery. Core flood experiments have tested the capacity of sophorolipids (SPLs) to boost oil recovery, where 27.27% of waste oil was recovered. The ability for microbial improved oil recovery applications was thus verified by the SPLs (Elshafie et al. 2015, Oliveira et al. 2015). This was found in other works like Minucelli et al. (2017) who observed that the sophorolipids by *Candida bombicola* ATCC22214 showed the addition of sophorolipids quickly and within a short period of time had an impact in promoting biological access to fast degradation substrates from oil waste. The outcome of biosurfactant on the biodegradation of petroleum compounds was explored by Gargouri et al. (2015). Their study showed that yeast *Candida* and *Trichosporon* had the potential to effectively degrade varied series of hydrocarbons and metabolized n-alkanes.

5.3 *Pharmaceutical Industry*

Biosurfactants have many applications, most of them being related to the pharmaceutical field (Fracchia 2015). Mukherjee et al. (2006) stated that the use of biosurfactants in pharmaceutical application have a good range such as antimicrobial activity, anti-cancer activity, anti-adhesive agents, antiviral activity and gene delivery. Biosurfactants are the usually preferred surface-active molecules to be used in this field because of their lower toxicity and safe characteristics when compared to synthetic surfactants (Ohadi et al. 2020). This section will state the use of biosurfactants in the pharmaceutical field.

Firstly, biosurfactants' antimicrobial activity is of use of in the pharmaceutical field. Rodrigues et al. (2006b) research highlighted antimicrobial activity in two biosurfactants that were obtained from probiotic bacteria which are *Lactococcus lactis* 53 and *Streptoccoccus thermophilis* A, against some of the bacterial and yeast strains isolated from explanted voice prostheses. They found out that both biosurfactants have a high antimicrobial activity even at low concentrations against *Candida tropicalis* GB 9/9. In another study, Gharaei-Fathabad (2011) reported that the biosurfactant that is produced by marine *Bacillus circulans* had a potent antimicrobial activity against Gram positive and Gram negative pathogens as well as Semi pathogen microbial strains including the MDR strain. In addition, Fukuoka et al. (2007b) identify that the antimicrobial activity of MELs has been assigned to damage caused in the bilayer cell membranes of microorganisms, and MEL-A and MEL-B patterns are reported to have high antimicrobial activity, mostly against

Gram-positive bacteria. Lately, Resende et al. (2019) studied the effectiveness of antimicrobial activity in toothpaste formulation by using biosurfactants which produced by *Pseudomonas aeruginosa* UCP 0992 (PB), *Bacillus metylotrophicus* UCP 1616 (BB) and *Candida bombicola* URM 3718 (CB) and chitosan. As a result, the toothpaste formulations were effective at inhibiting biofilm formed by *S. mutans*. The presence of chitosan in the formulations also showed it's importance in enhancing the inhibition of dental biofilm.

Karlapudi et al. (2020) reported that their latest isolated biosurfactant produced by *Acinetobacter indicus* M6 has the potential to be a good anti-proliferate against lung cancer cells. The anti-cancer activity of biosurfactant against lung cancer cells was assessed in terms of cell viability at different concentrations. As a result, the percentage of lung cancer viable cells decrease with increasing biosurfactant concentrations and incubation time. On another hand, exposure of PC 12 cells (Pheochromocytoma, chromaffin cell tumor of the adrenal medulla) to Mannosylerythritol lipid (MEL) increased the activity of acetylcholine esterase and interrupted the cell cycle at the G1 phase which can result in the overgrowth of neurites and partial cellular differentiation. The authors suggest that MEL induces neuronal differentiation in PC 12 cells and use glycolipids as novel reagents for the treatment of cancer cells (Krishnaswamy et al. 2008).

Thirdly, biosurfactants have been found to inhibit the adhesion of pathogenic organisms to solid surfaces or to infection sites (Rodrigues et al. 2006a). They showed that pre-coating the vinyl urethral catheter by running the surfactin solution through them before inoculation with media resulted in the decrease in the amount of biofilm formed by *Salmonella typhimurium*, *Salmonella enterica*, *E. coli* and *Proteus mirabilis*. Recently, Rodrigues et al. (2006a) demonstrated that when rinsing flow chambers which are designed to monitor microbial adhesion with a rhamnolipid solution, the rate of deposition and adhesion was dramatically reduced for some bacterial and yeast strains isolated from explanted voice prostheses to silicone rubber. Therefore, the authors believe that this rhamnolipid may be useful as a biodetergent solution for prostheses cleaning, prolonging their lifetime and directly benefiting laryngectomized patients.

Fourthly, biosurfactant is considered the best in pharmaceutical field application because it can perform effectively in antiviral activity (Biniarz et al. 2017). Krishnaswamy et al. (2008) stated that incidence of HIV in women are high so there is need for a female controlled, efficacious and safe vaginal topical microbicide. Therefore, the authors found out that sophorolipids surfactants from *C. bombicola* and its structural analogues such as sophorolipid diacetate ethyl ester is a very potent spermicidal and virucidal agent. Moreover, sophorolipids are also reported to perform against the human immunodeficiency virus (Shah et al. 2005). Our world has recently been shocked by the pandemic caused by Covid-19. Subramaniam et al. (2020) suggested that biosurfactants could be used against Covid-19. As a result, Covid-19 virus can be neutralized by the antibodies which directly prevent infection but it will remain high few weeks after infection but then typically begin to wane (Mahalaxmi et al. 2020). Hence, biosurfactant has high immunosuppressive potential and can be used as a novel treatment molecule in most of the immune diseases but for completely degrade the virus is still under study.

Lastly, gene delivery is gene transfection whereas the lipofection using cationic liposomes is considered to be a method to transport foreign gene to the target cells without any side effects (Zhang et al. 2010). Kitamato et al. (2002) stated that differing with commercially available cationic liposomes, biosurfactant supported liposomes show growing potency of gene transfection. Furthermore, biosurfactants have been used for gene transfection, as ligands for binding immunoglobulins, as adjuvants for antigens and also as inhibitors for fibrin clot formation and activators of fibrin clot lysis (Rodrigues et al. 2006).

5.4 Cosmeceutical Industry

Biosurfactants have been suggested to substitute chemical compounds in cosmetic products due to its emulsification, foaming, water binding capacity, spreading and wetting properties effect on viscidity and on product consistency (Gharaei-Fathabad 2011). Most of the biosurfactants are part of fatty acids, neutral lipids, glycolipids and lipopeptides (Gupta et al. 2019). The most widely used glycolipids biosurfactant in cosmetics are sophorolipids, rhamnolipids and mannosylerythritol lipids (MELs) (Lourith and Kanlayavattanakul 2009). Besides, biosurfactants such as rhamnolipids, are approved by US EPA as safe for use in cosmetics (Nitschke and Costa 2007). This section will clarify the use of those three glycolipids biosurfactant in the cosmetic industry.

Recently, biosurfactant have found a role due to their skin friendly properties in the cosmetic industry. Originally, sophorolipids obtained from the yeast Candida species, mainly from the strain *Candida Bomicola* (Ashby et al. 2006). Sophorolipids can be mixed with ethylene oxide and used as skin moisturizers (Gharaei-Fathabad 2011) due to its good skin compatibility and excellent moisturizing properties (Lourith and Kanlayavattanakul 2009). Furthermore, sophorolipids are useful as bactericidal agents and used in the treatment of acne, dandruff and body odours (Lourith and Kanlayavattanakul 2009). For instance, Kao Co. Ltd. have produced sophorolipids commercially as humectants as well as for skin application and contain 'Sopholiance S' as the active ingredient that is present in the make-up cosmetics 'Sofina' and 'Soliance' (Furuta et al. 2004). Sophorolipids are also present in pencil-shaped lip rouge, lip cream and eye shadow (Kawano et al. 1981).

Next, rhamnolipids are commonly produced by *Pseudomonas* sp. mainly by *Pseudomonas aeruginosa* during the cultivation on glucose, glycerol or triglycerides which will produce as monorhamnolipid and dirhamnolipid (Lang 2002). Rhamnolipids are natural emulsifiers and can be used as substitute to chemical substrates like sodium lauryl sulfate, sodium dodecyl sulfate and sodium laureth sulfate in a comestic composition (Farmer and Alibek 2020). Hence, Farmer and Alibek (2020), reported that rhamnolipids can be formulated to increase moisture retention, minimize the appearance of wrinkles and increase smoothness of skin. Thus, because of it skin compatibility and immensely low skin irritation (Haba et al. 2003), commercial skin care cosmetics were launched in numerous dosage forms and also cosmetics that contain rhamnolipids have been patented and used as anti-wrinkle and anti-ageing products (Desanto 2008). In addition, rhamnolipids have also been assimilated due to their antimicrobial action in some different cosmetic formulations such as deodorants, nail care products and toothpastes (Muhammad and Mahsa 2014).

Lastly, one of the most commonly used biosurfactant is Mannosylerythritol lipids (MELs), which is produced by the yeasts of Pseudozyma, Ustilago and Schizonella species (Yoshida et al. 2015). Morita et al. first reported that MEL-A can moisturise the dry and damaged skin by comparing the properties between biosurfactant and natural ceramides. Natural ceramides have great hydrating properties that have been applied in several skincare formulations, even though, the commercial use of it is of high cost due to their limited obtainability and advanced extraction-purification methods (Kitamoto et al. 2009). Therefore, MELs have been discovered to be used in skincare formulations due to their ceramide-like properties (Kitamoto et al. 2009, Kitagawa et al. 2008). Hence, MEL are widely used in the production of several cosmetic products such as lipsticks, lipmakers, eye shades, soap, sprays, powders, nail care, body massage oils and accessories (Ueno et al. 2007, Villeneuve 2007). In addition, the use of surfactin derivative lipopeptides has incredibly benefited the Japanese cosmetic industry because it gives a good impact in anti-wrinkle cosmetics and cleansing products (Gupta et al. 2019). Nowadays, MELs for cosmetic applications are commercially available as SurfMellow® produced by Toyobo Co., Ltd. which is available at Japan and USA (Salek and Uston 2019).

5.5 Bioremediation

Bioremediation is a method used to boost the degradation of environmental pollutants by microorganisms. In bioremediation, for example, the most important attention of biosurfactants was studied in the removal of heavy metal and hydrocarbons from polluted sites (Karlapudi et al. 2018). Due to its unique functionality and eco-friendly methodology, biosurfactant by yeast is used commercially in a significant application in the bioremediation process. Thus, biosurfactants tend to be important candidates for the substitution of synthetic surfactants in the field of bioremediation, due to their lower toxicity, high degradation, and in situ processing possibilities (Olasanmi and Thring 2018). Biosurfactant has been used to emulsify and spread hydrocarbons in the bioremediation of polluted water and soil due to its action on the oil-water interface, thereby improving the degradation of such compounds in the environment. Heavy metals also appear in the soil in addition to organic contaminants and are known to be inorganic contaminants with possible hazard to humans (Souza et al. 2014). Metal ions may exist in rocks, sand, and soil as fixed or soluble minerals, or water or vapors as dissolved ions. Inorganic or organic molecules may also be bound to metals or even bound to air particles. Surfactants may and have been used to fix soils polluted with metals and oils by desorption, solubilization, and dispersion of soil pollutants, enabling the elimination, aggregation, or reuse of soils (Fenibo et al. 2019b, Mao et al. 2015).

Several authors have documented the potential use of biosurfactants produced by different yeasts in bioremediation. Meneses et al. (2017), assessed the invention of biosurfactants by *Aureobasidium thailandense* LB011. The biosurfactant has demonstrated better efficiency than the chemical surfactant sodium dodecyl sulfate

(SDS) in oil dispersion studies, hence its possible bioremediation purpose. The biomolecule decreased the surface tension of water up to 31.2 mN/m. *Candida lipolytica* biosurfactant accomplishes an acceptable outcome in petroleum and heavy metal treatment at contaminated sites. The result showed that the biosurfactants was capable of 80% oil distribution effectiveness and could remove about 70% of motor oil from polluted. For the heavy metal treatment, the crude biomolecule was revealed to eliminate 30% to 40% of Copper and Lead from standard sand. It was also demonstrated how long-term stable agents can be marketed by making the production and use of biosurfactant more workable in the existing marketplace for petroleum-derived chemical surfactants (Santos et al. 2017).

The biosurfactant developed as glycolipids by *Meyerozma guilliermondii* yeast was found to solubilize 15.9% of cadmium in sewage sludge (anaerobic conditions). It is known that sophorolipids can help solubilize metals, so it is desirable to produce them in bioleaching reactors by co-inoculating them with acidophilic bacteria. In the case of cadmium, where a solution of 2% biosurfactant provided by yeast was added, the best metal removal has been achieved. *M. guilliermondii* was accomplished to solubilize 15.9 percent of this element with a pH modified to 2.0 (Camargo et al. 2018).

The accomplishment of biosurfactants on the biodegradation of diesel hydrocarbons was researched by Babaei and Habibi (2018). They revealed that the presence of biosurfactant sophorolipid-producing yeast *Candida ctenulata* KP324968 could boost biodegradation by approximately 82.1 percent after 6 days at a biodegradation rate of 0.378 g gcell^{-1}h^{-1}. The activity was performed well with active culture performance using optimal conditions, such as pH at 4.7, 204 rpm, the addition of a diesel concentration of 93.4 g L^{-1}, an *in situ* process. In, conclusion, diesel removal from the high concentration effluent was developed very well. Biosurfactants by *Candida sphaericia* UCP0995, also presented can eliminate iron, zinc, and lead from polluted soil with a reduction of these heavy metals of 95%, 90%, and 79%, respectively (Kieliszek et al. 2017, Luna et al. 2016b).

6. Conclusion

Biosurfactants from yeasts are good alternatives over chemical surfactants due to their low toxicity, exceptional physicochemical and biodegradability. In light with that, there are many studies reported, that support biosurfactants from yeast for their high potential applications in the industries such as food, pharmaceutical, cosmetics, petroleum and for bioremediation. Adoption of new and efficient upstream and downstream processing methods, process optimization, and research on use of alternative raw materials is needed to upkeep the advancement of state-of-the-art technologies for biosurfactants production using the yeast platform.

Acknowledgement

The authors would like to thank Research Management Center at Universiti Teknologi Malaysia (UTM) for support through grant No. Q.J130000.3051.02M09.

References

Abouseoud, M., R. Maachi, A. Amrane, S. Boudergua and A. Nabi. 2008. Evaluation of different carbon and nitrogen sources in production of biosurfactant by *Pseudomonas fluorescens*. Desalination 223(1-3): 143–151.

Accorsini, F.R., M.J.R. Mutton, E.G.M. Lemos and M. Benincasa. 2012. Biosurfactants production by yeasts using soybean oil and glycerol as low cost substrate. Braz. J. Microbiol. 43(1): 116–125.

Adamczak, M. and W. odzimierz Bednarski. 2000. Influence of medium composition and aeration on the synthesis of biosurfactants produced by *Candida antarctica*. Biotechnol. Lett. 22(4): 313–316.

Alcantara, V.A., I.G. Pajares, J.F. Simbahan, N.R. Villarante and M.L.D. Rubio. 2010. Characterization of biosurfactant from *Saccharomyces cerevisiae* 2031 and evaluation of emulsification activity for potential application in bioremediation. Philipp. Agric. Sci. 93(1): 22–30.

Alcantara, V.A., I.G. Pajares, J.F. Simbahan and M.D. Rubio. 2012. Substrate dependent production and isolation of an extracellular biosurfactant from *Saccharomyces cerevisiae* 2031. Philipp. J. Sci. 141: 13–24.

Alejandro, C.S., H.S. Humberto and J.F. María. 2011. Production of glycolipids with antimicrobial activity by *Ustilago maydis* FBD12 in submerged culture. Afr. J. Microbiol. Res. 5: 2512–2523.

Ali, L.H. and W.S. Ali. 2019. Production and antibacterial activity of biosurfactant from *Saccharomyces cerevisiae*. The 1st International Scientific Conference on Pure Science, 1–7. doi:10.1088/1742-6596/1234/1/012080.

Alizadeh-Sani, M., H. Hamishehkar, A. Khezerlou, M. Azizi-Lalabadi, Y. Azadi, E. Nattagh-Eshtivani, M. Fasihi, A. Ghavami, A. Aynehchi and A. Ehsani. 2018. Bioemulsifiers derived from microorganisms: applications in the drug and food industry. Adv. Pharm. Bull. 8(2): 191–199.

Alwaely, W.A., A.K. Ghadban and I.M. Alrubayae. 2019. Production and properties of biosurfactant from the local isolation of *Candida* spp. Drug Invent. Today. 12(5): 948–953.

Amaral, P.F.F., J.M. Da Silva, M. Lehocky, A.M.V. Barros-Timos, M.A.Z. Coelho, I.M. Marrucho and J.A.P. Countiho. 2006. Production and characterization of a bioemulsifier from *Yarrowia lipolytica*. Process Biochem. 41(8): 1894–1898.

Amaral, P.F., M.A.Z. Coelho, I.M. Marrucho and J.A. Coutinho. 2010. Biosurfactants from yeasts: characteristics, production and application. pp. 236–249. *In*: Biosurfactants. Springer, New York, NY.

Ashby, R.D., D.K.Y. Solaiman and T.A. Foglia. 2006. The use of fatty acid esters to enhance free acid sophorolipid synthesis. Biotechnol. Lett. 28: 253–260.

Babaei, F. and A. Habibi. 2018. Fast biodegradation of diesel hydrocarbons at high concentration by the sophorolipid-producing yeast *Candida catenulata* KP324968. J. Mol. Microbiol. Biotechnol. 28(5): 240–254.

Baki, A.S., U.K. Mohammad, A.A. Farouq, A. Gambo, M.A. Yahaya, U.S. Ahmad and S. Zaid. 2017. Evaluation of fungus from traditionally fermented cow milk and their application in the production of biosurfactant. Clin. Biotechnol. Microbiol. 1(5): 198–206.

Batista, R.M., R.D. Rufino, J.M. Luna, J.E.G. de Souza and L.A. Sarubbo. 2010. Effect of medium components on the production of a biosurfactant from *Candida tropicalis* applied to the removal of hydrophobic contaminants in soil. Water Environ. Res. 82(5): 418–425.

Batool, R., S. Ayub and I. Akbar. 2017. Isolation of biosurfactant producing bacteria from petroleum contaminated sites and their characterization. Soil Environ. 36(1): 35–44.

Bednarski, W., M. Adamczak, J. Tomasik and M. Płaszczyk. 2004. Application of oil refinery waste in the biosynthesis of glycolipids by yeast. Bioresour. Technol. 95(1): 15–18.

Bertrand, B., F. Martínez-Morales, N.S. Rosas-Galván, D. Morales-Guzmán and M.R. Trejo-Hernández. 2018. Statistical design, a powerful tool for optimizing biosurfactant production: A review. Colloids Interfaces 2(3): 1–18.

Besson, F. and G. Michel. 1992. Biosynthesis of iturin and surfactin by *Bacillus subtilis*: Evidence for amino acid activating enzymes. Biotechnol. Let. 14(11): 1013–1018.

Bhardwaj, G., S.S. Cameotra and H.K. Chopra. 2013a. Biosurfactants from fungi: a review. J. Pet Environ. Biotechnol. 4(6): 1–6.

Bhardwaj, G., S.S. Cameotra and H.K. Chopra. 2013b. Utilization of oleo-chemical industry by-products for biosurfactant production. AMB Express. 3(1): 1–5.

Biniarz, P., M. Łukaszewicz and T. Janek. 2017. Screening concepts, characterization and structural analysis of microbial-derived bioactive lipopeptides: a review. Crit. Rev. Biotechnol. 37: 393–410.

Bueno, J.L., P.A.D. Santos, R.R. da Silva, I.S. Moguel, A. Pessoa Jr, M.V. Vianna, F.C. Pagnocca, L.D. Sette and D.B. Gurpilhares. 2019. Biosurfactant production by yeasts from different types of soil of the South Shetland Islands (Maritime Antarctica). J. Appl. Microbiol. 126(5): 1402–1413.

Bushnell, L.D. and H.F. Haas. 1941. The utilization of certain hydrocarbons by microorganisms. J. Bacteriol. 41(5): 653.

Camargo, F.P., A.J.D. Menezes, P.S. Tonello, A.C.A. Dos Santos and I.C.S. Duarte. 2018. Characterization of biosurfactant from yeast using residual soybean oil under acidic conditions and their use in metal removal processes. FEMS Microbiol. Lett. 365(10): 1–8.

Campos, J.M., T.L. Stamford and L.A. Sarubbo. 2014. Production of a bioemulsifier with potential application in the food industry. Appl. Biochem. Biotechnol. 172(6): 3234–3252.

Campos, J.M., T.L. Stamford, R.D. Rufino, J.M. Luna, T.C.M. Stamford and L.A. Sarubbo. 2015. Formulation of mayonnaise with the addition of a bioemulsifier isolated from *Candida utilis*. Toxicol. Rep. 2: 1164–1170.

Campos, J.M., T.L.M. Stamford and L.A. Sarubbo. 2019. Characterization and application of a biosurfactant isolated from *Candida utilis* in salad dressings. Biodegradation 30(4): 313–324.

Casas, J. and F. García-Ochoa. 1999. Sophorolipid production by *Candida bombicola*: Medium composition and culture methods. J. Biosci. Bioeng. 88(5): 488–494.

Chandran, P. and N. Das. 2010. Biosurfactant production and diesel oil degradation by yeast species *Trichosporon asahii* isolated from petroleum hydrocarbon contaminated soil. Int. J. Eng. Sci. Technol. 2: 6942–6953.

Chandran, P. and N. Das. 2011. Characterization of sophorolipid biosurfactant produced by yeast species grown on diesel oil. Int. J. Sci. Nat. 2(1): 63–71.

Chaprão, M.J., I.N. Ferreira, P.F. Correa, R.D. Rufino, J.M. Luna, E.J. Silva and L.A. Sarubbo. 2015. Application of bacterial and yeast biosurfactants for enhanced removal and biodegradation of motor oil from contaminated sand. Electron. J. Biotechnol. 18(6): 471–479.

Ciesielska, K., I.N. Van Bogaert, S. Chevineau, B. Li, S. Groeneboer, W. Soetaert, Y. Van de Peer and B. Devreese. 2014. Exoproteome analysis of *Starmerella bombicola* results in the discovery of an esterase required for lactonization of sophorolipids. J. Proteom. 98: 159–174.

Dejwatthanakomol, C., J. Anuntagool, M. Morikawa and J. Thaniyavarn. 2016. Production of biosurfactant by *Wickerhamomyces anomalus* PY189 and its application in lemongrass oil encapsulation. Sci. Asia 42: 252–258.

Desai, J.D. and I.M. Banat. 1997. Microbial production of surfactants and their commercial potential. Microbiol. Mol. Biol. Rev. 61(1): 47–64.

Desanto, K. 2008. Rhamnolipid-based formulations. World Patent WO 2008013899 A3.

Deshpande, M. and L. Daniels. 1995. Evaluation of sophorolipid biosurfactant production by *Candida bombicola* using animal fat. Bioresour. Technol. 54(2): 143–150.

Dhivya, H., S. Balaji, R. Madhan and S. Akila. 2014. Production of amphiphilic surfactant molecule from *Saccharomyces cerevisiae* Mtcc 181 and its protagonist in nanovesicle synthesis. Int. J. Pharma. Sci. 3(11): 16–23.

Eldin, A.M., Z. Kamel and N. Hossam, 2019. Isolation and genetic identification of yeast producing biosurfactants, evaluated by different screening methods. Microchem. J. 146: 309–314.

Elshafie, A.E., S.J. Joshi, Y.M. Al-Wahaibi, A.S. Al-Bemani, S.N. Al-Bahry, D. Al-Maqbali and I.M. Banat. 2015. Sophorolipids production by *Candida bombicola* ATCC 22214 and its potential application in microbial enhanced oil recovery. Front Microbiol. 6: 1324.

Fakruddin, M.D. 2012. Biosurfactant: Production and application. J. Pet. Environ. Biotechnol. 3(4): 1–5.

Farmer, S. and K. Alibek. 2020. Yeast-Based Masks for Improved Skin, Hair and Scalp Health. U.S. Patent App. No. 16 752 844.

Fenibo, E.O., G.N. Ijoma, R. Selvarajan and C.B. Chikere. 2019a. Microbial Surfactants: The Next Generation Multifunctional Biomolecules for Applications in the Petroleum Industry and Its Associated Environmental Remediation. Microorganisms. 7(11): 581.

Fenibo, E.O., S.I. Douglas and H.O. Stanley. 2019b. A review on microbial surfactants: Production, classifications, properties and characterization. Adv. Appl. Microbiol. 18(3): 1–22.

Fontes, G.C., P.F.F. Amaral, M. Nele and M.A.Z. Coelho. 2010. Factorial design to optimize biosurfactant production by *Yarrowia lipolytica*. J. Biomed. Biotechnol. 2010: 1–8.

Fracchia, L., J.J. Banat, M. Cavallo and I.M. Banat. 2015. Potential therapeutic applications of microbial surface-active compounds. AIMS Bioeng. 2(3): 144–162.

Fukuoka, T., T. Morita, M. Konishi, T. Imura and D. Kitamoto. 2007a. Characterization of new types of mannosylerythritol lipids as biosurfactants produced from soybean oil by a basidiomycetous yeast, *Pseudozyma shanxiensis*. J. Oleo Sci. 56(8): 435–442.

Fukuoka, T., T. Morita, M. Konishi, T. Imura, H. Sakai and D. Kitamoto. 2007b. Structural characterization and surface-active properties of a new glycolipid biosurfactant, mono-acylated mannosylerythritol lipid, produced from glucose by *Pseudozyma antarctica*. Appl. Microbiol. Biotechnol. 76: 801–810.

Furuta, T., K. Igarashi and Y. Hirata. 2004. U.S. Patent App. No. 10 481 507.

Gargouri, B., N. Mhiri, F. Karray, F. Aloui and S. Savadi. 2015. Isolation and characterization of hydrocarbon-degrading yeast strains from petroleum contaminated industrial wastewater. Biomed. Res. Int.

George, S. and K. Jayachandran. 2013. Production and characterization of rhamnolipid biosurfactant from waste frying coconut oil using a novel *Pseudomonas aeruginosa* D. J. Appl. Microbiol. 114(2): 373–383.

Gharaei-Fathabad, E. 2011. Biosurfactants in pharmaceutical industry: a mini-review. Am. J. Drug Discov. Dev. 1(1): 58–69.

Guilmanov, V., A. Ballistreri, G. Impallomeni and R.A. Gross. 2002. Oxygen transfer rate and sophorose lipid production by *Candida bombicola*. Biotechnol. Bioeng. 77(5): 489–494.

Gupta, P.L., M. Rajput, T. Oza, U. Trivedi and G. Sanghvi. 2019. Eminence of microbial products in cosmetic industry. Nat. Prod. Bioprospect. 1–12.

Haba, E., A. Pinazo, O. Jauregui, M.J. Espuny, M.R. Infante and A. Manresa. 2003. Physicochemical characterization and antimicrobial properties of rhamnolipids produced by *Pseudomonas aeruginosa* 47T2 NCBIM 40044. Biotechnol. Bioeng. 81(3): 316–322.

Hammami, W., C. Labbé, F. Chain, B. Mimee and R.R. Bélanger. 2008. Nutritional regulation and kinetics of flocculosin synthesis by *Pseudozyma flocculosa*. Appl. Microbiol. Biotechnol. 80(2): 307.

Heryani, H. and M.D. Putra. 2017. Dataset on potential large scale production of biosurfactant using *Bacillus* sp. Data in Brief 13: 196–201.

Hewald, S., U. Linne, M. Scherer, M.A. Marahiel, J. Kämper and M. Bölker. 2006. Identification of a gene cluster for biosynthesis of mannosylerythritol lipids in the basidiomycetous fungus *Ustilago maydis*. Appl. Environ. Microbiol. 72(8): 5469–5477.

Hu, X., C. Wang and P. Wang. 2015. Optimization and characterization of biosurfactant production from marine *Vibrio* sp. strain 3B-2. Front. Microbiol. 6: 1–13.

Jahan, R. A.M. Bodratti, M. Tsianou and P. Alexandridis. 2020. Biosurfactants, natural alternatives to synthetic surfactants: physicochemical properties and applications. Adv. Colloid Int. Sci. 275: 1–22.

Jezierska, S., S. Sylwia, S. Claus and I. Van Bogaert. 2018. Yeast glycolipid biosurfactants. Febs. Let. 592(8): 1312–1329.

Jezierska, S., S. Claus, R. Ledesma-Amaro and I. Van Bogaert. 2019. Redirecting the lipid metabolism of the yeast *Starmerella bombicola* from glycolipid to fatty acid production. J. Ind. Microbiol. Biotechnol. 46(12): 1697–1706.

Jimoh, A.A. and J. Lin. 2020. Biotechnological applications of *Paenibacillus* sp. D9 lipopeptide biosurfactant produced in low-cost substrates. Appl. Biochem. Biotechnol. 191: 921–941.

Kapadia, S.G. and B.N. Yagnik. 2013. Current trend and potential for microbial biosurfactants. Asian J. Exp. Biol. Sci. 4(1): 1–8.

Karlapudi, A.P., T.C. Venkateswarulu, J. Tammineedi, L. Kanumuri, B.K. Ravuru, V. ramu Dirisala and V.P. Kodali. 2018. Role of biosurfactants in bioremediation of oil pollution—a review. Petroleum 4(3): 241–249.

Karlapudi, A.P., T.C. Venkateswarulu, K. Srirama, R.K. Kota, I. Mikkili and V.P. Kodali. 2020. Evaluation of anti-cancer, anti-microbial and anti-biofilm potential of biosurfactant extracted from an Acinetobacter M6 strain. J. King Saud Univ. Sci. 32(1): 223–227.

Kawano, J., T. Suzuki, S. Inoue and S. Hayashi. 1981. Powered compressed cosmetic material. US Patent 4 305 931.

Khan, M.S.A., B. Singh and S.S. Cameotra. 2014. Biological applications of biosurfactants and strategies to potentiate commercial production. Biosurfactants 159: 269.

Kieliszek, M., A.M. Kot, A. Bzducha-Wróbel, S. BŁażejak, I. Gientka and A. Kurcz. 2017. Biotechnological use of *Candida* yeasts in the food industry: a review. Fungal Biol. Rev. 31(4): 185–198.

Kim, H.S., J.W. Jeon, B.H. Kim, C.Y. Ahn, H.M. Oh and B.D. Yoon. 2006. Extracellular production of a glycolipid biosurfactant, mannosylerythritol lipid by *Candida* sp. SY16 using fed-batch fermentation. Appl. Microbiol. Biot. 70: 391–6.

Kitagawa, M., M. Suzuki, S. Yamamoto, A. Sogabe, D. Kitamoto, T. Imura, T. Fukuoka and T. Morita. 2008. Skin Care Cosmetic and Skin and Agent for Preventing Skin Roughness Containing Biosurfactants, EP 1 964 546 B1.

Kitamoto, D., H. Isoda and T. Nakahara. 2002. Functions and potential applications of glycolipid biosurfactant from energy saving materials to gene delivery carriers. J. Biosci. Bioeng. 943: 187–201.

Kitamoto, D., T. Morita, T. Fukuoka, Ma. Konishi and T. Imura. 2009. Self-assembling properties of glycolipid biosurfactants and their potential applications. Curr. Opin. Colloid Interface Sci. 14: 315–328.

Konishi, M., M. Fujita, Y. Ishibane, Y. Shimizu, Y. Tsukiyama and M. Ishida. 2016. Isolation of yeast candidates for efficient sophorolipids production: their production potentials associate to their lineage. Biosci. Biotechnol. Biochem. 80(10): 2058–2064.

Krishnaswamy, M., G. Subbuchettiar, T.K. Ravi and S. Panchaksharam. 2008. Biosurfactants properties, commercial production and application. Curr. Sci. 94: 736–747.

Kulakovskaya, T.V., A.S. Shashkov, E.V. Kulakovskaya and W.I. Golubev. 2004. Characterization of an antifungal glycolipid secreted by the yeast *Sympodiomycopsis paphiopedili*. FEMS Yeast Res. 5(3): 247–252.

Kurz, M., C. Eder, D. Isert, Z. Li, E.F. Paulus, M. Schiell, L. Toti, L. Vertesy, J. Wink and G. Seibert. 2003. Ustilipids, acylated β-D-mannopyranosyl D-erythritols from *Ustilago maydis* and *Geotrichum candidum*. J. Antibiot. 56(2): 91–101.

Lang, S. 2002. Biological amphiphiles (microbial biosurfactants). Curr. Opin. Colloid Interface Sci. 7: 12–20.

Lima, Á.S. and R.M. Alegre. 2009. Evaluation of emulsifier stability of biosurfactant produced by *Saccharomyces lipolytica* CCT-0913. Braz. Arch. Biol. Technol. 52(2): 285–290.

Lima, F.A., O.S. Santos, A.W.V. Pomella, E.J. Ribeiro and M.M. de Resende. 2020. Culture medium evaluation using low-cost substrate for biosurfactants lipopeptides production by *Bacillus amyloliquefaciens* in pilot bioreactor. J. Surfactants Deterg. 23(1): 91–98.

Lima, R.A., R.F.S. Andrade, D.M. Rodríguez, H.W.C. Araújo, V.P. Santos and G.M. Campos-Takaki. 2017. Production and characterization of biosurfactant isolated from *Candida glabrata* using renewable substrates. Afr. J. Microbiol. Res. 11(6): 237–244.

Lira, I.R.A.D.S., E.M.D.S. Santosa, A.A.S. Filhoa, C.B.B. Fariasb, J.M.C. Guerrab, L.A. Sarubboa and J.M. de Lunaa. 2020. Biosurfactant production from *Candida guilliermondii* and evaluation of its toxicity. Chem. Eng. Trans. 79: 457–462.

Liu, K., Y.M. Sun, J. Cao, J.R. Wang, Lu and H. Xu. 2020. Rational design, properties, and applications of biosurfactants: a short review of recent advances. Curr. Opin. Colloid Interface Sci. 45: 57–67.

Lourith, N. and M. Kanlayavattanakul. 2009. Natural surfactants used in cosmetics: glycolipids. Int. J. Cosmet. Sci. 31(4): 255–261.

Luepongpattana, S., S. Jindamarakot, S. Thaniyavarn and J. Thaniyavarn. 2014. Screening of biosurfactant production yeast and yeast-like fungi isolated from the coastal areas of Koh Si Chang. The 26th Annual Meeting of the Thai Society for Biotechnology and International Conference, 468–477.

Luna, J.M., A. Santos Filho, R.D. Rufino and L.A. Sarubbo. 2016a. Production of biosurfactant from *Candida bombicola* URM 3718 for environmental applications. Chem. Eng. Trans. 49: 583–588.

Luna, J.M, R.D. Rufino and L.A. Sarubbo. 2016b. Biosurfactant from *Candida sphaerica* UCP0955 exhibiting a heavy metal remediation properties. Process Saf. Environ. Prot. 102: 558–556.

Mahalaxmi, I., J. Kaavya, M.D. Subramaniam, S.B. Lee, A.A. Dayem, S.G. Cho and V. Balachandar. 2020. COVID-19: an update on diagnostic and therapeutic approaches. This article was published and explain the SARS-CoV-2 outbreak, treatment and other related approaches towards COVID-19. BMB Reports 53: 191–205.

Maneerat, S. and K. Phetrong. 2007. Isolation of biosurfactant-producing marine bacteria and characteristics of selected biosurfactant. Songklanakarin J. Sci. Technol. 29(3): 781–791.

Mao, X., R. Jiang, W. Xiao and J. Yu. 2015. Use of surfactants for the remediation of contaminated soils: a review. J. Hazard. Mater. 285: 419–435.

Marcelino, P.R.F., G.F.D. Peres, R. Terán-Hilares, F.C. Pagnocca, C.A. Rosa, T.M. Lacerda, J.C. dos Santos and S.S. da Silva. 2019. Biosurfactants production by yeasts using sugarcane bagasse hemicellulosic hydrolysate as new sustainable alternative for lignocellulosic biorefineries. Ind. Crops Prod. 129: 212–223.

Meneses, D.P., E.J. Gudiña, F. Fernandes, L.R. Goncalves, L.R. Rodrigues and S. Rodrigues. 2017. The yeast-like fungus *Aureobasidium thailandense* LB01 produces a new biosurfactant using olive oil mill wastewater as an inducer. Microbiol. Res. 204: 40–47.

Mimee, B., C. Labbé, R. Pelletier and R.R. Bélanger. 2005. Antifungal activity of flocculosin, a novel glycolipid isolated from *Pseudozyma flocculosa*. Antimicrob. Agents Chemother. 49(4): 1597–1599.

Minucelli, T., R.M. Ribeiro-Viana, D. Borsato, G. Andrade, M.V.T. Cely, M.R. Oliveira, C. Baldo and M.A.P.C. Celligoi. 2017. Sophorolipids production by *Candida bombicola* ATCC 22214 and its potential application in soil bioremediation. Waste Biomass Valori. 8(3): 743–753.

Morita, T., M. Konishi, T. Fukuoka, T. Imura and D. Kitamoto. 2008. Production of glycolipid biosurfactants, mannosylerythritol lipids, by *Pseudozyma siamensis* CBS 9960 and their interfacial properties. J. Biosci. Bioeng. 105(5): 493–502.

Morita, T., M. Kitagawa, M. Suzuki, S. Yamamoto, A. Sogabe, S. Yanagidani, T. Imura, T. Fukuoka and D. Kitamoto. 2009. A yeast glycolipid biosurfactant, mannosylerythritol lipid, shows potential moisturizing activity toward cultured human skin cells: the recovery effect of MEL-A on the SDS-damaged human skin cells. J. Oleo Sci. 58: 639–642.

Morita, T., H. Koike, H. Hagiwara, E. Ito, M. Machida, S. Sato, H. Habe and D. Kitamoto. 2014. Genome and transcriptome analysis of the basidiomycetous yeast *Pseudozyma antarctica* producing extracellular glycolipids, mannosylerythritol lipids. PloS One 9(2): e86490.

Muhammad, I.M. and S.S. Mahsa. 2014. Rhamnolipids: well-characterized glycolipids with potential broad applicability as biosurfactants. Ind. Biotechnol. 10: 285–291.

Mukherjee, S., P. Das and R. Sen. 2006. Towards commercial production of microbial surfactants. Trends Biotechnol. 24: 509–515.

Neboh, H.A., G.O. Abu and L. Uyigue. 2016. Utilization of agro-industrial wastes as substrates for biosurfactant production. Int. J. Geog. Environ. Mngmt. 2(1): 109–116.

Nitschke, M. and S.G. Costa. 2007. Biosurfactants in Food Industry. Trends Food Sci. Technol. 18: 252.

Nwaguma, I.V., C.B. Chikere and G.C. Okpokwasili. 2019a. Effect of cultural conditions on biosurfactant production by *Candida* sp. isolated from the sap of *Elaeis guineensis*. Biotechnol. J. Int. 23(3): 1–14.

Nwaguma, I.V., C.B. Chikere and G.C. Okpokwasili. 2019b. Isolation and molecular characterization of biosurfactant-producing yeasts from saps of *Elaeis guineensis* and *Raphia africana*. Microbiol. Res. J. Int. 29(4): 1–12.

Ohadi, M., A. Shahravan, N. Dehghannoudeh, T. Eslaminejad, I.M. Banat and G. Dehghannoudeh. 2020. Potential use of microbial surfactant in microemulsion drug delivery system: a systematic review. Drug Des. Devel. Ther. 14: 541.

Olasanmi, I.O. and R.W. Thring. 2018. The role of biosurfactants in the continued drive for environmental sustainability. Sustainability 10(12): 4817.

Oliveira, M.D., A. Magri, C. Baldo, D. Camilios-Neto, T. Minucelli and M.A.P.C. Celligoi. 2015. Review: sophorolipids a promising biosurfactant and its applications. Int. J. Adv. Biotechnol. Res. 6(2): 161–174.

Ongena, M. and P. Jacques. 2008. Bacillus lipopeptides: versatile weapons for plant disease biocontrol. Trends Microbiol. 16(3): 115–125.

Osman, M.S.Z., A.Z.U.R.A.I.E.N. Ibrahim, Japper-Jaafar and S. Shahir. 2019. Biosurfactants and its prospective application in the petroleum industry. J. Sustain. Sci. Manag. 14(3): 125–140.

Panilaitis, B., G.R. Castro, D. Solaiman and D.L. Kaplan. 2007. Biosynthesis of emulsan biopolymers from agro-based feedstocks. J. Appl. Microbiol. 102(2): 531–537.

Perfumo, A., I.M. Banat and R. Marchant. 2018. Going green and cold: biosurfactants from low temperature environments to biotechnology applications. Trends Biotechnol. 36: 277–289.

Pessôa, M.G., K.A.C. Vespermann, B.N. Paulino, M.C.S. Barcelos, G.M. Pastore and G. Molina. 2019. Newly isolated microorganisms with potential application in biotechnology. Biotechnol. Adv. 37: 319–339.

Price, N.P., K.J. Ray, K.E. Vermillion, C.A. Dunlap and C.P. Kurtzman. 2012. Structural characterization of novel sophorolipid biosurfactants from a newly identified species of *Candida* yeast. Carbohydr. Res. 348: 33–41.

Punrata, T., J. Thaniyavarna, S.C. Napathorna, J. Anuntagoolb and S. Thaniyavarna. 2020. Production of a sophorolipid biosurfactant by *Wickerhamomyces anomalus* MUE24 and its use for modification of rice flour properties. Science Asia 46(1): 11–18.

Qiao, N. and Z. Shao. 2010. Isolation and characterization of a novel biosurfactant produced by hydrocarbon-degrading bacterium *Alcanivorax dieselolei* B-5. J. Appl. Microbial. 108(4): 1207–1216.

Rawat, G., A. Dhasmana and V. Kumar. 2020. Biosurfactants: the next generation biomolecules for diverse applications. J. Environ. Sustain. 1–17.

Resende, A.H.M., J.M. Farias, D.D. Silva, R.D. Rufino, J.M. Luna, T.C.M. Stamford and L.A. Sarubbo. 2019. Application of biosurfactants and chitosan in toothpaste formulation. Colloids Surf. B Biointerfaces 181: 77–84.

Ribeiro, B.G., J.M.C. Guerra and L.A. Sarubbo. 2020a. Potential food application of a biosurfactant produced by *Saccharomyces cerevisiae* URM 6670. Front. Bioeng. Biotechnol. 8: 1–13.

Ribeiro, B.G, B.O. de Veras, J. dos Santos Aguiar, J.M.C. Guerra and L.A. Sarubbo. 2020b. Biosurfactant produced by *Candida utilis* UFPEDA1009 with potential application in cookie formulation. Electron. J. Biotechnol. 46: 14–21.

Rodrigues, L., I.M. Banat, J. Teixeira and R. Oliveira. 2006a. Biosurfactants: potential applications in medicine. J. Antimicrob. Chemother. 57(4): 609–618.

Rodrigues, L., H. Van Der Mei, I.M. Banat, J. Teixeira and R. Oliveira. 2006b. Inhibition of microbial adhesion to silicone rubber treated with biosurfactant from *Streptococcus thermophilus* A. FEMS Immunol. Med. Microbiol. 46(1): 107–112.

Roy, A. 2017. Review on the biosurfactants: properties, types and its applications. J. Fundam. Renew. Energy Appl. 8: 1–14.

Saha, P. and K.V.B. Rao. 2017. Biosurfactants—A Current Perspective on Production and Applications. Nat. Environ. Pollut. Technol. 16(1): 181–188.

Saharan, B.S., R.K. Sahu and D. Sharma. 2011. A review on biosurfactants: Fermentation, current developments and perspectives. Genet. Eng. Biotechnol. J. 2011(1): 1–14.

Sajid, M., M.S.A. Khan, S.S. Cameotra and A.S. Al-Thubiani. 2020. Biosurfactants: Potential applications as immunomodulator drugs. Immunol. Lett. 223: 71–77.

Sałek, K. and S.R. Euston. 2019. Sustainable microbial biosurfactants and bioemulsifiers for commercial exploitation. Process Biochem. 85: 143–155.

Sandrin, C., F. Peypoux and G. Michel. 1990. Coproduction of surfactin and iturin A, lipopeptides with surfactant and antifungal properties, by *Bacillus subtilis*. Biotechnol. Appl. Biochem. 12(4): 370–375.

Santos, D.K.F., R.D. Rufino, J.M. Luna, V.A. Santos and L.A. Sarubbo. 2016. Biosurfactants: Multifunctional biomolecules of the 21st century. Int. J. Mol. Sci. 17(3): 1–31.

Santos, D.K., A.H. Resende, D.G. de Almeida, R.D.C.F. Soares da Silva, R.D. Rufino, J.M. Luna, I.M. Banat and L.A. Sarubbo. 2017. Candida lipolytica UCP0988 biosurfactant: potential as a bioremediation agent and in formulating a commercial related product. Front. Microbiol. 8: 767.

Sarubbo, L.A., C.B.B. Farias and G.M. Campos-Takaki. 2007. Co-utilization of canola oil and glucose on the production of a surfactant by *Candida lipolytica.* Current Microbiol. 54: 68–73.

Sarubbo, L.A., R.B. Rocha Jr, J.M. Luna, R.D. Rufino, V.A. Santos and I.M. Banat. 2015. Some aspects of heavy metals contamination remediation and role of biosurfactants. Chem. Ecol. 31(8): 707–723.

Sarubbo, L.A., J.M. Luna, R.D. Rufino and P. Brasileiro. 2016. Production of a low-cost biosurfactant for application in the remediation of sea water contaminated with petroleum derivates. Chem. Eng. Trans. 49: 523–528.

Sekhon, K.K., S. Khanna and S.S. Cameotra. 2012. Biosurfactant production and potential correlation with esterase activity. J. Pet. Environ. Biotechnol. 3(133): 1–10.

Sen, R. 1997. Response surface optimization of the critical media components for the production of surfactin. J. Chem. Technol. Biotechnol. 68(3): 263–270.

Sen, S., S.N. Borah, A. Bora and S. Deka. 2017. Production, characterization, and antifungal activity of a biosurfactant produced by *Rhodotorula babjevae* YS3. Microb. Cell Fact. 16(1): 1–14.

Shah, V., G.F. Doncel, T. Seyoum, K.M. Eaton, I. Zalenskaya, R. Hagver, A. Azim and R. Gross. 2005. Sophorolipids, microbial glycolipids with anti-human immunodeficiency virus and spermimmobilizing activities. Antimicrob. Agents Chemother. 49: 4093–100.

Sharma, P., S. Sangwan and H. Kaur. 2019. Process parameters for biosurfactant production using yeast *Meyerozyma guilliermondii* YK32. Environ. Monit. Assess. 191(9): 1–13.

Silva, I.A., B.O. Veras, B.G. Ribeiro, J.S. Aguiar, J.M.C. Guerra, J.M. Luna and L.A. Sarubbo. 2020. Production of cupcake-like dessert containing microbial biosurfactant as an emulsifier. PeerJ. 8: 1–23.

Silva, R.D.C.F., D.G. Almeida, R.D. Rufino, J.M. Luna, V.A. Santos and L.A. Sarubbo. 2014. Applications of biosurfactants in the petroleum industry and the remediation of oil spills. Int. J. Mol. Sci. 15(7): 12523–12542.

Singh, P. and B.N. Tiwary. 2016. Isolation and characterization of glycolipid biosurfactant produced by a *Pseudomonas otitidis* strain isolated from Chirimiri coal mines, India. Bioresour. Bioprocess. 3(1): 1–16.

Souza, E.C., T.C. Vessoni-Penna and R.P. de Souza Oliveira. 2014. Biosurfactant-enhanced hydrocarbon bioremediation: An overview. Int. biodeterior. biodegradation 89: 88–94.

Souza, K.S.T., E.J. Gudiña, R.F. Schwan, L.R. Rodrigues, D.R. Dias and J.A. Teixeira. 2018. Improvement of biosurfactant production by *Wickerhamomyces anomalus* CCMA 0358 and its potential application in bioremediation. J. Hazard. Mater. 346: 152–158.

Subramaniam, M.D., D. Venkatesan, M. Iyer, S. Subbarayan, V. Govindasami, A. Roy and N.S. Kumar. 2020. Biosurfactants and anti-inflammatory activity: A potential new approach towards COVID-19. Curr. Opin. Environ. Sci. Health.

Thaniyavarn, J., T. Chianguthai, P. Sangvanich, N. Roongsawang and K. Washio. 2008. Production of sophorolipid biosurfactant by *Pichia anomala*. Biosci. Biotechnol. Biochem. 72: 2061–2068.

Ueno, Y., N. Hirashima, Y. Inoh, T. Furuno and M. Nakanishi. 2007. Characterization of biosurfactant-containing liposomes and their efficiency for gene transfection. Biol. Pharm. Bull. 30: 169.

Van Bogaert, I.N., M. Demey, D. Develter, W. Soetaert and E.J. Vandamme. 2009. Importance of the cytochrome P450 monooxygenase CYP52 family for the sophorolipid-producing yeast *Candida bombicola*. FEMS Yeast Res. 9(1): 87–94.

Vecino, X., J.M. Cruz, A.B. Moldes and L.R. Rodrigues. 2017. Biosurfactants in cosmetic formulations: trends and challenges. Critical Rev. Biotechnol. 37: 911–923.

Vijayakumar, S. and V. Saravanan. 2015. Biosurfactants-Types, sources and applications. Res. J. Microbiol. 10(5): 181–192.

Villeneuve, P. 2007. Lipases in lipophilization reactions. Biotechnol. Adv. 25(6): 515–536.

Yeh, M.S., Y.H. Wei and J.S. Chang. 2005. Enhanced production of surfactin from *Bacillus subtilis* by addition of solid carriers. Biotechnol. Prog. 21(4): 1329–1334.

Yoshida, S., M. Koitabashi, J. Nakamura, T. Fukuoka, H. Sakai, M. Abe, D. Kitamoto and H. Kitamoto. 2015. Effects of biosurfactants, mannosylerythritol lipids, on the hydrophobicity of solid surfaces and infection behaviours of plant pathogenic fungi. J. Appl. Microbiol. 119: 215–224.

Zhang, Y., H. Li, J. Sun, J. Gao, W. Liu, B. Li and J. Chen. 2010. DC-Chol/DOPE cationic liposomes: a comparative study of the influence factors on plasmid pDNA and siRNA gene delivery. Int. J. Pharm. 390(2): 198–207.

Zinjarde, S.S. and A. Pant. 2002. Emulsifier from a tropical marine yeast, *Yarrowia lipolytica* NCIM 3589. J. Basic Microbiol. 42(1): 67–73.

11

Insights into Production and Applications of Microbial Lipopeptides

Swasti Dhagat and *Satya Eswari Jujjavarapu**

1. Introduction to Lipopeptides

Lipopeptides are amphiphilic molecules that are synthesized as secondary metabolites by bacteria, yeast and actinomycetes. They consist of lipids connected to short linear chains or cyclic amino acids via an amide or ester bond. The fatty acid has a variable length and the amino acids are of D-configuration rather than L possibly to resist the action of proteases. These molecules are self-assembled into various structures by microbial species by non-ribosomal peptide synthetase (NRPS). A single microbial species can produce various isoforms of lipopeptides that differ in one or more amino acids and fatty acid moieties. They are surface-active agents which lower the surface and interfacial tension between two phases which can be either liquid-liquid or liquid-solid, respectively. Due to their surfactant-like properties lipopeptides have antimicrobial and haemolytic properties and hence they are used in chemical, agricultural, food and pharmaceutical industries (Eswari et al. 2019a). Generally, they are produced by Gram-positive bacteria. The excessive use of antibiotics to control pathogens has resulted in the development of antibiotic resistant strains of bacteria and fungi. Due to the unique mechanism of these lipopeptides against pathogens they are suitable alternative antimicrobial agents of this era. They have a wide-ranging antimicrobial activity and are important biomolecules in medicinal chemistry (Beltran-Gracia et al. 2017, Eswari et al. 2019d).

Department of Biotechnology, National Institute of Technology Raipur (C.G.) 492010, India.
Email: sdhagat.phd2016.bt@nitrr.ac.in
* Corresponding author: satyaeswarij.bt@nitrr.ac.in, eswari_iit@yahoo.co.in

2. Types and Classification of Lipopeptides

Microbial lipopeptides are classified based on their size, structural similarities, charge and hydrophobicity. Based on topology of peptide chain, they are broadly classified into two types: linear and cyclic (Beltran-Gracia et al. 2017). Linear lipopeptides can be easily synthesized chemically and hence are of great interest in pharmaceutical industries. Cyclic lipopeptides, on the other hand, have great oral bioavailability and antibacterial potential. Some of the classes of microbial lipopeptides are daptomycin, surfactin, iturin, fengycin, polymyxins, viscosin, pseudofactin, fusaricidin and paenibacterin. All the classes of lipopeptides have different mechanisms of action as antimicrobial and anticancer agents. Some of the major classes of lipopeptides are discussed here.

2.1 *Daptomycin*

Daptomycin is a cyclic lipopeptide with an average molecular weight of 1620.7 g/mol in which the decanoyl lipid is attached to a peptide of 13 amino acids (Fig. 1(A)). It is synthesized by a Gram-positive bacterium, *Streptomyces roseosporus* and acts against Gram-positive bacteria (Jujjavarapu and Dhagat 2018, Jujjavarapu et al. 2018). Due to its antimicrobial properties it has been clinically approved by Food and Drug Administration (FDA), USA as an antibiotic under the trade name "Cubicin". Daptomycin has been used to treat skin and soft tissue infections, osteomyelitis and endocarditis. It is also used as a final strategy for methicillin-resistant *Staphylococcus aureus* (MRSA) infections (Patel et al. 2015).

The mechanism of action of daptomycin as an antibiotic is still not clear but few hypotheses have been proposed for the same. Daptomycin, due to its anionic

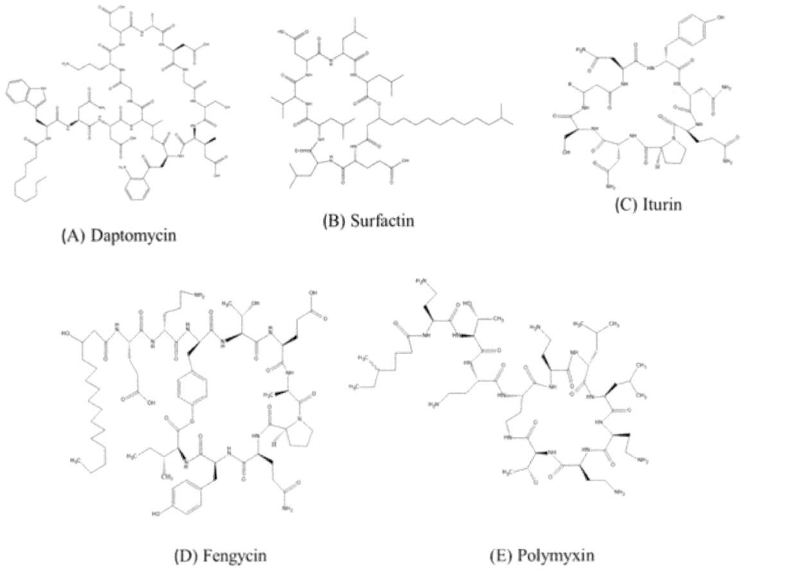

(A) Daptomycin

(B) Surfactin

(C) Iturin

(D) Fengycin

(E) Polymyxin

Figure 1. Structures of daptomycin (A), surfactin (B), iturin (C), fengycin (D) and polymyxin (E).

nature, disrupts the membrane potential of bacteria causing efflux of potassium ions and depolarization with the help of calcium ions and creates pores in the bacterial cell membrane (Beltran-Gracia et al. 2017). Due to the loss of potassium ions the synthesis of DNA, RNA and proteins are arrested leading to cell death. It also causes the redistribution and disruption of enzymes required in the synthesis of cell wall and cell division. This results in small, membrane lesions affecting cell morphology (Jujjavarapu et al. 2018). Daptomycin might also inhibit the synthesis of lipotechoic acid which is the proteoglycan component of cell wall of Gram-positive bacteria. Therefore, daptomycin prevents formation of cell wall leading to cell lysis (Beltran-Gracia et al. 2017).

2.2 Surfactin

Surfactin comprise a major class of lipopeptides with a broad range of antibacterial, antifungal, antiviral, anti-mycoplasma and antitumor activities and also supress inflammatory responses and have a very high surface active property (Kim et al. 1998, Hamley 2015). Surfactin has a molecular weight of 1036.3 g/mol and consists of a cyclic heptapeptide bonded to a β-hydroxy fatty acid of C13-15 fatty acyl chain (Shao et al. 2015) to form a lactone ring structure (Seydlová et al. 2011) (Fig. 1(B)). Different isoforms of surfactins varying according to the order of amino acid and size of lipid moiety coexist in a cell depending on the bacterial strain and culture conditions (Seydlová et al. 2011). Surfactin permeate lipid membranes after dimerization and form ion channels in lipid bilayer (Meena and Kanwar 2015a, Beltran-Gracia et al. 2017).

The antifungal activity of surfactin is by inhibiting glucan synthase which is an important enzyme required for callose synthesis in fungal cells. Surfactin inserts its peptide moiety into lipid bilayer and forms aggregate at the membrane. This creates pores in the cell membrane resulting in its breakage and cell lysis (Jujjavarapu and Dhagat 2018, Jujjavarapu et al. 2018, Eswari et al. 2019c). Surfactin also acts as antiviral agent against enveloped viruses by acting as viral fusion inhibitor (Yuan et al. 2018). Surfactin also exhibits anti-cancer activity against various cell lines, such as LoVo (human colon carcinoma) cells, Ehlrich carcinoma cells and human breast cancer cells (MCF-7, T47D and MDA-MB-231) (Jujjavarapu and Dhagat 2018, Jujjavarapu et al. 2018). It can also be used for the production of nanoparticles for drug delivery applications (Eswari et al. 2018). Due to the haemolytic activity of surfactin it has limited medical applications (Beltran-Gracia et al. 2017).

2.3 Iturin

Iturin is a small cyclic lipopeptide with a molecular weight of approximately 1043.2 g/mol. The peptide part of iturin consists of a ring of seven amino acids with a hydrophobic tail of β-amino fatty acid chain with 14 to 17 carbon (Jujjavarapu and Dhagat 2018, Jujjavarapu et al. 2018, Eswari et al. 2019c) (Fig. 1(C)). Due to the variations in amino acids iturins are classified as iturin A, iturin C, iturin D, iturin E, bacillomycin D, bacillomycin F, bacillomycin Lc, mycosubtilin and mojavensin (Ali et al. 2014). Iturin is produced by all strains of *Bacillus subtilis*. Iturin A forms ion-conducting potassium channels in lipid bilayer of cells and they can be utilized as biocontrol agents against plant pathogens (Patel et al. 2015). Iturin also causes leakage of potassium ions by disturbing cytoplasmic membrane of yeast cells resulting in cell death. They interact with sterol components of fungal cell membrane and act as antifungal agents. Due to their physiochemical and biological properties they are used in oil, food and pharmaceutical industries. The disadvantage of iturin is its haemolytic activity.

2.4 *Fengycin*

Fengycin is synthesized by various strains of *Bacillus subtilis* and some strains of *Paenibacillus*. It has a molecular weight of 1463.7 g/mol and has antifungal activity against filamentous fungi (Deleu et al. 2008). This family of lipopeptide consists of fengycins and plipastatin. They are lipodecapeptides with lactone ring in β-hydroxy fatty acid chain (either saturated or unsaturated) of 14 to 18 carbon atoms (Eswari et al. 2019c) (Fig. 1(D)). The cyclization is between the phenolic (-OH) side chain of Tyr3 and C-terminal COOH group of Ile20 (Pathak et al. 2012).

Fengycin acts on plasma membrane of fungi. The affinity of fengycin for fungal membrane consisting of phosphatidylcholine is more than bacterial membrane containing (phosphatidylethanolamine: phosphatidylglycerol) (Sur et al. 2018). They induce the production of reactive oxygen species (ROS) and cause cell death by mitochondria dependent apoptosis. They have mild haemolytic activity and hence can be used as potent anti-tumor agents (Yin et al. 2013, Beltran-Gracia et al. 2017). The mechanism of action of fengycin is because of both its amphiphilic nature and its affinity for lipid bilayers. At low concentration, fengycin exists as monomer in the lipid bilayer and does not disrupt the phospholipid bilayer. At medium concentration, accumulation of fengycin inside the lipid bilayer leads to self-aggregation and modification of phospholipid bilayer. This leads to the formation of ion channels and pores in the cell membrane (Bechinger et al. 1993, Deleu et al. 2008).

2.5 *Polymyxin*

Polymyxin is a cyclic lipopeptide synthesized by Gram-positive soil microbe *Paenibacillus polymyxa* with a molecular weight of 1203.5 g/mol (Choi et al. 2009). It is made up of a heptapeptide core as the main cyclic component, a linear tripeptide and a fatty acid tail (Dewick 2002, Velkov et al. 2010, Eswari et al. 2019b) (Fig. 1(E)). The presence of Dab (5 L-α,γ-diaminobutyric acid) residue makes polymyxin polycationic at pH 7.4. Due to the difference in amino acid side chains and fatty acids, polymyxins are classified into polymyxin A-E, M, S and T. Polymyxins exhibit antibacterial activity against a wide range of Gram-negative bacteria (*Escherichia coli*, *Pseudomonas aeruginosa*, *Acinetobacter baumannii*, *Klebsiella* sp., *Citrobacter* sp., *Enterobacter* sp., *Haemophilus influenza*, *Salmonella* sp., *Shigella* sp.) whereas polymyxins M and T are also active against some Gram-positive bacteria (Gales et al. 2001, Hogardt et al. 2004, Niks et al. 2004). Polymyxins, as topical applications, are already being used to treat infections caused by enterobacteria, pseudomonads and *Acinetobacter*. Polymyxin B and polymyxin E are permitted to be used as the last line of therapy against multi-drug resistant Gram-negative bacilli.

The antibacterial activity of polymyxin is by disrupting both the inner and outer membrane of Gram-negative cell. The site of action of polymyxin is lipid A of lipopolysaccharide (LPS) layer present on the outer membrane of bacteria. The cationic Dab residues present on the polymyxin molecules interacts with the phosphate groups of lipid A which are negatively charged. Divalent cations such as calcium (Ca^{2+}) and magnesium (Mg^{2+}) ions are required to stabilize lipopolysaccharide and are present in the cell membrane. L-Dab on polymyxin has higher affinity for divalent cations than LPS and thus create disturbance in the outer membrane of bacterial cell. This increases the permeability of cell membrane leading to leakage of cell contents and cell death (Newton 1956, Davis et al. 1971, Schindler and Osborn 1979).

3. Microbial Production of Lipopeptides

Lipopeptides are synthesized as a fermentation product by bacteria, yeast and actinomycetes. Microbial lipopeptides are produced from renewable sources and are biodegradable compared to their chemical counterparts which are produced from non-renewable sources, such as petrochemicals and hence are recalcitrant and non-biodegradable. They can also be produced by fermentation of inexpensive substrates, such as agricultural residues or carbohydrate-rich wastes and therefore their cost of production is reduced. Due to their production from green sources they are non-toxic and safe for various applications. Production using microorganisms also shortens the time of fermentation.

3.1 Bacterial Production

Among all the bacterial species, *Bacillus* sp. comprises a major genus of lipopeptide producing bacteria (Table 1). *Bacillus* sp. produces almost all the major families of lipopeptides, especially cyclic lipopeptides that have antibacterial and antifungal properties. Bacteria belonging to *Bacillus* species are easy to cultivate, are convenient producers of lipopeptides at industrial level and produce non-toxic by-products. But the production of lipopeptides by *Bacillus* is very low. *Bacillus* spp. are the efficient producers of surfactin, fengycin and iturin. They are also capable of producing lichenysin, pumilacidin, bacircine, daitocin and halo- and iso-halobactin. These lipopeptides have unusual β-amino acids and N-methylated and hydroxyl amino acids.

Another family of bacteria, *Paenibacillus*, also produces large numbers of cyclic and linear lipopeptides. Fengycin, polymyxin, fusaricidin, tridecaptin and paenibacterin are some of the lipopeptides produced by *Paenibacillus* family. All these lipopeptides have potent antibiotic activities. Polymyxin consists of 15 variants, one of which is octapeptin. Octapeptins have a variable fatty acyl group of C8 to C10 and have antimicrobial activity against Gram-positive and Gram-negative bacteria including multi-drug resistant *Pseudomonas aeruginosa* and *Escherichia coli*. Tridecaptins are linear lipopeptides with 13 amino acids. Paenibacterin also consists of 13 amino acids along with a C15 fatty acyl chain. *Brevibacillus laterosporus* produces tauramadine which consists of five amino acids linked to iso-methyl-octadecanoic acid. It also produces kurstakins, trideaptins and cerexins.

Table 1. Bacterial production of major classes of lipopeptides.

Lipopeptide	Producer bacteria	References
Daptomycin	*Streptomyces roseosporus*	Baltz et al. (2006)
Surfactin	*Bacillus subtilis*	Al-Ajlani et al. (2007)
	Bacillus amyloliquefaciens WH1	Gao et al. (2013)
	Bacillus tequilensis	Pathak et al. (2014)
	Bacillus polyfermenticus	Kim et al. (2009)
	Bacillus pumilus	Slivinski et al. (2012)
Iturin	*Bacillus subtilis*	Dang et al. (2019)
	Bacillus amyloliquefaciens	
	Bacillus licheniformis	
	Bacillus thuringiensis	
	Bacillus methyltrophicus	
	Bacillustequilensis	Dunlap et al. (2019)
Fengycin	*Bacillus subtilis*	Wei et al. (2010)
	Bacillus amyloliquefaciens	Hanif et al. (2019)
	Bacillus tequilensis	Pathak et al. (2014)
	Bacillus thuringiensis	Roy et al. (2013)
	Paenibacillus sp.	Beltran-Gracia et al. (2017)
Polymyxin	*Paenibacillus polymyxa*	Yu et al. (2018)
	Paenibacillus amylolyticus	DeCrescenzo et al. (2007)
Viscosin	*Pseudomonas fluorescens*	Alsohim et al. (2014)
Fusaricidin	*Paenibacillus polymyxa*	Li and Chen (2019)
Paenibacterin	*Paenibacillus thiaminolyticus* OSY-SE	Huang and Yousef (2014)
Pseudofactin	*Pseudomonas fluorescens* BD-5	Janek et al. (2010)
Lichenysin	*Bacillus licheniformis*	Madslien et al. (2013)
Serrawettin W1	*Serratiamarcescens*	Matsuyama et al. (2011)

Pseudomonas also produces numerous cyclic lipopeptides possessing antimicrobial and anticancer activities. Some of the lipopeptides produced by *Pseudomonas* spp. are syringomycin, syringopeptin, tolaasin, viscosin, pseudofactin, amphisin and putisolysin. Syringomycins and syringopeptines are lipodepsipeptides and are toxins that induce necrosis. Viscosin has nine amino acids whereas amphisin has eleven amino acids in their peptide moiety. Tolaasin consist of 19–25 amino acids with 3-hydroxydecanoic acid and putisolysin has 12 amino acids with hexanoic lipid tail. Plusbacins, tripropeptins and empedopeptin are cyclic lipopeptides and differ in the nature of fatty acid moiety and first three amino acids. Few of the linear lipopeptides produced by *Pseudomonas* are syringofactin and corrugatin. Corrugatin contains β-hydroxyhistidine which is a rare amino acid and acts as a ligand for ferric ions.

3.2 Production by Actinomycetes

Actinomycetes belong to a class of Gram-positive bacteria which have certain characteristics like fungi such as formation of conidia and hyphae but, unlike fungi, they have murine in their cell wall. Actinomycetes are oval-shaped facultative anaerobes whereas bacteria are spherical- or rod-shaped aerobes, anaerobes or facultative aerobes. Actinomycetes produce a large number of antimicrobial (antibacterial and antifungal) compounds. Some of the lipopeptides produced by actinomycetes are listed in Table 2. One of the most common lipopeptide, daptomycin, is produced by an Actinobacteria, *Streptomyces roseosporus*.

Actinomyces and *Streptomyces* produce a number of antibiotic molecules, namely, friulimicins, amphomycins and laspartomycins. Friulimicin B, produced by *Actinoplanes friuliensis*, is undergoing clinical trials. Ramoplanin is produced by *Actinoplanes* sp. It is a glycolipodepsipeptide and hence acts as an antibiotic against multi-drug resistant *Staphylococcus aureus*, *Enterococcus* sp. and *Clostridium difficile*. Other actinomycetes that produce lipopeptides are *Nocardiopsis* sp., *Streptomyces* sp., *Nesterenkonis* sp. and *Micromonospora* sp.

Table 2. Lipopeptides produced by various strains of actinomycetes.

Lipopeptide	Producer actinomycetes	References
Arylomycin A6 and arylomycin A5	*Streptomyces parvus* HCCB10043	Rao et al. (2013)
Arylomycin A and arylomycin B	*Streptomyces* sp. Tü 6075	Schimana et al. (2002)
Friulimicin A, B, C, and D	*Actinoplanes friuliensis*	Aretz et al. (2000); Vertesy et al. (2000)
Rakicidin A and B	*Micromonospora* sp.	McBrien et al. (1995)
TAN-1511 A, B and C	*Streptosporangium amethystogenes*	Takizawa et al. (1995)
Peptidolipins B-F	*Nocardia* sp.	Wyche et al. (2012)
Neopeptins A and B	*Streptomyces* sp. K-710	Satomi et al. (1982)

4. Process Optimization for the Production of Lipopeptides

Lipopeptides have a wide range of applications in medical, pharmaceutical and agricultural sectors. But the major limitation of microbial production of lipopeptides is a high production cost and low yield of the product. Process optimization, especially media optimization, is an important parameter for industrial production of lipopeptides as it increases production rates and yields (Suryawanshi et al. 2019b, Suryawanshi et al. 2020b). Lipopeptide production and yields are influenced by concentrations of C, N, P, Mg, Na, Fe, Mn and Zn ions along with medium used, process conditions and environmental parameters.

4.1 Selection of Suitable Production Medium for the Production of Lipopeptides

The optimum production medium helps in determining final yield and quality of product. Various components in medium that require optimization are carbon and nitrogen sources and trace metals.

4.1.1 Carbon Source

Carbon sources affect metabolism of bacteria and overall production cost. The quantity and nature of the carbon source is an important consideration for lipopeptide production. Low quantity of carbon will be insufficient for the growth of bacteria whereas high concentration might hinder bacterial growth by substrate inhibition or lead to formation of undesired by-products such as lactates, acetates or 2,3-butanediols (Glaser et al. 1995, Ramos et al. 2000). The nature of a carbon source is important as microorganisms might not be able to degrade complex substrates.

The type of carbon source is also dependent on the type of application of lipopeptide. For example, lipopeptides used for bioremediation, microbial enhanced oil recovery and as detergent formulations can be produced using agro-industrial wastes as substrates. These low cost substrates can be used to produce low-value high-volume lipopeptides (Mukherjee et al. 2006). Agro-industrial wastes for lipopeptide production include soybean meal, soy hydrolysate, corn steep liquor, cashew fruit juice, molasses, rice, fish, potato, and cassava wastes. Pre-defined or chemically defined media are suited for high-value low-volume applications of lipopeptides in therapeutic and pharmaceutical sectors. The advantage of using chemically defined media is that the relation between biomass, substrate consumption and product formation can be easily established. It also provides consistency among batches, eases scale-up process and reduces the need of additional processes such as pre-treatment of medium components.

4.1.2 Nitrogen Source

The ratio of C:N is very significant and depends on bacterial species used for the production and mode of cultivation (Fonseca et al. 2007). The nitrogen salt determines aerobic-to-anaerobic respiration activity in *Bacillus subtilis*. A diauxic consumption pattern is observed when dual nitrogen source such as NH_4NO_3 is used. In such type of respiration, ammonia is consumed first followed by nitrate. The mechanism of such a type of respiration can be explained as follows. The presence of excess ammonium ion in medium represses nitrate permease which prevents uptake of nitrate. When oxygen is depleted in the medium, the consumption of ammonium decreases and consumption of nitrate begins. Under anaerobic conditions, nitrate acts as final electron acceptor as well as nitrogen source by being converted to ammonia (Hoffmann et al. 1995). The conditions of oxygen depletion and nitrate limitation increase lipopeptide yield (Davis et al. 1999). The inorganic compounds used as nitrogen source are ammonium nitrate, ammonium sulphate, glutamic sodium, urea and sodium nitrate. The cheaper raw materials such as soybean flour, peptone and casein acid hydrolysate can also be used as nitrogen sources (Liu et al. 2012).

4.1.3 Trace Elements or Divalent Ions

Trace elements are the source of divalent ions. The importance of trace elements have been highlighted by metabolic flux analysis derived from metabolic pathways of products (Satya Eswari and Venkateswarlu 2016). Metal cations such as, Fe^{2+}, Mg^{2+}, Mn^{2+} and K^+ increase lipopeptide production by *Bacillus subtilis* (Cooper et al. 1981, Sheppard and Cooper 1991, Wei et al. 2007, Gancel et al. 2009). Fe^{2+}

stimulates bacterial growth and is a cofactor for many enzymes required for the synthesis of lipopeptides (Sen 1997, Wei and Chu 1998). Mn^{2+} promotes nitrate reductase synthesis which is required for assimilation of nitrate during oxygen-depleted environments (Huang et al. 2015).

4.2 *Optimization of Media Components and Operational Conditions*

The determination of concentration of carbon, nitrogen and trace elements to obtain high yields of lipopeptides by one-factor-at-a-time experiments is a tedious process. These experiments also do not study the interaction between different media components. Hence statistical optimization tools are employed to evaluate the concentrations of all media components.

The optimization methods screen most influential medium components and determine the optimum concentration of each media component. This results in a medium with high product yield and selectivity. The optimization tools such as artificial neural network (ANN), differential evolution (DE), genetic algorithm (GA), particle swarm optimization (PSO) improve lipopeptide production (Eswari et al. 2013, Peng et al. 2014, Chen et al. 2015a, Eswari et al. 2016, Eswari and Venkateswarlu 2016, Venkateswarlu and Jujjavarapu 2019). Optimization methods have generally been performed in shake flasks and not have been validated in bioreactors. Some factors such as oxygen transfer rates, aeration and agitation rates are significantly different in shake flasks and bioreactors. Hence, the expected productivity by optimization techniques is not guaranteed in bioreactor-scale operations.

Operational conditions such as pH, temperature, aeration rate and concentration of dissolved oxygen affect the growth of cells and lipopeptide production. The optimum operational conditions vary from one microbial strain to another. *Bacillus subtilis* produces surfactin at optimum temperature range of 25°C to 37°C whereas temperature required by thermophilic *Bacillus* spp. is greater than 40°C (Zhao et al. 2013). The optimum pH range for high lipopeptide production is 3.0–8.0.

Oxygen transfer is an important parameter for scale-up production and process optimization in surfactin production (Chen et al. 2015b). The conditions of oxygen transfer define whether the product is a mixture of lipopeptides or consists of only surfactin. Production of surfactin increases with increased mass transfer coefficient (kLa). Foam formation decreases oxygen transfer and reduces kLa (Wei et al. 2004). Therefore, this parameter needs to be taken care of during production of lipopeptides by fermentation (Fahim et al. 2012). Aeration results in excessive foaming. Use of antifoam during lipopeptide production is not recommended as they alter the physiology of bacteria and downstream processing. Bioreactors that are preferred for lipopeptide production are bubbleless membrane aerated bioreactor, bioreactor with foam collector, three-phase inverse fluidized bed bioreactor and rotating discs biofilm reactor (Chtioui et al. 2012, Coutte et al. 2013, Fahim et al. 2013). One of the possible alternatives to overcome this problem can be the use of anaerobic bioreactors. The design of bioreactor is still a challenge to obtain high productivities of lipopeptides for industrial processes.

5. Applications of Lipopeptides

Due to their amphiphilic nature and detergent-like properties, lipopeptides can be used for many industrial, pharmaceutical and biomedical applications.

5.1 Antimicrobial Agent

Due to their unique structure lipopeptides are suitable alternative for many clinical and biomedical applications (Mandal et al. 2013b). Daptomycin, under the trade name of Cubicin, was the first lipopeptide approved by FDA, USA in 2003. Since then it has been used to treat skin and blood infections caused by Gram-positive microorganisms (Nakhate et al. 2013). The bactericidal activity of lipopeptide is increased by the addition of lipid moiety of suitable length, preferably with 10–12 carbon atoms. As the length of lipid tail increases (C14–16) the lipopeptides also exhibit antifungal activity (Mandal et al. 2013b).

Surfactin exhibits antibacterial activity whereas fengycin and iturin are potential antifungal agents (Ongena and Jacques 2008, Gordillo and Maldonado 2012). Surfactin has excellent membrane-active properties and hence have great potential in health care and biotechnology-based applications. Surfactin, produced by *Bacillus circulans*, are active against multi-drug resistant strains of *Alcaligenes faecalis*, *Escherichia coli*, *Proteus vulgaris*, *Pseudomonas aeruginosa* and methicillin-resistant *Staphylococcus aureus* (Eswari and Yadav 2019). Lipopeptides produced by *Streptomyces* also have pharmaceutical applications (Sharma et al. 2014).

5.2 Antiviral Activity

Surfactin inactivates enveloped viruses more efficiently than non-enveloped virus (Seydlová et al. 2011) which is also dependent on the length of carbon atoms in acyl chain of the lipopeptide (Singla et al. 2014) and increases with the increase in hydrophobicity of fatty acid (Seydlová et al. 2011). The antiviral activity of surfactin is because of the physicochemical interaction between amphiphilic surfactin and the lipid membrane of viruses. Surfactin permeates itself in lipid bilayer of viruses. This disintegrates the viral envelope containing viral proteins which are required for adsorption and penetration of virus into target cells (Kracht et al. 1999). Surfactin exhibits antiviral properties against Herpes simplex virus (HSV-1 and HSV-2), Murine encephalomyocarditis virus, Semliki Forest virus, vesicular stomatitis virus, Simian immunodeficiency virus and Feline calcivirus, and inactivate cell-free viruses of Bursal disease virus, Porcine parvovirus, Newcastle disease virus and Pseudo rabies virus.

5.3 Anti-cancer Activity

Lipopeptides produced by *Bacillus subtilis* have antitumor properties against many different cells with varying mechanism of action. Due to their amphiphilic nature, the fatty acid moiety forms hydrophobic interaction with acyl chain of phospholipids present on cell membrane (Liu et al. 2010). Surfactin has a wide range of antitumor activity whereas that of iturin and fengycin is limited. Iturin inhibits proliferation of renal carcinoma (A498), alveolar adenocarcinoma (A549), colon adenocarcinoma

(HCT-15), hepatocellular carcinoma (HepG2), breast cancer cells (MDA-MB-231 and MCF-7) urinary bladder cancer cells (BIU87) and chronic myelogenous leukemia cells (K562). Fengycin blocks the proliferation of 95D (non-small cell lung cancer) cells in nude mice (Yin et al. 2013).

Surfactin exhibited anti-cancer activity against MCF-7, T47D and MDA-MB-231 breast cancer cell lines possibly by increasing cell cycle arrest at G0/G1 phase (Lee et al. 2012, Duarte et al. 2014). The apoptosis of Bcap-37 breast cancer cells were observed after treatment with surfactin-like lipopeptide. This was because of increased membrane fluidization caused by decrease in amount of fatty acids resulting in reduction in degree of unsaturated fatty acids (Liu et al. 2010). Other mechanisms by which surfactin exhibits anti-cancer activity is by increasing intracellular calcium concentration (Ca^{2+}) and inducing apoptosis by arresting cells at G2/M phase (Cao et al. 2009) (Fig. 2).

Surfactin interferes with cell cycle phase transition by inhibiting activity of G2-specific kinase, cyclin B1/p34cdc and by accumulating p53 (tumor suppressor) and p21Waf1/Cip1 (cyclin kinase inhibitor) (Cao et al. 2009). Surfactin also induces apoptosis by the generation of reactive oxygen species (ROS) and/or ROS-JNK-mediated mitochondrial/caspase pathway. All these lead to the release of cytochrome c (cyt c) and caspase-cascade reaction (Cao et al. 2010). Surfactin also represses PI3K/Akt and ERK signalling pathways which are important for cell proliferation, apoptosis and cell cycle and regulates NF-κB, AP-1, PI3K/Akt and ERK pathways simultaneously (Park et al. 2013). It inhibits 12-o-tetradecanoylphorbol-13-acetate

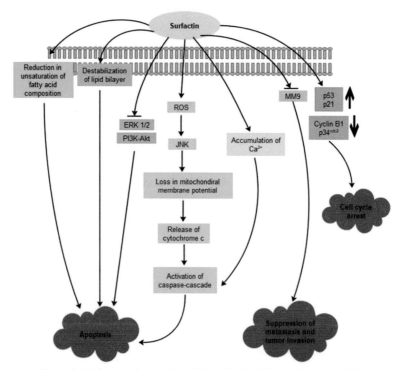

Figure 2. Various mechanisms by which surfactin exhibits anti-cancer activity.

(TPA)-induced migration, invasion and colony formation of breast cancer cells by down-regulating expression of matrix metalloproteinase-9 (MMP-9). The reduced TPA-induced activation, activating protein-1 (AP-1) and nuclear localization of nuclear factor-kappa B (NF-κB) leads to repressed PI3K/Akt and extracellular signal-regulated kinase (ERK) signalling pathways (Gudiña et al. 2013, Wu et al. 2017).

5.4 Anti-adhesive Agents

Bacteria propagate over surfaces by adhering to surface and form biofilm. Lipopeptides, especially surfactin, inhibit adhesion of microbes on surfaces and prevent biofilm formation. Because of their anti-adhesive property they find application as a coating on biomedical devices. The electrostatic repulsion between bacterial cells and anionic surfactin is responsible for its anti-adhesive properties (Zeraik and Nitschke 2010). Surfactin coated vinyl urethral catheters and stainless steel implants have shown to have reduced biofilm formation (Mireles II et al. 2001). Surfactin inhibits biofilm formation by *Staphylococcus aureus* and *Escherichia coli* on polystyrene surfaces by 97% and 90%, respectively (Rivardo et al. 2009).

5.5 Immunomodulators

Lipopeptides act as immunomodulators and interact with pattern recognition receptors such as Toll-like receptors (TLRs) and induce immune response (Steinhagen et al. 2011, Kelesidis 2014). Surfactin was found to suppress proliferation of T-cells, increase regulator T cells and down regulates TNF (tumor necrosis factor)-α, IFN (interferon)-Υ and activated CD8+ T cells. This alteration in immune response led to decrease in Type 1 diabetes mellitus resulting from autoimmunity (Gao et al. 2013). Daptomycin exhibits immunomodulatory properties by suppressing cytokine production (Tirilomis 2014).

Lipopeptides, in combination with conventional antigens, also act as promising immune adjuvants and are non-toxic and non-pyrogenic (Mittenbühler et al. 1997, BenMohamed et al. 2002, Moyle and Toth 2008, Toth et al. 2008). A mixture of polymyxin B, colistin and ovalbumin increased humoral immune responses (Yoshino et al. 2013). The formulation of WH1 fungin and hepatitis B surface antigen has shown to be a promising vaccine when administered to mice (Pan et al. 2014, Patel et al. 2015). Lipopeptides also act as potential macrophage activators and stimulate polyclonal B-lymphocytes. Conjugation of lipopeptides with low molecular weight antigens increases antibody-mediated immune response and stimulates cellular immune response *in vivo* (Bessler et al. 1997).

5.6 Food Industry

Lipopeptides are used in food industries due to their antimicrobial and anti-adhesive properties. Cyclic lipopeptides are preferred over linear lipopeptides as their cyclic peptides make them resistant to proteases. Due to their high emulsifying properties they can be utilized to process raw materials. For example, surfactin is used in baking as it helps in maintaining texture, volume and stability and also helps to emulsify fat to prevent formation of fat globules (Mandal et al. 2013a). Surfactin is also present as an additive in a Japanese fermented soybean dish, natto. Its application as a food additive is limited as its acceptable daily intake is not defined to consider it as non-toxic. But as it is produced by *Bacillus subtilis* which belongs to a category of generally regarded as safe (GRAS) organism the use of surfactin in food industry is promising (Juola et al. 2014).

Surfactin also reduces biofilm formed by food pathogens on food processing equipment. Adsorption of surfactin on stainless steel reduces pathogenic adhesion and biofilm formation (de Araujo et al. 2016). Incorporation of surfactin into food to replace some ingredients and improve the quality has also been used as a potential application of surfactin. The application of surfactin producing bacteria highly improved the texture of cookies when compared with chemical emulsifiers (Zouari et al. 2016). Lipopeptides produced by Enterobacteriaceae have also been used in food industries due to their high emulsification properties and viscosity at acidic pH (Mandal et al. 2013a). Lipopeptides are better food preservatives to avoid food spoilage during processing. They also control the growth of microorganisms responsible for food spoilage (Meena and Kanwar 2015b).

5.7 Agriculture

Use of chemicals as pesticides and insecticides on plants leads to contamination of groundwater, development of resistance in pathogens and health risks to humans. Hence there is a need to search for environmental friendly pesticides against plant pathogens. Application of lipopeptides is an eco-friendly approach to overcome both the issues. Different lipopeptides have unique characteristics to interact with plants. Lipopeptides from *Bacillus* sp. can be employed as biocontrol agents against plant disease as they have inhibitory spreading, antagonistic and immune-stimulating effects (Geissler et al. 2019). For example, surfactin is not used in agriculture *per se* but it has synergistic effects whereas iturin and fengycin have strong antifungal activities and can be used as biopesticides against plant pathogens (Ongena and Jacques 2008). Surfactin supports colonization of root tissues and wets the surface of plant roots to promote supply of nutrients. It also helps in biofilm formation which is essential for swarming on plant tissues (Cawoy et al. 2015). Biofilms formed on *Arabidopsis* roots with the help of surfactin exhibits antibacterial activity against *Pseudomonas syringae* (Bais et al. 2004).

Surfactin has shown to protect wheat against *Zymnoseptoria tritici* with 70% efficiency by stimulating jasmonic acid- and salicyclic acid-signalling pathways. These acids regulate plant defences against biotic stresses and hence surfactin stimulate immune system of plants (Le Mire et al. 2018). Antifungal activities have also been demonstrated by only surfactin producing bacteria, surfactin and iturin producers and surfactin, iturin and fengycin producers. Some species of fungi such as *Botrytis cinerea* and *Aspergillus flavus* are only inhibited by *Bacillus* strains producing all three types of lipopeptides simultaneously (Paraszkiewicz et al. 2017).

5.8 Bioremediation

Lipopeptides have detergent-like properties and hence they can be used for enhanced oil recovery and degradation of oil. Lipopeptides produced by *Escherichia fergusonii*

KLU01 have found to be resistant to heavy metals such as manganese, zinc, iron, copper, nickel and lead. This lipopeptide can be used for remediation of heavy metal and degradation of hydrocarbons (Sriram et al. 2011). The oil displacement abilities of lipopeptides produced by *Bacillus* sp. were very high which is comparable to commercial dispersants and synthetic surfactants. This lipopeptide was also stable at extreme salt concentrations, pH and temperature (Khondee et al. 2015).

Transportation of crude oil over a long distance from fields to refineries becomes limited due to high viscosity of heavy oils. The applications of lipopeptides reduces viscosity of heavy oils and improves miscibility. This increases mobility of oil and ease its transportation (Lima et al. 2011). Hence lipopeptides offer a cost-effective approach to heat, blend and lubricate crude oil to ease its handling and transportation. Lipopeptides (either produced *in situ* or added *ex situ*) decrease oil viscosity, promotes oil flow and immobilize oil trapped in reservoirs. Therefore, they can be used in oil reservoirs to increase their recovery. Other Lipopeptide producing microorganisms with a potential to enhance oil recovery are *Bacillus subtilis*, *Bacillus siamensis*, *Fusarium* sp. BS-8 (Qazi et al. 2013, Varadavenkatesan and Murty 2013, Pathak and Keharia 2014), *Bacillus subtilis* B30 (Al-Wahaibi et al. 2014), *Acinetobacter baylyi* and *Inquilinus limosus* (Saimmai et al. 2013, Zou et al. 2014). The most preferred lipopeptide for microbial enhanced oil recovery is surfactin due to its better soil recovery efficiency and interfacial activity.

6. Challenges in Microbial Production of Lipopeptides

Production of lipopeptides by microorganisms poses many challenges for large-scale industrial purposes and hence their industrial production is highly restricted. One of the major challenges is the cost of production of lipopeptides by fermentation process. Currently, lipopeptides are produced by expensive substrates. Hence, there is a need to search for low cost raw materials or biomass that reduces the cost of fermentation process. This problem can also be overcome by use of agricultural wastes or wastes containing high carbohydrate content as substrates (Suryawanshi et al. 2019a, Suryawanshi et al. 2020a). Also, separation and purification of lipopeptides in liquid medium add to the cost due to simultaneous production of various types of surface active molecules. Due to the surface-active properties of lipopeptides, they cause formation of foam during fermentation process. The overflow of foam results in loss of cells, nutrients and products. Therefore, controlling foam formation is also an important challenge for industrial production. A possible solution for this problem can be optimization of reactors and processes.

The lipopeptide production ability of wild-type microbial strains is very low. Two possible ways by which production of lipopeptides can be increased are by either selection of wild-type strains with high yields or by genetically engineering strains to increase specific productivity of lipopeptides. Genetic modification of strains becomes challenging as the genes responsible for expression of proteins for lipopeptide biosynthesis are very large in size and cloning them is difficult. Hence, effective engineering of lipopeptide synthesis is a problem. Some of the techniques for increasing the production of lipopeptides are optimization of culture medium with various optimization tools like ANN, GA, etc., alteration of fermentation process parameters and modification of producer strains (Zhao et al. 2017). Appropriate scale-up technologies need to be establishes to produce lipopeptides at industrial scale.

Acknowledgement

The authors are grateful to National Institute of Technology Raipur, India for providing the necessary facilities to carry out their research.

References

Al-Ajlani, M.M., M.A. Sheikh, Z. Ahmad and S. Hasnain. 2007. Production of surfactin from *Bacillus subtilis* MZ-7 grown on pharmamedia commercial medium. Microbial Cell Factories 6(1): 17.

Al-Wahaibi, Y., S. Joshi, S. Al-Bahry, A. Elshafie, A. Al-Bemani and B. Shibulal. 2014. Biosurfactant production by *Bacillus subtilis* B30 and its application in enhancing oil recovery. Colloids and Surfaces B: Biointerfaces 114: 324–333.

Ali, S., S. Hameed, A. Imran, M. Iqbal and G. Lazarovits. 2014. Genetic, physiological and biochemical characterization of *Bacillus* sp. strain RMB7 exhibiting plant growth promoting and broad spectrum antifungal activities. Microbial Cell Factories 13(1): 144.

Alsohim, A.S., T.B. Taylor, G.A. Barrett, J. Gallie, X.X. Zhang, A.E. Altamirano-Junqueira, L.J. Johnson, P.B. Rainey and R.W. Jackson. 2014. The biosurfactant viscosin produced by *Pseudomonas fluorescens* SBW 25 aids spreading motility and plant growth promotion. Environmental Microbiology 16(7): 2267–2281.

Aretz, W., J. Meiwes, G. Seibert, G. Vobis and J. Wink. 2000. Friulimicins: novel lipopeptide antibiotics with peptidoglycan synthesis inhibiting activity from *Actinoplanes friuliensis* sp. nov. The Journal of Antibiotics 53(8): 807–815.

Bais, H.P., R. Fall and J.M. Vivanco. 2004. Biocontrol of *Bacillus subtilis* against infection of Arabidopsis roots by *Pseudomonas syringae* is facilitated by biofilm formation and surfactin production. Plant Physiology 134(1): 307–319.

Baltz, R.H., P. Brian, V. Miao and S.K. Wrigley. 2006. Combinatorial biosynthesis of lipopeptide antibiotics in *Streptomyces roseosporus*. Journal of Industrial Microbiology and Biotechnology 33(2): 66–74.

Bechinger, B., M. Zasloff and S.J. Opella. 1993. Structure and orientation of the antibiotic peptide magainin in membranes by solid-state nuclear magnetic resonance spectroscopy. Protein Science 2(12): 2077–2084.

Beltran-Gracia, E., G. Macedo-Raygoza, J. Villafaña-Rojas, A. Martinez-Rodriguez, Y.Y. Chavez-Castrillon, F.M. Espinosa-Escalante and M. Beltran-Garcia. 2017. Production of lipopeptides by fermentation processes: Endophytic bacteria, fermentation strategies and easy methods for bacterial selection. Fermentation Processes 199–222.

BenMohamed, L., S.L. Wechsler and A.B. Nesburn. 2002. Lipopeptide vaccines—yesterday, today, and tomorrow. The Lancet Infectious Diseases 2(7): 425–431.

Bessler, W., L. Heinevetter, K.-H. WiesmÜller, G. Jung, W. Baier, M. Huber, A. Lorenz, U. Esche, K. Mittenbühler and P. Hoffmann. 1997. Bacterial cell wall components as immunomodulators—I. Lipopeptides as adjuvants for parenteral and oral immunization. International Journal of Immunopharmacology 19(9-10): 547–550.

Cao, X., A. Wang, R. Jiao, C. Wang, D. Mao, L. Yan and B. Zeng. 2009. Surfactin induces apoptosis and G2/M arrest in human breast cancer MCF-7 cells through cell cycle factor regulation. Cell Biochemistry and Biophysics 55(3): 163.

Cao, X.-h., A.-h. Wang, C.-l. Wang, D.-z. Mao, M.-f. Lu, Y.-q. Cui and R.-z. Jiao. 2010. Surfactin induces apoptosis in human breast cancer MCF-7 cells through a ROS/JNK-mediated mitochondrial/caspase pathway. Chemico-Biological Interactions 183(3): 357–362.

Cawoy, H., D. Debois, L. Franzil, E. De Pauw, P. Thonart and M. Ongena. 2015. Lipopeptides as main ingredients for inhibition of fungal phytopathogens by *Bacillus subtilis/amyloliquefaciens*. Microbial Biotechnology 8(2): 281–295.

Chen, F., H. Li, Z. Xu, S. Hou and D. Yang. 2015a. User-friendly optimization approach of fed-batch fermentation conditions for the production of iturin A using artificial neural networks and support vector machine. Electronic Journal of Biotechnology 18(4): 273–280.

Chen, W.-C., R.-S. Juang and Y.-H. Wei. 2015b. Applications of a lipopeptide biosurfactant, surfactin, produced by microorganisms. Biochemical Engineering Journal 103: 158–169.

Choi, S.-K., S.-Y. Park, R. Kim, S.-B. Kim, C.-H. Lee, J.F. Kim and S.-H. Park. 2009. Identification of a polymyxin synthetase gene cluster of *Paenibacillus polymyxa* and heterologous expression of the gene in *Bacillus subtilis.* Journal of Bacteriology 191(10): 3350–3358.

Chtioui, O., K. Dimitrov, F. Gancel, P. Dhulster and I. Nikov. 2012. Rotating discs bioreactor, a new tool for lipopeptides production. Process Biochemistry 47(12): 2020–2024.

Cooper, D., C. Macdonald, S. Duff and N. Kosaric. 1981. Enhanced production of surfactin from *Bacillus subtilis* by continuous product removal and metal cation additions. Applied and Environmental Microbiology 42(3): 408–412.

Coutte, F., D. Lecouturier, V. Leclère, M. Béchet, P. Jacques and P. Dhulster. 2013. New integrated bioprocess for the continuous production, extraction and purification of lipopeptides produced by *Bacillus subtilis* in membrane bioreactor. Process Biochemistry 48(1): 25–32.

Dang, Y., F. Zhao, X. Liu, X. Fan, R. Huang, W. Gao, S. Wang and C. Yang. 2019. Enhanced production of antifungal lipopeptide iturin A by *Bacillus amyloliquefaciens* LL3 through metabolic engineering and culture conditions optimization. Microbial Cell Factories 18(1): 68.

Davis, D., H. Lynch and J. Varley. 1999. The production of surfactin in batch culture by *Bacillus subtilis* ATCC 21332 is strongly influenced by the conditions of nitrogen metabolism. Enzyme and Microbial Technology 25(3-5): 322–329.

Davis, S.D., A. Iannetta and R.J. Wedgwood. 1971. Activity of colistin against *Pseudomonas aeruginosa*: inhibition by calcium. Journal of Infectious Diseases 124(6): 610–612.

de Araujo, L.V., C.R. Guimarães, R.L. da Silva Marquita, V.M. Santiago, M.P. de Souza, M. Nitschke and D.M.G. Freire. 2016. Rhamnolipid and surfactin: Anti-adhesion/antibiofilm and antimicrobial effects. Food Control 63: 171–178.

DeCrescenzo Henriksen, E., D. Phillips and J.D. Peterson. 2007. Polymyxin E production by *P. amylolyticus*. Letters in Applied Microbiology 45(5): 491–496.

Deleu, M., M. Paquot and T. Nylander. 2008. Effect of fengycin, a lipopeptide produced by *Bacillus subtilis*, on model biomembranes. Biophysical Journal 94(7): 2667–2679.

Dewick, P.M. 2002. Medicinal Natural Products: A Biosynthetic Approach. John Wiley & Sons.

Duarte, C., E.J. Gudiña, C.F. Lima and L.R. Rodrigues. 2014. Effects of biosurfactants on the viability and proliferation of human breast cancer cells. AMB Express 4(1): 40.

Dunlap, C., M. Bowman and A.P. Rooney. 2019. Iturinic lipopeptide diversity in the *Bacillus subtilis* species group–important antifungals for plant disease biocontrol applications. Frontiers in Microbiology 10: 1794.

Eswari, J. and M. Yadav. 2019. New perspective of drug discovery from herbal medicinal plants: *Andrographis paniculata* and *Bacopa monnieri* (terpenoids) and novel target identification against *Staphylococcus aureus*. South African Journal of Botany 124: 188–198.

Eswari, J.S., M. Anand and C. Venkateswarlu. 2013. Optimum culture medium composition for rhamnolipid production by *Pseudomonas aeruginosa* AT10 using a novel multi-objective optimization method. Journal of Chemical Technology & Biotechnology 88(2): 271–279.

Eswari, J.S. and C. Venkateswarlu. 2016. Multiobjective simultaneous optimization of biosurfactant process medium by integrating differential evolution with artificial neural networks. Indian Journal of Chemical Technology (IJCT) 23(5): 335–344.

Eswari, J.S., M. Anand and C. Venkateswarlu. 2016. Optimum culture medium composition for lipopeptide production by *Bacillus subtilis* using response surface model-based ant colony optimization. Sadhana 41(1): 55–65.

Eswari, J.S., S. Dhagat and P. Mishra. 2018. Biosurfactant assisted silver nanoparticle synthesis: A critical analysis of its drug design aspects. Advances in Natural Sciences: Nanoscience and Nanotechnology 9(4): 045007.

Eswari, J.S., S. Dhagat and R. Sen. 2019a. Biosurfactants, bioemulsifiers, and biopolymers from thermophilic microorganisms. pp. 87–97. Thermophiles for Biotech Industry. Springer.

Eswari, J.S., S. Dhagat and M. Yadav. 2019b. Antibacterial lipopeptides. pp. 49–78. Computer-Aided Design of Antimicrobial Lipopeptides as Prospective Drug Candidates. CRC Press.

Eswari, J.S., S. Dhagat and M. Yadav. 2019c. Antifungal lipopeptides. pp. 79–106. Computer-Aided Design of Antimicrobial Lipopeptides as Prospective Drug Candidates. CRC Press.

Eswari, J.S., S. Dhagat and M. Yadav. 2019d. Lipopeptides and computer-aided drug design. pp. 1–22. Computer-Aided Design of Antimicrobial Lipopeptides as Prospective Drug Candidates. CRC Press.

Fahim, S., K. Dimitrov, F. Gancel, P. Vauchel, P. Jacques and I. Nikov. 2012. Impact of energy supply and oxygen transfer on selective lipopeptide production by *Bacillus subtilis* BBG21. Bioresource Technology 126: 1–6.

Fahim, S., K. Dimitrov, P. Vauchel, F. Gancel, G. Delaplace, P. Jacques and I. Nikov. 2013. Oxygen transfer in three phase inverse fluidized bed bioreactor during biosurfactant production by *Bacillus subtilis*. Biochemical Engineering Journal 76: 70–76.

Fonseca, R., A. Silva, F. De França, V. Cardoso and E. Sérvulo. 2007. Optimizing carbon/nitrogen ratio for biosurfactant production by a *Bacillus subtilis* strain. pp. 471–486. Applied Biochemistry and Biotechnology. Springer.

Gales, A.C., A.O. Reis and R.N. Jones. 2001. Contemporary assessment of antimicrobial susceptibility testing methods for polymyxin B and colistin: review of available interpretative criteria and quality control guidelines. Journal of Clinical Microbiology 39(1): 183–190.

Gancel, F., L. Montastruc, T. Liu, L. Zhao and I. Nikov. 2009. Lipopeptide overproduction by cell immobilization on iron-enriched light polymer particles. Process Biochemistry 44(9): 975–978.

Gao, Z., X. Zhao, S. Lee, J. Li, H. Liao, X. Zhou, J. Wu and G. Qi. 2013. WH1 fungin a surfactin cyclic lipopeptide is a novel oral immunoadjuvant. Vaccine 31(26): 2796–2803.

Geissler, M., K.M. Heravi, M. Henkel and R. Hausmann. 2019. Lipopeptide biosurfactants from *Bacillus* species. pp. 205–240. Biobased Surfactants. Elsevier.

Glaser, P., A. Danchin, F. Kunst, P. Zuber and M.M. Nakano. 1995. Identification and isolation of a gene required for nitrate assimilation and anaerobic growth of *Bacillus subtilis*. Journal of Bacteriology 177(4): 1112–1115.

Gordillo, A. and M.C. Maldonado. 2012. Purification of peptides from *Bacillus* strains with biological activity. Chromatography and its Applications 11: 201–225.

Gudiña, E.J., V. Rangarajan, R. Sen and L.R. Rodrigues. 2013. Potential therapeutic applications of biosurfactants. Trends in Pharmacological Sciences 34(12): 667–675.

Hamley, I.W. 2015. Lipopeptides: from self-assembly to bioactivity. Chemical Communications 51(41): 8574–8583.

Hanif, A., F. Zhang, P. Li, C. Li, Y. Xu, M. Zubair, M. Zhang, D. Jia, X. Zhao and J. Liang. 2019. Fengycin produced by *Bacillus amyloliquefaciens* FZB42 inhibits *Fusarium graminearum* growth and mycotoxins biosynthesis. Toxins 11(5): 295.

Hoffmann, T., B. Troup, A. Szabo, C. Hungerer and D. Jahn. 1995. The anaerobic life of *Bacillus subtilis*: cloning of the genes encoding the respiratory nitrate reductase system. FEMS Microbiology Letters 131(2): 219–225.

Hogardt, M., S. Schmoldt, M. Götzfried, K. Adler and J. Heesemann. 2004. Pitfalls of polymyxin antimicrobial susceptibility testing of *Pseudomonas aeruginosa* isolated from cystic fibrosis patients. Journal of Antimicrobial Chemotherapy 54(6): 1057–1061.

Huang, E. and A.E. Yousef. 2014. Paenibacterin, a novel broad-spectrum lipopeptide antibiotic, neutralises endotoxins and promotes survival in a murine model of *Pseudomonas aeruginosa*-induced sepsis. International Journal of Antimicrobial Agents 44(1): 74–77.

Huang, X., J.n. Liu, Y. Wang, J. Liu and L. Lu. 2015. The positive effects of Mn^{2+} on nitrogen use and surfactin production by *Bacillus subtilis* ATCC 21332. Biotechnology & Biotechnological Equipment 29(2): 381–389.

Janek, T., M. Łukaszewicz, T. Rezanka and A. Krasowska. 2010. Isolation and characterization of two new lipopeptide biosurfactants produced by *Pseudomonas fluorescens* BD5 isolated from water from the Arctic Archipelago of Svalbard. Bioresource Technology 101(15): 6118–6123.

Jujjavarapu, S.E. and S. Dhagat. 2018. *In silico* discovery of novel ligands for antimicrobial lipopeptides for computer-aided drug design. Probiotics and Antimicrobial Proteins 10(2): 129–141.

Jujjavarapu, S.E., S. Dhagat and V. Kurrey. 2018. Identification of novel ligands for therapeutic lipopeptides: daptomycin, surfactin and polymyxin. Current Drug Targets 19(13): 1589–1598.

Juola, M., K. Kinnunen, K.F. Nielsen and A. von Right. 2014. Surfactins in natto: the surfactin production capacity of the starter strains and the actual surfactin contents in the products. Journal of Food Protection 77(12): 2139–2143.

Kelesidis, T. 2014. The interplay between daptomycin and the immune system. Frontiers in Immunology 5: 52.

Khondee, N., S. Tathong, O. Pinyakong, R. Müller, S. Soonglerdsongpha, C. Ruangchainikom, C. Tongcumpou and E. Luepromchai. 2015. Lipopeptide biosurfactant production by chitosan-immobilized *Bacillus* sp. GY19 and their recovery by foam fractionation. Biochemical Engineering Journal 93: 47–54.

Kim, K., S.Y. Jung, D.K. Lee, J.-K. Jung, J.K. Park, D.K. Kim and C.-H. Lee. 1998. Suppression of inflammatory responses by surfactin, a selective inhibitor of platelet cytosolic phospholipase A2. Biochemical Pharmacology 55(7): 975–985.

Kim, K.M., J.Y. Lee, C.K. Kim and J.S. Kang. 2009. Isolation and characterization of surfactin produced by *Bacillus polyfermenticus* KJS-2. Archives of Pharmacal Research 32(5): 711–715.

Kracht, M., H. Rokos, M. Özel, M. Kowall, G. Pauli and J. Vater. 1999. Antiviral and hemolytic activities of surfactin isoforms and their methyl ester derivatives. The Journal of Antibiotics 52(7): 613–619.

Le Mire, G., A. Siah, M.-N. Brisset, M. Gaucher, M. Deleu and M.H. Jijakli. 2018. Surfactin protects wheat against *Zymoseptoria tritici* and activates both salicylic acid-and jasmonic acid-dependent defense responses. Agriculture 8(1): 11.

Lee, J.H., S.H. Nam, W.T. Seo, H.D. Yun, S.Y. Hong, M.K. Kim and K.M. Cho. 2012. The production of surfactin during the fermentation of cheonggukjang by potential probiotic *Bacillus subtilis* CSY191 and the resultant growth suppression of MCF-7 human breast cancer cells. Food Chemistry 131(4): 1347–1354.

Li, Y. and S. Chen. 2019. Fusaricidin produced by *Paenibacillus polymyxa* WLY78 induces systemic resistance against fusarium wilt of cucumber. International Journal of Molecular Sciences 20(20): 5240.

Lima, T.M., A.F. Fonseca, B.A. Leao, A.H. Mounteer, M.R. Tótola and A. Borges. 2011. Oil recovery from fuel oil storage tank sludge using biosurfactants. J. Bioremed. Biodegrad. 2(12): 10.4172.

Liu, X., X. Tao, A. Zou, S. Yang, L. Zhang and B. Mu. 2010. Effect of themicrobial lipopeptide on tumor cell lines: apoptosis induced by disturbing the fatty acid composition of cell membrane. Protein & Cell 1(6): 584–594.

Liu, X., B. Ren, H. Gao, M. Liu, H. Dai, F. Song, Z. Yu, S. Wang, J. Hu and C.R. Kokare. 2012. Optimization for the production of surfactin with a new synergistic antifungal activity. PloS One 7(5): e34430.

Madslien, E., H. Rønning, T. Lindbäck, B. Hassel, M. Andersson and P. Granum. 2013. Lichenysin is produced by most *Bacillus licheniformis* strains. Journal of Applied Microbiology 115(4): 1068–1080.

Mandal, S.M., A.E. Barbosa and O.L. Franco. 2013a. Lipopeptides in microbial infection control: scope and reality for industry. Biotechnology Advances 31(2): 338–345.

Mandal, S.M., S. Sharma, A.K. Pinnaka, A. Kumari and S. Korpole. 2013b. Isolation and characterization of diverse antimicrobial lipopeptides produced by Citrobacter and Enterobacter. BMC Microbiology 13(1): 152.

Matsuyama, T., T. Tanikawa and Y. Nakagawa. 2011. Serrawettins and other surfactants produced by *Serratia*. pp. 93–120. Biosurfactants. Springer.

McBrien, K.D., R.L. Berry, S.E. Lowe, K.M. Neddermann, I. Bursuker, S. Huang, S.E. Klohr and J.E. Leet. 1995. Rakicidins, new cytotoxic lipopeptides from *Micromonospora* sp. fermentation, isolation and characterization. The Journal of Antibiotics 48(12): 1446–1452.

Meena, K.R. and S.S. Kanwar. 2015a. Lipopeptides as the antifungal and antibacterial agents: applications in food safety and therapeutics. BioMed Research International.

Meena, K.R. and S.S. Kanwar. 2015b. Lipopeptides as the antifungal and antibacterial agents: applications in food safety and therapeutics. BioMed Research International 2015: 473050.

Mireles II, J.R., A. Toguchi and R.M. Harshey. 2001. Salmonella enterica Serovar Typhimurium swarming mutants with altered biofilm-forming abilities: surfactin inhibits biofilm formation. Journal of Bacteriology 183(20): 5848–5854.

Mittenbühler, K., M. Loleit, W. Baier, B. Fischer, E. Sedelmeier, G. Jung, G. Winkelmann, C. Jacobi, J. Weckesser and M.H. Erhard. 1997. Drug specific antibodies: T-cell epitope-lipopeptide conjugates are potent adjuvants for small antigens *in vivo* and *in vitro*. International Journal of Immunopharmacology 19(5): 277–287.

Moyle, P.M. and I. Toth. 2008. Self-adjuvanting lipopeptide vaccines. Current Medicinal Chemistry 15(5): 506–516.

Mukherjee, S., P. Das and R. Sen. 2006. Towards commercial production of microbial surfactants. Trends in Biotechnology 24(11): 509–515.

Nakhate, P., V. Yadav and A. Pathak. 2013. A review on Daptomycin; the first US-FDA approved. Lipopeptide antibiotics. Journal of Scientific and Innovative Research 2(5): 970–980.

Newton, B. 1956. The properties and mode of action of the polymyxins. Bacteriological Reviews 20(1): 14.

Niks, M., J. Hanzen, D. Ohlasova, D. Rovná, A. Purgelová, Z. Szövényiová and A. Vaculikova. 2004. Multiresistant nosocomial bacterial strains and their "*in vitro*" susceptibility to chloramphenicol and colistin. Klinicka mikrobiologie a infekcni lekarstvi 10(3): 124.

Ongena, M. and P. Jacques. 2008. Bacillus lipopeptides: versatile weapons for plant disease biocontrol. Trends in Microbiology 16(3): 115–125.

Pan, H., X. Zhao, Z. Gao and G. Qi. 2014. A surfactin lipopeptide adjuvanted hepatitis B vaccines elicit enhanced humoral and cellular immune responses in mice. Protein and Peptide Letters 21(9): 901–910.

Paraszkiewicz, K., P. Bernat, P. Siewiera, M. Moryl, L.S. Paszt, P. Trzciński, Ł. Jałowiecki and G. Płaza. 2017. Agricultural potential of rhizospheric *Bacillus subtilis* strains exhibiting varied efficiency of surfactin production. Scientia Horticulturae 225: 802–809.

Park, S.Y., J.-H. Kim, Y.J. Lee, S.J. Lee and Y. Kim. 2013. Surfactin suppresses TPA-induced breast cancer cell invasion through the inhibition of MMP-9 expression. International Journal of Oncology 42(1): 287–296.

Patel, S., S. Ahmed and J.S. Eswari. 2015. Therapeutic cyclic lipopeptides mining from microbes: latest strides and hurdles. World Journal of Microbiology and Biotechnology 31(8): 1177–1193.

Pathak, K.V., H. Keharia, K. Gupta, S.S. Thakur and P. Balaram. 2012. Lipopeptides from the banyan endophyte, *Bacillus subtilis* K1: mass spectrometric characterization of a library of fengycins. Journal of the American Society for Mass Spectrometry 23(10): 1716–1728.

Pathak, K.V., A. Bose and H. Keharia. 2014. Characterization of novel lipopeptides produced by *Bacillus tequilensis* P15 using liquid chromatography coupled electron spray ionization tandem mass spectrometry (LC–ESI–MS/MS). International Journal of Peptide Research and Therapeutics 20(2): 133–143.

Pathak, K.V. and H. Keharia. 2014. Application of extracellular lipopeptide biosurfactant produced by endophytic *Bacillus subtilis* K1 isolated from aerial roots of banyan (*Ficus benghalensis*) in microbially enhanced oil recovery (MEOR). 3 Biotech 4(1): 41–48.

Peng, W., J. Zhong, J. Yang, Y. Ren, T. Xu, S. Xiao, J. Zhou and H. Tan. 2014. The artificial neural network approach based on uniform design to optimize the fed-batch fermentation condition: application to the production of iturin A. Microbial Cell Factories 13(1): 54.

Qazi, M.A., M. Subhan, N. Fatima, M.I. Ali and S. Ahmed. 2013. Role of biosurfactant produced by *Fusarium* sp. BS-8 in enhanced oil recovery (EOR) through sand pack column. International Journal of Bioscience, Biochemistry and Bioinformatics 3(6): 598.

Ramos, H.C., T. Hoffmann, M. Marino, H. Nedjari, E. Presecan-Siedel, O. Dreesen, P. Glaser and D. Jahn. 2000. Fermentative metabolism of *Bacillus subtilis*: physiology and regulation of gene expression. Journal of Bacteriology 182(11): 3072–3080.

Rao, M., W. Wei, M. Ge, D. Chen and X. Sheng. 2013. A new antibacterial lipopeptide found by UPLC-MS from an actinomycete *Streptomyces* sp. HCCB10043. Natural Product Research 27(23): 2190–2195.

Rivardo, F., R. Turner, G. Allegrone, H. Ceri and M. Martinotti. 2009. Anti-adhesion activity of two biosurfactants produced by *Bacillus* spp. prevents biofilm formation of human bacterial pathogens. Applied Microbiology and Biotechnology 83(3): 541–553.

Roy, A., D. Mahata, D. Paul, S. Korpole, O.L. Franco and S.M. Mandal. 2013. Purification, biochemical characterization and self-assembled structure of a fengycin-like antifungal peptide from *Bacillus thuringiensis* strain SM1. Frontiers in Microbiology 4: 332.

Saimmai, A., S. Udomsilp and S. Maneerat. 2013. Production and characterization of biosurfactant from marine bacterium *Inquilinus limosus* KB3 grown on low-cost raw materials. Annals of Microbiology 63(4): 1327–1339.

Satomi, T., H. Kusakabe, G. Nakamura, T. Nishio, M. Uramoto and K. Isono. 1982. Neopeptins A and B, new antifungal antibiotics. Agricultural and Biological Chemistry 46(10): 2621–2623.

Satya Eswari, J. and C. Venkateswarlu. 2016. Dynamic modeling and metabolic flux analysis for optimized production of rhamnolipids. Chemical Engineering Communications 203(3): 326–338.

Schimana, J., K. Gebhardt, A. Hoeltzel, D.G. Schmid, R. Suessmuth, J. Mueller, R. Pukall and H.-P. Fiedler. 2002. Arylomycins A and B, new biaryl-bridged lipopeptide antibiotics produced by *Streptomyces* sp. Tü 6075. The Journal of Antibiotics 55(6): 565–570.

Schindler, M. and M. Osborn. 1979. Interaction of divalent cations and polymyxin B with lipopolysaccharide. Biochemistry 18(20): 4425–4430.

Sen, R. 1997. Response surface optimization of the critical media components for the production of surfactin. Journal of Chemical Technology & Biotechnology: International Research in Process, Environmental and Clean Technology 68(3): 263–270.

Seydlová, G., R. Cabala and J. Svobodová. 2011. Biomedical engineering, trends, research and technologies. Surfactin—Novel Solutions for Global Issues 13: 306–330.

Shao, C., L. Liu, H. Gang, S. Yang and B. Mu. 2015. Structural diversity of the microbial surfactin derivatives from selective esterification approach. International Journal of Molecular Sciences 16(1): 1855–1872.

Sharma, D., S.M. Mandal and R.K. Manhas. 2014. Purification and characterization of a novel lipopeptide from *Streptomyces amritsarensis* sp. nov. active against methicillin-resistant *Staphylococcus aureus*. AMB Express 4(1): 1–9.

Sheppard, J.D. and D.G. Cooper. 1991. The response of *Bacillus subtilis* ATCC 21332 to manganese during continuous-phased growth. Applied Microbiology and Biotechnology 35(1): 72–76.

Singla, R.K., H.D. Dubey and A.K. Dubey. 2014. Therapeutic spectrum of bacterial metabolites. Indo Global Journal of Pharmaceutical Sciences 2(2): 52–64.

Slivinski, C.T., E. Mallmann, J.M. de Araújo, D.A. Mitchell and N. Krieger. 2012. Production of surfactin by *Bacillus pumilus* UFPEDA 448 in solid-state fermentation using a medium based on okara with sugarcane bagasse as a bulking agent. Process Biochemistry 47(12): 1848–1855.

Sriram, M.I., S. Gayathiri, U. Gnanaselvi, P.S. Jenifer, S.M. Raj and S. Gurunathan. 2011. Novel lipopeptide biosurfactant produced by hydrocarbon degrading and heavy metal tolerant bacterium *Escherichia fergusonii* KLU01 as a potential tool for bioremediation. Bioresource Technology 102(19): 9291–9295.

Steinhagen, F., T. Kinjo, C. Bode and D.M. Klinman. 2011. TLR-based immune adjuvants. Vaccine 29(17): 3341–3355.

Sur, S., T.D. Romo and A. Grossfield. 2018. Selectivity and mechanism of fengycin, an antimicrobial lipopeptide, from molecular dynamics. The Journal of Physical Chemistry B 122(8): 2219–2226.

Suryawanshi, N., S. Jujjavarapu and S. Ayothiraman. 2019a. Marine shell industrial wastes–an abundant source of chitin and its derivatives: constituents, pretreatment, fermentation, and pleiotropic applications-a revisit. International Journal of Environmental Science and Technology 1–22.

Suryawanshi, N., S. Naik and J.S. Eswari. 2019b. Extraction and optimization of exopolysaccharide from *Lactobacillus* sp. using response surface methodology and artificial neural networks. Preparative Biochemistry and Biotechnology 49(10): 987–996.

Suryawanshi, N., S. Ayothiraman and J.S. Eswari. 2020a. Ultrasonication mode for the expedition of extraction process of chitin from the maritime shrimp shell waste. Indian Journal of Biochemistry and Biophysics (IJBB 57(4): 431–438.

Suryawanshi, N., J. Sahu, Y. Moda and J.S. Eswari. 2020b. Optimization of process parameters for improved chitinase activity from *Thermomyces* sp. by using artificial neural network and genetic algorithm. Preparative Biochemistry & Biotechnology 1–11.

Takizawa, M., T. Hida, T. Horiguchi, A. Hiramoto, S. Harada and S. Tanida. 1995. Tan-1511 A, B and C, microbial lipopeptides with G-CSF and GM-CSF inducing activity. The Journal of Antibiotics 48(7): 579–588.

Tirilomis, T. 2014. Daptomycin and its immunomodulatory effect: consequences for antibiotic treatment of methicillin-resistant *Staphylococcus aureus* wound infections after heart surgery. Frontiers in Immunology 5: 97.

Toth, I., P. Simerska and Y. Fujita. 2008. Recent advances in design and synthesis of self-adjuvanting lipopeptide vaccines. International Journal of Peptide Research and Therapeutics 14(4): 333–340.

Varadavenkatesan, T. and V.R. Murty. 2013. Production of a lipopeptide biosurfactant by a novel *Bacillus* sp. and its applicability to enhanced oil recovery. International Scholarly Research Notices.

Velkov, T., P.E. Thompson, R.L. Nation and J. Li. 2010. Structure–activity relationships of polymyxin antibiotics. Journal of Medicinal Chemistry 53(5): 1898–1916.

Venkateswarlu, C. and S.E. Jujjavarapu. 2019. Stochastic Global Optimization Methods and Applications to Chemical, Biochemical, Pharmaceutical and Environmental Processes. Elsevier.

Vertesy, L., E. Ehlers, H. Kogler, M. Kurz, J. Meiwes, G. Seibert, M. Vogel and P. Hammann. 2000. Friulimicins: Novel lipopeptide antibiotics with peptidoglycan synthesis inhibiting activity from *Actinoplanes friuliensis* sp. nov. The Journal of Antibiotics 53(8): 816–827.

Wei, Y.-H. and I.-M. Chu. 1998. Enhancement of surfactin production in iron-enriched media by *Bacillus subtilis* ATCC 21332. Enzyme and Microbial Technology 22(8): 724–728.

Wei, Y.H., L.F. Wang and J.S. Chang. 2004. Optimizing iron supplement strategies for enhanced surfactin production with *Bacillus subtilis*. Biotechnology Progress 20(3): 979–983.

Wei, Y.-H., C.-C. Lai and J.-S. Chang. 2007. Using Taguchi experimental design methods to optimize trace element composition for enhanced surfactin production by *Bacillus subtilis* ATCC 21332. Process Biochemistry 42(1): 40–45.

Wei, Y.-H., L.-C. Wang, W.-C. Chen and S.-Y. Chen. 2010. Production and characterization of fengycin by indigenous *Bacillus subtilis* F29-3 originating from a potato farm. International Journal of Molecular Sciences 11(11): 4526–4538.

Wu, Y.-S., S.-C. Ngai, B.-H. Goh, K.-G. Chan, L.-H. Lee and L.-H. Chuah. 2017. Anticancer activities of surfactin and potential application of nanotechnology assisted surfactin delivery. Frontiers in Pharmacology 8: 761.

Wyche, T.P., Y. Hou, E. Vazquez-Rivera, D. Braun and T.S. Bugni. 2012. Peptidolipins B–F, antibacterial lipopeptides from an ascidian-derived *Nocardia* sp. Journal of Natural Products 75(4): 735–740.

Yin, H., C. Guo, Y. Wang, D. Liu, Y. Lv, F. Lv and Z. Lu. 2013. Fengycin inhibits the growth of the human lung cancer cell line 95D through reactive oxygen species production and mitochondria-dependent apoptosis. Anti-cancer Drugs 24(6): 587–598.

Yoshino, N., M. Endo, H. Kanno, N. Matsukawa, R. Tsutsumi, R. Takeshita and S. Sato. 2013. Polymyxins as novel and safe mucosal adjuvants to induce humoral immune responses in mice. PLoS One 8(4): e61643.

Yu, Z., Z. Sun, J. Yin and J. Qiu. 2018. Enhanced production of polymyxin E in Paenibacillus polymyxa by replacement of glucose by starch. BioMed Research International.

Yuan, L., S. Zhang, Y. Wang, Y. Li, X. Wang and Q. Yang. 2018. Surfactin inhibits membrane fusion during invasion of epithelial cells by enveloped viruses. Journal of Virology 92(21).

Zeraik, A.E. and M. Nitschke. 2010. Biosurfactants as agents to reduce adhesion of pathogenic bacteria to polystyrene surfaces: effect of temperature and hydrophobicity. Current Microbiology 61(6): 554–559.

Zhao, H., D. Shao, C. Jiang, J. Shi, Q. Li, Q. Huang, M.S.R. Rajoka, H. Yang and M. Jin. 2017. Biological activity of lipopeptides from *Bacillus*. Applied Microbiology and Biotechnology 101(15): 5951–5960.

Zhao, P., C. Quan, L. Jin, L. Wang, J. Wang and S. Fan. 2013. Effects of critical medium components on the production of antifungal lipopeptides from *Bacillus amyloliquefaciens* Q-426 exhibiting excellent biosurfactant properties. World Journal of Microbiology and Biotechnology 29(3): 401–409.

Zou, C., M. Wang, Y. Xing, G. Lan, T. Ge, X. Yan and T. Gu. 2014. Characterization and optimization of biosurfactants produced by Acinetobacter baylyi ZJ2 isolated from crude oil-contaminated soil sample toward microbial enhanced oil recovery applications. Biochemical Engineering Journal 90: 49–58.

Zouari, R., S. Besbes, S. Ellouze-Chaabouni and D. Ghribi-Aydi. 2016. Cookies from composite wheat–sesame peels flours: Dough quality and effect of *Bacillus subtilis* SPB1 biosurfactant addition. Food Chemistry 194: 758–769.

12

Microbial Biosurfactants
Sources, Classification, Properties and Mechanism of Interaction

P Saranraj,[1,]* *R Z Sayyed,*[2] *P Sivasakthivelan,*[3]
M Durga Devi,[4] *Abdel Rahman Mohammad Al Tawaha*[5]
and *S Sivasakthi*[6]

1. Introduction

Surfactants are among the most versatile products of Chemical industry. Surfactant is an abridgment of the term surface-active chemical compounds. A surfactant is a substance present in a system, having the properties of adsorbing onto the surface or interface of the system and altering the interfacial energy of those surfaces. The term interface denotes the difference between the two immiscible phases, and the term surface denotes the interface where one phase is gas which is generally air. The interfacial energy is the minimum amount of work required to create that interface. In order to work out interfacial surface tension between two phases, interfacial energy per unit area is to be measured. It is the minimum amount of work required to create a unit area of the interface or to expand it by unit area. The physical phenomenon is

[1] Department of Microbiology, Sacred Heart College (Autonomous), Tirupattur – 635 601, Tamil Nadu, India.
[2] Department of Microbiology, PSGVP Mandal's Arts, Science and Commerce College, Shahada, Maharashtra, India 425409.
[3] Department of Agricultural Microbiology, Faculty of Agriculture, Annamalai University, Annamalai Nagar – 608 002, Tamil Nadu, India.
[4] Department of Biochemistry, Sacred Heart College (Autonomous), Tirupattur – 635 601, Tamil Nadu, India.
[5] Department of Biological Sciences, Al - Hussein Bin Talal University, Maan, Jordan.
[6] Department of Microbiology, Shanmuga Industries Arts and Science College, Tiruvannamalai, Tamil Nadu, India.
* Corresponding author: microsaranraj@gmail.com

additionally a measure of the difference in the nature of two phases meeting at the surface. The interfacial energy per unit area required to make the extra amount of that interface is the product of interfacial surface tension and increase in the area of the interface. A surfactant is, therefore, a substance which at a low concentration adsorbs at some or all interfaces in the system, and significantly changes the amount of work required to expand those interfaces (Rosen and Kunjappu 2012). Surfactants are surface-active agents with wide-ranging properties including the lowering of surface and interfacial tensions of liquids (Christofi and Ivshina 2002). Biosurfactants are biological compounds that exhibit high surface-active properties (Georgiou et al. 1992, Kuiper et al. 2004). Surfactants are characteristically organic compounds containing both hydrophobic groups which form an integral part of tails and therefore, their heads are composed of hydrophilic groups. The surfactant molecule contains both a water-insoluble which is an oil-soluble component and a water-soluble component. They are surface-active amphiphilic agents possessing both hydrophilic as well as hydrophobic moieties that reduce the surface and interfacial tensions by accumulating at the interface between two fluids such as oil and water, signifying that surfactants moreover assist the solubility of polar compounds in organic solvents (Fig. 1). They are of synthetic or biological origin. Due to their interesting properties like lower toxicity levels, a higher degree of biodegrading ability, high foaming potential, and optimal activity at extreme conditions like temperatures, pH levels, and salinity, these qualities have been attracting the attention of the scientific and industrial communities. Increasing public awareness about environmental pollution influences the research and development of novel technologies that help in the clean-up of organic and inorganic contaminants like hydrocarbons and metals accumulated in society. An alternative and eco-friendly method of bioremediation technology for contaminated sites with these pollutants is the use of biosurfactants and biosurfactant-producing microorganisms. The diversity of biosurfactants makes them a vital group of compounds for potential use during a big variety of commercials and biotechnological applications (Pacwa Płociniczak et al. 2011).

The special properties of biosurfactants are that they allow their use and partial replacement of artificially or chemically synthesized surfactants in industrial operations. Chemically synthesized biosurfactants are mostly derived from oil and are widely utilized in cosmetics. Biosurfactants in aqueous solutions and hydrocarbon mixtures reduce surface tension, Critical Micelle Concentration (CMC), and interfacial tension (Rahman et al. 2002). Currently, Microbial biosurfactants have gained considerable interest in recent years due to their low toxicity, biodegradable

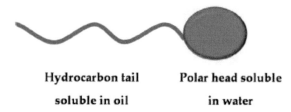

Hydrocarbon tail   **Polar head soluble**

soluble in oil   **in water**

Figure 1. Surfactant molecule with apolar (hydrophobic) and polar moieties (hydrophilic) (Source: Danyelle Khadydja Santos et al. 2016).

nature, and diversity. Their range of potential industrial applications includes enhanced oil recovery, crude oil drilling, lubricants, surfactant-aided bioremediation of water-insoluble pollutants, health care, and food processing (Sullivan 1998). Other areas that tap and use the potentials of biosurfactants are cosmetics and the formulations of soaps, foods, and both dermal and transdermal drug delivery systems.

2. Microbial Biosurfactant

Microorganisms produce a wide range of secondary metabolites namely surfactants, generally called Biosurfactants. Microbial Biosurfactants are low molecular weight surface-active compounds widely produced by bacteria, fungi, and yeast. The term "Surfactants" was first coined by Antara products in 1950 which covered all the products having surface activity, wetting agents, detergents, foaming agents, emulsifiers, and dispersants. Microbial surfactants are produced by a variety of microbes that possess different structures chemically and physically (Amaral et al. 2010). Microbial surface-active compounds are diverse molecules produced by different microorganisms in various structures and are mainly classified by their chemical structure and their microbial origin. They are made up of a hydrophilic moiety, comprising an acid, peptide cations, or anions, monosaccharides, disaccharides or polysaccharides, and a hydrophobic moiety of unsaturated or saturated hydrocarbon chains of fatty acids. The hydrophilic (polar) part of the biosurfactants is usually mentioned as 'head' and therefore the hydrophobic part (non-polar) is understood as 'tail' (Karanth et al. 1999). These structures confer a good range of properties, including the power to lower surface tension and interfacial surface tension of liquids and to make micelles and microemulsions between two different phases (Smyth et al. 2010).

Microorganisms utilize a spread of organic compounds because of the source of carbon and energy for his or her growth. When the carbon source is an insoluble substrate such as hydrocarbon, microbes facilitate their diffusion into the cell by producing a spread of drugs, the biosurfactants. Some bacteria excrete ionic surfactants, which emulsifies hydrocarbon substrates in the growth medium (Guerra Santos et al. 1986). Some microbes are capable of adapting their cell wall, by synthesizing lipopolysaccharides or non-ionic surfactants in their cell wall. The exact reason why some microorganisms produce surfactants is unclear (Deziel et al. 2000).

According to Danyelle Khadydja Santos et al. (2016), microorganisms use a group of carbon sources and energy for growth. The mix of carbon sources with insoluble substrates favor the intracellular diffusion and production of various substances. Microorganisms like bacteria, filamentous fungi, and yeasts can produce biosurfactants with different molecular structures and surface activities (Campos et al. 2013). In recent decades, there has been a rise in scientific interest regarding the isolation of microorganisms that produce tensioactive molecules with good surfactant characteristics, like a coffee Critical Micelle Concentration (CMC), low toxicity, and high emulsifying activity (Silva et al. 2014).

Literature describes the genera *Pseudomonas* and *Bacillus* as great biosurfactant producers (Silva et al. 2014). However, most biosurfactants of a bacterial origin are inadequate for use in the food industry due to their possible pathogenic nature

(Shepherd et al. 1995). *Candida bombicola* and *Candida lipolytica* are among the foremost commonly studied yeasts for the assembly of biosurfactants. A key advantage of using yeasts, like *Yarrowia lipolytica*, baker's yeast, and *Kluyveromyces lactis*, resides in their "Generally considered safe" (GRAS) status. Organisms with GRAS status do not offer the risks of toxicity or pathogenicity, which allows their use in the food and pharmaceutical industries (Campos et al. 2013). Biosurfactants have many environmental applications like bioremediation and dispersion of oil spills, enhanced oil recovery, and transfer of petroleum. Biosurfactants are potentially employed in the areas related to food, cosmetics, health care industries, and the cleaning of toxic chemicals of both industrial and agricultural origin. Biosurfactants produced by various microorganisms together with their properties are listed in Table 1.

3. Classification of Biosurfactant

Biosurfactants were categorized according to their microbial origin and chemical structure. Rosenberg and Ron (1999) classified biosurfactants into two categories based on their Molecular weight. They are (i) High molecular weight molecules (high mass) and (ii) Low molecular weight molecules (Low mass). Low mass biosurfactants include Glycolipids, Lipopeptides, Lipoproteins, Fatty acids, and Phospholipids, whereas High mass surfactants include Polymeric and Particulate surfactants. Bacteria are considered the major producers of biosurfactant among microbes. The production of biosurfactants from microbes depends on the condition in which it is fermented, environmental factors, and availability of nutrients. The High molecular weight microbial surfactants are generally polyanionic heteropolysaccharides containing both polysaccharides and proteins, the low molecular weight microbial surfactants are often Glycolipids. The production of microbially mediated biosurfactants differs from the environment of the growing microorganism based on its nutritional status. Microbial cells with high cell surface hydrophobicity are also surfactants and they themselves play a natural role in the growth of microbial cells on hydrocarbon, sulfur which are water-insoluble substrates. Cell adhesion, cell aggregation, emulsification, flocculation, dispersion, and desorption phenomena were mediated by the exocellular surfactants (Karanth et al. 1999).

The polar grouping was the base in classifying the artificially synthesized surfactants, whereas the biosurfactants were categorized by their chemical composition and their microbial origin. In general, their structure includes a hydrophilic moiety consisting of amino acids or peptides anions or cations; monosaccharides, disaccharides, or polysaccharides; and a hydrophobic moiety consisting of unsaturated, saturated, or fatty acids. The different groups of biosurfactants include Glycolipids, lipoproteins and Lipopeptides, Phospholipids and Fatty acids, Polymeric surfactants, and Particulate surfactants. A number of reports on the synthesis of biosurfactants by hydrocarbon-degrading microorganisms were recorded along with data stating that water-soluble compounds such as Glucose, Sucrose, Glycerol, or Ethanol were also capable of producing the same (Cooper and Goldenberg 1987, Hommel et al. 1994, Passeri 1992). The biosurfactant-producing microbes are distributed among a wide group of genera.

Table 1. Microorganisms involved in production of major Biosurfactants.

Biosurfactant class	Microorganisms
Glycolipids	
Rhamnolipids	*Pseudomonas aeruginosa*
Trehalose lipids or Trehalolipids	*Mycobacterium* sp., *Nocardia* sp., *Corynebacterium* sp., *Rhodococcus erythropolis*, *Rhodococcus ruber* and *Arthrobacter* sp.
Sophorolipids	*Torulopsis bombicola*, *Torulopsis apicola*, *Torulopsis petrophilum*, *Wickerhamiella domericqiae* and *Candida bogoriensis*
Mannosylerythritol lipids	*Candida antarctica*, *Psdozyma antarctica* and *Psdozyma rugulosa*
Cellobiolipids	*Ustilago zeae* and *Ustilago maydis*
Lipopeptides and Lipoproteins	
Surfactin	*Bacillus subtilis*
Subitisin	*Bacillus subtilis*
Lichenysin	*Bacillus licheniformis*
Iturnin	*Bacillus subtilis*
Fengycin	*Bacillus subtilis*
Viscosin	*Pseudomonas fluorescens*
Arthrofactin	*Pseudomonas* sp.
Amphisin	*Pseudomonas syringae*
Putisolvin	*Pseudomonas putida*
Serrawettin	*Serratia marcescens*, *Serratia liquefaciens* and *Serratia rubidaea*
Gramicidin	*Brevibacterium brevis*
Polymyxin	*Bacillus polymyxa*
Peptide lipid	*Bacillus licheniformis*
Fatty acids, Phospholipids and Neutral lipids	
Fatty acids	*Corynebacterium lepus*
Corynomicolic acids	*Corynebacterium insidibasseosum*
Phospholipids	*Rhodococcus erythropolis*, *Thiobacillus thiooxidans*, *Corynebacterium lepus*, *Micrococcus* sp., *Acinetobacter* sp. and *Candida* sp.
Neutral lipids	*Thiobacillus thiooxidans*
Ornithine lipids	*Pseudomonas* sp., *Thiobacillus* sp., *Gluconobacter* sp. and *Micrococcus* sp.
Polymeric Biosurfactants	
Emulsan	*Acinetobacter calcoaceticus*
Apoemulsan	*Acinetobacter venetianus*
Alasan	*Acinetobacter calcoaceticus* and *Acinetobacter radioresistens*
Liposan	*Candida lipolytica*
Biodispersan	*Acinetobacter calcoaceticus*
Lipomanan	*Candida tropicalis*
Carbohydrate – Protein - Lipid	*Pseudomonas fluorescens* and *Debaryomyces polymorphis*
Protein PA	*Pseudomonas aeruginosa*
Aminoacid – Lipids	*Bacillus* sp.
Particulate Biosurfactants	
Vesicles and Fimbriae	*Acinetobacter calcoaceticus*
Whole cells	Variety of bacteria

Source: Desai and Banat (1997), Muthusamy et al. (2008).

3.1 Glycolipids

Glycolipids are carbohydrates like monosaccharides, disaccharides, trisaccharides and tetrasaccharides that include glucose, mannose, galactose, glucuronic acid, rhamnose and galactose sulphate combined with long-chain aliphatic acids or hydroxy aliphatic acids. Popular biosurfactants are Glycolipids which are carbohydrates in combination with long-chain aliphatic acids or hydroxy aliphatic acids which are linked by ether or an ester group. The best-known Glycolipids are Rhamnolipids, Trehalolipids and Sophorolipids. Other types of Glycolipids have been reported in the literature such as Glycoglycerolipid (Nakata 2000), Sugar based bioemulsifiers (Kim et al. 2000), Cellobiolipids, Mannosylerythritol lipid A (Van Hoogmoed et al. 2000) and many different Hexose lipids (Golyshin et al. 1999).

3.1.1 Rhamnolipids

Certain species of Pseudomonas are characterized to produce large amounts biosurfactant containing one or two molecules of rhamnose linked to one or two molecules of hydroxydecanoic acid (Benincasa et al. 2004, Nitschke et al. 2005, Monteiro et al. 2007, Pornsunthorntawee et al. 2008). In Rhamnolipids, the molecules of Rhamnose are linked to one or two molecules of β-hydroxydecanoic acid, are the best-studied Glycolipids. The first report on the Production of rhamnose containing glycolipids was reported in *Pseudomonas aeruginosa* by Jarvis and Johnson (1949). L-Rhamnosyl-L-rhamnosyl-β-hydroxydecanoyl-β-hydroxydecanoate (Fig. 2) and L-rhamnosyl-β-hydroxydecanoyl-β-hydroxydecanoate, denoted as Rhamnolipid 1 and 2, respectively, are the principal Glycolipids produced by *Pseudomonas aeruginosa* (Edward and Hayashi 1965, Hisatsuka et al. 1977). The formation of rhamnolipid types 3 and 4 containing one b-hydroxydecanoic acid with one and two rhamnose units, respectively (Syldatk et al. 1985), methyl ester derivatives of Rhamnolipids 1 and 2 (Hirayamo and Kato 1982), and Rhamnolipids with alternative fatty acid chains have also been reported (Parra et al. 1989, Rendell et al. 1990). Edward and Hayashi (1965) have reported formation of Glycolipid, type R-1 containing two rhamnose and two hydroxydecanoic units by *Pseudomonas aeruginosa*. A second kind of rhamnolipid (R-2) containing one rhamnose unit was reported by Rendell et al. (1990). These mutants develop when the growth medium was supplemented with rhamnolipid. Rhamnolipids from *Pseudomonas* spp. have been demonstrated to lower the interfacial tension against n-hexadecane to 1 mN/m and the surface tension to 25 to 30 mN/m (Lang and Wagner 1987, Parra et al. 1989). They also emulsify alkanes and stimulate the growth of *Pseudomonas aeruginosa* on

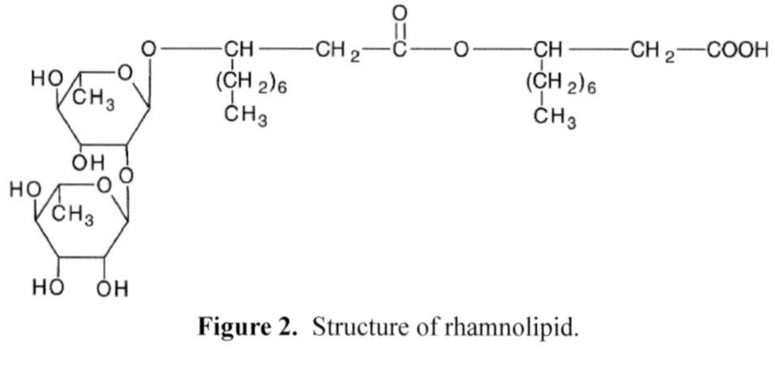

Figure 2. Structure of rhamnolipid.

hexadecane (Hisatsuka et al. 1977). Gas chromatographic analysis of hydroxyl fatty acids rhamnolipid produced by *Pseudomonas aeruginosa* showed that positions of the fatty acids in the lipid moiety were variable (Monteiro et al. 2007).

3.1.2 *Trehalose Lipids or Trehalolipids*

Several structural types of microbial trehalose lipid biosurfactants have been reported and the structure or Trehalolipids or Trehalose lipid is given in Fig. 3. Disaccharide trehalose linked at C-6 and C-6' to mycolic acids is associated with most species of *Mycobacterium* sp., *Nocardia* sp. and *Corynebacterium* sp. (Lang and Wagner 1987, Cooper et al. 1981). Mycolic acids are long-chain, α-branched-β-hydroxy fatty acids. Microorganisms produce Trehalolipids in different sizes and structures of mycolic acid and they also differ in the number of carbon atoms and degrees of unsaturation (Lang and Wagner 1987, Desai and Banat 1997). Trehalose dimycolate produced by *Rhodococcus erythropolis* and *Arthrobacter* sp. has been extensively studied (Kretschmer et al. 1982). *Rhodococcus erythropolis* also synthesizes a novel anionic trehalose lipid (Ristau and Wagner 1993). Trehalose lipids from *Rhodococcus erythropolis* and *Arthrobacter* sp. reduced the surface and interfacial tensions in the culture broth from 25 to 40 and 1 to 5 mN/m, respectively (Lang and Wagner 1987, Li et al. 1984). Philp (2002) reported the production of trehalose lipids from alkanotrophic *Rhodococcus ruber* on gaseous alkanes propane and butane.

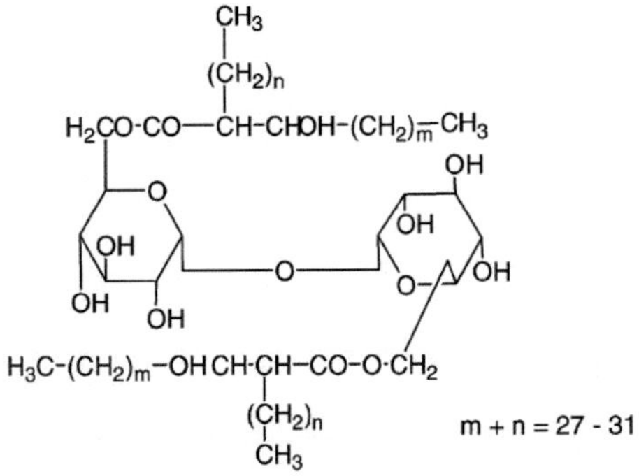

Figure 3. Trehalose lipids or trehalolipids.

3.1.3 *Sophorolipids*

Sophorolipids consist of a dimeric carbohydrate sophorose attached with a long-chain hydroxyl fatty acid and are mainly produced by yeasts such as *Torulopsis bombicola*, *Torulopsis apicola* (Tullock et al. 1967) and *Wickerhamiella domericqiae* (Chen et al. 2006) and consists of a dimeric carbohydrate sophorose linked to a long-chain hydroxyl fatty acid by glycosidic linkage. Biosurfactants are made up of a mixture of six to nine various hydrophobic sophorosides. Similar mixtures of water-soluble sophorolipids from several yeasts have also been reported (Hommel et al. 1987). Cutler and Light (1979) showed that *Candida bogoriensis* produces glycolipids in which sophorose is linked to docosanoic acid diacetate. Sophorolipids have the capacity to lower the surface tension of water from 72.8 mN/m to 40–30 mN/m, with a CMC of 40 to 100 mg/L (Van Bogaert et al. 2007). Sophorolipids was produced by *Torulopsis petrophilum* in water-insoluble substrates like alkanes and vegetable oils (Cooper and Paddock 1983). These sophorolipids, which were chemically identical to those produced by *Torulopsis bombicola,* did not emulsify alkanes or vegetable oils. Sophorolipids were not produced, when *Torulopsis petrophilum* is grown on a glucose yeast extract medium, instead, an effective protein-containing alkane emulsifying agent was formed (Cooper and Paddock 1983). These findings seem to contradict the conventional statement that surfactants as well as microbial emulsifiers are produced to enhance the uptake of water-insoluble substrates. Sophorolipids lower the surface and interfacial tension but they are not effective emulsifying agents (Cooper and Paddock 1984). Lactonic and acidic sophorolipids lowered the interfacial tension between n-hexadecane and water from 40 to 5 mN/m and exhibited stability towards changes in pH and temperature (Lang et al. 1989). Mostly Sophorolipids occur as a mixture of macrolactones and free acid form. It is reported that the lactone form of the Sophorolipid is required, or preferable, for various applications (Hu and Ju 2001). These biosurfactants are made up of a mixture of about six to nine different hydrophobic Sophorolipids (Fig. 4).

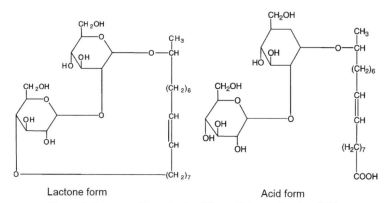

Lactone form   Acid form

Figure 4. Structure of lactonized and free-acid forms of sophorolipids.

3.1.4 Mannosylerythritollipids

Mannosylerythritollipids are the glycolipid biosurfactant consists of a sugar called mannosylerythritol and are synthesized by yeast like *Candida antarctica* (Crich et al. 2002) and *Candida* sp. (Kim et al. 1999). The fatty acid component of biosurfactant was determined to be hexanoic, dodecanoic, tetradecanoic or tetradecenoic acids (Kim et al. 1999). Mannosylerythritol lipids (Fig. 5) synthesized by *Candida* sp. lowered the surface tension of water to 29 dyne/cm at CMC of 10 mg/L and the minimum Interfacial tension was 0.1 dyne/cm against Kerosene (Kim et al. 1999). Fukuoka et al. (2007) have characterized the surface active properties of a new glycolipid biosurfactant, mono acylated mannosylerythritol lipid produced by *Psdozyma antarctica* and *Psdozyma rugulosa.*

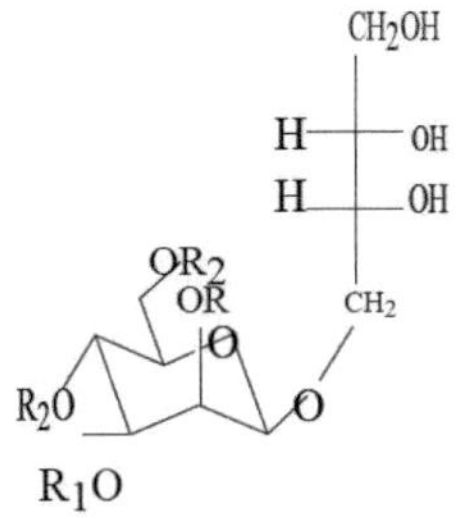

Figure 5. Mannosylerythritollipids.

3.2 *Lipopeptides and Lipoproteins*

Cyclic lipopeptides including decapeptide antibiotics (gramicidins) and lipopeptide antibiotics (polymyxins) are produced by *Bacillus brevis* (Marahiel et al. 1977) and *Bacillus polymyxa* (Suzuki et al. 1965), respectively, which possess surface-active properties. Ornithine-containing lipids from *Pterostylis rubescens* (Yamane 1987) and *Thiobacillus thiooxidans* (Knoche and Shiveley 1972), cerilipin, an ornithine and taurine containing lipids from *Gluconobacter cerinus* (Tahara et al. 1976), and lysine-containing lipids from *Agrobacterium tumefaciens* (Tahara et al. 1976) also exhibit excellent biosurfactant activity. An aminolipid biosurfactant called serratamolide has been isolated from *Serratia marcescens* (Mutsuyama et al. 1985). Biosurfactants increased cell hydrophilicity by blocking the hydrophobic sites on the surface of the cells when negatively mutated by serratamolide (Bar Ness et al. 1988).

3.2.1 *Surfactin*

Surfactin, a cyclic lipopeptide is one of the most effective powerful biosurfactants known so far, which was first reported in *Bacillus subtilis* (Fig. 6) (Kappeli et al. 1979). Because of its exceptional surfactant activity it is named as Surfactin (Rosenberg et al. 1979). Surfactins can lower the surface tension from 72 mN/m to 27.9 mN/m (Zosim et al. 1982) and have a CMC of 0.017 g/L (Cirigliano and Carman 1984). The surfactin groups of compounds are shown to be a cyclic lipoheptapeptides which contain a β-hydroxy fatty acid in its side chain (Cooper et al. 1981). Recent studies indicate that surfactin shows potent antiviral, antimycoplasma, antitumoral, anticoagulant activities as well as inhibitors of enzymes (Hisatsuka et al. 1971, Zosim et al. 1982). Although, such properties of surfactins qualify them for potential applications in medicine or biotechnology, they have not been exploited extensively till date.

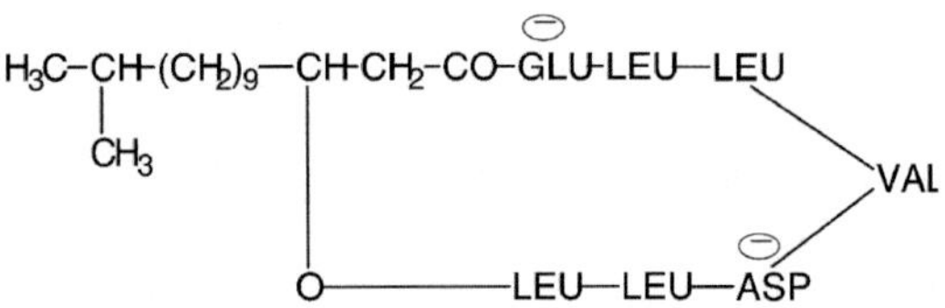

Figure 6. Structure of Surfactin.

3.2.2 Lichenysin

Lichenysin, produced by *Bacillus licheniformis* exhibits similar structure and physiochemical properties to that of Surfactin (McInerney et al. 1990). *Bacillus licheniformis* also produce several other surface-active agents that act synergistically and exhibit excellent temperature, pH and salt stability (McInerney et al. 1990). Lichenysin A produced by *Bacillus licheniformis* strain is characterized to contain a long-chain β-hydroxy fatty acid molecule (Yakimov et al. 1996). The Lichenysin produced by *Bacillus licheniformis* is capable of lowering the surface tension of water to 27 mN/m and the interfacial tension between water and n-hexadecane to 0.36 mN/m. Lichenysin is stable even under a wide range of temperature fluctuation, pH variation and NaCl concentration which promotes the dispersion of colloidal 3-silicon carbide and aluminum nitride slurries more efficiently than any other chemical agents (Horowitz and Currie 1990). It has also been reported that lichenysin is a more efficient cation chelator compared with Surfactin (Grangemard et al. 2001).

3.2.3 Iturnin

Iturin A, the first compound discovered of the Iturin group is a best-known member isolated from a *Bacillus subtilis* strain taken from the soil (Peypoux et al. 1978). The subsequent isolation from other strains of *Bacillus subtilis* of five other Lipopeptides such as Iturin AL, Mycosubtilin, Bacillomycin L, D, F, and LC (or Bacillopeptin), all having a common pattern of chemical constitution, led to the adoption of the generic name of "Iturins" for this group of Lipopeptides (Kajimura et al. 1995). The compounds of Iturin groups are cyclic lipoheptapeptides which contain a β-amino fatty acid in its side chain. Lipopeptides belonging to the iturin family are potent antifungal agents that can also be used as biopesticides for plant protection (Vater et al. 2002, Romero et al. 2007, Mizumoto et al. 2007).

3.2.4 Fengycin

Fengycin is a lipodecapeptide containing β-hydroxy fatty acid in its side chain which comprises of C15 to C17 variants which have a characteristic Ala-Val dimorphy at its 6th position in its peptide ring (Vater et al. 2002). Wang et al. (2004) have demonstrated the identification of Fengycin homologues produced by *Bacillus subtilis* by using Electrospray Ionization Mass Spectrometry (ESI-MS) technique.

3.2.5 Viscosin

Viscosin is majorly produced by *Pseudomonas fluorescens*. The bacterium can adhere to broccoli heads as a wetting agent and cause decay of the wounded and unwounded florets of broccoli. Mutants which is deficient in viscosin obtained by transposon mutagenesis affect the wounded broccoli florets but they are devoid of the ability to decay unwounded ones unlike the wild type of bacterium. Mating of these mutants with their triparent with their corresponding wild type clones and the helper *Escherichia coli* (with the mobilizable plasmid pPK2013) yielded transconjugants. Their linkage maps indicated that a 25 kb chromosomal DNA after transcription and translation forms three proteins as a synthetase complex and it is required to produce Viscosin. A probe was made from the DNA region hybridized with DNA fragments of other plant-pathogenic *Pseudomonas* to varying degrees (Braun et al. 2001).

3.2.6 Arthrofactin

Arthrofactin produced by *Pseudomonas* sp. MIS38, is the most potent cyclic lipopeptide-type biosurfactant ever reported. Three genes termed *ArfA*, *ArfB* and *ArfC* form the Arthrofactin synthetase gene cluster and encode *ArfA*, *ArfB* and *ArfC* which assemble to form a unique structure. *ArfA*, *ArfB* and *ArfC* contain two, four, and five functional modules, respectively. A module is defined as the unit that catalyzes the incorporation of a specific amino acid into the peptide product. The arrangement of the modules of a peptide synthetase is usually colinear with the amino acid sequence of the peptide. The modules can be further subdivided into different domains that are characterized by a set of short conserved sequence motifs. Each module bears a condensation domain [C] (responsible for formation of peptide bond between two consecutively bound amino acids), adenylation domain [A] (responsible for amino acid recognition and adenylation at the expense of ATP) and thiolation domain [T] (serves as an attachment site of 4-phosphopantetheine cofactor and a carrier of thioesterified amino acid intermediates). However, none of the 11 modules possess the epimerization domain [E] responsible for the conversion of amino acid residues from L to D form. Moreover, two thioesterase domains are tandemly located at the C-terminal end of *ArfC*. *ArfB* is the gene absolutely essential for arthrofactin production as its disruption impaired this act (Roongsawan et al. 2003).

3.2.7 Amphisin

Amphisin is produced by *Pseudomonas* sp. It has both biosurfactant and antifungal properties and brings about the inhibition of plant pathogenic fungi. The two component regulatory system *GacA/GacS* (*GacA* is a response regulator and *GacS* is a sensor kinase) controls the amphisin synthetase gene (*amsY*) (Koch et al. 2002). The surface motility of this bacterium requires the production of this biosurfactant as is indicated by the mutants defective in the genes *gacS* and *amsY*. Amphisin synthesis is regulated by *gacS* gene as the *gacS* mutant regains the property of surface motility upon the introduction of a plasmid encoding the heterologous wild-type *gacS* gene from *Pseudomonas syringae* (Andersen et al. 2003).

3.2.8 Putisolvin

Pseudomonas putida produces two surface active cyclic lipopeptides designated as Putisolvins I and II. The ORF (open reading frame) encoding the synthesis of the Putisolvins bears amino acid homology to various lipopeptide synthetases (Kuiper et al. 2004). Putisolvins are produced by a Putisolvin synthetase designated as *psoA*. Three heat shock genes *dnaK*, *dnaJ* and *grpE* positively regulate the biosynthesis of Putisolvin (Dubern et al. 2005). The *ppuI-rsaL-ppuR* quorum sensing system controls putisolvin biosynthesis. *ppuI* and *ppuR* mutants exhibit decreased Putisolvin production whereas *rsaL* mutants show enhanced Putisolvin production (Dubern et al. 2006).

3.2.9 Serrawettin

A group of Gram negative bacteria such as *Serratia* produces surface active cyclodepsipeptides known as Serrawettin W1, W2 and W3 (Matsuyama et al. 1986, Matsuyama et al. 1989). Varied strains of *Serratia marcescens* produces different

Serrawettins, e.g., Serrawettin W1 was produced by *Serratia marcescens* strains 274 and *Serratia marcescens* ATCC 13880 or *Serratia marcescens* NS 38. Serrawettin W2 was produced by *Serratia marcescens* strain NS 25 and W3 was produced by *Serratia marcescens* strain NS 45. Besides this *Serratia liquefaciens* produce Serrawettin W2. Temperature dependent synthesis of novel lipids–Rubiwettin R1 and RG1 was recorded in *Serratia rubidaea* (Matsuyama et al. 1990).

3.3 Fatty Acids, Phospholipids and Neutral Lipids

Large quantities of fatty acid and phospholipid surfactants are produced by bacterial and yeast species when grown on n-alkanes (Robert et al. 1989). Rich vesicles are produced by *Acinetobacter* sp. (Fig. 7), Phosphatidylethanolamine (Kappeli et al. 1979), which form clear microemulsions of alkanes in water. The quantitative production of phospholipids has been recorded in a few species of *Aspergillus* (Kappeli et al. 1979) and *Thiobacillus thiooxidans* (Beeba and Umbreit 1971). *Arthrobacter* strain (Wayman et al. 1984) and *Pseudomonas aeruginosa* (Robert et al. 1989) accumulate up to 40 to 80% (wt/wt) of such lipids when cultivated on hexadecane and olive oil, respectively. Phosphatidylethanolamine produced by *Rhodococcus* sp. Reduction in interfacial tension between water and hexadecane to less than 1 mN/m and a CMC of 30 mg/liter is notices when *Erythropolis* is grown on n-alkane (Kretschmer et al. 1982).

When grown on alkanes few hydrocarbon-degrading microbes produce extracellular free fatty acids that exhibit good surfactant activity. The fatty acid biosurfactants are saturated fatty acids in the range of C12 to C14 and complex fatty acids containing hydroxyl groups and alkyl branches (Mac Donald et al. 1981, Kretschmer et al. 1982). It was shown that *Arthrobacter* strain (Wayman et al. 1984) and *Pseudomonas aeruginosa* (Robert et al. 1989) accumulated up to 40–80% (w/w) of such lipids when cultivated on Hexadecane and olive oil respectively.

$$H_2C - O - \overset{\overset{\displaystyle O}{\|}}{C} - R_1$$
$$HC - O - \overset{\overset{\displaystyle O}{\|}}{C} - R_2$$
$$H_2C - O - \overset{\overset{\displaystyle O}{\|}}{\underset{\underset{\displaystyle O^-}{|}}{P}} - O - CH_2 - CH_2 - \overset{+}{N}H_3$$

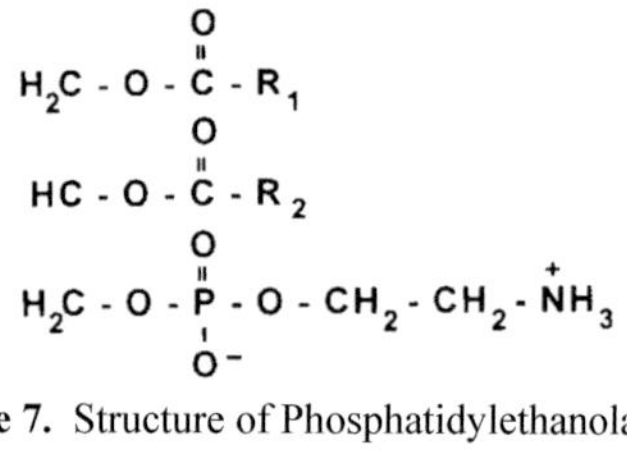

Figure 7. Structure of Phosphatidylethanolamine.

3.4 Polymeric Biosurfactants

Polymeric biosurfactants are high molecular weight biopolymers, which show certain properties like high viscosity, tensile strength and resistance to shear. The following are examples of different classes of polymeric biosurfactants.

3.4.1 Emulsan

Acinetobacter calcoaceticus produces a potent extracellular polymeric bioemulsifier called Emulsan (Rosenberg et al. 1979) which is characterized as a polyanionic amphipathic heteropolysaccharides (Fig. 8). The heteropolysaccharide backbone

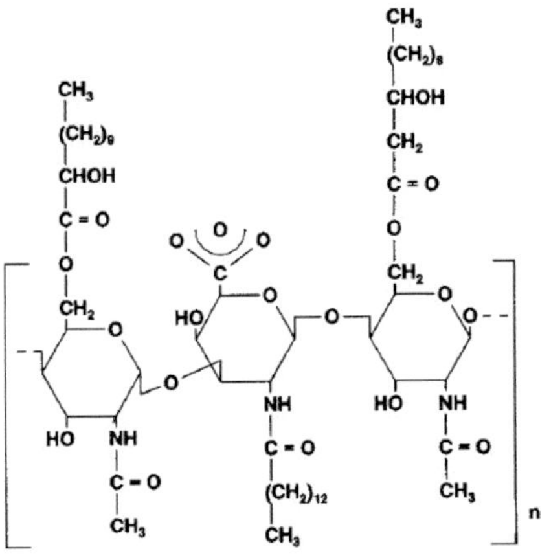

Figure 8. Structure of Emulsan.

consists of repeating units of trisaccharide of N-acetyl-d-galactosamine, N-acetylgalactosamineuronic acid and an unidentified N-acetylamino sugar (Zukereberg et al. 1979). Apoemulsan is the product obtained after the removal of a protein fraction which exhibits low emulsifying activity on hydrophobic substrates such as n-hexadecane. Major proteins associated with the emulsan complex is a cell surface esterase (Bach et al. 2003).

3.4.2 *Apoemulsan*

Apoemulsan is an extracellular, polymeric lipoheteropolysaccharide produced by *Acinetobacter venetianus*. Deproteinized Emulsan (apoemulsan, 103 kDa) consists of d-galactosamine, l-galactosamine uronic acid (pKa, 3.05) and a diamino, 2-desoxy n-acetylglucosamine (Mac Donald et al. 1981). It retained emulsifying activity towards certain hydrocarbon substrates but was unable to emulsify relatively non-polar, hydrophobic, aliphatic materials (Nitschke and Costa 2007, Singh et al. 2007). It is now known that polymers are synthesized from Wzy pathway. However, there also appears a differing report which claims that the process is based on presence of Polysaccharide-copolymerase (PCP) (Makkar and Cameotra 2002, Van Hamme et al. 2006). Singh et al. (2007) proved that synthesis of this polymer was dependant on Wzy pathway where, PCP protein controlled the length of the polymer. This was proved by inducing defined point mutations in the proline-glycine-rich region of apoemulsan PCP protein (Wzc). Five out of eight mutants produced higher weight BE than that of the wild type while four had modified biological properties. It has been suggested that emulsifying activity and release of polymer is mediated via esterase gene (34.5 kD). A study carried out by Das and Mukherjee (2005) proved that lipase is responsible for enhanced emulsification properties.

3.4.3 *Alasan*

Alasan is an anionic alanine-containing heteropolysaccharide protein biosurfactant produced by *Acinetobacter radioresistens* and it was reported to solubilize and degrade polyaromatic hydrocarbons (Barkay et al. 1999). Alasan is a surface-active component with 35.77 kD protein called AlnA. This surface active protein AlnA have a high amino acid sequence homology to *Escherichia coli* Outer membrane protein A (OmpA), but however OmpA does not possess any emulsifying activity (Toren et al. 2002). Three of Alasan proteins were purified from *Acinetobacter radioresistens* are having molecular masses of 16, 31 and 45 kD and it was demonstrated that the 45 kD protein had the highest specific emulsifying activity, 11% higher than the intact Alasan complex (Toren et al. 2002). The 16- and 31-kD proteins gave relatively low emulsifying activities, but they were significantly higher than that of apo-alasan (Toren et al. 2002).

3.4.4 Liposan

Candida lipolytica produce an extracellular water soluble emulsifier called Liposan which is composed of 83% (v/w) carbohydrate and 17% (w/v) protein. The carbohydrate portion is a heteropolysaccharides consisting of glucose, galactose, galactosamine and galacturonic acid (Singh et al. 2007).

3.4.5 Biodispersan

Biodispersan is an extracellular, anionic polysaccharide produced by *Acinetobacter calcoaceticus* which acts as a dispersing agent for water insoluble solids (Das and Mukherjee 2005). It is non-dialyzable, with an average molecular weight of 51,400 and contains four reducing sugars, namely, glucosamine, 6-methylaminohexose, galactosamine uronic acid and an unidentified amino sugar (Singh et al. 2007). Rich protein was also secreted along with the extracellular polysaccharides. Protein defective mutants produced an equal or enhanced biodispersion as compared to the parent strain (Bach et al. 2003).

3.5 Particulate Biosurfactants

Partition of hydrocarbons by Extracellular membrane vesicles leads to the formation of a microemulsion which plays a vital role in the uptake of alkane by microbial cells. *Acinetobacter* sp. strain with a vesicle diameter of 20 to 50 nm and a buoyant density of 1.158 g/cm^3 was made up of phospholipid, protein, and lipopolysaccharide. The membrane vesicles contain about 5 times as much phospholipid and about 350 times as many polysaccharides as does the outer membrane of the same organism. Hydrocarbon-degrading and pathogenic bacteria are attributed to surface activity by several cell surface components, which include different structures as follows, M protein and lipoteichoic acid in Group A *Streptococci*, protein A in *Staphylococcus aureus*, layer A in *Aeromonas salmonicida*, Prodigiosin in *Serratia* spp., Gramicidins in *Bacillus brevis* spores, and thin fimbriae in *Acinetobacter calcoaceticus* (Roggiani and Dubnau 1993).

4. Properties of Biosurfactants

Biosurfactants are of increasing demand in commercial use because of the wide spectrum of available sources. There are many advantages of microbial biosurfactants compared to their chemically synthesized counterpart. Microbial surfactants are selected based on varied parameters such as surface movement, resilience to pH, adaptation to varied temperature and ionic quality, biodegradability, low poisonous nature, emulsifying and demulsifying capacity, and antiadhesive activity. The main characteristic features of biosurfactants and a short description of each property are given below.

4.1 Surface and Interface Activity

Efficiency and effectiveness are basic characteristics of a best surfactant. Efficiency is measured by the CMC, whereas effectiveness is related to surface and interfacial tensions (Barros et al. 2007). The CMC of biosurfactants lies in between 1 to 2000 mg/L, whereas interfacial (oil/water) and surface tensions are respectively 1 and 30 mN/m. A good surfactant lowers the surface tension of water from 72 to 35 mN/m and the interfacial tension of water or hexadecane from 40 to 1 mN/m (Mullian 2005). Surfactin produced by *Bacillus subtilis* reduces the surface tension of water to 25 mN/m and interfacial tension of water or hexadecane to < 1 mN/m (Cooper et al. 1981). Rhamnolipids from *Pseudomonas aeruginosa* decreases the surface tension of water to 26 mN/m and the interfacial tension of water/hexadecane to < 1 mN/m (Hisatsuka et al. 1971). The Sophorolipids from *Torulopsis bombicola* have been reported to reduce the surface tension to 33 mN/m and the interfacial tension to 5 mN/m (Cooper and Cavalero 2003). In general, biosurfactants are more effective and efficient and their CMC is about 10–40 times lower than that of chemical surfactants, i.e., less surfactant is necessary to get a maximum decrease in surface tension (Desai and Banat 1997).

Surfactant helps in decreasing surface strain and the interfacial pressure. Surfactin produced by *Bacillus subtilis* can lessen the surface tension of water to 25 mN m^{-1} and interfacial strain water or hexadecane to under 1 mN m^{-1} (Cavalero and Cooper 2003). *Pseudomonas aeruginosa* produces Rhamnolipids which diminished the surface tension of water to 26 mN m^{-1} and interfacial strain of water or hexadecane to esteem under 1 mN m^{-1} (Chakrabarti 2012). Biosurfactants are more powerful and effective and their CMC is around a few times lower than chemical surfactants, i.e., for maximal decline on surface strain, less surfactant is fundamental (Das and Mukherjee 2007).

4.2 Temperature, pH and Ionic Strength Tolerance

Most of the biosurfactants and their surface activities are not influenced by varied external factors such as temperature and pH. McInerney et al. (1990) reported that Lichenysin from *Bacillus licheniformis* and another biosurfactant produced by *Arthrobacter protophormiae* was not affected by temperature (30°C to 100°C), pH (2.0–12.0) and by Sodium chloride (10% NaCl concentration whereas 2% NaCl is enough to inactivate synthetic surfactants) and Calcium concentrations (up to 25 g/L). A lipopeptide of *Bacillus subtilis* was stable even after autoclaving (121°C/20 min) and after 6 months at –18°C. No changes in the surface activity were noticed from pH 5 to 11 and in the concentration of Sodium chloride up to 20% (Nitschke and Pastore 1990). Extremophiles biosurfactant has gained importance in recent years for their commercial utilization. Since, industrial procedures include extremes of temperature, pH and weight, it is important to separate novel microbial items that ready to work under these conditions (Cooper et al. 1981).

4.3 Biodegradability

Microbial derived compounds degrade easily compared to the Synthetic surfactants. Biosurfactants are easily degraded by microorganisms in water and soil, making these compounds adequate for bioremediation of toxic substances, oil spills, and waste treatment (Desai and Banat 1997). Synthetic surfactants cause various environmental issues and thus, biodegradable biosurfactants from marine microorganisms were concerned for the biosorption of ineffectively solvent polycyclic sweet-smelling hydrocarbon, phenanthrene contaminated in aquatic surfaces (Gautam and Tyagi 2006). Marine algae, *Cochlodinium* by utilizing the biodegradable biosurfactant Sophorolipids with the removal efficiency of 90% of every 30 minutes treatment (Gharaei Fathabad 2011).

4.4 Low Toxicity

Microbial surfactants are generally considered as low or nontoxic products and therefore are appropriate for many industries. A low degree of toxicity allows the use of biosurfactants in foods, cosmetics and pharmaceutical industries. A low level of toxicity is essential for environmental applications. Biosurfactants can be produced from industrial wastes. There is less work recorded with respect to the poisonous nature of biosurfactants. Poremba et al. (1991) showed that the elevated level of toxicity in chemically-derived surfactant which displayed an LC50 against *Photobacterium phosphoreum* and was found to be 10 times lower than that of Rhamnolipids. The low toxicity profile of biosurfactant, Sophorolipids from *Candida bombicola* made them helpful in nourishment ventures (Hatha et al. 2007).

By analyzing the toxicity of synthetic surfactants and commercial dispersants, it was found that most biosurfactants degraded faster, except synthetic sucrose-stearate which showed structure homology to Glycolipids and degraded quickly than the biogenic Glycolipids. It was also reported that biosurfactants showed higher EC50 (effective concentration to decrease 50% of test population) values than synthetic dispersants (Poremba et al. 1991). A biosurfactant from *Pseudomonas aeruginosa* was tested along with a synthetic surfactant (Marlon A-350) which is mostly used in the industry, for its toxicity and mutagenic properties. The results revealed a higher level of toxicity and mutagenic effect in the chemical-derived surfactant and non-toxic and non-mutagenic in the biosurfactant (Flasz et al. 1998).

4.5 Specificity

Biosurfactants are multiple molecules with specific groups with varied functions such as the detoxification of different pollutants and deemulsification of industrial emulsion and in the fields of food, pharmaceuticals, and cosmetics with a specific action.

4.6 Biocompatibility and Digestibility

Properties such as Biological compatibility and digestibility of microbial biosurfactants allow the use of biomolecules in many industries, especially the food, pharmaceutical, cosmetics.

4.7 Emulsion Forming and Emulsion Breaking

Stable emulsion can be produced with a life span of months and year. Higher molecular-mass biosurfactants are in general better emulsifiers than the low-molecular-mass biosurfactants. Sophorolipids from *Torulopsis bombicola* have been shown to reduce surface tension, but are not good emulsifiers. By contrast, Liposan does not reduce the surface tension, but has been used successfully to emulsifyedible oils. Polymeric surfactants offer additional advantages because they coat droplets of oil, thereby forming the stable emulsion. This property is especially useful for making oil/water emulsion for cosmetics and food.

Biosurfactants act as emulsifiers or de-emulsifiers. An emulsion can be depicted as a heterogeneous framework, comprising of oneimmiscible fluid scattered in another as beads, whose distance acrossby and large surpasses 0.1 mm. Emulsions are two types: oil-in-water or water-in-oil emulsions. They have a minimal stability which might be balanced out by added substances, for example, biosurfactants and can be kept up as steady emulsions for a considerable length of time to years (Hu and Ju 2001).

Liposan, which is a water soluble emulsifier synthesized by *Candida lipolytica* has been used with edible oils to form stable emulsions. Liposan are commonly used in the cosmetic and food industries for producing stable oil/water emulsions (Campos et al. 2013). Polymeric surfactants offer additional advantages because they coat droplets of oil, thereby forming stable emulsions. This property is especially useful for making oil/water emulsions for cosmetics and food.

4.8 Chemical Diversity

The chemical diversity of naturally produced Biosurfactants offers a wide selection of surface-active agents with properties closely related to specific applications.

4.9 Antiadhesive Agent

A biofilm can be depicted as a group of microbes or other organic matter that have aggregated on any surface (Hatha et al. 2007). The initial step of biofilm foundation is bacterial adherence over the surface was influenced by different components including sort of microorganism, hydrophobicity and electrical charges of surface, ecological conditions and capacity of microorganisms to deliver extracellular polymers that assistance cells to grapple to surfaces (Jadhav et al. 2011). The biosurfactants can be utilized as a part of changing the hydrophobicity of the surface which thus influences the bond of microorganisms over the surface. A surfactant from *Streptococcus thermophilus* backs off the colonization of other thermophilic strains of *Streptococcus* over the steel which are in charge of fouling. So also, a biosurfactant from *Pseudomonas fluorescens* hindered the connection of *Listeria monocytogenes* onto steel surface (Konishi et al. 2008).

5. Mechanism of Interaction of Biosurfactants

Biosurfactants are microbial amphiphilic polymers and polyphilic polymers that tend to interact with the phase boundary between two phases in a heterogeneous system, defined as the interface. For all interfacial systems, it is known that organic molecules from the aqueous phase tend to immobilize at the solid interface. There they eventually form a film known as a conditioning film, which will change the properties (wettability and surface energy) of the original surface (Neu 1996). In an analogy to organic conditioning films, biosurfactants may interact with the interfaces and affect the adhesion and detachment of bacteria. In addition, the substratum surface properties determine the composition and orientation of the molecules conditioning the surface during the first hour of exposure. After about 4 hrs, a certain degree of uniformity is reached and the composition of the adsorbed material becomes substratum independent (Neu 1996).

Owing to the amphiphilic nature of biosurfactants, not only hydrophobic but a range of interactions are involved in the possible adsorption of charged biosurfactants to interfaces. Most natural interfaces have an overall negative or, rarely, positive charge. Thus, the ionic conditions and the pH are important parameters if interactions of ionic biosurfactants with interfaces are to be investigated (Craig et al. 1993). Gottenbos et al. (2001) demonstrated that positively charged biomaterial surfaces exert an antimicrobial effect on adhering Gram negative bacteria, but not on Gram positive bacteria. In addition, the molecular structure of a surfactant will influence its behaviour at interfaces. In describing the surface-active approach, an effort is made to elaborate on the possible theoretical locations and orientations of the biosurfactants. Nevertheless, it must be kept in mind that the situation in natural systems is far more complex and requires the consideration of many additional parameters.

6. Conclusion

Surfactants have long been among the most versatile of process chemicals. Their market is extremely competitive, and manufacturers will have to expand their arsenal to develop products for the 1990s and beyond. In this regard, biosurfactants are promising candidates. Biosurfactants are unique biomolecules with a variety of functions that are fast becoming a more efficient and greener alternative to their predecessor chemical surfactants. The unique properties of biosurfactants allow their use and possible replacement of chemically synthesized surfactants in a great number of industrial operations. During the last 2–3 decades a wide variety of microorganisms have been reported to produce numerous types of biosurfactants. Their biodegradability and lower toxicity gives them an advantage over their chemical counterparts and therefore may make them suitable for replacing chemicals. The emulsifying activity of the biosurfactant is another branch having a full scope that they can be used as emulsion forming agents for hydrocarbons and oils giving stable emulsions. While many types of biosurfactants are in use, no single biosurfactant is suitable for all potential applications. To date, biosurfactants are unable to compete economically with chemically synthesized compounds in the market, mainly due to their high production costs and the lack of comprehensive toxicity testing. Biosurfactants are used by many industries and one could easily say that there is almost no modern industrial operation where properties of surfaces and surface active agents are not exploited. The potential application of biosurfactants in industries is also a reality.

References

Amaral, P.F., M.A. Coelho, I.M. Marrucho and J.A. Coutinho. 2010. Biosurfactants from yeasts: Characteristics, production and application. Adv. Exp. Med. Bio. 672: 236–249.

Andersen, J.B., B. Koch, T.H. Nielsen, D. Sorensen and M. Hansen. 2003. Surface motility in *Pseudomonas* sp. DSS73 is required for efficient biological containment of the root pathogenic microfungi *Rhizoctonia solani* and *Pythium ultimum.* Microbio. 149: 37–46.

Bach, H., Y. Berdichevsky and D. Gutnick. 2003. An exocellular protein from the oil-degrading microbe *Acinetobacter venetianus* RAG-1 enhances the emulsifying activity of the polymeric bioemulsifier emulsan. Appl. Environ. Microbiol. 69: 2608–2615.

Bar Ness, R., N. Avrahamy, T. Matsuyama and M. Rosenberg. 1988. Increased cell surface hydrophobicity of a *Serratia marcescens* NS 38 mutant lacking wetting activity. J. Bacteriol. 170: 4361–4364.

Barkay, T., S. Navon-Venezia and E.Z. Ron. 1999. Enhancement of solubilization and biodegradation of polyaromatic hydrocarbons by the bioemulsifier alasan. Appl. Environ. Microbiol. 65: 2697–2702.

Barros, F.F.C., C.P. Quadros, M.R. Maróstica and G.M. Pastore. 2007. Surfactina: Propriedades químicas, tecnológicas e funcionais para aplicações em alimentos. Quím. Nova. 30: 1–14.

Beeba, J.L. and W.W. Umbreit. 1971. Extracellular lipid of *Thiobacillus thiooxidans.* J. Bacteriol. 108: 612–615.

Benincasa, M., A. Abalos and I. Oliveria. 2004. Chemical structure, surface properties and biological activities of the biosurfactant produced by *Pseudomonas aeruginosa* LBI from soapstock. Antonie van Leeuwenhoek 85: 1–8.

Braun, P.G., P.D. Hildebrand, T.C. Ells and D.Y. Kobayashi. 2001. Evidence and characterization of a gene cluster required for the production of Viscosin, a Lipopeptide biosurfactant, by a strain of *Pseudomonas fluorescens.* Can. J. Micro. 47: 294–301.

Campos, J.M., T.L.M. Stamford, L.A. Sarubbo, J.M. Luna, R.D. Rufino and I.M. Banat. 2013. Microbial biosurfactants as additives for food industries. Biotechnol. Prog. 29: 1097–1108.

Cavalero, D.A. and D.G. Cooper. 2003. The effect of medium composition on the structure and physical state of Sophorolipids produced by *Candida bombicola* ATCC 22214. J. Biotechnol. 103: 31–41.

Chakrabarti, S. 2012. Bacterial biosurfactant: Characterization, antimicrobial and metal remediation properties. Ph.D. Thesis, National Institute of Technology, Mumbai, India.

Chen, J., X. Song and H. Zhang. 2006. Production, structure elucidation and anticancer properties of Sophorolipid from *Wickerhamiella domercqiae.* Enzyme Microb. Technol. 39: 501–506.

Christofi, N. and I.B. Ivshina. 2002. Microbial surfactants and their use in field studies of soil remediation. J. Appl. Micro. 93: 915–929.

Cirigliano, M. and G. Carman. 1984. Isolation of a Bioemulsifier from *Candida lipodytica.* Appl. Environ. Microbiol. 48: 747–750.

Cooper, D.G., C.R. MacDonald, S.J.B. Duff and N. Kosaric. 1981. Enhanced production of Surfactin from *Bacillu subtilis* by continuous product removal and metal cation additions. Appl. Environ. Microbiol. 42: 408–412.

Cooper, D.G. and D.A. Paddock. 1983. *Torulopsis petrophilum* and surface activity. Appl. Environ. Microbiol. 46: 1426–1429.

Cooper, D.G. and D.A. Paddock. 1984. Production of a biosurfactant from *Torulopsis bombicola.* Appl. Environ. Microbiol. 47: 173–176.

Cooper, D.G. and B.G. Goldenberg. 1987. Surface active agents from two *Bacillus* species. Appl. Environ. Micro. 53: 224–229.

Cooper, D.G. and D.A. Cavalero. 2003. The effect of medium composition on the structure and physical state of Sophorolipids produced by *Candida bombicola* ATCC 22214. J. Biotechnol. 103: 31–41.

Craig, V.S.J., B.W. Ninham and R.M. Pashley. 1993. Effect of electrolytes on bubble coalescence. Nature 364: 317–319.

Crich, D., M.A. De la Mora and R. Cruz. 2002. Synthesis of the Mannosyl erythritol lipid MEL A; confirmation of the configuration of the meso-erythritol moiety. Tetrahedron 58: 35–44.

Cutler, A.J. and R.J. Light. 1979. Regulation of hydroxydocosanoic and Sophoroside production in *Candida bogoriensis* by the level of glucose and yeast extract in the growth medium. J. Biol. Chem. 254: 1944–1950.

Das, K. and A.K. Mukherjee. 2005. Characterization of biochemical properties and biological activities of biosurfactants produced by *Pseudomonas aeruginosa* mucoid and non-mucoid strains. Appl. Microbiol. Biotechnol. 69: 192–199.

Das, K. and A.K. Mukherjee. 2007. Crude petroleum-oil biodegradation efficiency of *Bacillus subtilis* and *Pseudomonas aeruginosa* strains isolated from petroleum oil contaminated soil from north-east India. Biores. Tech. 98: 1339–1345.

Desai, J.D. and I.M. Banat. 1997. Microbial production of surfactants and their commercial potential. Microbiol. Mol. Biol. Rev. 61: 47–64.

Deziel, E., F. Lepine, S. Milot and R. Villemur. 2000. Mass spectrometry monitoring of Rhamnolipids from a growing culture of *Pseudomonas aeruginosa* Strain 57RP. Biochimica et Biophysica Acta 1485: 145–152.

Dubern, J.F., E.L. Lagendijk, B.J.J. Lugt enberg and G.V. Bloemberg. 2005. The heat shock genes *dna*K, *dna*J, and *grp*E are involved in regulation of Putisolvin biosynthesis in *Pseudomonas putida* PCL1445. J. Bacteriol. 187: 5967–5976.

Dubern, J.F., B.J.J. Lugt enberg and G.V. Bloemberg. 2006. The ppuI-rsaL-ppuR quorum sensing system regulates biofilm formation of *Pseudomonas putida* PCL 1445 by controlling biosynthesis of the cyclic lipopeptides Putisolvins I and II. Journal of Bacteriology 188(8): 2898–2906.

Edward, J.R. and J.A. Hayashi. 1965. Structure of a Rhamnolipid from *Pseudomonas aeruginosa*. Arch. Biochem. Biophys. 111: 415–421.

Flasz, A., C.A. Rocha, B. Mosquera and C. Sajo. 1998. A comparative study of the toxicity of a synthetic surfactant and one produced by *Pseudomonas aeruginosa* ATCC 55925. Med. Sci. Res. 26: 181–185.

Fukuoka, T., T. Morita and T. Konishi. 2007. Characterization of new Glycolipid biosurfactants, tri-acylated mannosylerythritol lipids, produced by pseudozyma yeasts. Biotechnol. Lett. 29: 1111–1118.

Gautam, K.K. and V.K. Tyagi. 2006. Microbial surfactants: A review. J. Oleo Sci. 55: 155–166.

Georgiou, G., S.C. Lin and M.M. Sharma. 1992. Surface active compounds from microorganisms. Biotech. 10: 60–65.

Gharaei Fathabad, E. 2011. Biosurfactants in pharmaceutical industry. Ameri. J. Drug Disco. Develop. 1(1): 58–69.

Golyshin, P.M., H.L. Fredrickson and L. Giuliano. 1999. Effect of novel biosurfactants on biodegradation of polychlorinated biphenyls by pure and mixed bacterial cultures. Microbiologica. 22: 257–267.

Gottenbos, B., D. Grijpma and C. Vander Mei. 2001. Antimicrobial effects of positively charged surfaces on adhering Gram positive and Gram negative bacteria. J. Antimicrob. Chemother. 48: 7–13.

Grangemard, I., J. Wallach and R. Maget Dana. 2001. Lichenysin: a more efficient cation chelator than Surfactin. Appl. Biochem. Biotechnol. 90: 199–210.

Guerra Santos, L., O. Kappeli and A. Fiechter. 1986. Dependence of *Pseudomonas aeruginosa* continuous culture biosurfactant production on nutritional and environmental factors. Appl. Microbio. Biotech. 24: 443–448.

Hatha, A.A.M., G. Edward and K.S.M.P. Rahman. 2007. Microbial biosurfactants—Review. J. Mar. Atmos. Res. 3: 1–17.

Hirayama, T. and I. Kato. 1982. Novel Rhamnolipids from *Pseudomonas aeruginosa*. FEBS Lett. 139: 81–85.

Hisatsuka, K., T. Nakahara, N. Sano and K. Yamada. 1971. Formation of rhamnolipid by *Pseudomonas aeruginosa*: Its function in hydrocarbon fermentations. Agric. Biol. Chem. 35: 686–692.

Hisatsuka, K., T. Nakahara, Y. Minoda and K. Yamada. 1977. Formation of protein like activator for n-alkane oxidation and its properties. Agric. Biol. Chem. 41: 445–450.

Hommel, R.K., O. Stuwer, W. Stuber, D. Haferburg and H.P. Kleber. 1987. Production of water soluble surface active exolipids by *Torulopsis apicola*. Appl. Microbiol. Biotechnol. 26: 199–205.

Hommel, R.K., L. Weber, A. Weiss, U. Himelreich, O. Rilke and H.P. Kleber. 1994. Production of Sophorose lipid by *Candida apicola* grown on glucose. J. Biotechnol. 33: 147–155.

Horowitz, S. and J.K. Currie. 1990. Novel dispersants of silicon carbide and aluminium nitrate. J. Dispersion Sci. Technol. 11: 637–659.

Hu, Y. and J.K. Ju. 2001. Purification of lactonic sophorolipids by crystallization. J. Biotechnol. 87: 263–272.

Jadhav, M., S. Kalme, D. Tamboli and S. Govindwar. 2011. Rhamnolipid from *Pseudomonas desmolyticum* NCIM-2112 and its role in the degradation of Brown 3REL. J. Basic Microbiol. 51: 385–396.

Jarvis, F.G. and M.J. Johnson. 1949. A glycolipid produced by *Pseudomonas aeruginosa*. J. Am. Chem. Soc. 71: 4124–4126.

Kajimura, Y., M. Sugiyama and M. Kaneda. 1995. Bacillopeptins, new cyclic lipopeptide antibiotics from *Bacillus subtilis* FR-2. J. Antibiot. (Tokyo) 48: 1095–1103.

Kappeli, O., R. Shah and W.R. Finnerty. 1979. Partition of alkane by an extracellular vesicle derived from hexadecane grown *Acinetobacter*. J. Bacteriol. 140: 707–712.

Karanth, N.G.K., P.G. Deo and N.K. Veenanadig. 1999. Microbial biosurfactant and their importance. Cur. Sci. 77: 116–126.

Khadydja, D., F. Santos, Raquel D. Rufino, Juliana M. Luna, Valdemir A. Santos and Leonie A. Sarubbo. 2016. Biosurfactants: Multifunctional biomolecules of the 21st century. Int. J. Mol. Sci. 17(401): 1–31.

Kim, H.S., B.D. Yoon and D.H. Choung. 1999. Characterization of a biosurfactant, mannosylerythritol lipid produced from *Candida* sp. SY16. Appl. Microbiol. Biotechnol. 52: 713–721.

Kim, H.S., E.J. Lim and S.O. Lee. 2000. Purification and characterization of biosurfactants from *Nocardia* sp. L-417. Biotechnol. Appl. Biochem. 31: 249–253.

Knoche, H.W. and J.M. Shiveley. 1972. The structure of an Ornithine containing lipid from *Thiobacillus thiooxidans*. J. Biol. Chem. 247: 170–178.

Koch, B., T.H. Nielsen, D. Sorensen, J.B. Andersen and C. Christophersen. 2002. Lipopeptide production in *Pseudomonas* sp. DSS73 is regulated by components of sugar beet seed exudate *via* the Gac two-component regulatory system. Appl. Environ. Microbio. 68: 4509–4516.

Konishi, M., T. Fukuoka, T. Morita, T. Imura and D. Kitamoto. 2008. Production of new types of Sophorolipids by *Candida batistae*. J. Oleo. Sci. 57: 359–369.

Kretschmer, A., H. Bock and F. Wagner. 1982. Chemical and physical characterization of interfacial-active lipids from *Rhodococcus erythropolis* grown on *n*-alkane. Appl. Environ. Microbiol. 44: 864–870.

Kuiper, I., L. Ellen, R.P. Lagendijk, P.D. Jeremy, E.M.L. Gerda, E.T. Jane, J.L.L. Ben and V.B. Guido. 2004. Characterization of two *Pseudomonas putida* lipopeptide biosurfactants, Putisolvin I and II, which inhibit biofilm formation and break down existing biofilms. Mol. Microbio. 51(1): 97–113.

Lang, S. and F. Wagner. 1987. Structure and properties of biosurfactants. pp. 21–47. *In*: Kosaric, N., W.L. Cairns and N.C.C. Gray (eds.). Biosurfactants and Biotechnology. Marcel Dekker, Inc., New York, N.Y.

Lang, S., E. Katsiwela and F. Wagner. 1989. Antimicrobial effects of biosurfactants. Fat. Sci. Technol. 91: 363–366.

Li, Z.Y., S. Lang, F. Wagner, L. Witte and V. Wray. 1984. Formation and identification of interfacial-active glycolipids from resting microbial cells of *Arthrobacter* sp. and potential use in tertiary oil recovery. Appl. Environ. Microbiol. 48: 610–617.

Mac Donald, C.R., D.G. Cooper and J.E. Zajic. 1981. Surface-active lipids from *Nocardia erythropolis* grown on hydrocarbons. Appl. Environ. Microbiol. 41: 117–123.

Makkar, R. and S.S. Cameotra. 2002. An update on the use of unconventional substrates for biosurfactant production and their application. Appl. Microbiol. Biotechnol. 58: 428–434.

Marahiel, M., W. Denders, M. Krause and H. Kleinkauf. 1977. Biological role of gramicidin S in spore functions. Studies on gramicidin-S negative mutants of *Bacillus brevis* 9999. Eur. J. Biochem. 99: 49–52.

Matsuyama, T., K. Kameda and I. Yano. 1986. Two kinds of bacterial wetting agents: aminolipid and glycolipid. Proc. Japan Soc. Mass Spec. 11: 125–128.

Matsuyama, T., M. Sogawa and Y. Nakagawa. 1989. Fractal spreading growth of *Serratia marcescens* which produces surface-active exolipids. FEMS Microbio. Lett. 61: 243–246.

Matsuyama, T., K. Keneda, I. Ishizuka, T. Toida and I. Yano. 1990. Surface active novel glycolipid and linked 3-hydroxy fatty acids produced by *Serratia rubidaea*. J. Bact. 172(6): 3015–3022.

McInerney, M.J., M. Javaheri and D.P. Nagle. 1990. Properties of the biosurfactant produced by *Bacillus liqueniformis* strain JF-2. J. Microbiol. Biotechnol. 5: 95–102.

Mizumoto, S., M. Hirai and M. Shoda. 2007. Enhanced Iturin a production by *Bacillus subtilis* and its effect on suppression of the plant pathogen *Rhizoctonia solani*. Appl. Microbiol. Biotechnol. 75: 1267–1274.

Monteiro, S.A., G.L. Sassaki and L.M. De Souza. 2007. Molecular and structural characterization of the biosurfactant produced by *Pseudomonas aeruginosa* DAUPE 614. Chem. Phys. Lip. 147: 1–13.

Muthusamy, K., S. Gopalakrishnan, Th.K. Ravi and P. Sivachidambaram. 2008. Biosurfactants: Properties, commercial production and application. Cur. Sci. 94(25): 739–747.

Mutsuyama, T., M. Fujita and I. Yano. 1985. Wetting agent produced by *Serratia marcescens*. FEMS Microbiol. Lett. 28: 125–129.

Nakata, K. 2000. Two glycolipids increase in the bioremediation of halogenated aromatic compounds. J. Biosci. Bioeng. 89: 577–581.

Neu, T. 1996. Significance of bacterial surface active compounds in interaction of bacteria with interfaces. Microbiol. Rev. 60: 151–166.

Nitschke, M. and G.M. Pastore. 1990. Production and properties of a surfactant obtained from *Bacillus subtilis* grown on cassava wastewater. Bioresour. Technol. 97: 336–341.

Nitschke, M., S.G. Costa and J. Contiero. 2005. Rhamnolipid surfactants: an update on the general aspects of these remarkable biomolecules. Biotechnol. Prog. 21: 1593–1600.

Nitschke, M. and S.G.V.A.O. Costa. 2007. Biosurfactants in food industry. Trends Food Sci. Technol. 18: 252–259.

Parra, J.L., J. Guinea, M.A. Manresa, M. Robert, M.E. Mercade, F. Comelles and M.P. Bosch. 1989. Chemical characterization and physicochemical behaviour of biosurfactants. J. Am. Oil Chem. Soc. 66: 141–145.

Passeri, A. 1992. Marine biosurfactants—Production, characterization and biosynthesis of anionic glucose lipid from marine bacterial strain MM1. Appl. Microbiol. Biotechnol. 37: 281–286.

Peypoux, F., F. Besson and G. Michel. 1978. Structure de Iturine C de *Bacillus subtilis*. Tetrahedron. 38: 1147–1152.

Philp, J.C., M.S. Kuyukina and I.B. Ivshina. 2002. *Alkanotripic rhodococcus* ruber as a biosurfactant producer. Appl. Microbiol. Biotechnol. 59: 318–324.

Płociniczak, P.M., G.A. Płaza, Z. Piotrowska Seget and S.S. Cameotra. 2011. Environmental applications of biosurfactants: recent advances. Int. J. Mol. Sci. 1: 633–654.

Poremba, K., W. Gunkel, S. Lang and F. Wagner. 1991. Toxicity testing of synthetic and biogenic surfactants on marine microorganisms. Environ. Toxicol. Water Qual. 6: 157–163.

Pornsunthorntawee, O., P. Wongpanit and S. Chavadej. 2008. Structural and physicochemical characterization of crude biosurfactant produced by *Pseudomonas aeruginosa* SP4 isolated from petroleum-contaminated soil. Bioresour. Technol. 99: 1589–1595.

Rahman, K.S.M., I.M. Banat, T.J. Rahman, T. Thayumanavan and P. Lakshmanaperumalsamy. 2002. Bioremediation of gasoline contaminated soil by bacterial consortium amended with poultry litter, coir pith and rhamnolipid biosurfactant. Biores. Tech. 81: 25–32.

Rendell, N.B., G.W. Taylor, M. Somerville, H. Todd, R. Wilson and J. Cole. 1990. Characterization of *Pseudomonas* rhamnolipids. Biochim. Biophys. Acta 1045: 189–193.

Ristau, E. and F. Wagner. 1993. Formation of novel anionic trehalosetetraesters from *Rhodococcus erythropolis* under growth limiting conditions. Biotechnol. Lett. 5: 95–100.

Robert, M., M.E. Mercade, M.P. Bosch, J.L. Parra, M.J. Espuny, M.A. Manresa and J. Guinea. 1989. Effect of the carbon source on biosurfactant production by *Pseudomonas aeruginosa* 44T. Biotechnol. Lett. 11: 871–874.

Roggiani, M. and D. Dubnau. 1993. ComA, a phosphorylated response regulator protein of *Bacillus subtilis*, binds to the promoter region of *srfA*. J. Bacteriol. 175: 3182–3187.

Romero, D., A. De Vicente and J.L. Olmos. 2007. Effect of lipopeptides of antagonistic strains of *Bacillus subtilis* on the morphology and ultrastructure of the cucurbit fungal pathogen *Podosphaera fusca*. J. Appl. Microbiol. 103: 969–976.

Roongsawang, N., K. Hase, M. Haruki, T. Imanaka and M. Morikawa. 2003. Cloning and characterization of the gene cluster encoding arthrofactin synthetase from *Pseudomonas* sp. MIS38. Chem. Biol. 10: 869–880.

Rosen, M.J. and J.T. Kunjappu. 2012. Characteristic features of surfactants. *In*: Milton, J.T. and J. Rosen (eds.). Surfactants and Interfacial Phenomena (Chapter 1). Hoboken, NJ, USA. John Wiley & Sons, Inc.

Rosenberg, E., A. Zuckerberg, C. Rubinovitz and D.L. Gutinck. 1979. Emulsifier *Arthrobacter* RAG-1: Isolation and emulsifying properties. Appl. Environ. Microbiol. 37: 402–408.

Rosenberg, E. and E.Z. Ron. 1999. High and low molecular mass microbial surfactants. Appl. Microbio. Biotech. 52(2): 154–162.

Shepherd, R., J. Rockey, I.W. Shutherland and S. Roller. 1995. Novel bioemulsifier from microorganisms for use in foods. J. Biotech. 40: 207–217.

Silva, R.C.F.S., A.G. Almeida, J.M. Luna, R.D. Rufino, V.A. Santos and L.A. Sarubbo. 2014. Applications of biosurfactants in the petroleum industry and the remediation of oil spills. Int. J. Mol. Sci. 15: 12523–12542.

Singh, A., J.D. Van Hamme and O.P. Ward. 2007. Surfactants in microbiology and biotechnology. Biotechnol. Adv. 25: 99–121.

Smyth, T.J., A. Perfumo, R. Marchant, I.M. Banat, M. Chen, R.K. Thomas and N.J. Parry. 2010. Directed microbial biosynthesis of deuterated biosurfactants and potential future application to other bioactive molecules. Appl. Microbio. Biotech. 87(4): 1347–1354.

Sullivan, E.R. 1998. Molecular genetics of biosurfactant production. Curr. Opin. Biotechnol. 9: 263–269.

Suzuki, T., K. Hayashi, K. Fujikawa and K. Tsukamoto. 1965. The chemical structure of Polymyxin E. The identies of polymyxin E1 with colistin A and polymyxin E2 with colistin B. J. Biol. Chem. 57: 226–227.

Syldatk, C., S. Lang, F. Wagner, V. Wray and L. Witte. 1985. Chemical and physical characterization of four interfacial-active rhamnolipids from *Pseudomonas* sp. DSM 2874 grown on n-alkanes. Zeitschrift Naturforschung. 40: 51–60.

Tahara, Y., Y. Yamada and K. Kondo. 1976. A new lipid; the ornithine and taurine-containing 'cerilipin'. Agric. Biol. Chem. 40: 243–244.

Toren, A., G. Segal and E.Z. Ron. 2002. Structure—function studies of the recombinant protein Bioemulsifier Aln A. Environ. Microbiol. 4: 257–261.

Tullock, P., A. Hill and J.F.T. Spencer. 1967. A new type of marocyclic lactone from *Torulopsis apicola*. J. Chem. Soc. Chem. Commun. 21: 584–586.

Van Bogaert, I.N., K. Saerens and C. De Muynck. 2007. Microbial production and application of sophorolipids. Appl. Microbiol. Biotechnol. 76: 23–34.

Van Hamme, J.D., A. Singh and O.P. Ward. 2006. Physiological aspect. Part-1 in a series of papers devoted to surfactants in microbiology and biotechnology. Biotechnol. Adv. 24: 604–620.

Van Hoogmoed, C.G., M. Van der Kuijl Booij and H.C. Van der Mei. 2000. Inhibition of *Streptococcus mutans* NS adhesion to glass with and without a salivary conditioning film by biosurfactant-releasing *Streptococcus mitis* strains. Appl. Environ. Microbiol. 66: 659–663.

Vater, J., B. Kablitz and C. Wilde. 2002. Matrix assisted laser desorption ionization-time of flight mass spectrometry of lipopeptide biosurfactants in whole cells and culture filtrates of *Bacillus subtilis* C-1 isolated from petroleum sludge. Appl. Environ. Microbiol. 68: 6210–6219.

Wang, J., J. Liu and X. Wang. 2004. Application of electrospray ionization mass spectrometry in rapid typing of Fengycin homologues produced by *Bacillus subtilis*. Lett. Appl. Microbiol. 39: 98–102.

Wayman, M., A.D. Jenkins and A.G. Kormady. 1984. Biotechnology for oil and fat industry. J. Am. Oil Chem. Soc. 61: 129–131.

Yakimov, M.M., H.L. Fredrickson and K.N. Timmis. 1996. Effect of heterogeneity of hydrophobic moieties on surface activity of Lichenysin A, a lipopeptide biosurfactant from *Bacillus licheniformis* BAS50. Biotechnol. Appl. Biochem. 23: 13–18.

Yamane, T. 1987. Enzyme technology for the lipid industry: an engineering overview. J. Ameri. Oil Chem. Soci. 64(12): 1657–1662.

Zosim, Z., D.L. Guntick and E. Rosenberg. 1982. Properties of hydrocarbon in water emulsion. Biotechnol. Bioeng. 24: 281–292.

Zukerberg, A., A. Diver and Z. Peeri. 1979. Emulsifier of *Arthrobacter* RAG-1: chemical and physical properties. Appl. Environ. Microbiol. 37: 414–420.

13

Promising Strategies for Economical Production of Biosurfactants
The Green Molecules

Rupali Sawant,[1] *Anushka Devale,*[1] *Shilpa Mujumdar,*[1,*]
Karishma Pardesi[2] and *Yogesh Shouche*[3]

1. Introduction

Surfactants are economically significant products due to their variety of industrial applications. Commercially available surfactants are mostly synthesised from petroleum derivatives. Due to extensive application of the surfactants in various sectors there is an increasing concern about their effect on the environment. Therefore, there is a necessity for a search for natural surfactants as an alternative to existing synthetic surfactants. A number of natural compounds with surface active properties are synthesised by living organisms. Compounds of a microbial origin that exhibit surfactant properties (emulsification capacity and a reduction in surface tension) are called biosurfactants (Eshrat Gharaei-Fathabad 2011, Sandeep 2017, Sorbino et al. 2014).

Biosurfactants (BS) and bioemulsifiers (BE) are amphiphilic surface active molecules. The presence of hydrophobic and hydrophilic moieties within these molecules allows them to aggregate at interfaces such as liquid/liquid, gas/liquid

[1] Department of Microbiology, Modern College of Arts, Science and Commerce, Shivajinagar, Pune 411005, Maharashtra, India; Emails: devalemicro@moderncollegepune.edu.in; hodmicro@moderncollegepune .edu.in
[2] Department of Microbiology, Savitribai Phule Pune University, Ganeshkhind, Pune-411007, Maharashtra, India; Email: karishma@unipune.ac.in
[3] National Centre for Cell Science, National Centre for Microbial Resource, Pune 411007, Maharashtra, India; Email: yogesh@nccs.res.in
* Corresponding author: sawantmicro@moderncollegepune.edu.in

or solid/liquid. Such properties are responsible for emulsifying, foaming, cleansing and dispersing characteristics (Makkar et al. 2011). Therefore, these surface active molecules play an important role in various fields like bioremediation, biodegradation, oil recovery, food, pharmaceutics, and many other industrial applications (Banat et al. 2014, Singh et al. 2019). BS and BE are efficient and eco-friendly substitutes to their synthetic counterparts.

Biosurfactants are generally of a low molecular weight microbial products composed of sugars, amino acids (hydrophilic moieties), saturated and unsaturated fatty acids (hydrophobic moieties) and functional groups such as carboxylic acids, e.g., glycolipids and lipopeptides (Sivapathasekaran et al. 2017, Uzoigwe et al. 2015). They are amphiphilic in nature and therefore, can dissolve in both polar and nonpolar solvents. Biosurfactants are known for their distinctive surface activity which involves lowering the surface and interfacial tension between different phases (liquid-air, liquid-liquid, and liquid-solid). They also possesses low critical micelle concentration (CMC) and form stable emulsions. They can act as saturating, foaming and solubilizing agents in different industrial processes (Rahman and Gapke 2008, Uzoigwe et al. 2015).

Bioemulsifiers are high molecular weight biopolymers or exopolysaccharides. These are multifaceted mixtures of heteropolysaccharides, lipopolysaccharides, lipoproteins and proteins (Uzoigwe et al. 2015). Like biosurfactants, these molecules can efficiently emulsify two immiscible liquids such as hydrocarbons or other hydrophobic substrates even at low concentrations but are less effective at surface tension reduction (Sandeep 2017, Shah et al. 2016). They are also involved in the solubilization of poorly-soluble substrates, thus increasing their access and availability for biodegradation. Bioemulsifiers are able to stabilize emulsions and this property has increased their use in numerous industries such as cosmetics, food, pharmaceutical and petroleum (Uzoigwe et al. 2015).

Diverse groups of microorganisms such as bacteria, fungi and yeasts are reported to produce different types of biosurfactants. The type and the quantity of the biosurfactant produced is dependent on the nature of the organism and the media constituents in which the organism is cultivated (Helmy et al. 2011, Singh et al. 2019).

2. Need for Low Cost Substrates

The market value of surfactants is very high and the need of these molecules is ever increasing as they constitute a part of many products used by us in our day-to-day life such as soap and detergents, cosmetics, pharmaceutical products, food and beverages (Banat et al. 2014, Prasad 2015, Singh et al. 2019). The market for biosurfactants is projected to grow at a rate of 4.9% in terms of value, from USD 15.02 billion in 2019 to reach USD 21.84 billion by 2027 (Natural Surfactants Market Analysis, By Product 2020). The market price of synthetic surfactants is around 1–3 USD/ kg, whereas for biosurfactants it is around 2–20 USD/kg, and in the case of some of the purified microbial surfactants the cost even high, e.g., surfactin (98% purity) available from the Sigma Chemical Company costs USD 13.94/mg (Dhanarajan and Sen 2014, Henkel et al. 2012, Randhawa et al. 2014, Singh et al. 2019). Thus, there is a necessity to lower the costs of biosurfactants so as to compete with the synthetic surfactants in the market. With the consideration of the current environmental legislations, in spite of huge market demand, the production of these molecules is limited especially because of the high production cost (Singh et al. 2019). Therefore, there is an urge to explore low cost substrates which will facilitate the large scale economical production of BS/BE to satisfy the current market demand.

3. Potential Low Cost Substrates

Large numbers of economically cost effective substrates, especially waste material from various industries and agricultural sectors, have been explored by researchers as a substrate for cultivating BS/BE producing organisms. Use of these wastes as substrate would also help in addressing the issues related to management of industrial and agricultural waste (Singh et al. 2019). Table 1 summarizes composition of different low cost substrates from the agricultural and food sector. Many cheap substrates from different sectors such as, agriculture, food, oil based industries, dairy and distillery, are being reported to be good sources of carbon and are used for the cultivation of BS/BE producing organisms (Banat et al. 2014). The following substrates can be used as promising substantial media for the microbial production of surface active agents.

3.1 Agriculture and Food Sector

Agricultural and food waste contains high amount of carbohydrates, lipids and therefore, have great potential for use as a carbon source, and for the growth of microbial cells and hence for the production of biosurfactants (Banat et al. 2014). Agricultural waste such as sugar cane and beet molasses, bagasse of sugarcane, wheat straw, rice bran and straws, hull of soy, corn cob, barley bran husk and pulp, coconut husk, sunflower seed shells, jackfruit seed powder, yellow cashew bagasse, potato and orange peel, rice mill polishing residue, cassava flour wastewater, cereals and pulses processed waste water are rich in starch content are also used for the production of surface active agents by the microorganisms (Banat et al. 2014, Henkel et al. 2012, Makkar et al. 2011, Prasad et al. 2015, Rane et al. 2017, Tan and Li 2018, Waghmode et al. 2014). *Pseudomonas aeruginosa* cultivated on waste canola oil as its carbon source and sodium nitrate as the nitrogen source, led to the production of 3585.31 ± 66.24 mg/L of rhamnolipid biosurfactant (Armendáriz et al. 2019). Palm oil an agricultural refinery waste was used as a carbon source for the production of rhamnolipid biosurfactant produced by *Pseudomonas aeruginosa* PAO1. Results demonstrated that *Pseudomonas aeruginosa* PAO1 can produce 430 mg/L of rhamnolipid production (Radzuan et al. 2016). Rhamnolipid was also produced by *Pseudomonas fluorescens* cultivated on whey tofu supplemented with nutrients and salt (Suryanti et al. 2017). Hydrolysed vine trimming shoots were used as a carbon source for the production biosurfactant by *Bacillus tequilensis* ZSB10, resulting in the yield of 1520 mg/L of extracellular biosurfactant (Cortés-camargo et al. 2016). Corn steep liquor was used as a low cost substrate obtained from agricultural waste for the production of biosurfactant. Results demonstrated a higher yield and also a drop in raw material cost per unit of biosurfactant from 47–12 USD/kg (Ebadipour et al. 2015).

Table 1. Composition of different low cost substrates from agricultural and food sector.

Low cost substrates	Carbohydrates	Proteins	Nitrogen	Cellulose	Hemicellulose	Lignin	Miscellaneous	References
Molasses	70	3	0.4–2	-	-	-	8–14% Ash 2 lipids	Banat et al. (2014), Eggleston et al. (2017), Sellami et al. (2016)
Bagasse of sugarcane	-	-	-	50	25	25	-	Hajiha and Sain (2015)
Rice straw and wheat straw	-	-	-	30–45	20–25	15–20	Minor organic compounds	Boschma and Kwant (2013)
Rice bran	-	-	-	15	31	11	Fibre and other components	Sunphorka et al. (2012)
Barley bran	26 Starch	17					3% Ash and Amino acids	Zheng et al. (2011)
Barley husk	11 Starch	4	-	39	12	22	4% Fat	Kohli et al. (2013)
Barley pulp	7 Starch	28	-	30	-	-	10% Oil and other components	Kaskatepe et al. (2017)
Coconut husk	-	-	-	23	12	40	3% Ash	De Farias Silva et al. 2016
Sunflower seed shell/hull	-	5	-	43	16	22	5% Lipids, 3% Minerals	Evon et al. (2007)
Jackfruit seed powder	43	16	-	-	-	-	1% Ash, 4% Fibre and Minerals	Muhammed et al. (2017)
Potato peel	69	8	1.3	-	-	-	6% Ash and 3% Fat	Arapoglou et al. 2010
Orange peel	14	-	-	-	-	6	7% Ash, 2% Fat and 8% Fibre	Ahmed et al. (2015)

Note : (-) Data not available

3.2 Oil and Oil Based Industrial Sector

Vegetable oils are commonly used all over the world for the cooking purpose. Therefore, the accumulation of large volumes of the waste oil is also increasing (Md Badrul Hisham et al. 2019). Appropriate disposal of the waste oil is an global environmental challenge (Banat et al. 2014, Md Badrul Hisham et al. 2019, Pathania and Jana 2019). Regulators have therefore accentuated on the management of the waste oil through their conversion into value added products (Pathania and Jana 2020). The Green alternative for such products is the use of these products as raw material or substrate for the economical production of commercially important bioproducts such as biosurfactants (Banat et al. 2014). Production of rhamnolipids biosurfactants can be achieved by using different vegetable oils such as soybean oil, sunflower oil, rapeseed oil, palm oil, fish oil, coconut oil, olive oil and glycerol. Production can also be done by low cost oil waste substrates like fatty acids, waste frying oil, olive oil production effluents, glycerol containing waste and soap stock (Banat et al. 2014, Henkel et al. 2012, Makkar et al. 2011, Tan and Li 2018). Waste olive oil and sunflower oil, waste from cotton seeds, soyabean, palm oil and corn oil refineries were studied as a substrate for the production of rhamnolipids (Banat et al. 2014, Nogueira et al. 2020, Prasad et al. 2015). A renewable feedstock, i.e., used cooking oil was used for the production of lipopeptide biosurfactant from *Bacillus* sp. HIP3 for removal of heavy metals (Md Badrul Hisham et al. 2019). A bacterial strain *Bacillus pseudomycoides* BS6 isolated from edible oil contaminated soil, produced a novel cyclic lipopeptide biosurfactant with good interfacial characters by using soybean oil waste as the sole source of carbon for energy (Li et al. 2016). Biosurfactant was also produced by *Pseudomonas aeruginosa* from kitchen waste oil, crude oil and waste cooking oil, as a low cost carbon source (Chen et al. 2018, Patowary et al. 2017, Sharma et al. 2015).

3.3 Dairy Sector

Dairy industries produce a variety of sugar (lactose) and protein containing by-products and waste, such as curd and cheese whey, whey wastewater. These by-products and waste can serve as a source of carbon and nitrogen for the growth of biosurfactant producing microorganisms (Banat et al. 2014, Henkel et al. 2012, Prasad et al. 2015, Satpute et al. 2017, Tan and Li 2018). Permeate from whey ultrafiltration composed of lactose, vitamins and minerals was used as a culture medium for the production of biosurfactant surfactin by *Bacillus methylotrophicus* and *Bacillus pumilus*. This study demonstrated the production of good quality biosurfactants along with recovery by membrane separation process (MSP) as an efficient method for purification of biosurfactants (Decesaro et al. 2020). Milk whey was optimized as a medium for the production of biosurfactant from *Penicillium sclerotiorum* UCP1361 (Truan et al. 2020). Whey waste was used as a cost effective medium for the production of biosurfactants from lactic acid bacteria (Alkan et al. 2019). Formulated curd whey used as a medium for the production of biosurfactants using *Pseudomonas aeruginosa* PBS29 produced 3670 ± 20 mg/L of biosurfactant. This study, therefore recommends use of formulated curd whey medium for the large scale production of biosurfactants (Poonguzhali et al. 2018).

3.4 Distillery Sector

Distillers dried grains with solubles (DDGS) is a by-product of alcohol fermentation by using starch rich grains. DDGS is rich in crude protein, fat, fiber, vitamins and minerals. Therefore, in recent years it has been used as a substrate for the production value added products by microbial fermentation (Iram et al. 2020). Rice based Distillers dried grains with solubles (rDDGS) was used as the carbon source for the production of novel rhamnolipid by *Pseudomonas aeruginosa* SS14. This study reported a yield of 14870 mg/L of rhamnolipid (Borah et al. 2019). Distillers' grains (DGS) were used as the sole carbon source by *Bacillus amyloliquefaciens* MT45 for the production of surfactin (Zhi et al. 2017). Distillery wastewater and spent wash can be used for the production of surface active agents especially rhamnolipids (Banat et al. 2014, Henkel et al. 2012, Makkar et al. 2011, Prasad 2015).

3.5 Microorganisms Used for BS Production on Low Cost Media

There are many reports on production of BS on different kinds of low cost media. Mainly rhamnolipids and gycolipids are produced by many organisms when grown on low cost substrates. In the case of bacteria, different species of *Bacillus* and *Pseudomonas* are found dominant in production of BS on low cost media, where as few fungi are also reported to produce BS on these media. Table 2 summarizes different biosurfactants produced by different bacteria as well as fungi.

4. Promising Approaches

In the past few decades developments in the applications of biosurfactants have urged a necessicity to economize the production of biosurfactants in order to compete with the synthetic counterparts. In view of this, various approaches such as, solid state fermentation, the co-production of BS with other bioproducts, use of statistical designs, immobilization technique, use of mutants and nanoparticles, can be used as promising strategies. These approaches are described as follows.

4.1 Solid State Fermentation (SSF)

Solid state fermentation is a technique which uses solid material as substrates with either the absence or near absence of free water (Cerda et al. 2019). Solid state fermentation is gaining importance as a promising strategy for the production of biosurfactants because it involves a low operational cost as compared to the submerged fermentation technique (Cerda et al. 2019, Singh et al. 2019). Many studies had reported the production of various surfactants such as surfactin, sophorolipids, lipopeptides, rhamnolipids by using low cost substrates for SSF (Banat et al. 2021, Cerda et al. 2019, Costa et al. 2018, Rodríguez et al. 2020, Singh et al. 2019). Figure 1 represents different cost effective substances used in SSF.

The use of the statistical approach of RSM-CCD under SSF enhanced rhamnolipid production by *Serratia rubidaea* SNAU02 from mahua oil cake. The rhamnolipid obtained was further tested for it's ability as a biocontrol agent (Nalini and Parthasarathi 2017). SSF carried out using mixed olive cake flour and olive leaf residue flour, as a substrate for the production of lipopeptide biosurfactant from

Table 2. Biosurfactants produced by various groups of microorganisms using low cost substrates.

Microbial group	Biosurfactant type	Low cost substrate used	References
Bacteria			
Bacillus subtilis	Lipopeptide Surfactin Fengycin Iturin	Automobile oil, Cashew apple juice, Cassava flour, Whey, Molasses	Asgher et al. (2020) Verma et al. (2020) Henkel et al. (2012) Modabber et al. (2020) Hu et al. (2019)
Bacillus licheniformis	Lichenysin	Molasses	Coronel-Leon et al. (2015)
Bacillus brevis	-	-	Moua et al. (2016)
Bacillus amyloliquefaciens MT45	Surfactin	Distiller's grains rDDGS	Zhi et al. (2017)
Bacillus velezensis ASN1	Lipopeptide	Waste office paperhydrolysate	Nair et al. (2019)
Bacillus methylotrophicus DCS1	-	Potato starch	Hmidet et al. (2018)
Pseudomonas aeruginosa SS14	Rhamnolipid	Rice based distiller's dried grain solubles (rDDGS)	Borah et al. (2019)
Pseudomonas aeruginosa	Rhamnolipid	Soyabean oil refinery waste, Lignocellulosic biomass	Bagheri et al. (2017), Eslami and Hajfarajollah (2020), Henkel et al. (2012)
Pseudomonas fluorescens	Rhamnolipid	Whey tofu	Suryanti et al. (2017)
Ochrobactrum intermedium	Glycolipid	Hexadecane	Ferhat et al. (2017)
Enterobacter aerogenes *Serratia marcescens* SNAU02	Rhamnolipid Rhamnolipid	Lignocellulosic biomass Mahua oil cake	Arumugam and Shereen (2019) Nalini and Parthasarathi (2017)
Stenotrophomonas acidaminiphila TW3	Glycolipid	Palm oil soapstocks	Onlamool et al. (2020)
Lactic Acid bacteria:			
Weissella cibaria PN3	Glycolipid	Soyabean oil	Subsanguan et al. (2020)
Pediococcus dextrinicus SHU1593	Lipoprotein	-	Ghasemi et al. (2019)
Fungi			
Mucor polymorphosphorus	-	Soyabean waste oil and corn steep liqour	Souza et al. (2016)
Penicillium sclerotiorum UCP1361	-	Barley and milk whey	Truan et al. (2020)

Table 2 Contd. ...

Table 2 Contd. ...

Microbial group	Biosurfactant type	Low cost substrate used	References
Aspergillus niger		Wheat bran and Corn cob	Kreling and Simon (2020)
Yeast			
Candida tropicalis	-	Sugarcane molasses, corn steep liquor and waste fruing oil	Almeida et al. (2017)
Candida lipolytica			Bertrand et al. (2018)
Candida bombicola	Glycolipid	Molasses, corn steep liquor and soyabean waste frying oil	Luna et al. (2016)
Candida guilliermondii	-	Molasses, corn steep liquor and waste frying oil	Sarubbo et al. (2016)
Yarrowia lipolytica	-	-	Bertrand et al. (2018)

Note : (-) Data not available

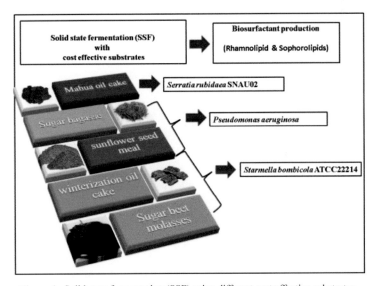

Figure 1. Solid state fermentation (SSF) using different cost effective substrates.

Bacillus subtilis SPB1 yielded 30.67 mg of biosurfactant per gram of the solid substrate (Zouari et al. 2015). SSF was used, as a tool for obtaining the cost effective production of biosurfactant from *Bacillus subtilis* SPB1. With the use of millet as a low cost substrate and by optimizing the production conditions such as temperature, inoculumn size and moisture content, with statistical methods of Plackett-Burman design and response surface methodology, maximum yield of the biosurfactant was obtained. The biosurfactant obtained showed good antimicrobial activity against multi drug resistant microorganisms (Ghribi et al. 2012). Figure 2 represents experimental designing for the solid state fermentation process.

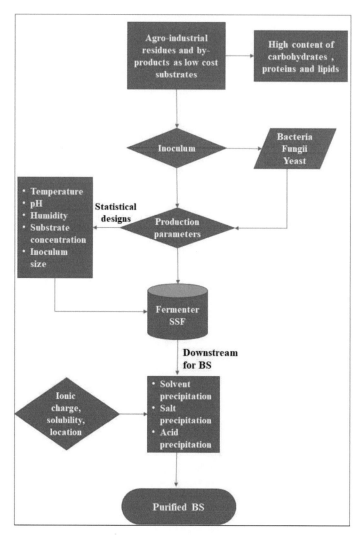

Figure 2. Process and parameters that affect on solid state fermentation (SSF).

4.2 *Co-production of Biosurfactants with Other Bioproducts*

Co-production or concomitant production, is a technique of cultivating microorganisms for obtaining two different bioproducts from the single bioprocess (Singh et al. 2019). The use of expensive substrates, low yield, foaming are some of the challenges in the large scale production of BS (Hmidet et al. 2018). Therefore, co-production of BS with other value added bioproducts using low cost substrates will contribute to enhance the production of BS and reduce the production cost, especially when the products obtained are used in the same industry or have some complementary properties (Hmidet et al. 2018, Makkar et al. 2011, Singh et al. 2019). Various studies of the co-production of biosurfactants with other bioproducts of industrial importance such as lipases, proteases, pectinases and PHA are been reported (Kavuthodi et al.

2015, Kreling and Simon 2020, Makkar et al. 2011, Nicolo et al. 2014, Singh et al. 2019). A recent study has reported co-production of biosurfactant rhamnolipid and PHA using *E. aerogenes* from low cost substrate *C. inophyllum* oil cake as a sole source of carbon (Arumugam and Shereen 2019). Concomitantly produced lipases and biosurfactants using agro industrial raw material such as wheat bran and corn cob, as a culture media for growth of *Aspergillus niger* exhibited bioremediation activity (Kreling and Simon 2020). Co-production of biosurfactant and microbial lipids was successfully done by using *Bacillus velezensis* ASN$_1$ from pre-treated waste office paper as a substrate (Nair et al. 2019). Lipopeptide biosurfactant and alkaline amylase were co-produced by *Bacillus methylotrophicus*, the study reported production of unique amylase with high stability, also the biosurfactant produced exhibited high stability at different physiological conditions (Hmidet et al. 2019). Statisfactory yield of PHA and biosurfactant was achieved in a single integrated fermentation process. Co-production of PHA and biosurfactant was obtained from *Pseudomonas mediterranea,* by using crude glycerol (by-product of biodiesel production) as a substrate. The yield achieved was 1100 mg/L of PHA and 720 mg/L of biosurfactant (Nicolo et al. 2014). Co-production of rhamnolipid and ethanol was achieved by using *Saccharomyces cerevisiae*, *Pseudomonas aeruginosa* and crude enzyme complexes (CECs) from exploded sugarcane bagasse (Lopes et al. 2017). Thus, co-production of biosurfactants with other industrially important bioproducts such as lipases, pectinases and PHA, help in reducing the process economics, thus making the production of the biosurfactant and these bioproducts cost effective.

4.3 *Statistical Designs for Biosurfactant Production*

Statistical designs have proved to be a promising tool for enhancing the production of biosurfactants. The most popular tools used for the optimization studies of BS production are factorial designs and Response Surface Methodology (RSM). By definition *'Factorial design is a type of research methodology that allows for the investigation of the main and interaction effects between two or more independent variables and on one or more outcome variable(s)'* (Nordstokke and Colp 2014). Factorial designs are experimental tools, wherein multiple factors are varied simultaneously or in combination. This helps us to examine the effect of two or more independent variables at the same time. It also helps us to determine the interactions among the variables (Lavrakas 2008). Production of biosurfactant is dependent on the culture media and culture conditions. Use of appropriate and adequate nutrients especially, carbon and nitrogen, influences the biosurfactant production. Therefore, optimization studies on culture media and conditions using experimental designs are essential for enhancing the biosurfactant production. Statistical designs aid in screening variety of carbon nitrogen sources, various culture conditions also provide insights to the significant concentrations and interactions amongst the variables under the study (Bertrand et al. 2018). The factorial designs such as two-level factorial design, the Plackett–Burman Design (PBD), and the Taguchi design are implemented for the biosurfactant production. Most commonly implemented RSM for the studies on enhancing biosurfactant production are the Central Composite Design (CCD) and the Box–Behnken Design (BBD) (Bertrand et al. 2018). In a study, the yield of biosurfactant produced by *Bacillus subtilis* RSL-2, was optimized by RSM-central composite design (CCD) on sugarcane molasses as the sole nutrient source. High yield as 12340 mg/L ± 100 mg/L of biosurfactant showing excellent surface activity as well as thermal and colloidal stability was obtained (Verma et al. 2020). Increase in the yield of glycolipid biosurfactant produced by *Stenotrophomonas acidaminiphila* TW3 from palm oil as the carbon source and $NaNO_3$ as the nitrogen source was achieved by the combination of CCRD and RSM (Onlamool et al. 2020). RSM was used for the optimization of biosurfactant production by using the mutant strain of *Bacillus subtilis* from automobile oil as a low cost substrate (Asgher et al. 2020). Optimization of critical media components for the production of biosurfactant by *Achromobacter xylos* strain GSR21 was successfully achieved by implementation of RSM (Reddy et al. 2018). Central composite rational design (CCRD) was demonstrated to be a suitable tool for the optimization of production conditions for biosurfactants, produced by *Candida tropicalis* UCP0996 using low cost substrates such as sugarcane molasses, corn steep liquor and waste frying oil (Almeida et al. 2017). Increase in the yield of rhamnolipid biosurfactant produced by *Pseudomonas aeruginosa* OG1 was obtained on cost effective substrates such as waste frying oil and chicken feather peptone by using Box-Behnken design (Ozdal and Gurkok 2017). RSM helped in the development of a polynomial model for maximizing the production of biosurfactant produced by *Bacillus brevis* (Moua et al. 2016). Studies demonstrated that CCRD used for study of low cost substrates soybean waste oil and corn steep liquor for the biosurfactant production by *Mucor polymorphosphorus* proved adequate for identifying the optimum production conditions (Souza, P.M. et al. 2016). Cost effective mediums containing molasses was optimized for the production of biosurfactant by RSM using *Bacillus licheniformis*. Production of 3200 mg/L was obtained with medium containing 107.82 gm/L of molasses, 6.47 gm/L of $NaNO_3$ and 9.7 gm/L of K_2HPO_4/KH_2PO_4 (Coronel-Leon et al. 2015). Thus, the use of statistical approaches for biosurfactant production help in determining the relationship and effect of factors under study on biosurfactant production and also result in enhancing the yield of the biosurfactants.

4.4 Advances in Bioprocess Technology such as Media Modifications and Immobilization

4.4.1 Media Modification

The large scale production of biosurfactant with reduced production costs can be achieved by bioprocess optimization. For the production of biosurfactant, culture media and culture conditions play a very significant role. Components of the culture media directly influence the biosurfactant metabolism and the production. Especially, carbon and nitrogen sources, used for cell growth and reproduction also contribute in biosurfactant synthesis, as building blocks. For instance, sugars contribute in formation of polar portions whereas oils contribute in the formation of the non-polar portions of the biosurfactants (Bertrand et al. 2018, Helmy et al. 2011, Singh et al. 2019). Characterization of biosurfactants produced from *Lactobacillus* strains revealed that when sugar cane molasses were used as substrate, glycoprotein biosurfactants were produced whereas when glycerol was used as a substrate glycolipid biosurfactants were produced (Mouafo et al. 2018). In a study effect of various carbon sources such as, glycerol, glucose and waste cooking oil (WCO) was determined on the synthesis of rhamnolipids, while the nitrogen source was kept constant. Characterization of biosurfactants synthesized revealed that glucose and glycerol based biosurfactants were di-substituted rhamnolipids with difference in their hydrophobic moieties, whereas biosurfactants synthesized from WCO was a lipopeptide (Ehinmitola et al. 2018). Amino acids enhance the activity and production of biosurfactants (Bertrand et al. 2018, Helmy et al. 2011). The effect of 19 different amino acids (Histidine, Threonine, Valine, Isoleucine, Leucine, Phenylalanine, Arginine, Proline, Methionine, Tryptophan, Alanine, Glycine, Glutamine-Cysteine, Hydroxyproline, Valine, Asparagine, Proline, Glutamine, Serine, and Glutamic acid, Hydroxyproline, Tyrosine) on biosurfactants production by *Pediococcus acidilactici* F70 was determined by single factor experiment. The result demonstrated an increase in the yield of the biosurfactants with addition of the amino acids. The yield of Glutamine produced biosurfactants was highest in the supernatant, while the yield of Arginine produced biosurfactants was highest on the cell surface biosurfactants, as compared to control groups. Further studies on main amino acids promoting biosurfactants production by Plackett-Burman design revealed that Alanine, Proline and Leucine showed substantial effect on the production of biosurfactants (Dong et al. 2020). Thus, the constituents of the growth medium used for the biosurfactants production play a crucial role in the type and the quantity of the biosurfactant produced (Helmy et al. 2011, Singh et al. 2019).

4.4.2 Immobilization

Immobilization of microbial cells can be defined as, *"the physical confinement or localization of intact cells to a certain region of space; without loss of desired biological activity"* (Willaert 2011). Main challenges in continuous culture and down stream processing of the biosurfactants production are the washout of the cells from reactors, effects of altering reactor conditions and undesirable metabolites on the cells and foam induced by free cells (Singh et al. 2019). Thus, use of immobilized cells or resting cells for the large scale production of the biosurfactants can prove an promising tool for enhancement in the production of the biosurfactant, as the growth and the product formation phases are separated. It will facilitate the downstream process and also the continuity in the bioprocess will be achieved (Helmy et al. 2011, Singh et al. 2019). Lactic acid bacteria strain *Weissella cibaria* PN3, producing glycolipid biosurfactant from 2% soybean oil were immobilized on commercial porous carrier and were reused up to nine cycles for biosurfactants production (Subsanguan et al. 2020). Biosurfactants producing *Pseudomonas aeruginosa* encapsulated in calcium alginate could maintain viability and ability to produce biosurfactants (Dehghannoudeh et al. 2019). Immobilization of biosurfactants producing *Ochrobactum intermedium* was beneficial for recovery of the biosurfactants (Ferhat et al. 2017). Rhamnolipid producing *Pseudomonas aeruginosa* MR01 cultivated on inexpensive soybean oil refinery wastes entrapped by sol-gel technique and using tetraethyl orthosilicate were reused for consecutive fermentation batches and in flow fermentation process (Bagheri et al. 2017). Thus, cell immobilization offers advantages such as, high cell density, high stability, eliminate washout of the cells and increase reaction time.

Immobilization increases the viability and resistance of the cells, also the cells can be reused multiple times (Bagheri et al. 2017, Singh et al. 2019).

4.5 *Use of Bacterial Mutants*

With the view to reduce production cost for biosurfactants, one of the aspects that can be exploited is the selection, adaption and engineering of the microbial cells for obtaining higher yields (Helmy et al. 2011). Exploring bioengineering by the use of strategies such as genetic engineering and the random mutation for strain improvement, it is experimentally proved by the researchers that an increase in the production and enhancement in the activity of the biosurfactant can be achieved (Asgher et al. 2020, Bouassida et al. 2018, Eslami and Hajfarajollah 2020). Genetic engineering has been extensively used as a tool for generating genetically modified strains (Eslami and Hajfarajollah 2020). Amongst the rhamnolipid (RL) producing strains, *Pseudomonas aeruginosa* has been reported to produce the highest yield of RL's. *Pseudomonas aeruginosa* was therefore, used as a model by researchers to obtain knowledge of the genes involved in the biosynthesis of RL's (Chong and Li 2017). It has been reported that 3-(Hydroxyalkanoyloxy)alkanoic acid (HAA; synthesized from β-hydroxydecanoyl-ACP) and dTDP-l-rhamnose are two important precursors for biosynthesis of rhamnolipid (Chong and Li 2017, Eslami and Hajfarajollah 2020). Though many bacteria are capable of synthesizing β-hydroxydecanoyl-ACP and dTDP-lrhamnose, the enzymes necessary to synthesize HAA and mono and di-rhamnolipids are solely produced by *Pseudomonas* sp. and *Burkholderia* sp. (Chong and Li 2017). Therefore, it is possible to produce non-pathogenic recombinant strains, by introducing genes necessary for RL synthesis in these organisms by metabolic engineering, thereby enhancing the production of biosurfactant rhamnolipid (Fig. 3). Another strategy which is widely used is random

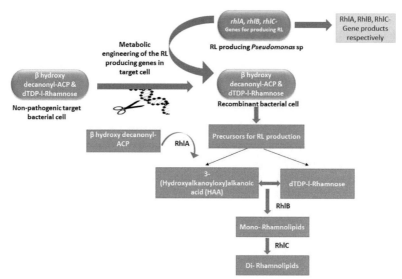

Figure 3. Synthesis of rhamnolipid from recombinant non-pathogenic strains generated by metabolic engineering.

mutagenesis using chemical mutagenic agents and radiations (Chong and Li 2017, Eslami and Hajfarajollah 2020).

Synthesis of lipopeptide biosurfactant was effectively improved after random chemical mutagenesis in *Bacillus subtilis* strain as compared to the wild type. Chemical mutagenesis was performed by exposing the strain to ethidium bromide. Results showed that higher yield was obtained from the mutant (4200 mg/L) than the wild type (yield obtained for the wild type was 2100 mg/L). Lipopeptide obtained from mutant also exhibited improved emulsification and oil displacement activity. Biosurfactant production by the mutant was further optimized by RSM and the results demonstrated that maximum biosurfactant was produced at pH 7, temperature 35°C, inoculum size 3 mL and at incubation period of 120 h (Asgher et al. 2020). In a study biosurfactants producing a strain of *Bacillus subtilis* was subjected to random mutagenesis by using a combination of UV irradiation and nitrous acid treatment. The mutant obtained produced higher concentration of biosurfactant as compared to the wild type (Bouassida et al. 2018).

4.6 Use of Nanoparticles

Use of nanoparticles for enhancing biosurfactants production is yet another promising strategy (Singh et al. 2019). Research work in this aspect has demonstrated that, especially metal nanoparticles such as Fe added in the culture media, have promoted increased production of the biosurfactant (Modabber et al. 2020, Singh et al. 2018). Figure 4 represents the experimental planning of iron nanoparticles on production of biosurfactant (Modabber et al. 2020). Addition of metal nanoparticles in culture media probably fulfill the necessity of required metal ion, thereby promoting growth and enhancing biosurfactant production (Helmy et al. 2011, Singh et al. 2019). However the scarcity of literature limits the discussion of this approach.

5. Conclusions

The global market for the surfactants is ever increasing because of their various applications in diverse industrial sectors. In the past few decades, biosurfactants have emerged as a promising eco-friendly alternative to synthetic surfactants due to characteristics such as non-toxicity, biodegradability, biocompatibility, efficiency at low concentrations unlike synthetic surfactants. High production cost of biosurfactants can be lowered by using cost effective substrates from various agricultural and industrial sectors, especially biological waste generated from these sectors such as molasses. The use of low cost substrates reduces the production cost thereby making the biosurfactants competent in the global market as against the synthetic surfactants. Advances in biosurfactant production technology such as solid state fermentation and co-production of biosurfactants with other industrially important bioproducts such as lipases, pectinases and PHA, help in reducing the process economics, thus making the production cost effective. The use of statistical approaches for biosurfactant production helps in determining the relationship and effect of factors under study on biosurfactants production and also results in enhancing the yield of the biosurfactants. Immobilization of biosurfactant producing cells facilitate the reuse of the biomass and also eases the downstream processing

**Step 1: Isolation and Identification of
Biosurfactant Producing
Microorganisms**

Step 2: Preparation of iron nanoparticles

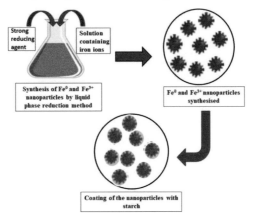

Step 3: Effect of the iron nanoparticles on surfactin production

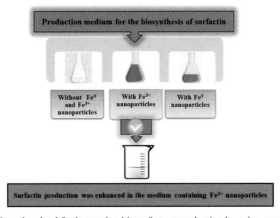

Figure 4. Steps involved for improving biosurfactant production by using nanoparticles.

for the product recovery, making the process economical. Thus, use of various low cost substrates and technologies such as solid state fermentation, immobilization of cells, modification in media constituents, strain improvement, use of nanoparticles, co-production of biosurfactants along with other economically important bioproducts coupled with statistical designs such as factorial designs and RSM can emerge as promising strategies for sustaining the global demand of biosurfactants.

References

Ahmed, I., M. Anjum, M. Azhar, Z. Akram and M. Tahir. 2015. Bioprocessing of citrus waste peel for induced pectinase production by *Aspergillus niger*; its purification and characterization. Journal of Radiation Research and Applied Sciences 3–9. https://doi.org/10.1016/j.jrras.2015.11.003.

Alkan, Z., Z. Ergİnkaya, G. Konuray and E.Ü. Turhan. 2019. Production of biosurfactant by lactic acid bacteria using whey as growth medium. Turkish Journal of Veterinary and Animal Sciences 676–683. https://doi.org/10.3906/vet-1903-48.

Almeida, D.G., R.C. Soares da Silva, J.M. Luna, R.D. Rufino, V.A. Santos and L.A. Sarubbo. 2017. Response Surface Methodology for Optimizing the Production of Biosurfactant by Candida tropicalis on Industrial Waste Substrates. Frontiers in Microbiology 8: 157. https://doi.org/10.3389/fmicb.2017.00157.

Arapoglou, D., T. Varzakas, A. Vlyssides and C. Israilides. 2010. Ethanol production from potato peel waste (PPW). Waste Management (New York, N.Y.) 30: 1898–1902. https://doi.org/10.1016/j.wasman.2010.04.017.

Armendáriz, B.P., C. Cal, E. Girgis, E. Kassis, L. Daniel and O. Martínez. 2019. Use of waste canola oil as a low-cost substrate for rhamnolipid production using *Pseudomonas aeruginosa*. AMB Express. https://doi.org/10.1186/s13568-019-0784-7.

Arumugam, A. and M.F. Shereen. 2019. Bioresource technology bioconversion of *Calophyllum inophyllum* oilcake for intensification of rhamnolipid and polyhydroxyalkanoates co-production by *Enterobacter aerogenes*. Bioresource Technology 296: 122321. https://doi.org/10.1016/j.biortech.2019.122321.

Asgher, M., M. Afzal, S. Ahmad and Q. Nimrah. 2020. Optimization of biosurfactant production from chemically mutated strain of *Bacillus subtilis* using waste automobile oil as low-cost substrate. Environmental Sustainability (0123456789). https://doi.org/10.1007/s42398-020-00127-9.

Bagheri Lotfabad, T., N. Ebadipour, R. Roostaazad, M. Partovi and M. Bahmaei. 2017. Two schemes for production of biosurfactant from *Pseudomonas aeruginosa* MR01: Applying residues from soybean oil industry and silica sol–gel immobilized cells. Colloids and Surfaces B: Biointerfaces 152: 159–168. https://doi.org/10.1016/j.colsurfb.2017.01.024.

Banat, I.M., S.K. Satpute, S.S. Cameotra, R. Patil and N.V. Nyayanit. 2014. Cost effective technologies and renewable substrates for biosurfactants' production. Frontiers in Microbiology. https://doi.org/10.3389/fmicb.2014.00697.

Banat, I.M., Q. Carboué, G. Saucedo-Castañeda and J. de Jesús Cázares-Marinero. 2021. Biosurfactants: The green generation of speciality chemicals and potential production using Solid-State fermentation (SSF) technology. Bioresource Technology 320: 124222. https://doi.org/https://doi.org/10.1016/j.biortech.2020.124222.

Bertrand, B., F. Mart, N. Sarela, D. Morales-guzm and R. Trejo-hern. 2018. Statistical design, a powerful tool for optimizing biosurfactant production: A review. Colloids Interfaces (8): 1–18. https://doi.org/10.3390/colloids2030036.

Borah, S.N., S. Sen, L. Goswami, A. Bora, K. Pakshirajan and S. Deka. 2019. Rice based distillers dried grains with solubles as a low cost substrate for the production of a novel rhamnolipid biosurfactant having anti-biofilm activity against *Candida tropicalis*. Colloids and Surfaces B: Biointerfaces 182: 110358. https://doi.org/https://doi.org/10.1016/j.colsurfb.2019.110358.

Boschma, S. and W.K. Kwant. 2013. Rice straw and wheat straw: potential feedstocks for the biobased economy. NL Agency Ministry of Economic Affairs, 6–30. Retrieved from http://english.rvo.nl/sites/default/files/2013/12/Straw report AgNL June 2013.pdf%0Awww.wageningenur.nl/fbr.

Bouassida, M., I. Ghazala, S. Ellouze-Chaabouni and D. Ghribi. 2018. Improved biosurfactant production by *Bacillus subtilis* SPB1 mutant obtained by random mutagenesis and its application in enhanced oil recovery in a sand system. Journal of Microbiology and Biotechnology 28(1): 95–104. https://doi.org/10.4014/jmb.1701.01033.

Cerda, A., A. Artola, R. Barrena, X. Font, T. Gea and A. Sánchez. 2019. Innovative production of bioproducts from organic waste through solid-state fermentation. Frontiers in Sustainable Food Systems 3(August): 1–6. https://doi.org/10.3389/fsufs.2019.00063.

Chen, C., N. Sun, D. Li, S. Long, X. Tang, G. Xiao and L. Wang. 2018. Optimization and characterization of biosurfactant production from kitchen waste oil using *Pseudomonas aeruginosa*. Environmental Science and Pollution Research 25(15): 14934–14943. https://doi.org/10.1007/s11356-018-1691-1.

Chong, H. and Q. Li. 2017. Microbial production of rhamnolipids: Opportunities, challenges and strategies. Microbial Cell Factories. https://doi.org/10.1186/s12934-017-0753-2.

Coronel-Leon, A.M., J. Marques and A.M. Bastida. 2015. Optimizing the production of the biosurfactant lichenysin and its application in biofilm control. Journal of Applied Microbiology 99–111. https://doi.org/10.1111/jam.12992.

Cortés-camargo, S., N. Pérez-rodríguez, R. Pinheiro, D.S. Oliveira, B.E. Barragán and J. Manuel. 2016. Production of biosurfactants from vine-trimming shoots using the halotolerant strain *Bacillus tequilensis* ZSB10. Industrial Crops and Products 79: 258–266.

Costa, J.A.V., H. Treichel, L.O. Santos and V.G. Martins. 2018. Solid-state fermentation for the production of biosurfactants and their applications. Chapter-16. pp. 357–372. *In*: Pandey, A., C. Larroche, C.R.B.T.-C.D. and B. Soccol (eds.). Current Developments in Biotechnology and Bioengineering. Elsevier. https://doi.org/https://doi.org/10.1016/B978-0-444-63990-5.00016-5.

De Farias Silva, C., M. Cabral, A.K. Abud and R. Almeida. 2016. Bioethanol production from coconut husk fiber. Ciência Rural 46. https://doi.org/10.1590/0103-8478cr20151331.

Decesaro, A., T. Strieder, M. Ângela, C. Cappellaro, A. Rempel, D. Zampieri and L. Maria. 2020. Biosurfactants production using permeate from whey ultrafiltration and bioproduct recovery by membrane separation process. Journal of Surfactants and Detergents. https://doi.org/10.1002/jsde.12399.

Dehghannoudeh, G., K. Kiani, M.H. Moshafi, N. Dehghannoudeh, M. Rajaee, S. Salarpour and M. Ohadi. 2019. Optimizing the immobilization of biosurfactant-producing *Pseudomonas aeruginosa* in alginate beads. Journal of Pharmacy and Pharmacognosy Research 7(6): 413–420.

Dhanarajan, G. and R. Sen. 2014. Cost analysis of biosurfactant production from a scientist's perspective. pp. 153–162. *In*: Biosurfactants. https://doi.org/10.1201/b17599-12.

Dong, Y., G. Shu, C. Dai, M. Zhang and H. Wan. 2020. Effect of amino acids on the production of biosurfactant by *Pediococcusacidilactici* F70. Acta Universitatis Cibiniensis. Series E: Food Technology 24(1): 129–138. https://doi.org/10.2478/aucft-2020-0011.

Ebadipour, N., T.B. Lotfabad, S. Yaghmaei and R. Roostaazad. 2015. Optimization of low-cost biosurfactant production from agricultural residues through the response surface methodology. Preparative Biochemistry and Biotechnology (April 2015): 37–41. https://doi.org/10.1080/10826068.2014.979204.

Eggleston, G., B. Legendre and M. Godshall. 2017. Sugar and other sweeteners. pp. 933–978. https://doi.org/10.1007/978-3-319-52287-6_15.

Ehinmitola, E.O., E.F. Aransiola and O.P. Adeagbo. 2018. Comparative study of various carbon sources on rhamnolipid production. South African Journal of Chemical Engineering 26: 42–48. https://doi.org/https://doi.org/10.1016/j.sajce.2018.09.001.

Eslami, P. and H. Hajfarajollah. 2020. Recent advancements in the production of rhamnolipid biosurfactants by *Pseudomonas aeruginosa*. RSV Advances 10: 34014–34032. https://doi.org/10.1039/d0ra04953k.

Evon, P., V. Vandenbossche, P.-Y. Pontalier and L. Rigal. 2007. Direct extraction of oil from sunflower seeds by twin-screw extruder according to an aqueous extraction process: Feasibility study and influence of operating conditions. Industrial Crops and Products 26: 351–359. https://doi.org/10.1016/j.indcrop.2007.05.001.

Ferhat, S., R. Alouaoui, A. Badis and N. Moulai-mostefa. 2017. Production and characterization of biosurfactant by free and immobilized cells from *Ochrobactrum intermedium* isolated from the soil of southern Algeria with a view to environmental application. Biotechnology & Biotechnological Equipment, 2818. https://doi.org/10.1080/13102818.2017.1309992.

Gharaei-Fathabad, E. 2011. Biosurfactants in pharmaceutical industry (A Mini-Review). American Journal of Drug Discovery and Development 1: 58–69. DOI: 10.3923/ajdd.2011.58.69.

Ghasemi, A., M. Moosavi-nasab, P. Setoodeh and G. Mesbahi. 2019. Biosurfactant production by lactic acid bacterium *Pediococcus dextrinicus* SHU1593 grown on different carbon sources: strain screening followed by product characterization. Scientific Reports 1–12. https://doi.org/10.1038/s41598-019-41589-0.

Ghribi, D., L. Abdelkefi-Mesrati, I. Mnif, R. Kammoun, I. Ayadi, I. Saadaoui and S. Chaabouni-Ellouze. 2012. Investigation of antimicrobial activity and statistical optimization of *Bacillus subtilis* SPB1 biosurfactant production in solid-state fermentation. Journal of Biomedicine and Biotechnology 2012: 373682. https://doi.org/10.1155/2012/373682.

Hajiha, H. and M. Sain. 2015. 17 – The use of sugarcane bagasse fibres as reinforcements in composites. pp. 525–549. *In*: Faruk, O. and M.B.T.-B.R. in C.M. Sain (eds.). Biofiber Reinforcements in Composite Materials. Woodhead Publishing, 2015. doi: 10.1533/9781782421276.4.525.

Helmy, Q., E. Kardena, N. Funamizu and Wisjnuprapto. 2011. Strategies toward commercial scale of biosurfactant production as potential substitute for it's chemically counterparts. International Journal of Biotechnology 12(1–2): 66–86. https://doi.org/10.1504/IJBT.2011.042682.

Henkel, M., M.M. Müller, J.H. Kügler, R.B. Lovaglio, J. Contiero, C. Syldatk and R. Hausmann. 2012. Rhamnolipids as biosurfactants from renewable resources: concepts for next-generation rhamnolipid production. Process Biochemistry 47(8): 1207–1219. https://doi.org/https://doi.org/10.1016/j.procbio.2012.04.018.

Hmidet, N., N. Jemil and M. Nasri. 2018. Simultaneous production of alkaline amylase and biosurfactant by *Bacillus methylotrophicus* DCS1: application as detergent additive. Biodegradation 8. https://doi.org/10.1007/s10532-018-9847-8.

Hu, F., Y. Liu and S. Li. 2019. Rational strain improvement for surfactin production: enhancing the yield and generating novel structures. Microbial Cell Factories (30): 1–13. https://doi.org/10.1186/s12934-019-1089-x.

Iram, A., D. Cekmecelioglu and A. Demirci. 2020. Distillers' dried grains with solubles (DDGS) and its potential as fermentation feedstock. Applied Microbiology and Biotechnology 104: 6115–6128. https://doi.org/10.1007/s00253-020-10682-0.

Kaskatepe, B., S. Yildiz, M. Gumustas and S. Ozkan. 2017. Rhamnolipid Production by Pseudomonas putida IBS036 and Pseudomonas pachastrellae LOS20 with Using Pulps. Current Pharmaceutical Analysis 13: 138–144. https://doi.org/10.2174/1573412912666161018144635.

Kavuthodi, B., S. Thomas and D. Sebastian. 2015. Co-production of pectinase and biosurfactant by the newly isolated strain *Bacillus subtilis* BKDS1. British Microbiology Research Journal 10(2): 1–12. https://doi.org/10.9734/bmrj/2015/19627.

Kreling, N.E. and V. Simon. 2020. Simultaneous production of lipases and biosurfactants in solid-state fermentation and use in bioremediation. Journal of Environmental Engineering 146(9): 1–10. https://doi.org/10.1061/(ASCE)EE.1943-7870.0001785.

Kohli, D., S. Garg and A.K. Jana. 2013. Thermal and morphological properties of chemically treated barley husk fiber. International Journal of Research in Mechanical Engineering & Technology 5762: 153–156. http://www.ijrmet.com/vol32/deepak.pdf.

Lavrakas, P.J. 2008. Encyclopedia of survey research methods (Vols. 1-0). Thousand Oaks, CA: Sage Publications, Inc. doi: 10.4135/9781412963947.

Li, J., M. Deng, Y. Wang and W. Chen. 2016. International biodeterioration & biodegradation production and characteristics of biosurfactant produced by *Bacillus pseudomycoides* BS6 utilizing soybean oil waste. International Biodeterioration & Biodegradation 112: S37–S62. https://doi.org/10.1016/j.ibiod.2016.05.002.

Lopes, V.D.S., J. Fischer, T.M.A. Pinheiro, B.V. Cabral, V.L. Cardoso and U. Coutinho Filho. 2017. Biosurfactant and ethanol co-production using *Pseudomonas aeruginosa* and *Saccharomyces cerevisiae* co-cultures and exploded sugarcane bagasse. Renewable Energy 109: 305–310. https://EconPapers.repec.org/RePEc:eee:renene:v:109:y:2017:i:c:p:305-310.

Luna, J.M., A. Santos Filho, R.D. Rufino and L.A. Sarubbo. 2016. Production of a biosurfactant from *Candida bombicolaurm* 3718 for environmental applications. Chemical Engineering Transactions 49: 583–588. DOI: 10.3303/CET1649098.

Makkar, R. ., S.S. Cameotra and I.M. Banat. 2011. Advances in utilization of renewable substrates for biosurfactant production. AMB Express 1(1): 1–19. https://doi.org/10.1186/2191-0855-1-5.

Md Badrul Hisham, N.H., M.F. Ibrahim, N. Ramli and S. Abd-Aziz. 2019. Production of biosurfactant produced from used cooking oil by *Bacillus* sp. HIP3 for heavy metals removal. Molecules 24(14). https://doi.org/10.3390/molecules24142617.

Modabber, G., A. Akhavan Sepahi, F. Yazdian and H. Rashedi. 2020. Surfactin production in the bioreactor: Emphasis on magnetic nanoparticles application. Engineering in Life Sciences 20(8). https://doi.org/10.1002/elsc.201900163.

Moua, F.E., M.M. Abo and M.E. Moharam. 2016. Optimization of biosurfactant production by *Bacillus brevis* using response surface methodology. Biotechnology Reports 9: 31–37. https://doi.org/10.1016/j.btre.2015.12.003.

Mouafo, T.H., A. Mbawala and R. Ndjouenkeu. 2018. Effect of different carbon sources on biosurfactants production by three strains of *Lactobacillus* spp. BioMed Research International 2018: 5034783. https://doi.org/10.1155/2018/5034783.

Muhammed, Y., M. Miah, S. Bhattacharjee, A. Sultana, S. Bhowmik, A. Sarker and A. Zaman. 2017. Evaluation of amino acid profile of jackfruit (*Artocarpus heterophyllus*) seed and its utilization for development of protein enriched supplementary food. Journal of Noakhali Science and Technology University 1: 77–84.

Nair, A.S., S. Al-bahry and N. Sivakumar. 2019. Co-production of microbial lipids and biosurfactant from waste office paper hydrolysate using a novel strain *Bacillus velezensis* ASN1. Biomass Conversion and Biorefinery. https://doi.org/10.1007/s13399-019-00420-6.

Nalini, S. and R. Parthasarathi. 2017. Optimization of rhamnolipid biosurfactant production from *Serratia rubidaea* SNAU02 under solid-state fermentation and its biocontrol efficacy against Fusarium wilt of eggplant. Annals of Agrarian Sciences. https://doi.org/10.1016/j.aasci.2017.11.002.

Natural Surfactants Market Analysis, By Product (Anionic, Nonionic, Cationic, Amphoteric), application (Detergent, Personal Care, Industrial and Institutional Cleaning, Oilfield Chemicals, Agricultural Chemicals), forecasts To 2027, Reports and Data, Category: Solvents and Surfactants • ID: RND_001934 • Format: PDF • Publish Date: July, 2020.

Nicolò, M.S., D. Franco, V. Camarda, R. Gullace, M.G. Rizzo, M. Fragala, G. Licciardello, A. Catara and S.P.P. Guglielmino. 2014. Integrated microbial process for bioconversion of crude glycerol from biodiesel into biosurfactants and PHAs. Chemical Engineering Transactions 38: 187–192. DOI: 10.3303/CET1438032.

Nogueira, I.B., D.M. Rodríguez, R.F. da Silva Andradade, A.B. Lins, A.P. Bione, I.G.S. da Silva and G.M. de Campos-Takaki. 2020. Bioconversion of agroindustrial waste in the production of bioemulsifier by *Stenotrophomonas maltophilia* UCP 1601 and application in bioremediation process. International Journal of Chemical Engineering 2020: 9434059. https://doi.org/10.1155/2020/9434059.

Nordstokke, D. and S.M. Colp. 2014. Factorial design. *In*: Michalos, A.C. (eds.). Encyclopedia of Quality of Life and Well-Being Research. Springer, Dordrecht. https://doi.org/10.1007/978-94-007-0753-5_982.

Onlamool, T., A. Saimmai, N. Meeboon and S. Maneerat. 2020. Enhancement of glycolipid production by *Stenotrophomonas acidaminiphila* TW3 cultivated in a low cost substrate. Biocatalysis and Agricultural Biotechnology 26(May): 101628. https://doi.org/10.1016/j.bcab.2020.101628.

Ozdal, M. and S. Gurkok. 2017. Optimization of rhamnolipid production by *Pseudomonas aeruginosa* OG1 using waste frying oil and chicken feather peptone. 3 Biotech 1–8. https://doi.org/10.1007/s13205-017-0774-x.

Pathania, A.S. and A.K. Jana. 2020. Improvement in production of rhamnolipids using fried oil with hydrophilic co-substrate by indigenous *Pseudomonas aeruginosa* NJ2 and characterizations. Applied Biochemistry and Biotechnology 191(3): 1223–1246. https://doi.org/10.1007/s12010-019-03221-9.

Patowary, K., R. Patowary, M.C. Kalita and S. Deka. 2017. Characterization of biosurfactant produced during degradation of hydrocarbons using crude oil as sole source of carbon. Frontiers in Microbiology 8: 1–14. https://doi.org/10.3389/fmicb.2017.00279.

Poonguzhali, P., S. Rajan and R. Parthasarathi. 2018. Utilization of potato peel and curd whey as the agro industrial substrates for biosurfactant production using *Pseudomonas aeruginosa* PBS29. Journal of Pharmacy 8(6): 54–62.

Prasad, B., H.P. Kaur and S. Kaur. 2015. Potential biomedical and pharmaceutical applications of microbial surfactants. World Journal of Pharmacy and Pharmaceutical Sciences 4(04): 1557–1575.

Radzuan, M.N., I. Banat and J. Winterburn. 2016. Production and characterization of rhamnolipid using palm oil agricultural refinery waste. Bioresource Technology. https://doi.org/10.1016/j.biortech.2016.11.052.

Rahman, P.K.S.M. and E. Gakpe. 2008. Production, characterisation and applications of biosurfactants—A review. Biotechnology 7: 360–370. DOI: 10.3923/biotech.2008.360.370.

Randhawa, K.K.S. and P.K.S.M. Rahman. 2014. Rhamnolipid biosurfactants—past, present, and future scenario of global market. Frontiers in Microbiology 5(SEP): 1–7. https://doi.org/10.3389/fmicb.2014.00454.

Rane, A.N., V.V. Baikar, D.V. Ravi Kumar and R.L. Deopurkar. 2017. Agro-industrial wastes for production of biosurfactant by *Bacillus subtilis* ANR 88 and its application in synthesis of silver and gold nanoparticles. Frontiers in Microbiology 8: 1–12. https://doi.org/10.3389/fmicb.2017.00492.

Reddy, G.S., K. Srinivasulu and B. Mahendran. 2018. Statistical optimization of medium components for biosurfactant production by. International Journal of Green Pharmacy 2018(4): 815–821.

Rodríguez, A., T. Gea, A. Sánchez and X. Font. 2020. Agro-wastes and inert materials as supports for the production of biosurfactants by solid-state fermentation. Waste Biomass Valor. https://doi.org/10.1007/s12649-020-01148-5.

Sandeep, L. 2017. Biosurfactant: Pharmaceutical perspective. Journal of Analytical & Pharmaceutical Research 4(3): 19–21. https://doi.org/10.15406/japlr.2017.04.00105.

Sarubbo, L.A., J.M. Luna, R.D. Rufino and P. Brasileiro. 2016. Production of a low-cost biosurfactant for application in the remediation of sea water contaminated with petroleum derivates. Chemical Engineering Transactions 49: 523–528. DOI: 10.3303/CET1649088.

Satpute, S.K., G.A. Płaza and A.G. Banpurkar. 2017. Biosurfactants reduction from renewable natural resources: Example of innovative and smart technology. Management Systems in Production Engineering 25(1): 46–54.

Sellami, M., F. Frikha, N. Miled, B. Lassaad and F. Ben Rebah. 2016. Agro-industrial waste based growth media optimization for biosurfactant production by *Aneurinibacillus migulanus*. Journal of Microbiology, Biotechnology and Food Sciences 05: 578–583. https://doi.org/10.15414/jmbfs.2016.5.6.578-583.

Shah, N., R. Nikam, S. Gaikwad, V. Sapre and J. Kaur. 2016. Biosurfactant: Types, detection methods, importance and applications. Indian Journal of Microbiology Research 3(1): 5. https://doi.org/10.5958/2394-5478.2016.00002.9.

Sharma, D., B.S. Saharan, N. Chauhan, S. Procha and S. Lal. 2015. Isolation and functional characterization of novel biosurfactant produced by *Enterococcus faecium*. SpringerPlus 4(1). https://doi.org/10.1186/2193-1801-4-4.

Singh, P., Y. Patil and V. Rale. 2018. Biosurfactant production: emerging trends and promising strategies. Journal of Applied Microbiology 126(1): 2–13.

Sivapathasekaran, C. and R. Sen. 2017. Origin, properties, production and purification of microbial surfactants as molecules with immense commercial potential. 54: 92–107.

Sobrinho, H., J. Luna, R. Rufino, A. Porto and L. Sarubbo. 2014. Biosurfactants: classification, properties and environmental applications. pp. 303–330. *In*: Govil, J.N. (ed.). Recent Developments in Biotechnology. Studium Press LLC, USA.

Souza, A.F., D.M. Rodriguez, D.R. Ribeaux, M.A.C. Luna, T.A. Lima, R.F.S. Andrade and G.M. Campos-takaki. 2016. Waste soybean oil and corn steep liquor as economic substrates for bioemulsifier and biodiesel production by *Candida lipolytica* UCP 0998. International Journal of Molecular Sciences 1–18. https://doi.org/10.3390/ijms17101608.

Souza, P.M., M. Freitas-silva, T. Alves, D. Lima, G.K.B. Silva, M.A.B. Lima and G.M. Campos-takaki. 2016. Factorial design based medium optimization for the improved production of biosurfactant by *Mucor polymorphosphorus*. International Journal of Current Microbiology and Applied Sciences 5(11): 898–905. http://dx.doi.org/10.20546/ijcmas.2016.511.103.

Subsanguan, T., N. Khondee and P. Nawavimarn. 2020. Reuse of immobilized *Weissella cibaria* PN3 for long-term production of both extracellular and biosurfactants, cell-bound glycolipid. Frontiers in Bioengineering and Biotechnology 8(July): 1–14. https://doi.org/10.3389/fbioe.2020.00751.

Sunphorka, S., W. Chavasiri, Y. Oshima and S. Ngamprasertsith. 2012. Protein and sugar extraction from rice bran and de-oiled rice bran using subcritical water in a semi-continuous reactor: optimization by response surface methodology. International Journal of Food Engineering 8. https://doi.org/10.1515/1556-3758.2262.

Suryanti, V., D.S. Handayani, S.D. Marliyana and S. Suratmi. 2017. Physicochemical properties of biosurfactant produced by *Pseudomonas fluorescens* grown on whey tofu. IOP Conference Series: Materials Science and Engineering 176: 12003. https://doi.org/10.1088/1757-899x/176/1/012003.

Tan, Y.N. and Q. Li. 2018. Microbial production of rhamnolipids using sugars as carbon sources. Microbial Cell Factories 1–13. https://doi.org/10.1186/s12934-018-0938-3.

Truan, L., N. Marques, A. Souza, D. Rubio-Ribeaux, A. Cine, R. Andrade, T. Silva, K. Okada and G. Takaki. 2020. Sustainable biotransformation of barley and milk whey for biosufactant production by *Penicillium Sclerotiorum* Ucp 1361. Chemical Engineering Transactions 79: 259–264. DOI:10.3303/CET2079044.

Uzoigwe, C., J.G. Burgess, C.J. Ennis and P.K.S.M. Rahman. 2015. Bioemulsifiers are not biosurfactants and require different screening approaches. Frontiers in Microbiology. https://doi.org/10.3389/fmicb.2015.00245.

Verma, R., S. Sharma, L. Mohan and L.M. Pandey. 2020. Experimental investigation of molasses as a sole nutrient for the production of an alternative metabolite biosurfactant. Journal of Water Process Engineering 38: 101632. https://doi.org/10.1016/j.jwpe.2020.101632.

Waghmode, S., C. Kulkarni. S. Shukla, P. Sursawant and C. Velhal. 2014. Low cost production of biosurfactant from different substrates and their comparative study with commercially available chemical surfactant. International Journal of Scientific & Technology Research 3: 146–14.

Willaert, R. 2011. Cell immobilization and its applications in biotechnology. pp. 313–367. https://doi.org/10.1201/b11490-13.

Zheng, X., L. Li and X. Wang. 2011. Molecular characterization of arabinoxylans from hull-less barley milling fractions. Molecules (Basel, Switzerland) 16: 2743–2753. https://doi.org/10.3390/molecules16042743.

Zhi, Y., Q. Wu and Y. Xu. 2017. Production of surfactin from waste distillers grains by co-culture fermentation of two *Bacillus amyloliquefaciens* strains. Bioresource Technology 235: 96–103. https://doi.org/https://doi.org/10.1016/j.biortech.2017.03.090.

Zouari, R., S. Ellouze-Chaabouni and D. Ghribi-aydi. 2015. Achievements in the life sciences optimization of *Bacillus subtilis* SPB1 biosurfactant production under solid-state fermentation using by-products of a traditional olive mill factory. ALS 8(2): 162–169. https://doi.org/10.1016/j.als.2015.04.007.

14

Biosurfactant Mediated Synthesis of Nanoparticles and their Applications

A F El-Baz,[1,]* R A El Kassas,[2] Shetaia, Yousseria M[2] and H A El Enshasy[3,4]

1. Introduction

Nanotechnology is the branch of science that evaluates and manages the atomic, molecular, and supramolecular scale of material. Although nanoscales have been utilized for hundreds of years nanotechnology (nanoscience) has only recently, around 40 years ago, been implemented. However, the modern use of nanoscale technology differs greatly from its historic use when it was used to produce colors on glass using different sized gold and silver particles. Nanoparticles possess a range of beneficial properties that are exploited by their users, including their greater tendency to undergo chemical reactions, higher management of light spectrum, lower mass, and higher strength. The enhanced properties of nanoparticles have given them scope in a range of fields such as surface science, organic chemistry, and molecular biology. Nanoparticles are particles of varying size with diameters ranging from 1 nanometer to up to 100 nanometers (Khan et al. 2019). Due to their minimal sizes, nanoparticles cannot be seen by optical microscopes as they are of a wavelength a lot smaller than that of visible light and therefore require electron microscopes to be seen.

[1] Department of Industrial Biotechnology, Genetic Engineering and Biotechnology Research Institute, University of Sadat City, Sadat, Egypt.
[2] Department of Microbiology, Faculty of Science, Ain Shams University, Cairo, Egypt.
[3] Institute of Bioproducts Development (IBD), Universiti Teknologi Malaysia (UTM), Skudai, Johor, Malaysia.
[4] Bioprocess Development Department, City for Scientific Research, New Burg Al Arab, Alexandria, Egypt.
* Corresponding author: ashraf.elbaz@gebri.usc.edu.eg; ashrafhawase@gmail.com

Nanoparticles have been shown to exhibit different characteristics compared to their larger counterparts (Buzea et al. 2007). This is due to the fact that any particle properties are dominated/dictated not beyond a few microscopic diameters of its surface. This difference in characteristics is often noticeable when nanoparticles are dispersed in a medium of different compositions as interface interactions become significant. Often, minimal point defects, which occur only at or around single lattice points, can be found in nanoparticles in comparison to larger particles of the same substance. Nanoparticles, also, allow for a range of dislocations that can be seen with the aid of electron microscopes. Individually, each nanoparticle displays various displacement mechanisms which along with their exceptional surface structure allows for mechanisms that vary from those of the mass material (Guo et al. 2013). The nanoparticles anisotropy variation of different properties along different axes results in an extensive change in properties. Thus, nanoparticles of gold and silver, for example, are intensively researched due to the tremendous spectra of implementations in different fields. Moreover, a sphere trigonometry of nanoparticles allows for higher cross-sections along with deeper colors of colloidal solutions (Guo et al. 2013). This presents the ability to shift the resonance wavelength, through the tuning of the particle's geometry, and thus allowing the particles to use in the fields of trace metal detection, molecular labeling, nano-technical application, and biomolecular assays (Mourdikoudis et al. 2018).

The top-down and the bottom-up paths are two common approaches for nanoparticle synthesis (Wang and Xia 2004). The prior procedure involves the progressive disintegration of the majority of substance into nanosized materials, while the latter approach involves the build-up of atoms or molecules into nano-meter sized molecular structures. For the biological and chemical synthesis of nanoparticles the latter approach "the bottom-up approach" is very popular.

Historically, nanoparticles have been synthesized, through different chemical methods. Such chemical methods result in the generation of environmental pollutants due to the diverse physical and chemical complicated procedures required for production, such as high pressure, energy, and temperature, resulting in the production of many toxic compounds into the environment (Korbekandi et al. 2012). Chemical reduction is one of the most common methods for nanoparticle synthesis. This method results in the production of metal atoms through the reduction of metal salts with the use of reducing chemical agents, the likes of ethylene glycol, citrate, hydrazine, and hydrides (Korbekandi et al. 2012). All prior mentioned chemicals pose both environmental and health risks. Moreover, such reduction processes bear both large financial consequences along with issues in scaling the process adequately.

1.1 *Microbial Biosynthesis of Nanoparticles*

Recently much interest has developed in "bionanotechnology" which explores the union of microbial biotechnology and nanotechnology. Robinson (2015) discussed the extensive biological synthesis of nanoparticles from pure metals "MeNPs" such as gold and silver. He also discussed how biosynthesis of nanoparticles has seen extensive exploration and advances as eco-friendly substitutes to the traditional historic chemical method. This biotechnological-based synthesis of nanoparticles in comparison to chemical synthesis is part of the new green generation, these are environmentally friendly substitutes to synthetic methods. The biosynthesis of such particles is also known as the "green-synthesis" or "green chemistry" method. The biological approach as opposed to the chemical approach does not require drastic circumstances such as high pressure and temperatures; and also has the benefit of not leaving behind harmful remains of the negative impact on the environment. The novel use of living systems for synthesizing nanoparticles not only carries the significant benefits of not harming the environment compared to the more traditional methods but is also easier and cheaper to carry out (Robinson 2015). In the last few years' biosynthesis of nanoparticles has originated from a large number of biological sources including microbes (bacteria, yeast, fungi, algae, cyanobacteria, actinomycetes), plant extracts, and biomolecules (Li et al. 2011). Also, several researchers have studied and published their work on the advantages of the products of biosynthesized metallic and bimetallic nanoparticles (Thakkar et al. 2010, Narayanan and Sakthivel 2010). Research has not only been carried out on producing metallic NPs but on the microorganisms involved in their synthesis. The nanoparticles produced were also analyzed microscopically to identify the resulting nanostructure (Quester et al. 2013, Płaza et al. 2014). These results hold a valuable key for future research in the field. This novel practice begins to allow nanoparticles to function within the biomedical realm in diverse and complex ways.

The biosynthesis process involves microorganisms grabbing onto target ions from solutions. Once the microorganisms latch onto the target ions, the reduction process begins, due to the microbial metabolic, enzymatic bio-reducing activity, reducing the metal to its elemental form (Robinson 2015). Upon completion of the biosynthesis, the process is classified as either extracellular or intracellular as per the location where the nanoparticle is formed. As previously discussed, the importance of the biosynthesis of nanoparticles by biological materials lies in their wide range of possible applications as well as the possibility to synthesize them from an enormous number of different microbial species. Predominantly, two established methods of environmentally friendly "green synthesis" methods of nanoparticle biosynthesis have been confirmed (Chopra 1969). One of the synthesis methods is through the use of biological sources whilst the other involves the use of metabolites synthesized by biological sources, i.e., the microorganism synthesis of the nanoparticles is carried out either intra- or extracellularly (Chopra 1969, Płaza et al. 2014). Intracellular synthesis occurs within the microbial cell, in such cases, the synthesized product formed through the reductive pathway of the cell wall is stored in the area between the inner cytoplasmic membrane and the bacterial outer membrane. However, in the case of extracellular synthesis, the dissolvable or the reducing cell wall enzymes are not produced internally, the extracellular enzyme production accordingly results in the metal reduction process happening externally (Chopra 1969). An example of an enzyme involved in the metal nanoparticle biosynthesis is nitrate reductase, the enzyme "NADH-" and "NADPH-dependent reductase" results in the production of the metallic form of the ion (i.e., Me^{+1} to Me^{0}). This reduction happens when NADH-dependent reductase, as an electron carrier, transfers an electron from NADH, causing the metal ion (Me^{+1}) reduction to its metallic form (Me^{0}) due to the ion gaining electrons. An example is the reduction of gold metal ions Au^{+3} to their metallic form Au^0 due to the metal gaining electrons (Chopra 1969). Sadowski (2010), detailed the use of *Bacillus licheniformis* in a resembling pathway for AgNPs synthesis. Table 1 describes several types of nanoparticles and the microorganisms from which they are synthesized.

Table 1. Nanoparticles, and the microorganisms from which they are synthesized.

Microbial strain	Nanoparticle types	Reference
Bacteria		
Bacillus subtilis	Silver	Pugazhenthiran et al. 2009
Bacillus licheniformis	Gold	Fortin and Beveridge 2000
Shewanella algae	Gold	Kalishwaralal et al. 2009
Lactobacillus	Gold	Husseiny et al. 2007
Clostridium thermoaceticum	Gold, silver, Au–Ag alloy	Nair and Pradeep 2002
Klebsiella aerogenes	Cadmium sulfide	Cunningham and Lundie 1993
Klebsiella pneumonia	Cadmium sulfide	Sweeney et al. 2004
Escherichia coli	Silver	Mokhtari et al. 2009
Desulfobacteraceae	Cadmium sulfide, Gold	Sweeney et al. 2004
Thermoanaerobacterethanolicus	Zinc sulfide	Labrenz et al. 2000
Magnetospirilliummagnetotacticum	Magnetite	Zhang et al. 1998
Rhodococcus	Gold	Ahmad et al. 2003c
Chlorella vulgaris	Gold	Ahmad et al. 2003b
Phaeodactylumtricornutum	Gold	Hosea et al. 1986
Rhodopseudomonascapsulata	Cadmium sulfide	Scarano and Morelli 2003
Rhodobactersphaeroides	Gold	He et al. 2007
Brevibacteriumcasei SRKP2	Lead sulfide	Bai et al. 2009
Brevibacteriumcasei SRKP2	Lead sulfide, Polyhydroxyalkanoates	Pandian et al. 2009
Yeast		
Candida glabrata	Cadmium sulfide	Reese and Winge 1988
Torulopsis sp.	Lead sulfide	Kowshik et al. 2002a,b
Schizosaccharomycespombe	Cadmium sulfide	Kowshik et al. 2002a,b
Trichosporonjirovecii	Cadmium sulfide	El-Baz et al. 2016
Fungi		
Fusariumsemitectum	Silver	Basavaraja et al. 2008
Fusariumoxysporum	Gold, silver, Au–Ag alloy, cadmium sulfide, zirconia	Mukherjee et al. 2002, Ahmad et al. 2003a, Senapati et al. 2005, Ahmad et al. 2002, Bansal et al. 2004
Colletotrichum sp.	Gold	Shankar et al. 2003
Aspergillus flavus, Phaenerochaete chrysosporium	Silver	Vigneshwaran et al. 2006
Verticillium	Gold, Silver	Mukherjee et al. 2001a, b
Monascuspurpureus	Silver	Elbaz et al. 2016
Aspergillus niger	Cadmium sulfide	Alsaggaf et al. 2020
Gluconacetobacterxylinus	Cellulose magnetite	Hassen et al. 2019
Penicillium fellutanum	Silver	Kathiresan et al. 2009

2. Surfactants

2.1 Surfactants, Different Structures

The generic name surfactants classifies a group of chemicals that decrease the interfacial surface energy between materials. Structurally, conventional surfactants consist of two components, "moieties" a polar or ionic head group, the "hydrophilic" component "moiety", as well as a "hydrophobic" component "moiety" this can be: -Straight hydrocarbon-Branched hydrocarbon-Fluorocarbon chain with 8–18 carbon atoms. It is the balance between these components that gives surfactants their unique properties and abilities. Such properties are derived by the amphiphile ability to lower free energy at the phase boundary resulting in a decrease in interface tension. Figure 1 adapted from (Chevalier 2002) presents a brief illustration of the different structures of surfactants. Small molecule amphiphile, which play an important role in the solubilization of substances of hydrophobic nature within aqueous systems are known as hydrotropes (Eastoe et al. 2011). Gemini surfactant are two conventional surfactant molecules connected chemically by a spacer bond, this bond varies in length and strength. The two hydrophobic tails vary in length with cationic, anionic, or non-ionic hydrophilic heads of surfactant which can self-assemble at low concentrations, is of preferred properties over conventional surfactants regarding interface surface activity. The stronger activity of such surfactant type can be explained easily, as two hydrophobic tails have double the properties compared to one hydrophobic tail, present within the conventional surfactants. Such superior properties allow for the use of a lesser amount of Gemini surfactant than that of conventional surfactants (Sekhon 2004). Bolaform surfactants, also known as bolaform amphiphiles (bolas) (Zana 2002) are two hydrophilic components that have mono-layered lipid membranes, joined by a hydrophobic spacer (Song et al. 2005). The ability of archaebacteria to survive in anaerobic surroundings, acidic pH, and harsh temperatures, has been attributed to the stability of the mono-layered lipid membranes (Benvegnu et al. 2004). Such surfactants have successfully been implemented for the delivery of drugs, genes, or ions, success has also been realized in biosensor devices (Shimizu et al. 2005). Diblock copolymers and small surfactant molecules self-association is found to be similar to that of monomers. The shape of the micellar aggregates, spherical, cylindrical, or vesicular, is dependent on the surrounding conditions (Alexandridis and Lindman 2000).

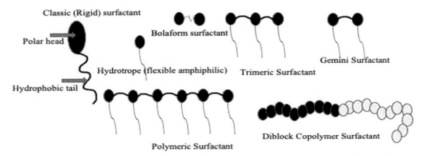

Figure 1. Illustration of different surfactant structures. Adapted from (Chevalier 2002).

Trimeric surfactants which have superior properties to monomeric and dimeric surfactants types are limited in number since they are difficult to synthesize (El-Said et al. 2018).

Polymeric surfactants as implied by their name are macromolecules of monomeric surfactants (Raffa et al. 2016).

2.2 *Surfactants General Properties, and Applications*

Surfactants have been implemented in a broad variety of applications due to their ability to lower the surface tension between substances. Application for surfactants are accordingly dependent on this characteristic where they have seen a large number of applications including laundry detergents, emulsifying agents, super-active agents, expanding agents, and propellants (Hirsch 2015). As detailed above surfactants are chemical compounds containing carbon-hydrogen bonds, i.e., organic compounds; that are characterized by the presence of hydrophobic and hydrophilic groups, i.e., they are amphiphilic compounds (Martin 2009). It is in accordance with the presence of both these groups, one insoluble in water the other soluble, that surfactants gain their ability to permeate water while at the same time act as an absorbent at interfaces between air and water or air and oil. For this reason, there are limitless applications for surfactants within chemical industries, with applications varying to those within the household, e.g., detergents to those for industrial cleaning. Surfactants are also used in the agroindustry, the paper and ink industry in self-care products, and even in pharmaceutics (Cheng et al. 2020). Surfactants are widely produced with recent estimates of its production globally exceeding 15 million tons per year. Half of the world production of surfactants is of soaps whilst other large-scale production of surfactants includes alkyl benzene sulfonates, lignin sulfonates, fatty alcohol ethoxylates, and alkylphenol ethoxylates (Martin 2009). Unfortunately, such great advantages come with the disadvantage of an unfavorable impact on not only the environment but on our wellbeing as well. Increased awareness of the environment and the continuous research for alternative natural resources have resulted in the production of new eco-responsible alternatives using "clean and/or sustainable" methodologies (Guenic et al. 2018).

2.3 *Chemical Surfactants Stabilizers for Nanoparticles*

Research was carried out on surfactants which when dissolved in water produce negatively charged particles "anions". Such surfactants are also known as anionic surfactants, e.g., tween 80 and SDS. Non-ionic surfactants that are neutrally charged, i.e., neither forming (positive charge) cations nor anions in water, e.g., Brij detergents, and polymer (Polyethylene Glycols (PEG) and polyvinylpyrrolidone (PVP)), were also researched. Neutral detergents have a negatively charged hydrophilic polar head, consisting of uncharged, head groups which include polyoxyethylene moieties, solubility being dependent on the binding of the hydrophilic parts to the water molecules (Chen and Wiktorowicz 1993).

Out of all investigated surfactants SDS, Tween 80 and PVP presented the most potential for further application as stabilizers for nanoparticle production. The results noted the aforementioned surfactants also strengthened and improved the modified Silver (Ag) nanoparticles antibacterial activity (Chen and Wiktorowicz 1993).

The team of researchers also focused on understanding the mechanical surface interaction system occurring between silver nanoparticles [Ag-NPs] the surfactants and the polymers that result in antibacterial ability. Unfortunately, Kvitek et al. (2008) worked to understand the interaction through which the ionic surfactants were adsorbed in the case of Ag-NPs they were unsuccessful. However, a different group of researchers had previously proposed a system of organization on the NP. They noted that an anionic surfactant, sodium dodecyl sulfate (SDS) could adsorb its hydrophilic group superficially on the silver nanoparticle, Ag-NPs, with a primary layer consisting of the surfactants' hydrophobic part reflecting outward orientation. This led to the formation of a secondary "counter-layer" aligned oppositely. The outcome of such formation involves the hydrophobic tails of surfactant penetrating between two thin like sheets of external-facing hydrophilic moiety. This surface arrangement increases cell-wall permeability or destructs Gram-positive bacteria cell wall, therefore increasing antibacterial activity of the Ag-NPs modified with surfactant "SDS" (Chen and Yeh 2002). The use of Ag, Ag ions, and AgNPs as an antibacterial as well as an antiviral substance is now established and accordingly extensively implemented (Nakamura et al. 2019).

3. Biosurfactants

3.1 Biosurfactants, Types, and Structures

Biosurfactants are surface-active biomolecules of biological origin, with the majority of such compounds originating from microorganisms. A large portion of the microorganisms that produce biosurfactants were isolated from polluted soils, wastewater, and sewage. These surface-active biomolecules have a similar structure and mode of action as that of chemical surfactants. Biosurfactants similar to chemical surfactants are amphiphilic biochemical compounds; their amphiphilic characteristic allows for their existence or separation at the interconnection between ionic and non-ionic media (Ramanathan et al. 2013). The biosurfactant is either synthesized within the cell "intracellular" or secreted outside the cell wall "extracellular". Biosurfactants, produced through fermentation or those bio-based present themselves as biomolecules with a hydrophilic head group and a hydrophobic tail (Nurfarahin et al. 2018). Structurally, conventional biosurfactants consist of a polar head group, a hydrophilic moiety, which can be a carbohydrate, cyclic peptide, amino acid, carboxylic acid, phosphate, or an alcohol (Ramanathan et al. 2013). Hydroxyl fatty acid moiety, long-chain fatty acid, or α-alkyl β-hydroxy fatty acid are examples of their hydrophobic tail. Table 2 below describes several types of biosurfactants and the microorganisms from which they are synthesized. The functionality of any given biosurfactant is determined by its structure. In other words, it is the structural characteristics such as location and size that determine their properties (Ramanathan et al. 2013). In accordance with the hydrophobic moiety, Morita et al. (2009) categorized biosurfactants into four types: polymers, fatty acids, lipopeptides, and glycolipids. On the other hand, Rosenberg and Ron (1999), categorized biosurfactants in accordance to their molecular weight

Table 2. Microorganism and major biosurfactant types (Płaza et al. 2014).

Biosurfactant producing microbes	Low molecular weight biosurfactant	
Pseudomonas spp., *Pseudomonas chlororaphis, Burkholderia* spp.	*Glycolipids*	Rhamnolipids
Candida bombicola/apicola, Torulopsispetrophilum, Candida sp., *Candida antartica, Candida botistae, Candida riodocensis, Candida stellata, Candida bogoriensis*		Sophorolipids
Mycobacterium fortium, Micromonospora sp., *M. smegmatis, M. paraffnicum, Nocardia* sp. *Rhodococcuserythropoli, Arthobacter* sp.		Trehalose esters
Rhodococcuserythropolis, Arthobacterparaffineu, Mycobacterium phlei, Nocardiaerythropolis		Trehalose mycolates
Pseudomonas flocculosa		Flocculosin
Bacillus brevis	*Lipopeptides and lipoproteins*	Gramicidins
Bacillus licheniformis		Peptide lipids
Bacillus licheniformis IM1307		Lichenysin G
Bacillus subtilis		Surfactin, subtilysin, subsporin, Iturin A, Bacillomycin L, Mycobacillin
Bacillus thuringiensis CMB26		Fengycin
Serratiamarcescens		Serrawettin
Streptomyces canus		Amphomycin
Streptomyces globocacience		Globomycin
Pseudomonas putida		Putisolvin I and II
Arthobacter sp.		Arthrofactin
Candida sp., *Corynebacterium* sp., *Micrococcus* sp., *Acinetobacter* sp., *Thiobacillusthiooxidans, Aspergillus* sp., *Pseudomonas* sp., *Mycococcus* sp., *Penicillium* sp., *Clavibactermichiganensis* subsp. *insidiosus*	*Phospholipids and fatty acids*	Phospholipids, Fatty acids
Biosurfactant producing microbes	**High molecular weight biosurfactant**	
Pseudomonas aeruginosa	*Polymeric biosurfactants*	Protein PA
Acinetobactercaloaceticus A		Biodispersan
Acinetobactercalcoaceticus		Alasan
Candida lipolytica		Liposan
Acinetobactercaloaceticus RAG-1, *Arethrobactercalcoaceticus*		Emulsan
Acinetobactercalcoaceticus, Cyanobacteria	*Particulate biosurfactants*	Fimbriae, whole cell
Acinetobacter sp. HO1-N, *A. calcoaceticus*		Membrane vesicles

naming two categories, those of high-molecular-weight (HMW), commonly known as bioemulsifiers and those of low-molecular-weight (LMW), referred to as biosurfactants (Uzoigwe et al. 2015). LMW, such as glycolipids, lipopeptides, phospholipids, and fatty acids, are often used in the process of lowering surface tension. HMW polymers, polymeric and particulate biosurfactants such as emulsan and alasan, are effective as emulsion-stabilizing agents due to their ability to firmly attach to surfaces (Ron and Rosenberg 2001).

In general, biosurfactants can be characterized based on:

1. The concentration of surfactants above which aggregates form and at which any surfactant added will form micelles, this is also known as their critical micelle concentration (CMC). The lower the CMC the higher the biosurfactants efficiency, i.e., to decrease surface tension lower biosurfactants concentration is necessary. The concentration can also affect physicochemical properties (Uzoigwe et al. 2015).

2. The balance between the concentrations and size of the hydrophilic and lipophilic group of the surfactant molecule also known as the hydrophilic-lipophilic balance (HLB). Those of low HLB are lipophilic and can stabilize water-in-oil emulsions, on the other hand, those of high HLB present higher water solubility and therefore form an emulsion of oil-in-water (Pacwa-Płociniczak et al. 2011).

3. Charge, biosurfactants originating from microbes are either anionic or neutral.

Biosurfactants prosper from a large range of benefits over their chemical counterparts. These benefits include their minimal effect on the environment and unlike their counterparts they are considered eco-friendly since they do not build up in the ecosystem as they are easily degradable. Biosurfactants are also biocompatible and as such allows for their uncensored application in beauty products, food and medicinal products (Santos et al. 2016); In addition, they are found in large quantities at minimal costs.

Biosurfactants can be used in the process of biological decomposition and decontamination of industrial wastewater, bioremediation of polluted soils, and controlling of oil-spillage. They also perform a crucial part in the stabilization of industrial emulsions (Santos et al. 2016). Lastly, medicinal products and nutritional applications can also be produced through the decontamination of certain pollutants using biosurfactants.

3.2 *Biosurfactants for the Stabilization and Synthesis of Nanoparticles (Microemulsion & Reverse Micelle)*

Dissolving surfactants in solvents of an organic nature result in the formation of spherical aggregates, where the hydrophilic head is directed inwards towards the center. The formed aggregates are found to be of a small size if formed in the absence of water. Large size aggregates are formed if water is introduced. Euliss et al. (2006) describe micelles to be nanoparticles formed by self-assembly of surfactants and lipids. The supramolecular formation of colloidal aggregates of micelles is dependent on the hydrophobicity of the amphiphilic molecules in water. Spherical shaped micelles show an increase in dimension upon the increase of surfactant concentration (Nelson et al. 1997). Nanoparticles can form micelle which make nanoparticles so interesting. However, controlling the properties including the morphology of these structures is of utmost importance. Capping agents, e.g., biosurfactants, in the synthesis of nanoparticles allows for homogenous distribution of the nanoparticle in the liquid medium. The use of biosurfactants for the stabilization and synthesis of controlled nanoparticles is therefore presenting itself as an eco-friendly method of production (Kulkarni et al. 2019). Figure 2, below, shows an illustration of the micelles size increase proportional to water content and the size of the nanoparticles (NP) controlled by the biosurfactants mediated process.

Water dissolves in the polar center resulting in a water-surfactant molar ratio "W". Per the amount of water if W is less than 15 the aggregates are named reverse micelles (Fig. 3a). On the other hand, if W is more than 15 they are known as microemulsions (Luisi et al. 1986). It is the hydrophilic-lipophilic balance value of the surfactant used that results in the formation of either a water impregnated micelles dispersed in oil "water-in-oil (W/O) microemulsion" (Fig. 3b), also known as reverse microemulsion, or in oil-impregnated micelles dispersed in water "oil-in-water (O/W) microemulsion" (Fig. 3c). Adapted from Kiran et al. 2011.

As researchers continue to understand and recognize the value of biologically active nanoparticles being synthesized through environmentally friendly methods; they have begun evaluating the process of synthesizing metal nanoparticles with the advantage of biosurfactants acting as a stabilizing agent, providing colloidal stability, preventing aggregation and unwanted growth, i.e., acting as capping agents. Biosurfactants were found to play an essential part in the aggregation and stabilization process of nanoparticles (Moo-Young and Moreira 2011), with potential in a broad variety of medical biotechnology implementations including use as; antibiotics, antifungals (Roy et al. 2013, Elshikh et al. 2017) and chemotherapy drugs (Dey et al. 2015) the introduction of genetic material into host cells "gene delivery" (Rodrigues et al. 2006) and healing of wounds (Ohadi et al. 2017a, 2017b, Gupta 2017). As such biologically synthesized surfactants have become a green alternative optimizing nanoparticle synthesis and stabilization. The most common method used involves

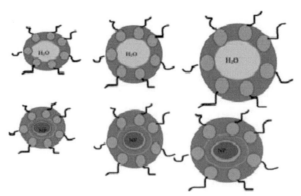

Figure 2. Illustration of the micelles size increases proportional to water content increase and the size of the nanoparticles (NP) controlled by the biosurfactants mediated process. Adapted from Kiran et al. 2011.

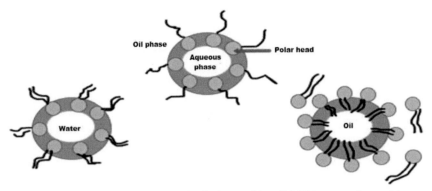

Figure 3a. Reverse micelle, (3b) Water in oil microemulsion, (3c) Oil in water microemulsion.

the absorption onto metallic nanoparticles (MeNPs) which stabilizes the surface of the nanoparticle whilst also ensuring that no subsequent aggregation occurs (Moo-Young and Moreira 2011). On the other hand, regarding the effect of the surfactant, the kind of surfactant adsorption is highly dependent on the type of surfactant as well as the density of the adsorbed coat. Researchers continue to examine the effect of the composition of biologically synthesized surfactants on the synthesis and characteristics of produced metal nanoparticles', yet no established studies have been completed. A large number of researchers have also examined the effect of both laboratory and commercially produced biosurfactants in stabilizing and modifying the nanoparticle synthesis (Reddy et al. 2009, Kiran et al. 2011, Farias et al. 2014).

3.2.1 Glycolipids Biosurfactants Stabilized Nanoparticles

3.2.1.1 Rhamnolipids

This class of glycolipids produced by a range of organisms mainly *Pseudomonas aeruginosa*, has come to be known as the most characterized surfactant of bacterial origin. They are characterized by a glycosyl head group with a 3-alkanoic acid fatty acid tail (Farias et al. 2014). AgNPs synthesis in water-in-oil microemulsion stabilized by commercial rhamnolipid was first discovered by Xie et al. 2006. The reducing agent, Sodium borohydride, also known as sodium tetrahydridoborate and sodium tetrahydroborate [NaBH], in rhamnolipid reverse micelles, was used for synthesis; dispersion in heptane occurred after the extraction of the micellar solution. The radiation absorption of the completed AgNPs was recorded using ultraviolet-visible spectroscopy, the AgNPs was then left to stabilize throughout two months (Farias et al. 2014). Further identification of the structure of the silver nanoparticles found the particle structures to be spherical and uniform.

Shortly after Xie et al. (2006) findings, Palanisamy and Raichur 2009, used commercial rhamnolipids to stabilize nickel oxide [NiO] nanorods (Płaza et al. 2014). The study used two methods for nanoparticle synthesis; two different microemulsion approaches. The first microemulsion solution involved commercial rhamnolipid dissolved in heptane, later nickel chloride hydrate [$NiCl_2 \cdot H_2O$] was added to the solution whilst it was continuously stirred. Upon completion both mixtures were added to one another and mixed; and a centrifuge was used to disassociate the nickel hydroxide precipitate resulting from the mixture. Ethanol was then used to wash off the heptane and biosurfactants that were still present. The nanorod, a single format of the nanoscale substance, in this case, nickel oxide a width of about 22 nm and 150–250 nm was formed after the drying of dissociated washed precipitate for 3 hrs at a temperature of 600-degree centigrade (Płaza et al. 2014). Moreover, pH was seen to be a determining factor in the nanorods morphology. The second method entailed adding ammonium hydroxide [NH_4OH] as an alternative to the nickel chloride solutions (Płaza et al. 2014).

Palanisamy and Raichur 2009, used a similar method to obtain spherically uniformly distributed nickel oxide nanoparticles. They reported the conventional use of rhamnolipid for the microemulsion production technique. Moreover, the study noticed that increasing the pH had an inverse proportional relationship to the size of the nanoparticles, where pHs ranging from 11 to 12.5 resulted in NPs sizes ranging from 47 to 86 nm (Kong et al. 2011). Also, other research has reported on the potential formation of a reverse micelle when rhamnolipids purified from *Pseudomonas aeruginosa* with either silver nitrate or sodium borohydride (in its aqueous form) were used to biosynthesize AgNPs (Kumar et al. 2010). In this study, the first step involves the production of reverse micelles through mixing rhamnolipids, n-butanol/ n-heptane, and aqueous silver nitrate. Similar to the first step, the second resulted in the formation of a reverse micelle through mixing equal volumes of rhamnolipid and aqueous borohydride as well as required solvents. Once again, the products of the two steps of the reactions were blended resulting in the formation of silver nanoparticles precipitate which disassociated from the mixture after being centrifuged. The product of the reaction was tested and reflected considerable broad-spectrum antibacterial activity (Kumar et al. 2010).

Narayanan et al. (2010) work involved the use of water-soluble rhamnolipids in the process of capping zinc sulfide, ZnS, nanoparticles. The research displayed the metal reactivity, dimensions as well as the synthesized ZnS nanoparticles' ability to react with light. The nanoparticles produced were both stable and water-soluble being synthesized in an aqueous environment without any organic solvents.

Worakitsiri et al. (2011), also researched using biologically synthesized surfactants for the production of metallic nanoparticles. Their research revolved around using a soft, easily removable rhamnolipids template to produce nanofibers and nanotubes of polyaniline (PANI). The synthesis began with the oxidative polymerization of aniline, the starting monomer, to which a dopant agent hydrochloric acid and an oxidant ammonium peroxydisulfate were added (Sapurina and Shishov 2012). A noticeable difference between polyaniline synthesis without and with the aid of rhamnolipid is the uniformity of the morphology, larger particles and thus high crystallinity and electrical conductivity of those synthesized with rhamnolipids (Sapurina and Shishov 2012). The controlled morphology and size allowed researchers to conclude that the addition of rhamnolipids to the synthesis process is beneficial in creating highly conductive polymeric nanoparticles.

Kumar et al. (2010) have recently made use of rhamnolipids produced by *Pseudomonas aeruginosa* for their research. The researchers used *Pseudomonas aeruginosa* to synthesize silver nanoparticles in an extracellular manner (Singh et al. 2018). Similar to prior research, nitrate reductase was added into the solution as a reducing agent for the silver ions. A conclusion was made that the nanoparticles were stabilized by the rhamnolipids that were in the culture supernatant. Other, similar, research found that nanoparticles produced by rhamnolipids were stable for two months due to the ability of the biosurfactants to stabilize the solution for two months at the same time preventing the formation of aggregates in the synthesis process (Singh et al. 2018). Moreover, later research by Saikia et al. 2013, further identified the importance of rhamnolipids in protecting silver nanoparticles against salt formation.

Reports followed the Hazra et al. 2012, investigation on rhamnolipid biosurfactant capped ZnS nanoparticles, structure, properties, compatibility, and toxicity to living cells. The capped nanoparticles were also investigated as nano-sized photocatalysts degrading azo dyes for implementation within the textile industry (Hazra et al. 2012). For this investigation rhamnolipids were once again used as both a capping and stabilizing agent in the solution, during the synthesis process of ZnS nanoparticles. They reported that within the textile dye industry the rhamnolipids usage resulted in less toxic ZnS nanoparticles with more degradation effectiveness (Hazra et al. 2012). The research noted the value of nanoparticles in the future of the generation of nano photocatalysts.

3.2.1.2 Sophorolipids

These are a class of glycolipids of a range of non-pathogenic yeast origin, structurally characterized by a dimeric sophorose linked to long-chain hydroxy fatty acids. This group has great potential as biosurfactants as they are biodegradable and are eco-friendly. Sophorolipids produced similar results to those obtained by rhamnolipids. Kasture et al. 2007, reported on the use of sophorolipids as a new capping agent for cobalt nanoparticles. *Starmerella bombicola,* yeast cells grown on oleic acid, produced sophorolipids (Huang et al. 2013). The capping of the cobalt nanoparticles afforded the sugar moiety exposure to the dissolvent granting it stability, making it water-redispersible. Baccile et al. (2013), suggests that exposure of the sophorose group found at the nanoparticles' surface might be the reason for the sophorolipids' biological advantageous properties (Płaza et al. 2014). The product of the biosynthesis was therefore a more resistant cobalt nanoparticle of paramagnetic characteristics. On an atomic level, the particles were polydisperse and consistently parted with an average dimension of approximately 50 nm.

Research continued by Kasture et al. (2008) for the production of silver nanoparticles at different temperatures using sophorolipids not only as a capping agent but as a reducing agent as well. The use of oleic acid or linoleic acid resulted in the production of two types of sophorolipids mixtures. Production of two different types of sophorolipids mixtures may be the result of these acids impacting the length of the fatty acid chain within the hydrophobic functional group (Elshafie et al. 2015). It was found that the silver nanoparticles synthesized at lower temperatures had larger particles with a high variety in size. Whereas at higher temperatures, the particle size decreased whilst their size distribution became narrower. The effect of the biosurfactants was confirmed when the linoleic acid solution synthesized larger zinc oxide particles whilst the oleic acid solution produced smaller particles; it was confirmed that biosurfactants play a direct role in both molecule size and distribution (Elshafie et al. 2015).

Moreover, iron oxide nanoparticles were also biosynthesized using sophorolipids where the acidic sophorolipids were found to play a role as surface complexing agents by keeping ions from interfering with the surfactant during the iron oxide nanoparticles synthesis. Sophorolipid derived nanoparticles were found to have exceptional stability in water and salt aqueous solutions, this was reflected by dynamic light scattering experiments (Płaza et al. 2014). A closer look at the morphology showed gathered average particle size in water to range from 10 nms to 30 nms for the sophorolipid-functionalized iron oxide nanoparticles. The maximum size in KCl-water solutions being no more than 30 nm. The findings further highlighted the importance of sophorolipids, due to its properties, for implementation in the bio-medical field.

3.2.1.3 Lipopeptides Biosurfactants Stabilized Nanoparticles

Lipopeptides consist of lipids joined to peptides and are synthesized as microbial secondary metabolites some of which exhibiting antimicrobial activity. These molecules have also been studied for their use in the synthesis process of biosurfactants. Surfactin, an example of bacterial lipopeptide, is used for the synthesis of two nanoparticles, AgNPs and AuNPs. Reddy et al. (2009), showed that surfactin can be used not only as a template for the synthesis process but also plays a role as a stabilizing agent. The team of researchers set out to demonstrate a mechanism where *Bacillus subtilis* originating surfactin mixed with aqueous $AuCl_4$ reduced by sodium borohydrate resulted in the production of stable gold nanoparticles. Foam fractionation was used to recover the surfactin from the culture supernatant. The recovered surfactin was then combined with the chloroaurate solution, this results in the solution's pale-yellow color turning to red purple (Reddy et al. 2009). The appearance of the red-purple color infers the oxidation of the metal and confirms the production of the gold nanoparticles. The environment of the synthesis was changed as it occurred in pH's of 5.7 and 9 with temperatures of 4°C and room temperature. Their results found stable nanoparticles were produced at a pH of 7 and 9 and remained intact for two months (Reddy et al. 2009); moreover, the gold nanoparticles aggregated within 24 hours at a pH of 5. The experiment also concluded a negative correlation between mean particle size and pH. The nanoparticles synthesized at 4°C were not as monodispersed or as uniform as those synthesized at room temperature (Reddy et al. 2009).

Further work by Reddy et al. 2011, resulted in the stabilized formation of silver nanoparticles using commercial surfactin. The experiment entailed mixing $AgNO_2$ and NaBH solution in the presence of surfactin to stabilize synthesis (Reynoso-García et al. 2018). As expected, the silver nanoparticles produced were stable for two months. Correlating to prior research results the mean particle size also decreased with increases in pH. The same conclusions they found with temperature in their prior work were also seen (Reynoso-García et al. 2018). Figure 4 displays the nanoparticles produced at different pH's and temperatures.

The precipitation method was used to prepare nanostructured zinc oxide with the aid of surfactin. Surfactin was used as a templating agent stabilizer in the rose-like arrangement (Reynoso-García et al. 2018). The thickness of the petal-like sections of the particle was measured to be between 9 and 13 nm. Moreover, the experiment found that the surfactin concentration largely affected morphology, mass, and consistency of the petal-like areas in the zinc oxide nanoparticles. A negative correlation was found between 'petal' thickness and surfactin concentration, with increases in concentration causing a fall in 'petal' thickness (Reynoso-García et al. 2018).

Singh et al. (2011) also used *Bacillus amyloliquifaciens* produced surfactin in the process of synthesizing stable cadmium sulfide nanoparticles (Singh et al. 2011). The amount of surfactin obtained from *Bacillus amyloliquifaciens* was calculated to be 160 mg/L (Singh et al. 2011). The produced surfactin played an important part in the production of the cadmium sulfide nanoparticles and was described to have stabilized and protected the nanoparticles. The synthesized nanoparticles showed stability for up to six months without significant changes to their structure.

4. Biosurfactants Stabilized Nanoparticles and Applications

The largest issue facing silver nanoparticle synthesis is related to inefficient stability and dispersion. Accordingly, aggregation does not occur, and hence only minimal antibacterial activity is realized (Gudiña et al. 2016). Ag nanoparticles' physical, chemical, and biological properties are highly dependent on the dimension, configuration, and their level of dispersion regulation. This then affects their application in the ecosystem as well as in the biotechnical and biomedical fields (Gudiña et al. 2016). It has also been determined that both rapid and green synthetic

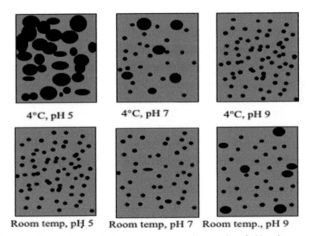

Figure 4. Illustration of transmission electron microscopy micrographs of pH and temperature effect on silver nanoparticle production (Reddy et al. 2009).

methods, which make use of various microorganisms and biosurfactants, may enhance the efficiency of Ag nanoparticle synthesis (Fenibo et al. 2019). Research continues to attempt to control particle size and monodispersity in the nanoparticle synthesis process by microbes. Investigations should be carried out on the influence of parameters such as concentration, temperature, and pH, used during the processing of the synthesis of Ag-NPs. The effect of such parameters on the dimension, shape, and distributions of sizes requires further research. Further research confirmed that specifically synthesized metal nanoparticles have a positive microbial antagonistic activity allowing them to be used effectively as antibacterial and antifungal agents. Broad-spectrum antibacterial activity and antimicrobial activity against *Candida albicans* was exhibited by silver nanoparticles synthesized using rhamnolipids originating from *Pseudomonas aeruginosa* BS-161R (Rodrigues 2015). As research continues in the nanoparticle synthesis field the development continues. The largest extent of research goes towards improving the bio-synthesis process and understanding the connections between the physical, chemical, and microbial antagonistic abilities of metal nanoparticles (Gudiña et al. 2016).

4.1 Medical and Surgical Applications

4.1.1 Antimicrobial Properties

Experimentation into surfactants and biosurfactants emphasized their important antimicrobial, anticarcinogenic activity as well as their ability to inhibit cell growth "antiproliferative effects" on tumor cells (Gudiña et al. 2016). Research on the use of surfactants from *Bacillus* reflected the ability of these surfactants to control bacterial, yeast cells, and fungal pathogens.

The necessity for antimicrobial agents in food products has led to the study of applications of oils for that purpose. The reactivity, volatile nature, and low solubility of oils in the aqueous phase restricts their application for such purpose. The increased stability of such oils upon their formulation in nanoemulsion systems not only prevents any reaction with other components but also increases their capacity as antimicrobial agents. Such formulations have the advantage of not affecting any general and specific characteristics of food (Donsì et al. 2012).

Even though the use of nanoparticles with biosurfactants as a microbial antagonistic requires extensive research, legislation imposed preventing the use of harmful chemicals and requiring the prevention emissions of gases from the greenhouse has increased the need for beneficial metallic nanoparticles. Such demand has resulted in the environmentally friendly, nontoxic biosynthesis process of metallic nanoparticles through the use of microorganisms or plant extracts rather than the conventional chemical and physical synthetic process (Barr and Dominici 2010).

4.1.2 Antibacterial Shielding of Medical Devices

Human implantable devices that require antimicrobial coatings are of two types. One of which are those completely implantable medical appliances, such as cardiac valves or dental implants, whilst the other are incompletely implantable medical appliances, such as catheters, intravenous catheters, or neurosurgical catheters. The disadvantage of the latter lies in the fact that they have greater susceptibility to bacterial growth, accordingly, increasing the likelihood of contamination. An important property for the antimicrobial coating of cardiovascular apparatuses is that to avoid the occurrence of thrombosis the coating has to be compatible with the patient's blood. "The application of a titanium oxide coating on implants is based on pore morphology, with calcium, silicon, phosphorus, and silver particle enrichment" (Xia et al. 2012). The coating plays an essential role in warding off bacterial attachment and growth which if not prevented would result in inflammation within the vicinity of the implants. Contamination of medical implants can be caused by a wide range of bacteria including *Streptococcus mutans, Streptococcus epidermis,* and *E. coli.* Research has shown nanocoating ability to boost the attachment and generation of osteoblast cell lines (Della Valle et al. 2012). Nanopolymers, in particular, can act as antibacterials and consequently can be used to prevent biofilm microbial growth of catheter (Samuel and Guggenbichler 2004, Galiano et al. 2008). The use of nano polymer coatings on neurosurgical catheters introduced invasively into the patient decreases the possibility of bacterial contamination and resulting complications. It has been noted that the nanopolymer production over a 6d period extensively reduced *S. aureus* growth (Wang et al. 2017).

4.1.3 Injury Dressings

Skin is the first line of natural protection for humans, protecting humans' internal organs from microbes and external objects whilst also managing to maintain the water and electrolyte vital and dynamic equilibrium. Wound or injury dressing provide protection due to its ability to duplicate the skin's ability to promote the generation and fibroblast migration, accelerate epithelial tissue formation, reduce the formation of marks or scars, preventing bacterial growth and inflammatory reactions (Yu et al. 2014). Within the medical practice, dressing is essential to allow the skin time to rebuild its natural protection line, while also speeding up the wound healing process. Dressing also prevents the occurrence of infection during the period where the skin is damaged and unable to carry out its natural task.

Skin injury is a result of trauma, burns as well as skin ulcers. Injured skin may get infected due to a broad spectrum of bacteria, e.g., *Staphylococcus, Streptococcus, E. coli,* and *Klebsiella* spp. Chronic wound infection might also be caused by several bacterial species along with antibiotic resistance.

Studies have reflected that implementing NPs with broad-spectrum antimicrobial properties plays an important function in decreasing bacterial proliferation and reproduction. Such studies showed that a fiber mat resulting from mixing AgNP and a combination of polyvinyl alcohol and chitosan can be used in injury curing (Li et al. 2013). Nanosilver surface of high specificity provides bacterial contact, resulting in a noticed inhibition of bacterial growth with a boosted wound healing speed.

4.1.4 Orthopedic and Trauma Applications

Polymethyl methacrylate, its modified form, and methyl methacrylate (MMA) are self-curing plastic at room temperature otherwise known as bone cement. Bone cement has a wide range of orthopedic applications, where it is frequently used in replacement surgery 'arthroplasty' for knees and hips to fix joint prostheses. The properties of bone cement allow it to block the space between implant and bone. Joint replacement surgery was shown to be followed by infection at a high rate of up to 3% (Kreutzer et al. 2005). Such infection occurs through a wide range of bacteria including *Staphylococcus aureus* as well as, methicillin-resistant *Staphylococcus aureus* (also widely known as MRSA), Gram positive *Staphylococcus epidermidis,* or *Acinetobacter baumannii* infections. Research on the loading of bone cement with antibiotics to effectively reduce the percentage of infection following joint replacement has reflected confusing results. Some research showing the ability of such antibiotic-loaded applications to potentially decrease or completely prevent such infection (Nowinski et al. 2012). Several other studies, on the other hand, show no significant reduction of infection rate (Zhou et al. 2015, Hinarejos et al. 2013). Fear of incurable postoperative infectious diseases that might lead to death, due to the rise of resistant bacterial strains, has led to more attention being paid to NPs applications in this field. NPs show no infection by certain antibiotic-resistant bacteria (Pelgrift and Friedman 2013). Both the Kirby–Bauer and the time-kill method have shown that mixing Ag NPs, at concentration as low as 0.05% with PMMA-based bone cement decreases surface biofilms formation preventing bacterial surface colonization (Miola et al. 2015). It can be concluded that nanosilver can significantly decrease infection related to surgery and that Ag NPs can be applied in place of antibiotics on bone cement (Prokopovich et al. 2015, Yun'an Qing et al. 2018).

4.1.5 Maxillofacial Prostheses

Dental plaque usually forms as a result of a reduction in pH during treatment with the parallel increase of bacterial growth. Nanocrystallization has been shown to decrease infection when mixed with dental materials. Examples include combining nanodiamond-functionalized amoxicillin with gutta-percha for root canal filling to prevent residual bacteria (Lee et al. 2015). Adding nano-titanium dioxide to maxillofacial prostheses increased antibacterial effects after being exposed to light. Prostheses, with no additives, exposed to such environmental conditions have shown biofilms development causing surrounding tissue inflammation (Aboelzahab et al. 2012). CuO and ZnO NPs coated brackets have successfully prevented *S. mutans* growth (Ramazanzadeh et al. 2015, Yassaei et al. 2020).

4.1.6 Microemulsion Drug Delivery System

Enhancing the bio-accessible oral administration of poor aqueous solubility drugs represent a challenge in the development of drug delivery systems (DDS). Microemulsions drug delivery systems (MDDS) consists of surfactants, cosurfactants, lipids, and/or co-solvents (Wu et al. 2017). Such systems cover a wide range of delivery systems, e.g., through the mouth "oral", nose "nasal", eye "ocular", skin "topical" and into veins through injections "intravenous". Surfactant molecules partition at the interface of a solution resulting in a lowered surface tension. Research shows that the ability of surfactant to adsorb, self-accumulate, and/or aggregate in an aqueous solution results in the production of micelle, lamellar, and crystalline structures. The disadvantage of preparing MDDS using surfactants include low level of solubility in lipids, reduced ability to load drugs, and increased possibility of irritation of the gastrointestinal GI system caused by increased amounts of surfactants up to 30–60% w/w. The use of biosurfactants in the preparation of these systems has helped decrease toxicity and GI (Anton et al. 2008). Another advantageous feature of such microemulsion systems is the fact that both hydrophilic and lipophilic active pharmaceutical ingredients (APIs) are soluble. The advantageous characteristics, superior self-assembling, and emulsifying traits, of using biosurfactants in MDDS preparation have led to extensive study to replace synthetic surfactants with microbial surfactants/surfactants to produce non-toxic microemulsion systems pharmaceuticals (Anton et al. 2008). Such research has resulted in biosurfactants replacing synthetic surfactants. For implementation as MDDS the most extensively studied biosurfactants are both glycolipids and lipopeptides (Ohadi et al. 2020).

For example, *Brevibacterium casei* MSA19, a sponge-associated marine bacterium produces a glycolipid biosurfactant when it utilizes industrial and agriculturally based industrial waste. Oilseed cake as a base, carbon source in the form of glucose, nitrogen source in the form of beef extract, metal in the form of $FeSO_4 7H_2O$, 2% NaCl were also added this is adjusted at pH 7.0 with a temperature of 30°C to optimize glycolipid biosurfactant production. *B. casei* MSA19 originating biosurfactant reflected a higher emulsification index when compared to the same for the synthetic surfactants SDS, Tween20, and Tween80. Silver nanoparticles can be produced through reverse microemulsion technique using the glycolipid biosurfactants for stabilization. The synthesized Ag nanoparticles remain stable and for 2 months (Kiran et al. 2010). Further investigations show nanoreactor microemulsions that utilize the glycolipid biosurfactant result in nanoparticles that are uniform and stable (Xie et al. 2006, Palanisamy and Raichur 2009, Kiran et al. 2010, Kumar et al. 2010, Ohadi et al. 2020). Rhamnolipid and sophorolipid, without the addition of a co-surfactant, resulted in stable and biocompatible microemulsions when mixed with lecithins (Nguyen et al. 2010). The two hydrophilic sugar groups in rhamnolipid resulted in a more hydrophilic biosurfactant. On the other hand, the unsaturated single long tail of fatty acid along with the groups of acetyl within sophorolipid biosurfactants resulted in hydrophobic characteristics (Nguyen et al. 2010).

Trehalose lipid was used for the synthesis of polymethyl methacrylate nanoparticles (nPMMA) using oil-in-water (O/W), dispersed phase modified atomized microemulsion procedure. To synthesis nPMMA a much smaller amount of trehalose lipid was needed compared to those required in other microemulsion polymerization systems. Morphologically the synthesized microemulsions varied where nano-spheres and needle-like non-calcinated particles were reported (Hazra et al. 2014, Ohadi et al. 2020).

4.1.7 Cell Culture Applications

Nanoemulsions are naturally composed of triacylglycerols: phospholipids, allowing cells (due to phospholipids being the main components of membranes) to accept them well. Their chemical composition is advantageous as it allows a replicable and controlled transfer of supplements dissolvable in oil to cell lines of human origin. Due to this, nanoemulsions have seen a great amount of interest in cell culture applications (Zülli et al. 2000, Ahmad et al. 2017).

4.2 Personal Care/Cosmetics Applications

Nanoemulsions have found a large range of use in both the personal and cosmetics applications. Such applications within the personal care industry include amongst others the production of skincare products, the likes of moisturizers, cream, and hair products. Nanoemulsions without alcohol additives can also be used to deliver fragrance or to extend the effect of the same in cosmetic/personal care products. The widespread application of nanoemulsions in these industries may be attributed to their transparency, fluidity, their remarkable hydrating ability as well as their extensive surface area which allows active components to penetrate the skin quickly. The fact Chellapa et al. (2016), Sonneville-Aubrun et al. (2004) showed that nanoemulsion can be sterilized by filtration, is of great advantage for both industries.

Simonnet et al. (2002, 2004) patented alkyl ether citrates, ethoxylated fatty ethers, or fatty esters of sugar-based nanoemulsions for use in cosmetic production. The ability of rhamnolipids to form O/W nanoemulsions qualifies them to be used in the toiletries industry (Bai and McClements 2016).

4.3 Agricultural Applications

The agrochemical industry largely depends on pesticides of maximum efficiency and with the least effect on the environment. Nanoemulsion presents itself as a solution to resolve the issue of a great number of pesticides not reaching the targeted plant site while decreasing the surfactant exhaustion. Within the agrochemical industry, the mentioned loss of pesticides is attributed to several factors such as run-off, rain fastness, volatilization, etc. Jiang et al. (2012) studied the production of a nanoemulsion composed of fatty-acid methyl esters and alkyl polyglucosides as a product for the glyphosate herbicide application. Alkyl polyglucosides were also used by Du et al. (2016) along with polyoxyethylene 3-lauryl ether as surfactants. Du et al. (2016) produced an oil/water nanoemulsion for water-insoluble pesticide β-cypermethrin delivery. The production of the nanoemulsion was executed using the oil phase methyl laurate.

4.4 Industrial Applications

4.4.1 Chloroform Detection using Ethoxylated Phytosterol-capped Gold Nanoparticles (BPS-GNPs)

Exposure to certain chloroform concentrations is of drastic effect to humans. Inhaling this chemical compound may lead to hepatic damage, convulsions, and even death due to respiratory failure. Methods to easily detect chloroform, widely found to be a contaminant of groundwater globally, had to be produced (Zhang et al. 2011, Yunus et al. 2012). A sensor was produced using biosurfactant ethoxylated phytosterol-capped gold nanoparticles (BPS-GNPs). Quick aggregation of these nanoparticles "BPS-GNPs" reflected the presence of chloroform, with UV-Spectra indicating minute amounts of chloroform. Small amounts of chloroform can be identified by observing, using no visual aid, the decrease of color of ethoxylated phytosterol-capped gold nanoparticles aqueous solution (Jia et al. 2017).

4.4.2 Tertiary Recovery/Enhanced Oil Recovery (EOR)

EOR is the only method of extracting certain crude oil from oil fields, this method allows for oil recovery from wells and reservoirs. Statistics indicate that this method of oil recovery allows for 30% to 60% more extraction compared to 20% to 40% using primary and secondary recovery methods. Oil recovery methods are evaluated and improved using rheology, i.e., describing and assessing the deformation and flow behavior of field materials, interfacial tension, wettability alteration "making the reservoir rock more water-wet" and core flooding experiments, i.e., studying the gas displacement process of EOR in a controlled environment. Kamal et al. (2017) noted that electrostatic repulsive interaction between nanoparticles and biosurfactants increased oil recovery. Their work showed that flooding fields with biopolymer following incubation with biosurfactant resulted in higher oil recovery. Accordingly, it can be stated that the use of microbes increases oil recovery, this is known as microbial enhanced oil recovery "MEOR".

4.4.3 Template Technology

The convenience of template technology lies in the ability to control morphology, particle size, and structure of the nano-sized particle production. This advantage allows for applications controlling specific property production (optical, magnetic, elastic, or chemical). The advantage of surfactants lies in the ability to come together in a supramolecular complex, which acts as templates to produce inorganic substances of nano-size (Holmberg 2004).

Nanoemulsion templates have been observed as a base, for application within the drug delivery industry, for a large number of nanoparticle preparation. This method requires the critical choice of specific nanoemulsion formulation methods and surfactants to allow required adaptation to the targeted drug and its administration path (Anton et al. 2008).

Mesoporous materials are valuable in many fields such as, e.g., gadgets of electronic nature or metal catalysts, and can be prepared by applying template technology. Hard template, the likes of mesoporous silica, have historically been the template of choice. Recently, however, more work has been carried out on the application of templates such as surfactants, and polymer. The advantages continue to be reproducibility and controlled delivery, along with the benefit of removing the template using a simple process (Xie et al. 2016). An example of such templates, used for mesoporous alumina synthesis, is a N-lauroyl-L-glutamic acid di-n-butylamine gel, this has worm-like channels, a large surface area with a small pore diameter of 3 nm (Cui et al. 2013).

The application of template technology has initiated an enormous amount of work such as targeting lipopeptides as a template to produce supramolecular gels for nano and micro-materials synthesis (Delbecq 2014).

References

Aboelzahab, A., A.M. Azad, S. Dolan and V. Goel. 2012. Mitigation of *Staphylococcus aureus*-mediated surgical site infections with IR photoactivated TiO$_2$ coatings on Ti implants. Advanced Healthcare Materials 1: 285–291.

Ahmad, A., P. Mukherjee, D. Mandal, S. Senapati, M.I. Khan, R. Kumar and M. Sastry. 2002. Enzyme mediated extracellular synthesis of CdS nanoparticles by the fungus, *Fusarium oxysporum*. Journal of the American Chemical Society 124: 12108–12109.

Ahmad, A., P. Mukherjee, S. Senapati, D. Mandal, M.I. Khan, R. Kumar and M. Sastry. 2003a. Extracellular biosynthesis of silver nanoparticles using the fungus *Fusarium oxysporum*. Colloids and Surfaces B: Biointerfaces 28: 313–318.

Ahmad, A., S. Senapati, M.I. Khan, R. Kumar, R. Ramani, V. Srinivas and M. Sastry. 2003b. Intracellular synthesis of gold nanoparticles by a novel alkalotolerant actinomycete, *Rhodococcus* species. Nanotechnology 14: 824.

Ahmad, A., S. Senapati, M.I. Khan, R. Kumar and M. Sastry. 2003c. Extracellular biosynthesis of monodisperse gold nanoparticles by a novel extremophilic actinomycete, *Thermomonospora* sp. Langmuir 19: 3550–3553.

Ahmad, M., J.A. Sahabjada, A. Hussain, M.A. Badaruddeen and A. Mishra. 2017. Development of a new rutin nanoemulsion and its application on prostate carcinoma PC3 cell line. Excli Journal 16: 810.

Alexandridis, P. and B. Lindman. 2000. Amphiphilic block copolymers: self-assembly and applications. Elsevier Press.

Alsaggaf, M.S., A.F. Elbaz, S. El-baday and S.H. Moussa. 2020. Anticancer and antibacterial activity of cadmium sulfide nanoparticles by *Aspergillus niger*. Advances in Polymer Technology Volume 2020 |Article ID 4909054.

Anton, N., J.P. Benoit and P. Saulnier. 2008. Design and production of nanoparticles formulated from nano-emulsion templates—a review. Journal of Controlled Release 128: 185–199.

Baccile, N., R. Noiville, L. Stievano and I. Van Bogaert. 2013. Sophorolipids-functionalized iron oxide nanoparticles. Physical Chemistry Chemical Physics 15: 1606–1620.

Bai, H.J. and Z.M. Zhang. 2009. Microbial synthesis of semiconductor lead sulfide nanoparticles using immobilized *Rhodobactersphaeroides*. Materials Letters 63: 764–766.

Bai, L. and D.J. McClements. 2016. Formation and stabilization of nanoemulsions using biosurfactants: Rhamnolipids. Journal of Colloid and Interface Science 479: 71–79.

Bansal, V., D. Rautaray, A. Ahmad and M. Sastry. 2004. Biosynthesis of zirconia nanoparticles using the fungus *Fusarium oxysporum*. Journal of Materials Chemistry 14: 3303–3305.

Barr, C.D. and F. Dominici. 2010. Cap and trade legislation for greenhouse gas emissions: public health benefits from air pollution mitigation. JAMA 303: 69–70.

Basavaraja, S., S.D. Balaji, A. Lagashetty, A.H. Rajasab and A. Venkataraman. 2008. Extracellular biosynthesis of silver nanoparticles using the fungus *Fusarium semitectum*. Materials Research Bulletin 43: 1164–1170.

Benvegnu, T., M. Brard and D. Plusquellec. 2004. Archaeabacteria bipolar lipid analogs: structure, synthesis, and lyotropic properties. Current Opinion in Colloid & Interface Science 8: 469–479.

Buzea, C., I.I. Pacheco and K. Robbie. 2007. Nanomaterials and nanoparticles: sources and toxicity. Biointerphases 2: MR17–MR71.

Chellapa, P., F.D. Ariffin, A.M. Eid, A.A. Almahgoubi, A.T. Mohamed, Y.S. Issa and N.A. Elmarzugi. 2016. Nanoemulsion for cosmetic application. European Journal of Biomedical and Pharmaceutical Sciences 3: 8–11.

Chen, S. and E.J. Wiktorowicz. 1993. Techniques in protein chemistry IV. RH Angeletti. Biomedicine & Pharmacotherapy 47(10): 469.

Chen, Y.H. and C.S. Yeh. 2002. Laser ablation method: use of surfactants to form the dispersed Ag nanoparticles. Colloids and Surfaces A: Physicochemical and Engineering Aspects 197(1-3): 133–139.

Cheng, K.C., Z.S. Khoo, N.W. Lo, W.J. Tan and N.G. Chemmangattuvalappil. 2020. Design and performance optimization of detergent product containing binary mixture of anionic-nonionic surfactants. Heliyon 6: p.e03861.

Chevalier, Y. 2002. New surfactants: new chemical functions and molecular architectures. Current Opinion in Colloid & Interface Science 7: 3–11.

Chopra, R. 1969. Glossary of Indian Medicinal Plants. Supplement. New Delhi: Council of Scientific & Industrial Research.

Cui, X., S. Tang and H. Zhou. 2013. Mesoporous alumina materials synthesized in different gel templates. Materials Letters 98: 116–119.

Cunningham, D.P. and L.L. Lundie. 1993. Precipitation of cadmium by *Clostridium thermoaceticum*. Applied and Environmental Microbiology 59(1): 7–14.

Delbecq, F. 2014. Supramolecular gels from lipopeptidegelators: Template improvement and strategies for the *in-situ* preparation of inorganic nanomaterials and for the dispersion of carbon nanomaterials. Advances in Colloid and Interface Science 209: 98–108.

Della Valle, C., L. Visai, M. Santin, A. Cigada, G. Candiani, D. Pezzoli, C.R. Arciola, M. Imbriani and R. Chiesa. 2012. A novel antibacterial modification treatment of titanium capable to improve osseointegration. The International Journal of Artificial Organs 35: 864–875.

Dey, G., R. Bharti, R. Sen and M. Mandal. 2015. Microbial amphiphiles: a class of promising new-generation anticancer agents. Drug Discovery Today 20: 136–146.

Donsi, F., Sessa and G. Ferrari. 2012. Effect of emulsifier type and disruption chamber geometry on the fabrication of food nanoemulsions by high pressure homogenization. Industrial & Engineering Chemistry Research 51: 7606–7618.

Du, H., C. Liu, Y. Zhang, G. Yu, C. Si and B. Li. 2016. Preparation and characterization of functional cellulose nanofibrils via formic acid hydrolysis pretreatment and the followed high-pressure homogenization. Industrial Crops and Products 94: 736–745.

Eastoe, J., M.H. Hatzopoulos and P.J. Dowding. 2011. Action of hydrotropes and alkyl-hydrotropes. Soft Matter 7: 5917–5925.

El-Baz, A.F., A.I. El-Batal, F.M. Abomosalam, A.A. Tayel, Y.M. Shetaia and S.T. Yang. 2016. Extracellular biosynthesis of anti-Candida silver nanoparticles using *Monascus purpureus*. J. Basic Microbiol. 56: 531–540.

El-Baz, A.F., N.M. Sorour and Y.M. Shetaia. 2016. *Trichosporonjirovecii*–mediated synthesis of cadmium sulfide nanoparticles. Journal of Basic Microbiology 56: 520–530.

El-Said, W.A., A.S. Moharram, E.M. Hussein and A.M. El-Khawaga. 2018. Synthesis, characterization, and applications of some new trimeric-type cationic surfactants. Journal of Surfactants and Detergents 21: 343–353.

Elshafie, A., S. Joshi, Y. Al-Wahaibi, A. Al-Bemani, S. Al-Bahry, D. Al-Maqbali and I. Banat. 2015. Sophorolipids production by *Candida bombicola* ATCC 22214 and its potential application in microbial enhanced oil recovery. Frontiers in Microbiology 6: 1324.

Elshikh, M., S. Funston, A. Chebbi, S. Ahmed, R. Marchant and I.M. Banat. 2017. Rhamnolipids from non-pathogenic *Burkholderiathailandensis* E264: physicochemical characterization, antimicrobial, and antibiofilm efficacy against oral hygiene related pathogens. New Biotechnology 36: 26–36.

Euliss, L.E., J.A. DuPont, S. Gratton and J. DeSimone. 2006. Imparting size, shape, and composition control of materials for nanomedicine. Chemical Society Reviews 35: 1095–1104.

Farias, C.B., A. Ferreira Silva, R. DinizRufino, J. Moura Luna, J.E. Gomes Souza and L.A. Sarubbo. 2014. Synthesis of silver nanoparticles using a biosurfactant produced in low-cost medium as stabilizing agent. Electronic Journal of Biotechnology 17: 122–125.

Fenibo, E.O., G.N. Ijoma, S. Ramganesh and C.B. Chikere. 2019. Microbial Surfactants: The Next Generation Multifunctional Biomolecules for Diverse Applications.

Fortin, D. and T.J. Beveridge. 2000. From biology to biotechnology and medical applications. Biomineralization. Wiley-VCH, Weinheim, pp. 7–22.

Galiano, K., C. Pleifer, K. Engelhardt, G. Brössner, P. Lackner, C. Huck, C. Lass-Flörl and A. Obwegeser. 2008. Silver segregation and bacterial growth of intraventricular catheters impregnated with silver nanoparticles in cerebrospinal fluid drainages. Neurological Research 30: 285–287.

Gudiña, E., J. Teixeira and L. Rodrigues. 2016. Biosurfactants produced by marine microorganisms with therapeutic applications. Marine Drugs 14: 38.

Guo, D., G. Xie and J. Luo. 2013. Mechanical properties of nanoparticles: basics and applications. Journal of Physics D: Applied Physics 47: 013001.

Gupta, S., N. Raghuwanshi, R. Varshney, I.M. Banat, A.K. Srivastava, P.A. Pruthi and V. Pruthi. 2017. Accelerated *in vivo* wound healing evaluation of microbial glycolipid containing ointment as a transdermal substitute. Biomedicine & Pharmacotherapy 94: 1186–1196.

Hassan, A., N.M. Sorour, A. El-Baz and Y. Shetaia. 2019. Simple synthesis of bacterial cellulose/magnetite nanoparticles composite for the removal of antimony from aqueous solution. International Journal of Environmental Science and Technology 16: 1433–1448.

Hazra, C., D. Kundu, A. Chaudhari and T. Jana. 2012. Biogenic synthesis, characterization, toxicity, and photocatalysis of zinc sulfide nanoparticles using rhamnolipids from *Pseudomonas aeruginosa* BS01 as capping and stabilizing agent. Journal of Chemical Technology & Biotechnology 88: 1039–1048.

Hazra, C., D. Kundu, A. Chatterjee, A. Chaudhari and S. Mishra. 2014. Poly (methyl methacrylate) (core)–biosurfactant (shell) nanoparticles: size controlled sub-100 nm synthesis, characterization, antibacterial activity, cytotoxicity, and sustained drug release behavior. Colloids and Surfaces A: Physicochemical and Engineering Aspects 449: 96–113.

He, J.Z., J.P. Shen, L.M. Zhang, Y.G. Zhu, Y.M. Zheng, M.G. Xu and H. Di. 2007. Quantitative analyses of the abundance and composition of ammonia-oxidizing bacteria and ammonia-oxidizing archaea of a Chinese upland red soil under long-term fertilization practices. Environmental Microbiology 9: 2364–2374.

He, S., Z. Guo, Y. Zhang, S. Zhang, J. Wang and N. Gu. 2007. Biosynthesis of gold nanoparticles using the bacteria *Rhodopseudomonascapsulata*. Materials Letters 61: 3984–3987.

Hinarejos, P., P. Guirro, J. Leal, F. Montserrat, X. Pelfort, M.L. Sorli, J.P. Horcajada and L. Puig. 2013. The use of erythromycin and colistin-loaded cement in total knee arthroplasty does not reduce the incidence of infection: a prospective randomized study in 3000 knees. JBJS 95: 769–774.

Hirsch, M. 2015. Surface active agents (Surfactants): Types and applications. Prospector Knowledge Center. https://knowledge.ulprospector.com/3106/pc-surface-active-agents-surfactants/> [Accessed 13 August 2020].

Holmberg, K. 2004. Surfactant-templated nanomaterials synthesis. Journal of Colloid and Interface Science 274: 355–364.

Hosea, M., B. Greene, R. Mcpherson, M. Henzl, M.D. Alexander and D.W. Darnall. 1986. Accumulation of elemental gold on the alga *Chlorella vulgaris*. Inorganica Chimica Acta 123: 161–165.

Huang, F., A. Peter and W. Schwab. 2013. Expression and characterization of CYP52 genes involved in the biosynthesis of sophorolipid and alkane metabolism from *Starmerellabombicola*. Applied and Environmental Microbiology 80(2): 766–776.

Husseiny, M.I., M. Abd El-Aziz, Y. Badr and M.A. Mahmoud. 2007. Biosynthesis of gold nanoparticles using *Pseudomonas aeruginosa*. Spectrochimica Acta Part A: Molecular and Biomolecular Spectroscopy 67: 1003–1006.

Jia, H., X. Leng, P. Huang, N. Zhao, J. An, H. Wu, Y. Song, Y. Zhu and H. Zhou. 2017. A sensitive and selective sensor for the detection of chloroform using biosurfactant ethoxylated phytosterol-capped gold nanoparticles. Materials Chemistry and Physics 202: 285–288.

Jiang, K., D.P. Lepak, J. Hu and J.C. Baer. 2012. How does human resource management influence organizational outcomes? A meta-analytic investigation of mediating mechanisms. Academy of Management Journal 55: 1264–1294.

Kalishwaralal, K., V. Deepak, S.R.K. Pandian and S. Gurunathan. 2009. Biological synthesis of gold nanocubes from *Bacillus licheniformis*. Bioresource Technology 100(21): 5356–5358.

Kamal, M.S., I.A. Hussein and A.S. Sultan. 2017. Review on surfactant flooding: phase behavior, retention, IFT, and field applications. Energy & Fuels 31(8): 7701–7720.

Kasture, M., S. Singh, P. Patel, P.A. Joy, A.A. Prabhune, C.V. Ramana and B.L.V. Prasad. 2007. Multiutility sophorolipids as nanoparticle capping agents: synthesis of stable and water dispersible Co nanoparticles. Langmuir 23(23): 11409–11412.

Kasture, M.B., P. Patel, A.A. Prabhune, C.V. Ramana, A.A. Kulkarni and B.L.V. Prasad. 2008. Synthesis of silver nanoparticles by sophorolipids: Effect of temperature and sophorolipid structure on the size of particles. Journal of Chemical Sciences 120(6): 515–520.

Kathiresan, K., S. Manivannan, M.A. Nabeel and B. Dhivya. 2009. Studies on silver nanoparticles synthesized by a marine fungus, *Penicillium fellutanum* isolated from coastal mangrove sediment. Colloids and Surfaces B: Biointerfaces 71(1): 133–137.

Khan, I., K. Saeed and I. Khan. 2019. Nanoparticles: Properties, applications, and toxicities. Arabian Journal of Chemistry 12(7): 908–931.

Kiran, G.S., A. Sabu and J. Selvin. 2010. Synthesis of silver nanoparticles by glycolipid biosurfactant produced from marine *Brevibacterium casei* MSA19. Journal of Biotechnology 148(4): 221–225.

Kiran, G.S., J. Selvin, A. Manilal and S. Sujith. 2011. Biosurfactants as green stabilizers for the biological synthesis of nanoparticles. Critical Reviews in Biotechnology 31: 354–364.

Kong, B., J.H. Seog, L.M. Graham and S.B. Lee. 2011. Experimental considerations on the cytotoxicity of nanoparticles. Nanomedicine 6: 929–941.

Korbekandi, H., S. Iravani and S. Abbasi. 2012. Optimization of biological synthesis of silver nanoparticles using *Lactobacillus casei* subsp. *casei*. Journal of Chemical Technology & Biotechnology 87: 932–937.

Kowshik, M., N. Deshmukh, W. Vogel, J. Urban, S.K. Kulkarni and K.M. Paknikar. 2002a. Microbial synthesis of semiconductor CdS nanoparticles, their characterization, and their use in the fabrication of an ideal diode. Biotechnology and Bioengineering 78: 583–588.

Kowshik, M., W. Vogel, J. Urban, S.K. Kulkarni and K.M. Paknikar. 2002b. Microbial synthesis of semiconductor PbS nanocrystallites. Advanced Materials 14: 815–818.

Kreutzer, J., M. Schneider, U. Schlegel, V. Ewerbeck and S.J. Breusch. 2005. Cemented total hip arthroplasty in Germany—an update. Zeitschrift fur Orthopadie und ihreGrenzgebiete 143: 48–55.

Kulkarni, P., R. Chakraborty and S. Chakraborty. 2019. Biosurfactant mediated synthesis of silver nanoparticles using *Lactobacillus brevis* (MTCC 4463) and their antimicrobial studies. International Journal of Pharmaceutical Science and Research 10: 1753–1759.

Kumar, C.G., S.K. Mamidyala, B. Das, B. Sridhar, G.S. Devi and M.S. Karuna. 2010. Synthesis of biosurfactant-based silver nanoparticles with purified rhamnolipids isolated from *Pseudomonas aeruginosa* BS-161R. J. Microbiol. Biotechnol. 20: 1061–1068.

Kvítek, L., A. Panáček, J. Soukupova, M. Kolář, R. Večeřová, R. Prucek, M. Holecova and R. Zbořil. 2008. Effect of surfactants and polymers on stability and antibacterial activity of silver nanoparticles (NPs). The Journal of Physical Chemistry C 112: 5825–5834.

Labrenz, M., G.K. Druschel, T. Thomsen-Ebert, Gilbert, S.A. Welch, K.M. Kemner, G.A. Logan, R.E. Summons, G. De Stasio, P.L. Bond and B. Lai. 2000. Formation of sphalerite (ZnS) deposits in natural biofilms of sulfate-reducing bacteria. Science 290: 1744–1747.

Le Guenic, S., L. Chaveriat, V. Lequart, N. Joly and P. Martin. 2019. Renewable surfactants for biochemical applications and nanotechnology. Journal of Surfactants and Detergents 22: 5–21.

Lee, D.K., S.V. Kim, A.N. Limansubroto, A. Yen, A. Soundia, C.Y. Wang, W. Shi, C. Hong, S. Tetradis, Y. Kim and N.H. Park. 2015. Nanodiamond–guttapercha composite biomaterials for root canal therapy. ACS Nano 9: 11490–11501.

Li, C., R. Fu, C. Yu, Z. Li, H. Guan, D. Hu, D. Zhao and L. Lu. 2013. Silver nanoparticle/chitosan oligosaccharide/poly (vinyl alcohol) nanofibers as wound dressings: a preclinical study. International Journal of Nanomedicine 8: 4131.

Li, X., H. Xu, Z.S. Chen and G. Chen. 2011. Biosynthesis of nanoparticles by microorganisms and their applications. Journal of Nanomaterials 2011: Article ID270974.

Luisi, P.L., L.J. Magid and J.H. Fendler. 1986. Solubilization of enzymes and nucleic acids in hydrocarbon micellar solution. Critical Reviews in Biochemistry 20: 409–474.

Martin, P.M. 2009. Handbook of deposition technologies for films and coatings: science, applications, and technology. William Andrew.

Miola, M., G. Fucale, G. Maina and E. Verné. 2015. Antibacterial and bioactive composite bone cements containing surface silver-doped glass particles. Biomedical Materials 10: 055014.

Mokhtari, N., S. Daneshpajouh, S. Seyedbagheri, R. Atashdehghan, K. Abdi, S. Sarkar, S. Minaian, H.R. Shahverdi and A.R. Shahverdi. 2009. Biological synthesis of very small silver nanoparticles by culture supernatant of *Klebsiella pneumonia*: The effects of visible-light irradiation and the liquid mixing process. Materials Research Bulletin 44: 1415–1421.

Moo-Young, M. and A. Moreira. 2011. Industrial Biotechnology and Commodity Products. Amsterdam: Elsevier.

Morita, T., T. Fukuoka, T. Imura and D. Kitamoto. 2009. Production of glycolipid biosurfactants by basidiomycetous yeasts. Biotechnol. Appl. Biochem. 53: 39–49.

Mourdikoudis, S., R. Pallares and N. Thanh. 2018. Characterization techniques for nanoparticles: comparison and complementarity upon studying nanoparticle properties. Nanoscale 10: 12871–12934.

Mukherjee, P., A. Ahmad, D. Mandal, S. Senapati, S.R. Sainkar, M.I. Khan, R. Ramani, R. Parischa, P.V. Ajayakumar, M. Alam and M. Sastry. 2001a. Bioreduction of AuCl4⁻ ions by the fungus, *Verticillium* sp., and surface trapping of the gold nanoparticles formed. Angewandte Chemie International Edition 40: 3585–3588.

Mukherjee, P., A. Ahmad, D. Mandal, S. Senapati, S.R. Sainkar, M.I. Khan, R. Parishcha, P.V. Ajaykumar, M. Alam, R. Kumar and M. Sastry. 2001b. Fungus-mediated synthesis of silver nanoparticles and their immobilization in the mycelial matrix: a novel biological approach to nanoparticle synthesis. Nano Letters 1: 515–519.

Mukherjee, P., S. Senapati, D. Mandal, A. Ahmad, M.I. Khan, R. Kumar and M. Sastry. 2002. Extracellular synthesis of gold nanoparticles by the fungus *Fusarium oxysporum*. ChemBioChem. 3: 461–463.

Nair, B. and T. Pradeep. 2002. Coalescence of nanoclusters and formation of submicron crystallites assisted by *Lactobacillus* strains. Crystal Growth & Design 2: 293–298.

Nakamura, S., M. Sato, Y. Sato, N. Ando, T. Takayama, M. Fujita and M. Ishihara. 2019. Synthesis and application of silver nanoparticles (Ag NPs) for the prevention of infection in healthcare workers. International Journal of Molecular Sciences 20: 3620.

Narayanan, K.B. and N. Sakthivel. 2010. Biological synthesis of metal nanoparticles by microbes. Advances in Colloid and Interface Science 156: 1–13.

Nelson, P.H., G.C. Rutledge and T.A. Hatton. 1997. On the size and shape of self-assembled micelles. The Journal of Chemical Physics 107: 10777–10781.

Nguyen, T.T., A. Edelen, B. Neighbors and D.A. Sabatini. 2010. Biocompatible lecithin-based microemulsions with rhamnolipid and sophorolipid biosurfactants: formulation and potential applications. Journal of Colloid and Interface Science 348: 498–504.

Nowinski, R.J., R.J. Gillespie, Y. Shishani, B. Cohen G. Walch and R. Gobezie. 2012. Antibiotic-loaded bone cement reduces deep infection rates for primary reverse total shoulder arthroplasty: a retrospective, cohort study of 501 shoulders. Journal of Shoulder and Elbow Surgery 21: 324–328.

Nurfarahin, A.H., M.S. Mohamed and L.Y. Phang. 2018. Culture medium development for microbial-derived surfactants production—an overview. Molecules 23: 1049.

Ohadi, M., G. Dehghannoudeh, M. Shakibaie, I.M. Banat, M. Pournamdari and H. Forootanfar. 2017a. Isolation, characterization, and optimization of biosurfactant production by an oil-degrading Acinetobacterjunii B6 isolated from an Iranian oil excavation site. Biocatalysis and Agricultural Biotechnology 12: 1–9.

Ohadi, M., H. Forootanfar, H.R. Rahimi, E. Jafari, M. Shakibaie, T. Eslaminejad and G. Dehghannoudeh. 2017b. Antioxidant potential and wound healing activity of biosurfactant produced by Acinetobacterjunii B6. Current Pharmaceutical Biotechnology 18: 900–908.

Ohadi, M., A. Shahravan, N. Dehghannoudeh, T. Eslaminejad, I.M. Banat and G. Dehghannoudeh. 2020. Potential use of microbial surfactant in microemulsion drug delivery system: a systematic review. Drug Design, Development, and Therapy 14: 541.

Pacwa-Płociniczak, M., G. Płaza, Z. Piotrowska-Seget and S. Cameotra. 2011. Environmental applications of biosurfactants: Recent advances. International Journal of Molecular Sciences 12: 633–654.

Palanisamy, P. and A.M. Raichur. 2009. Synthesis of spherical NiO nanoparticles through a novel biosurfactant mediated emulsion technique. Materials Science and Engineering: C 29: 199–204.

Pandian, S.R.K., V. Deepak, K. Kalishwaralal, J. Muniyandi, N. Rameshkumar and S. Gurunathan. 2009. Synthesis of PHB nanoparticles from optimized medium utilizing dairy industrial waste using *Brevibacterium casei* SRKP2: a green chemistry approach. Colloids and Surfaces B: Biointerfaces 74: 266–273.

Pelgrift, R.Y. and A.J. Friedman. 2013. Nanotechnology as a therapeutic tool to combat microbial resistance. Advanced Drug Delivery Reviews 65: 1803–1815.

Płaza, G., J. Chojniak and I. Banat. 2014. Biosurfactant mediated biosynthesis of selected metallic nanoparticles. International Journal of Molecular Sciences 15: 13720–13737.

Prokopovich, P., M. Köbrick, E. Brousseau and S. Perni. 2015. Potent antimicrobial activity of bone cement encapsulating silver nanoparticles capped with oleic acid. Journal of Biomedical Materials Research Part B: Applied Biomaterials 103: 273–281.

Pugazhenthiran, N., S. Anandan, G. Kathiravan, N.K.U. Prakash, S. Crawford and M. Ashokkumar. 2009. Microbial synthesis of silver nanoparticles by *Bacillus* sp. Journal of Nanoparticle Research 11: 1811.

Quester, K., M. Avalos-Borja and E. Castro-Longoria. 2013. Biosynthesis and microscopic study of metallic nanoparticles. Micron 54: 1–27.

Raffa, P., A.A. Broekhuis and F. Picchioni. 2016. Polymeric surfactants for enhanced oil recovery: A review. Journal of Petroleum Science and Engineering 145: 723–733.

Ramanathan, M., L. Shrestha, T. Mori, Q. Ji, J. Hill and K. Ariga. 2013. Amphiphile nanoarchitectonics: from basic physical chemistry to advanced applications. Physical Chemistry Chemical Physics 15: 10580.

Ramazanzadeh, B., A. Jahanbin, M. Yaghoubi, N. Shahtahmassbi, K. Ghazvini, M. Shakeri and H. Shafaee. 2015. Comparison of antibacterial effects of ZnO and CuO nanoparticles coated brackets against *Streptococcus mutans*. Journal of Dentistry 16: 200.

Reddy, A., C. Chen, S. Baker, C. Chen, J. Jean, C. Fan, H. Chen and J. Wang. 2009. Synthesis of silver nanoparticles using surfactin: A biosurfactant as stabilizing agent. Materials Letters 63: 1227–1230.

Reddy, A.S., Y.H. Kuo, S.B. Atla, C.Y. Chen, C.C. Chen, R.C. Shih, Y.F. Chang, J.P. Maity and H.J. Chen. 2011. Low-temperature synthesis of rose-like ZnO nanostructures using surfactin and their photocatalytic activity. Journal of Nanoscience and Nanotechnology 11: 5034–5041.

Reese, R. and D.R. Winge. 1988. Sulfide stabilization of the cadmium-gamma-glutamyl peptide complex of *Schizosaccharomycespombe*. Journal of Biological Chemistry 263: 12832–12835.

Reynoso-García, P., M. Güizado-Rodríguez, V. Barba, G. Ramos-Ortiz and H. Martínez-Gutiérrez. 2018. Stabilization of silver nanoparticles with a dithiocarbamate ligand and formation of nanocomposites by combination with polythiophene derivative nanoparticles. Advances in Condensed Matter Physics 2018: 1–9.

Robinson, P. 2015. Enzymes: principles and biotechnological applications. Essays in Biochemistry 59: 1–41.

Rodrigues, L., I.M. Banat, J. Teixeira and R. Oliveira. 2006. Biosurfactants: potential applications in medicine. Journal of Antimicrobial Chemotherapy 57: 609–618.

Rodrigues, L.R. 2015. Microbial surfactants: fundamentals and applicability in the formulation of nano-sized drug delivery vectors. Journal of Colloid and Interface Science 449: 304–316.

Ron, E.Z. and E. Rosenberg. 2001. Natural roles of biosurfactants: Mini review. Environmental Microbiology 3: 229–236.

Rosenberg, E. and E.Z. Ron. 1999. High-and low-molecular-mass microbial surfactants. Applied Microbiology and Biotechnology 52: 154–162.

Roy, A., D. Mahata, D. Paul, S. Korpole, O.L. Franco and S.M. Mandal. 2013. Purification, biochemical characterization, and self-assembled structure of a fengycin-like antifungal peptide from *Bacillus thuringiensis* strain SM1. Frontiers in Microbiology 4: 332.

Sadowski, Z. 2010. Biosynthesis and application of silver and gold nanoparticles. pp. 257–276. *In*: Perez, D.P. (ed.). Silver Nanoparticles. InTech Press.

Saikia, J.P., P. Bharali and B.K. Konwar. 2013. Possible protection of silver nanoparticles against salt by using rhamnolipid. Colloids and Surfaces B: Biointerfaces 104: 330–332.

Samuel, U. and J.P. Guggenbichler. 2004. Prevention of catheter-related infections: the potential of a new nano-silver impregnated catheter. International Journal of Antimicrobial Agents 23: 75–78.

Santos, D., R. Rufino, J. Luna, V. Santos and L. Sarubbo. 2016. Biosurfactants: Multifunctional biomolecules of the 21st century. International Journal of Molecular Sciences 17: 401.

Sapurina, I.Y. and M.A. Shishov. 2012. Oxidative polymerization of aniline: molecular synthesis of polyaniline and the formation of supramolecular structures. New Polymers for Special Applications 740: 272.

Sarah Le Guenic, Ludovic Chaveriat, Vincen Liquart, Nicolas Joly and Pattrick Martin. 2019. Renewable Surfactants for Biochemical Applications and Nanotechnology. J. of Surfactants and Detergents. 22(1): 5–21. https://doi.org/10.1002/jsde.12216.

Scarano, G. and E. Morelli. 2003. Properties of phytochelatin-coated CdS nanocrystallites formed in a marine phytoplanktonic alga (*Phaeodactylumtricornutum*, Bohlin) in response to Cd. Plant Science 165: 803–810.

Sekhon, B.S. 2004. Gemini (dimeric) surfactants. Resonance 9: 42–49.

Senapati, S., A. Ahmad, M.I. Khan, M. Sastry and R. Kumar. 2005. Extracellular biosynthesis of bimetallic Au–Ag alloy nanoparticles. Small 1: 517–520.

Shankar, S.S., A. Ahmad, R. Pasricha and M. Sastry. 2003. Bioreduction of chloroaurate ions by geranium leaves and its endophytic fungus yields gold nanoparticles of different shapes. Journal of Materials Chemistry 13: 1822–1826.

Shimizu, T., M. Masuda and H. Minamikawa. 2005. Supramolecular nanotube architectures based on amphiphilic molecules. Chemical Reviews 105: 1401–1444.

Simonnet, J.T., O. Sonneville, S. Legret and S.A. LOreal. 2002. Nanoemulsion based on ethylene oxide and propylene oxide block copolymers and its uses in the cosmetics, dermatological, and/or ophthalmological fields. U.S. Patent 6,464,990.

Simonnet, J.T., O. Sonneville, S. Legret and S.A. LOreal. 2004. Nanoemulsion based on sugar fatty esters or on sugar fatty ethers and its uses in the cosmetics, dermatological, and/or ophthalmological fields. U.S. Patent 6,689,371.

Singh, B., S. Dwivedi, A. Al-Khedhairy and J. Musarrat. 2011. Synthesis of stable cadmium sulfide nanoparticles using surfactin produced by *Bacillus amyloliquifaciens* strain KSU-109. Colloids and Surfaces B: Biointerfaces 85: 207–213.

Singh, H., J. Du, P. Singh and T. Yi. 2018. Extracellular synthesis of silver nanoparticles by *Pseudomonas* sp. THG-LS1.4 and their antimicrobial application. Journal of Pharmaceutical Analysis 8: 258–264.

Song, B., Z. Wang, S. Chen, X. Zhang, Y. Fu, M. Smet and W. Dehaen. 2005. The introduction of π–π stacking moieties for fabricating stable micellar structure: formation and dynamics of disklike micelles. Angewandte Chemie International Edition 44: 4731–4735.

Sonneville-Aubrun, O., J.T. Simonnet and F. L'alloret. 2004. Nanoemulsions: a new vehicle for skincare products. Advances in Colloid and Interface Science 108: 145–149.

Sweeney, R.Y., C. Mao, X. Gao, J.L. Burt, A.M. Belcher, G. Georgiou and B.L. Iverson. 2004. Bacterial biosynthesis of cadmium sulfide nanocrystals. Chemistry & Biology 11: 1553–1559.

Thakkar, K.N., S.S. Mhatre and R.Y. Parikh. 2010. Biological synthesis of metallic nanoparticles. Nanomedicine: Nanotechnology, Biology and Medicine 6: 257–262.

Uzoigwe, C., C.J. Burgess, C.J. Ennis and P.K. Rahman. 2015. Bioemulsifiers are not biosurfactants and require different screening approaches. Frontiers in Microbiology 6: 245.

Vigneshwaran, N., A.A. Kathe, P.V. Varadarajan, R.P. Nachane and R.H. Balasubramanya. 2006. Biomimetics of silver nanoparticles by white rot fungus, *Phaenerochaetechrysosporium*. Colloids and Surfaces B: Biointerfaces 53: 55–59.

Wang, L., C. Hu and L. Shao. 2017. The antimicrobial activity of nanoparticles: present situation and prospects for the future. International Journal of Nanomedicine 12: 1227.

Wang, Y. and Y. Xia. 2004. Bottom-Up and Top-Down approaches to the synthesis of monodispersed spherical colloids of low melting-point metals. Nano Letters 4: 2047–2050.

Worakitsiri, P., O. Pornsunthorntawee, T. Thanpitcha, S. Chavadej, C. Weder and R. Rujiravanit. 2011. Synthesis of polyaniline nanofibers and nanotubes via rhamnolipid biosurfactant templating. Synthetic Metals 161: 298–306.

Wu, Y.S., S.C. Ngai, B.H. Goh, K.G. Chan, L.H. Lee and L.H. Chuah. 2017. Anticancer activities of surfactin and potential application of nanotechnology assisted surfactin delivery. Frontiers in Pharmacology 8: 761.

Xia, W., K. Grandfield, A. Hoess, A. Ballo, Y. Cai and H. Engqvist. 2012. Mesoporous titanium dioxide coating for metallic implants. Journal of Biomedical Materials Research Part B: Applied Biomaterials 100: 82–93.

Xie, X., M.Q. Zhao, B. Anasori, K. Maleski, C.E. Ren, J. Li, B.W. Byles, E. Pomerantseva, G. Wang and Y. Gogotsi. 2016. Porous heterostructured MXene/carbon nanotube composite paper with high volumetric capacity for sodium-based energy storage devices. Nano Energy 26: 513–523.

Xie, Y., R. Ye and H. Liu. 2006. Synthesis of silver nanoparticles in reverse micelles stabilized by natural biosurfactant. Colloids and Surfaces A: Physicochemical and Engineering Aspects 279: 175–178.

Yassaei, S., A. Nasr, H. Zandi and M.N. Motallaei. 2020. Comparison of antibacterial effects of orthodontic composites containing different nanoparticles on *Streptococcus mutans* at different times. Dental Press Journal of Orthodontics 25: 52–60.

Yu, C., Z.Q. Hu and R.Y. Peng. 2014. Effects and mechanisms of a microcurrent dressing on skin wound healing: a review. Military Medical Research 1: 1–8.

Yun'an Qing, L.C., R. Li, G. Liu, Y. Zhang, X. Tang, J. Wang, H. Liu and Y. Qin. 2018. Potential antibacterial mechanism of silver nanoparticles and the optimization of orthopedic implants by advanced modification technologies. International Journal of Nanomedicine 13: 3311.

Yunus, I.S., Harwin, A. Kurniawan, D. Adityawarman and A. Indarto. 2012. Nanotechnologies in water and air pollution treatment. Environmental Technology Reviews 1: 136–148.

Zana, R. 2002. Dimeric and oligomeric surfactants. Behavior at interfaces and in aqueous solution: a review. Advances in Colloid and Interface Science 97: 205–253.

Zhang, B., H. Misak, P.S. Dhanasekaran, D. Kalla and R. Asmatulu. 2011, September. Environmental impacts of nanotechnology and its products. pp. 1–9. In Proceedings of the 2011 Midwest Section Conference of the American Society for Engineering Education.

Zhang, C., H. Vali, C.S. Romanek, T.J. Phelps and S.V. Liu. 1998. Formation of single-domain magnetite by a thermophilic bacterium. American Mineralogist 83: 1409–1418.

Zhou, Y., L. Li, Q. Zhou, S. Yuan, Y. Wu, H. Zhao and H. Wu. 2015. Lack of efficacy of prophylactic application of antibiotic-loaded bone cement for prevention of infection in primary total knee arthroplasty: results of a meta-analysis. Surgical Infections 16: 183–187.

Zülli, F., C. Liechti and F. Suter. 2000. Controlled delivery of lipophilic agents to cell cultures for *in vitro* toxicity and biocompatibility assays. International Journal of Cosmetic Science 22: 265–270.

Index